Methods in Enzymology

Volume 147
PEPTIDE GROWTH FACTORS
Part B

METHODS IN ENZYMOLOGY

EDITORS-IN-CHIEF

Sidney P. Colowick   Nathan O. Kaplan

Methods in Enzymology

Volume 147

Peptide Growth Factors

Part B

EDITED BY

David Barnes

DEPARTMENT OF BIOCHEMISTRY AND BIOPHYSICS
OREGON STATE UNIVERSITY
CORVALLIS, OREGON

David A. Sirbasku

DEPARTMENT OF BIOCHEMISTRY AND MOLECULAR BIOLOGY
UNIVERSITY OF TEXAS MEDICAL SCHOOL
HOUSTON, TEXAS

1987

ACADEMIC PRESS, INC.
Harcourt Brace Jovanovich, Publishers
Orlando   San Diego   New York   Austin
Boston   London   Sydney   Tokyo   Toronto

ACADEMIC PRESS, INC.
Orlando, Florida 32887

United Kingdom Edition published by
ACADEMIC PRESS INC. (LONDON) LTD.
24–28 Oval Road, London NW1 7DX

LIBRARY OF CONGRESS CATALOG CARD NUMBER: 54-9110

ISBN 0–12–182047–5 (alk. paper)

PRINTED IN THE UNITED STATES OF AMERICA

87 88 89 90   9 8 7 6 5 4 3 2 1

Table of Contents

Section I. Platelet-Derived Growth Factor

v

Section II. Angiogenesis, Endothelial and Fibroblast Growth Factors

Section III. Nerve and Glial Growth Factors

Section IV. Transferrin, Erythropoietin, and Related Factors

Section V. Techniques for the Study of Growth Factor Activity: Genetic Approaches and Biological Effects

Contributors to Volume 147

Article numbers are in parentheses following the names of contributors.
Affiliations listed are current.

JOHN M. ALETTA (18), *Department of Pharmacology, New York University School of Medicine, New York, New York 10016*

ROBERT H. ALLEN (21), *Division of Hematology, Department of Medicine, University of Colorado Health Sciences Center, Denver, Colorado 80262*

HARRY N. ANTONIADES (3), *Department of Nutrition, Harvard School of Public Health, Boston, Massachusetts 02115*

THOMAS J. BARIBAULT (16), *Division of Horticultural Research, CSIRO, Adelaide 5001, South Australia*

GRAEME I. BELL (17), *Howard Hughes Medical Institute, University of Chicago, Chicago, Illinois 60637*

MAURICE C. BONDURANT (30), *Department of Medicine, Division of Hematology, Vanderbilt University School of Medicine, Nashville, Tennessee 37232*

JEREMY P. BROCKES (19), *MRC Cell Biophysics Unit, London WC2B 5RL, England*

DAN CASSEL (38), *Department of Biology, Technion-Israel Institute of Technology, Haifa 32000, Israel*

HIDEO CHIBA (29), *Department of Food Science and Technology, Faculty of Agriculture, Kyoto University, Kyoto 606, Japan*

BRENT H. COCHRAN (6), *Center for Cancer Research and Department of Biology, Massachusetts Institute of Technology, Cambridge, Massachusetts 02139*

DENNIS D. CUNNINGHAM (14), *Department of Microbiology and Molecular Genetics, College of Medicine, University of California, Irvine, California 92717*

DERRICK DOMINGO (24), *Department of Cancer Biology, The Salk Institute for Biological Studies, San Diego, California 92138*

BO EK (1), *Ludwig Institute for Cancer Research, Biomedical Center, S-751 23 Uppsala, Sweden*

MICHAEL W. FANGER (16), *Department of Immunology, Dartmouth Medical Center, Hanover, New Hampshire 03755*

BRIAN FAULDERS (1), *Ludwig Institute for Cancer Research, Biomedical Center, S-751 23 Uppsala, Sweden*

D. GOSPODAROWICZ (9), *Cancer Research Institute, University of California Medical Center, San Francisco, California 94143*

STEVEN H. GREEN (18), *Department of Pharmacology, New York University School of Medicine, New York, New York 10016*

LLOYD A. GREENE (18), *Department of Pharmacology, New York University School of Medicine, New York, New York 10016*

GARY R. GROTENDORST (12), *Departments of Medicine and Biochemistry, Medical University of South Carolina, Charleston, South Carolina 29425*

RICHARD T. HAMILTON (39), *Department of Zoology, Iowa State University, Ames, Iowa 50011*

ANNET HAMMACHER (1), *Ludwig Institute for Cancer Research, Biomedical Center, S-751 23 Uppsala, Sweden*

MAUREEN A. HARRINGTON (36), *Comprehensive Cancer Center, University of Southern California, Los Angeles, California 90033*

SHUICHI HASHIZUME (27), *Morinaga Institute of Biological Science, Yokohama 230, Japan*

CARL-HENRIK HELDIN (1), *Ludwig Institute for Cancer Research, Biomedical Center, S-751 23 Uppsala, Sweden*

HARVEY R. HERSCHMAN (31), *Department of Biological Chemistry, and Laboratory of Biomedical and Environmental Sciences, University of California, Los Angeles School of Medicine, Los Angeles, California 90024*

FUMIAKI ITO (34), *Department of Biochemistry, Faculty of Pharmaceutical Sciences, Setsunan University, 45-1 Nagaotoge-cho, Hirakata-shi, Osaka 573-01, Japan*

ANN JOHNSSON (1), *EMBL, Postfach 1022.40, 6900 Heidelberg 1, Federal Republic of Germany*

MICHAEL E. KAMARCK (25), *Molecular Therapeutics Inc., West Haven, Connecticut 06516*

JERRY KAPLAN (22), *Department of Pathology, University of Utah School of Medicine, Salt Lake City, Utah 84132*

WILLIAM R. KIDWELL (37), *Growth Regulation Center, Laboratory of Tumor Immunology and Biology, National Cancer Institute, Bethesda, Maryland 20892*

MICHAEL KLAGSBRUN (8), *Departments of Biological Chemistry and Surgery, Children's Hospital and Harvard Medical School, Boston, Massachusetts 02115*

SIGRUN KORSCHING (15), *Division of Biology, California Institute of Technology, Pasadena, California 91125*

MARK J. KOURY (30), *Department of Medicine, Division of Hematology, Vanderbilt University School of Medicine, Nashville, Tennessee 37232*

KAZUHIKO KURODA (27), *Morinaga Institute of Biological Science, Yokohama 230, Japan*

JAYNE F. LESLEY (24), *Department of Cancer Biology, The Salk Institute for Biological Studies, San Diego, California 92138*

RAMON LIM (20), *Division of Neurochemistry and Neurobiology, Department of Neurology, University of Iowa, Iowa City, Iowa 52242*

STEVE LOVEJOY (28), *SRI International, Menlo Park, California 94025*

GEORGE J. MARKELONIS (26), *Department of Anatomy, University of Maryland School of Medicine, Baltimore, Maryland 21201*

ALAN McCLELLAND (25), *Molecular Therapeutics Inc., West Haven, Connecticut 06516*

K. A. McKEEHAN (13, 35), *W. Alton Jones Cell Science Center, Inc., Lake Placid, New York 12946*

W. L. McKEEHAN (13, 35), *W. Alton Jones Cell Science Center, Inc., Lake Placid, New York 12946*

MARK MERCOLA (6), *Department of Microbiology and Molecular Genetics, Harvard Medical School and the Dana-Farber Cancer Institute, Boston, Massachusetts 02115*

JOYCE F. MILLER (20), *Division of Neurochemistry and Neurobiology, Department of Neurology, University of Iowa Hospital, Iowa City, Iowa 52242*

HIROKI MURAKAMI (27), *Department of Food Science and Technology, Kyushu University, Fukuoka 812, Japan*

KENNETH E. NEET (16), *Department of Biochemistry, Case Western Reserve University, Cleveland, Ohio 44106*

MARIT NILSEN-HAMILTON (39), *Department of Biochemistry and Biophysics, Iowa State University, Ames, Iowa 50011*

TAE H. OH (26), *Department of Anatomy, University of Maryland School of Medicine, Baltimore, Maryland 21201*

PANAYOTIS PANTAZIS (3), *Department of Nutrition, Harvard School of Public Health, Boston, Massachusetts 02115*

LOREN PICKART (28), *Procyte, Redmond, Washington 98052*

W. JACKSON PLEDGER (7, 36), *Department of Cell Biology, Vanderbilt University Medical School, Nashville, Tennessee 37232*

ELAINE W. RAINES (5), *Department of Pathology, University of Washington, Seattle, Washington 98195*

HANS-GEORG RAMMENSEE (24), *Basel Institute for Immunology, 4005 Basel, Switzerland*

BARRETT ROLLINS (6), *Department of Microbiology and Molecular Genetics, Harvard Medical School and the Dana-Farber Cancer Institute, Boston, Massachusetts 02115*

LARS RÖNNSTRAND (1), *Ludwig Institute for Cancer Research, Biomedical Center, S-751 23 Uppsala, Sweden*

RUSSELL ROSS (5), *Department of Pathology, University of Washington, Seattle, Washington 98195*

PAUL ROTHENBERG (38), *Department of Pathology, Brigham and Women's Hospital, Harvard Medical School, Boston, Massachusetts 02115*

E. ROZENGURT (4), *Laboratory of Growth Regulation, Imperial Cancer Research Fund, London WC2A 3PX, England*

FRANK H. RUDDLE (25), *Department of Biology, Yale University, New Haven, Connecticut 06511*

ADRIANA RUKENSTEIN (18), *Department of Pharmacology, New York University School of Medicine, New York, New York 10016*

RUSSELL RYBKA (8), *Department of Surgery, Children's Hospital, Boston, Massachusetts 02115*

RYUZO SASAKI (29), *Department of Food Science and Technology, Faculty of Agriculture, Kyoto University, Kyoto 606, Japan*

CAROL SAUVAGE (24), *Department of Cancer Biology, The Salk Institute for Biological Studies, San Diego, California 92138*

STEPHEN T. SAWYER (30), *Department of Medicine, Division of Hematology, Vanderbilt University School of Medicine, Nashville, Tennessee 37232*

CHARLES D. SCHER (7), *Division of Oncology, Department of Pediatrics, The Children's Hospital of Philadelphia, University of Pennsylvania, Philadelphia, Pennsylvania 19104*

ROBERTA SCHULTE (24), *Department of Cancer Biology, The Salk Institute for Biological Studies, San Diego, California 92138*

JAMES SCOTT (17), *Molecular Medicine Group, MRC Clinical Research Centre, Harrow, Middlesex HA1, 3UJ, England*

MARK J. SELBY (17), *Hormone Research Institute, University of California, San Francisco, California 94143*

PAUL A. SELIGMAN (21), *Division of Hematology, Department of Medicine, University of Colorado Health Sciences Center, Denver, Colorado 80262*

NOBUYOSHI SHIMIZU (33, 34), *Department of Molecular Biology, Keio University School of Medicine, 35 Shinanomachi, Shinjuku-ku, Tokyo 160, Japan, and Department of Molecular and Cellular Biology, University of Arizona, Tucson, Arizona 85721*

YUEN SHING (8), *Department of Surgery, Children's Hospital, Boston, Massachusetts 02115*

JAI PAL SINGH (2), *Department of Atherosclerosis and Thrombosis, The Upjohn Company, Kalamazoo, Michigan 49001*

SANDRA SMITH (8), *Department of Surgery, Children's Hospital, Boston, Massachusetts 02115*

CHARLES D. STILES (6), *Department of Microbiology and Molecular Genetics, Harvard Medical School and the Dana-Farber Cancer Institute, Boston, Massachusetts 02115*

DANIEL S. STRAUS (32), *Biomedical Sciences Division and Department of Biology, University of California, Riverside, California 92521*

P. STROOBANT (4), *Medical and Molecular Biology Unit, Department of Biochemistry, University College and Middlesex School of Medicine, London W1P 6DB, England*

ROBERT SULLIVAN (8), *Department of Surgery, Children's Hospital, Boston, Massachusetts 02115*

HOWARD H. SUSSMAN (23), *Department of Pathology, Stanford University School of Medicine, Stanford, California 94305*

HANS THOENEN (15), *Department of Neurochemistry, Max-Planck-Institute for Psychiatry, D-8033 Martinsried, Federal Republic of Germany*

KENNETH A. THOMAS (10), *Department of Biochemistry and Molecular Biology, Merck Institute for Therapeutic Research, Merck Sharp and Dohme Research Laboratories, Rahway, New Jersey 07065*

JAMES A. THOMPSON (14), *Department of Microbiology and Molecular Genetics, College of Medicine, University of California, Irvine, California 92717*

SUSAN E. TONIK (23), *Department of Pathology, Stanford University School of Medicine, Stanford, California 94305*

IAN S. TROWBRIDGE (24), *Department of Cancer Biology, The Salk Institute for Biological Studies, San Diego, California 92138*

JOHN H. WARD (22), *Department of Medicine, University of Utah School of Medicine, Salt Lake City, Utah 84132*

ÅKE WASTESON (1), *Department of Cell Biology, University of Linköping, S-581 85 Linköping, Sweden*

M. D. WATERFIELD (4), *Medical and Molecular Biology Unit, Department of Biochemistry, University College and Middlesex School of Medicine, London W1P 6DB, England*

SUZANNE WENNERGREN (1), *Ludwig Institute for Cancer Research, Biomedical Center, S-751 23 Uppsala, Sweden*

BENGT WESTERMARK (1), *Department of Pathology, University Hospital, S-751 85 Uppsala, Sweden*

SHIN-ICHI YANAGAWA (29), *Department of Biophysics, Institute for Virus Research, Kyoto University, Kyoto 606, Japan*

BRUCE R. ZETTER (11), *Departments of Physiology and Surgery, Harvard Medical School, Children's Hospital, Boston, Massachusetts 02115*

JOHN ZULLO (6), *Department of Microbiology and Molecular Genetics, Harvard Medical School and the Dana-Farber Cancer Institute, Boston, Massachusetts 02115*

PETER ZUMSTEIN (6), *W. Alton Jones Cell Science Center, Lake Placid, New York, 12946*

Dedication

Sidney Colowick and Nate Kaplan were personally involved to a great degree in the development of Peptide Growth Factors, Parts A and B, Volumes 146 and 147 of *Methods in Enzymology.* Although their scientific interests and research directions were varied, they directed their interests toward hormone-related research from time to time throughout their careers. (See, for instance, the chapter on Growth Factor Stimulation of Sugar Uptake by Inman and Colowick in Volume 146.) From the early discussions regarding these volumes, Sidney's and Nate's comments, advice, and suggestions made our task easier, and their wisdom contributed greatly to the quality of the final product. Sadly, neither lived to see the completion of the work.

Other tributes to these extraordinary men have appeared in *Methods in Enzymology,* and we could add little that has not been expressed. The series itself may be the best demonstration of their remarkable scientific insight and comprehension. However, those who knew them realize that in spite of their enormous intellectual contributions it is the personal impact they had on colleagues, students, and friends that is their greatest legacy. These men and their families touched our lives profoundly. We dedicate these volumes to Sidney and Maryda, Nate and Goldie, and their families.

DAVID BARNES
DAVID A. SIRBASKU

Preface

Applications of new techniques and proliferation of investigators in the area of peptide growth factors have led to enormous expansion in understanding the mechanisms of action of these molecules in the past few years and have, in turn, raised new and intriguing questions concerning the relationship of peptide growth factors, receptors, and related molecules to normal and abnormal regulation of cell division and differentiation *in vivo*. Volumes 146 and 147 of *Methods in Enzymology* are a reflection of the recent advances. In these volumes we have attempted to cover methods specific for each growth factor in the areas of purification, immunoassay, radioreceptor assay, biological activity assay, and receptor identification and quantitation. Also included in both volumes are general methods for the study of mechanisms of action that are applicable to many of the factors.

Volume 146 includes techniques concerning epidermal growth factor, transforming growth factors alpha and beta, somatomedin C/insulin-like growth factors, and bone and cartilage growth factors. It also contains methods for quantitative cell growth assays and techniques for the study of growth factor-modulated protein phosphorylation and cell surface membrane effects. Volume 147 covers techniques concerning platelet-derived growth factor, angiogenesis, endothelial and fibroblast growth factors (heparin-binding growth factors), nerve and glial growth factors, transferrin, erythropoietin, and related factors. Also included are genetic approaches to growth factor action and additional methods for the study of biological effects of these molecules.

Procedures related to the basic aspects of research concerning epidermal, nerve, platelet-derived, and insulin-like growth factors have appeared in the Hormone Action series of *Methods in Enzymology,* and methods concerning most aspects of growth factors for lymphoid cells have appeared in the Immunochemical Techniques series. We have attempted to avoid major duplication of material appearing elsewhere in this series in the belief that volumes of moderate size providing recent and complimentary methodology are of more practical value to the researcher. For example, we have not attempted to cover growth factors for lymphoid cells, with the exception of transferrin, which appears to be a requirement for optimal growth of many cell types. Similarly, some approaches that are finding current intensive use in the field, such as recombinant DNA, hybridoma, nucleic acid, or peptide synthesis techniques, are covered per se in other volumes of *Methods in Enzymology*. Detailed techniques in these areas are included only in situations in which the authors or editors felt that they were sufficiently novel or specific to the area of growth factor research.

We thank the many contributors to this project, and hope these volumes will be useful to investigators in the field.

DAVID BARNES
DAVID A. SIRBASKU

METHODS IN ENZYMOLOGY

EDITED BY

Sidney P. Colowick and Nathan O. Kaplan

VANDERBILT UNIVERSITY
SCHOOL OF MEDICINE
NASHVILLE, TENNESSEE

DEPARTMENT OF CHEMISTRY
UNIVERSITY OF CALIFORNIA
AT SAN DIEGO
LA JOLLA, CALIFORNIA

METHODS IN ENZYMOLOGY

EDITORS-IN-CHIEF

Sidney P. Colowick and Nathan O. Kaplan

VOLUME XVIII. Vitamins and Coenzymes (Parts A, B, and C)
Edited by DONALD B. MCCORMICK AND LEMUEL D. WRIGHT

VOLUME XIX. Proteolytic Enzymes
Edited by GERTRUDE E. PERLMANN AND LASZLO LORAND

VOLUME XX. Nucleic Acids and Protein Synthesis (Part C)
Edited by KIVIE MOLDAVE AND LAWRENCE GROSSMAN

VOLUME XXI. Nucleic Acids (Part D)
Edited by LAWRENCE GROSSMAN AND KIVIE MOLDAVE

VOLUME XXII. Enzyme Purification and Related Techniques
Edited by WILLIAM B. JAKOBY

VOLUME XXIII. Photosynthesis (Part A)
Edited by ANTHONY SAN PIETRO

VOLUME XXIV. Photosynthesis and Nitrogen Fixation (Part B)
Edited by ANTHONY SAN PIETRO

VOLUME XXV. Enzyme Structure (Part B)
Edited by C. H. W. HIRS AND SERGE N. TIMASHEFF

VOLUME XXVI. Enzyme Structure (Part C)
Edited by C. H. W. HIRS AND SERGE N. TIMASHEFF

VOLUME XXVII. Enzyme Structure (Part D)
Edited by C. H. W. HIRS AND SERGE N. TIMASHEFF

VOLUME XXVIII. Complex Carbohydrates (Part B)
Edited by VICTOR GINSBURG

VOLUME XXIX. Nucleic Acids and Protein Synthesis (Part E)
Edited by LAWRENCE GROSSMAN AND KIVIE MOLDAVE

VOLUME XXX. Nucleic Acids and Protein Synthesis (Part F)
Edited by KIVIE MOLDAVE AND LAWRENCE GROSSMAN

VOLUME XXXI. Biomembranes (Part A)
Edited by SIDNEY FLEISCHER AND LESTER PACKER

VOLUME 146. Peptide Growth Factors (Part A)
Edited by DAVID BARNES AND DAVID A. SIRBASKU

VOLUME 147. Peptide Growth Factors (Part B)
Edited by DAVID BARNES AND DAVID A. SIRBASKU

VOLUME 148. Plant Cell Membranes (in preparation)
Edited by LESTER PACKER AND ROLAND DOUCE

VOLUME 149. Drug and Enzyme Targeting (Part B) (in preparation)
Edited by RALPH GREEN AND KENNETH J. WIDDER

VOLUME 150. Immunochemical Techniques (Part K: *In Vitro* Models of B
and T Cell Functions and Lymphoid Cell Receptors) (in preparation)
Edited by GIOVANNI DI SABATO

VOLUME 151. Molecular Genetics of Mammalian Cells (in preparation)
Edited by MICHAEL M. GOTTESMAN

VOLUME 152. Guide to Molecular Cloning Techniques (in preparation)
Edited by SHELBY L. BERGER AND ALAN R. KIMMEL

VOLUME 153. Recombinant DNA (Part D) (in preparation)
Edited by RAY WU AND LAWRENCE GROSSMAN

VOLUME 154. Recombinant DNA (Part E) (in preparation)
Edited by RAY WU AND LAWRENCE GROSSMAN

VOLUME 155. Recombinant DNA (Part F) (in preparation)
Edited by RAY WU

VOLUME 156. Biomembranes (Part P: ATP-Driven Pumps and Related
Transport: The Na,K-Pump) (in preparation)
Edited by SIDNEY FLEISCHER AND BECCA FLEISCHER

Section I

Platelet-Derived Growth Factor

[1] Purification of Human Platelet-Derived Growth Factor

By Carl-Henrik Heldin, Ann Johnsson, Bo Ek,
Suzanne Wennergren, Lars Rönnstrand, Annet Hammacher,
Brian Faulders, Åke Wasteson, and Bengt Westermark

Platelet-derived growth factor (PDGF) is the major mitogen in serum for connective tissue cells (for a recent review, see Ref. 1). The molecule · is an M_r 31,000 protein composed of two disulfide-bonded polypeptide chains, A and B.[2] PDGF is probably a heterodimer of one A chain and one B chain, although the possibility that homodimers occur has not been ruled out. Determination of a partial amino acid sequence of PDGF revealed that the B chain is almost identical to part of p28[sis], the transforming protein of simian sarcoma virus.[3,4] This indicates that a PDGF-like growth factor is involved in simian sarcoma virus-induced cell transformation. The A chain of PDGF showed approximately 60% homology to the B chain,[5] indicating that the gene sequences encoding the two PDGF chains have evolved from a common ancestral gene.

Small amounts of PDGF were first purified in 1979[6,7]; protocols for the purification of larger quantities of human PDGF from fresh platelets[8,9] or from outdated platelet concentrates[10–12] have subsequently been published. While essentially all studies so far have been performed using

[1] C.-H. Heldin, Å. Wasteson, and B. Westermark, *Mol. Cell. Endocrinol.* **39**, 169 (1985).
[2] A. Johnsson, C.-H. Heldin, B. Westermark, and Å. Wasteson, *Biochem. Biophys. Res. Commun.* **104**, 66 (1982).
[3] M. D. Waterfield, G. T. Scrace, N. Whittle, P. Stroobant, A. Johnsson, Å. Wasteson, B. Westermark, C.-H. Heldin, J. S. Huang, and T. F. Deuel, *Nature (London)* **304**, 35 (1983).
[4] R. F. Doolittle, M. W. Hunkapiller, L. E. Hood, S. G. Devare, K. C. Robbins, S. A. Aaronson, and H. N. Antoniades, *Science* **221**, 275 (1983).
[5] A. Johnsson, C.-H. Heldin, Å. Wasteson, B. Westermark, T. F. Deuel, J. S. Huang, D. H. Seeburg, E. Gray, A. Ullrich, G. T. Scrace, P. Stroobant, and M. D. Waterfield, *EMBO J.* **3**, 921 (1984).
[6] H. N. Antoniades, C. D. Scher, and C. D. Stiles, *Proc. Natl. Acad. Sci. U.S.A.* **76**, 1809 (1979).
[7] C.-H. Heldin, B. Westermark, and Å. Wasteson, *Proc. Natl. Acad. Sci. U.S.A.* **76**, 3722 (1979).
[8] C.-H. Heldin, B. Westermark, and Å. Wasteson, *Biochem. J.* **193**, 907 (1981).
[9] C. N. Chesterman, T. Walker, B. Grego, K. Chamberlain, M. T. W. Hearn, and F. J. Morgan, *Biochem. Biophys. Res. Commun.* **116**, 809 (1984).
[10] T. F. Deuel, J. S. Huang, R. T. Proffitt, J. U. Baenziger, D. Chang, and B. B. Kennedy, *J. Biol. Chem.* **256**, 8896 (1981).
[11] H. N. Antoniades, *Proc. Natl. Acad. Sci. U.S.A.* **78**, 7314 (1981).
[12] E. W. Raines and R. Ross, *J. Biol. Chem.* **257**, 5154 (1982).

METHODS IN ENZYMOLOGY, VOL. 147

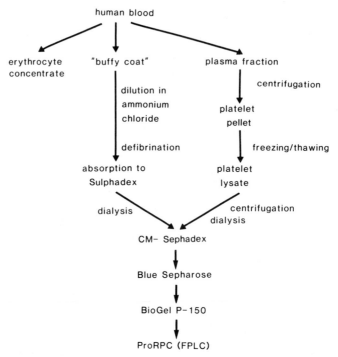

FIG. 1. Flow diagram for the preparation of human PDGF.

human PDGF, purification of PDGF from porcine platelets was recently reported.[13]

The procedure to purify PDGF from human platelets described in this communication is based on accumulated experience from PDGF purification in our laboratory, and has been used in its present form for the last 6 years (Fig. 1).

Assay for Growth-Promoting Activity

Growth-promoting activity was determined by a [³H]thymidine ([³H]TdR) incorporation assay using human foreskin fibroblasts (line AG 1523, obtained from the Human Genetic Cell Repository, Camden, NJ) as target cells.[14] The cells were routinely grown in Eagle's minimal essential medium supplemented with 10% newborn calf serum. For assay, confluent cell cultures were trypsinized and replated sparsely on 35-mm Petri

[13] P. Stroobant and M. D. Waterfield, *EMBO J.* **3,** 2963 (1984).
[14] C. Betsholtz and B. Westermark, *J. Cell. Physiol.* **118,** 203 (1984).

dishes (approximately 3×10^4 cells per dish). The medium was changed on day 4 to 2 ml of serum-free MCDB 105 medium, with a reduced calcium concentration (0.5 mM). Test samples were added on day 6 together with [³H]TdR (0.02 μCi/ml, 5 Ci/mmol) and as a carrier human serum albumin (100 μg/ml). Each 35-mm dish contained at this time approximately 1×10^5 cells. Forty-eight hours later the cell cultures were harvested. The medium of the cell cultures was removed and the cells were fixed in 5% (w/v) trichloroacetic acid (3–4 ml). After 5–10 min the trichloroacetic acid was removed and the culture dishes were rinsed extensively under running tap water and then air dried. The DNA was solubilized by the addition of 0.5 ml of 1% SDS, 0.3 M NaOH per dish. After about 15 min at room temperature the lysates were transferred to scintillation vials and subjected to liquid scintillation counting using 5 ml of Instagel (Packard) per sample. Maximal activity in the assay (5–10 × 10^3 cpm, 50–80% labeled nuclei) was obtained with 10 ng/ml of PDGF.

Purification of PDGF

Preparation of Platelet Lysate

Centrifugation of human blood according to the routine procedure at the blood banks yields an erythrocyte concentrate, a "buffy coat" fraction containing mainly leukocytes, and a plasma fraction. Due to their heterogeneity in size, the platelets are distributed between the buffy coat and the plasma fraction. For the purification of PDGF we have used both these fractions as starting material, which were made available to us through the courtesy of the Red Cross in Helsinki (G. Myllylä) and the Dept. of Virology in Uppsala (G. Alm), as by-products in their production of interferon.

The Plasma Fraction. The platelets of the plasma fraction were recovered by centrifugation and the pellets were stored at $-20°$. Platelets were lysed by four cycles of thawing and freezing in distilled water containing protease inhibitors [1 mM phenylmethylsulfonyl fluoride, 20 mM benzamidine, 100 mM ε-aminocaproic acid, 1% (v/v) aprotinin (Sigma)]. A volume of 4 ml per unit (one unit is derived from 400 ml of blood) was used; 2000–3000 units were processed simultaneously. The ionic strength was then increased by the addition of 1 M NaCl, 0.01 M sodium phosphate buffer, pH 7.4 (2 ml/unit). Thereafter the lysate was subjected to continuous flow centrifugation using a Beckman J2-21M centrifuge equipped with a JCF-2 rotor and a standard core (15,000 rpm, flow rate 60 ml/min). The supernatant was stored at $-20°$ in 3-liter aliquots, each derived from 500 units.

The Buffy Coat Fraction. For the purpose of recovering leukocytes for interferon production, the buffy coat fractions (approximately 10–20% of the original blood volume) are diluted with 4 volumes of 0.15 M ammonium chloride in order to disintegrate the erythrocytes, and then subjected to centrifugation. The resulting supernatant was used for PDGF production and stored in 10-liter volumes at $-20°$. Due to the large volume of this fraction, protease inhibitors were not added. Six 10-liter containers of platelet-enriched plasma were processed at a time (containing about 3000 g of protein). After thawing, $CaCl_2$ was added to a final concentration of 10 mM and the containers were agitated on a shaking platform at 4° overnight. Fibrin clots were removed by sieving and Sulphadex gel (300 ml/10 liters) was then added. [Sulphadex is a sulfated Sephadex G-50 medium gel (Pharmacia), synthesized as described.[15]] After shaking overnight the gel beads were allowed to sediment, and the unadsorbed fraction removed. The gel was poured into a column (60 × 5 cm) and washed with 0.5 M NaCl, 0.01 M phosphate buffer, pH 7.4 (3–4 liters). PDGF-containing material was eluted with 1.5–2.0 liters of 1.5 M NaCl, 0.01 M phosphate buffer, pH 7.4. The Sulphadex gel was discarded afterward.

CM-Sephadex Chromatography

Platelet lysate from either the platelet pellet (approximately 3 liters, containing about 80 g of protein) or Sulphadex eluate (two batches combined to 3–4 liters, approximately 8 g of protein) was dialyzed against 0.01 M phosphate buffer pH 7.4 (20 liters), followed by 0.08 M NaCl, 0.01 M phosphate buffer, pH 7.4 (2 × 20 liters). A small precipitate was removed by centrifugation at 20,000 g for 20 min. Forty grams of dry CM-Sephadex (C-25, Pharmacia) was added to the supernatant and equilibrated end over end overnight at 4°. The gel was then allowed to sediment, the unadsorbed fraction removed, and the gel poured into a column (40 × 5 cm). The column was washed with 0.01 M phosphate buffer, pH 7.4 (2 liters) and eluted with 0.5 M NaCl, 0.01 M phosphate buffer, pH 7.4 (600 ml).

Blue Sepharose

The eluate from the CM-Sephadex chromatography (approximately 200 mg of protein) was diluted with an equal volume of 0.01 M phosphate buffer, pH 7.4, and applied to a column of Blue Sepharose (Pharmacia; 2 × 7 cm) at a flow rate of 25 ml/hr. The column was washed with 1 M

[15] J. P. Miletic, G. J. Broze, Jr., and P. W. Majerus, *Anal. Biochem.* **105**, 304 (1980).

NaCl, 0.01 M phosphate buffer, pH 7.4 (200 ml), and eluted with 50% ethylene glycol, 1 M NaCl, 0.01 M phosphate buffer, pH 7.4 (60 ml).

BioGel P-150 Chromatography

The eluates from two Blue Sepharose runs were combined (approximately 20 mg of protein), dialyzed against 1 M acetic acid, and lyophilized. The sample was dissolved in 3 ml of 1 M acetic acid. After approximately 1 hr at room temperature it was centrifuged; a small insoluble pellet was obtained. The pellet was extracted once with 1 ml of 1 M acetic acid; after centrifugation the supernatant was combined with the first 3 ml portion and applied to a BioGel P-150 column (100–200 mesh, Bio-Rad; 150 × 2 cm). (P-100 has recently been used instead of P-150, with similar results.) The column was eluted with 1 M acetic acid at room temperature. The flow rate was about 6 ml/hr, and 4 ml fractions were collected. The elution patterns of growth-promoting activity and proteins are shown in Fig. 2.

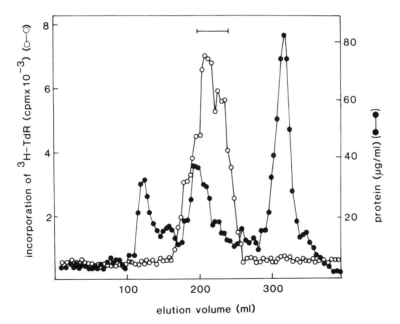

FIG. 2. Gel chromatography of PDGF partially purified through the Blue Sepharose step. Material originating from platelets of the plasma fraction was applied to a BioGel P-150 column eluted with 1 M acetic acid. The effluent fractions were analyzed for growth-promoting activity at a dilution of 1 : 2000, and for protein concentration.

Reversed-Phase Chromatography

The active material from the BioGel P-150 chromatography (approximately 1 mg of protein) was lyophilized and dissolved in 200 μl of 1 M acetic acid. It was then applied to a ProRPC column attached to an FPLC apparatus (Pharmacia) and operated at room temperature. The column was washed with 4 ml of 2 M guanidine–HCl, 1 M acetic acid, and then developed with a gradient (30 ml) from 0 to 28% of propanol in 2 M guanidine–HCl, 1 M acetic acid. One milliliter fractions were collected. The elution patterns are shown in Fig. 3. Active fractions were pooled, diluted with two volumes of 1 M acetic acid, and adsorbed to a SepPak[R] column (Waters). The column was washed with 1 M acetic acid, and PDGF was eluted with 70% acetonitrile, 0.1% trifluoroacetic acid. The material was lyophilized and stored in aliquots. Protein concentration of the purified product was determined by the Lowry method[16] and by amino acid analysis.

Properties of the Purified PDGF

The purified PDGF was analyzed by SDS–gel electrophoresis[17] and silver[18] staining (Fig. 4). Nonreduced PDGF consisted of several components in the M_r region 28,000–31,000; after reduction components of M_r: s 17,000 and lower were seen. A similar kind of heterogeneity has been reported from all other purification procedures described.[8-12] It is not known whether it was caused during the experimental handling, or had occurred already in the platelets. Addition of protease inhibitors during purification, however, reduced the extent of proteolysis, indicating that part of the proteolysis occurred during purification (not shown). Thus material derived from the plasma fraction was more homogeneous than that obtained from the buffy coat. It should be pointed out, however, that the specific activity of the two preparations was identical. The purity of the product was difficult to assess due to the heterogeneity. However, several lines of evidence indicate that the purified PDGF contains no or only very small amounts of non-PDGF-derived contaminants: (1) all stainable components of the nonreduced SDS–gel electropherogram were found to have mitogenic activity when they were extracted from the gel (not shown), (2) no stainable material was left at the M_r 30,000 region after reduction, and (3) only peptides derived from the A and B chains were

[16] O. H. Lowry, N. J. Rosebrough, A. L. Farr, and R. J. Randall, *J. Biol. Chem.* **193,** 265 (1951).

[17] G. Blobel and B. Dobberstein, *J. Cell Biol.* **67,** 835 (1975).

[18] J. H. Morrisey, *Anal. Biochem.* **117,** 307 (1981).

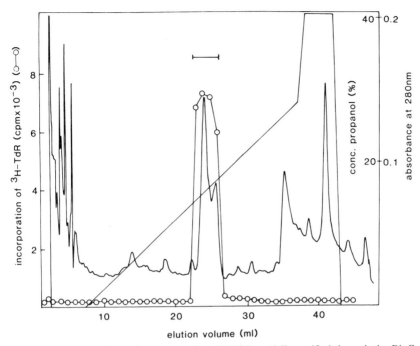

FIG. 3. Reversed-phase chromatography of PDGF partially purified through the BioGel P-150 step. Active fractions from a BioGel P-150 run (pooled as indicated in Fig. 2), of material originating from platelets of the buffy coat fraction, were applied to a ProRPC column attached to an FPLC apparatus. Effluent fractions were analyzed for growth-promoting activity at a dilution of 1 : 4000, and for protein concentration.

found when the material was reduced, fractionated, and the amino acid sequence determined.[5] Furthermore, when the same reversed-phase system was used in the purification of PDGF-like growth factors from certain human tumor cell lines,[19] pure products were obtained from more crude fractions applied to the column (unpublished data). The purity of these factors could be determined more accurately since they were more homogeneous than PDGF from platelets.

As can be seen in Fig. 5, PDGF gives half maximal growth-promoting activity on human foreskin fibroblasts at 3 ng/ml (0.1 nM).

Comments on the Purification Method

A major difficulty in the purification of PDGF has been the stickiness of the molecule. PDGF has a strong positive charge at neutral pH as well

[19] C.-H. Heldin, B. Westermark, and Å. Wasteson, *J. Cell. Physiol.* **105**, 235 (1980).

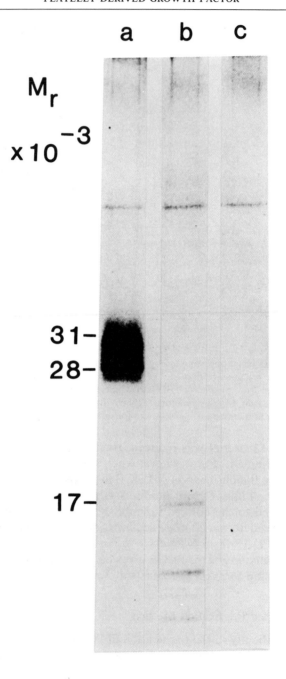

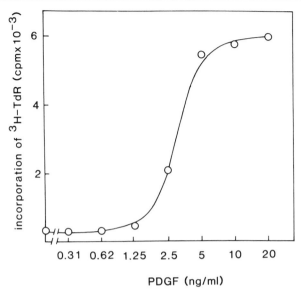

PDGF (ng/ml)

Fig. 5. Stimulation of [³H]TdR incorporation in human fibroblasts by pure PDGF.

as hydrophobic determinants. At neutral pH the yield of PDGF is better at high ionic strength (e.g., 1 M NaCl) than at low ionic strength. Purified preparations of PDGF are preferentially handled in solvents of low pH, e.g., 1 M acetic acid. PDGF can be stored in 1 M acetic acid at $-20°$ for at least several months, or at room temperature for at least a few days, without loss of activity.

The overall yield of PDGF in the present method is about 10%. The main loss occurs during the first part of the procedure, up until and including the Blue Sepharose step. Approximately 100 μg of PDGF was obtained per batch (from the platelets in the plasma fraction of 500 units, or 120 liters of diluted buffy coat fraction). The yield of PDGF in the purification procedure seems to be related to the scale of the preparation, better recoveries being obtained with large batch sizes.

Fig. 4. SDS–gel electrophoresis of PDGF. The active material from a ProRPC chromatogram of PDGF from the buffy coat fraction was pooled as indicated in Fig. 3 and analyzed by SDS–gel electrophoresis.[17] A gradient gel of 13–18% acrylamide was used and proteins were visualized by silver staining.[18] About 1 μg of PDGF, in nonreduced (a) or reduced (c) form, was applied to the gel. Note that the stainability of PDGF is better in the nonreduced form. A blank lane (b) is shown to illustrate some staining artifacts in the high M_r region of the gel.

Our procedure was worked out for fresh platelets. However, since the protein compositions of the starting material from the buffy coat fraction and that of platelet-rich plasma are rather similar, it is likely that this method would also be useful for the purification of PDGF from outdated platelet concentrates.

Ion-exchange chromatography using CM-Sephadex is a powerful purification step due to the unusually high isoelectric point of PDGF. As an alternative to batch elution with 0.5 M NaCl, 0.01 M phosphate buffer, pH 7.4, elution of the column with a gradient of NaCl was tried. Under these conditions the mitogenic activity eluted in a broad peak. Since this procedure therefore only gave a less than 2-fold better purification compared to the batch elution, the latter method was preferred, in order to save time and effort.

The affinity of PDGF for Blue Sepharose is probably due to a combination of hydrophobic and elecrostatic interactions, since the most efficient elution was obtained from a combination of 50% ethylene glycol and 1 M NaCl, 0.01 M phosphate buffer, pH 7.4. Some activity, usually less than 10%, was also found in the 1 M NaCl, 0.01 M phosphate buffer, pH 7.4, wash fraction. Hydrophobic matrices, such as octyl-Sepharose (Pharmacia), can also be used but was found to give a less reproducible yield.

Since gel chromatography matrices do not swell as much in 1 M acetic acid as in buffers of neutral pH, it may be difficult to obtain a good flow rate in a P-150 column eluted with 1 M acetic acid. We have therefore operated the column at room temperature. In addition, the following steps have proven to be important during the preparation of the column. The gel (BioGel P-150, 100–200 mesh) should be swollen at room temperature for at least 2 days. During packing, the column is first clamped and 1 M acetic acid is added to a height of 60 cm. The gel slurry (approximately 50% gel v/v) is then added and allowed to sediment, without flow, through the 1 M acetic acid solution for at least 4 hr. The flow is then started and the ʾcking continued according to the usual procedure.

The last purification step, on the reversed-phase column, has a high resolving power. We have used HPLC (RP8 column) and FPLC (ProRPC column) with equally good results. The solvent system containing 2 M guanidine–HCl, 1 M acetic acid was found to give the most reproducible yield among the solvent systems that we have tried. A rather good result was also obtained with 0.1% trifluoroacetic acid in conjunction with a gradient of acetonitrile. Pyridine/formic acid, as used previously,[2] was found to give a good recovery but led to chemical modification of PDGF.

The method described in this communication is used continuously in our laboratory to purify PDGF. It has proven to be a reliable method which give a highly purified product at a reasonable yield.

Acknowledgments

We express our sincere thanks to the Finnish Red Cross, Helsinki (G. Myllylä) and the Department of Virology, Uppsala (G. Alm) for the generous supply of platelets. Financial support was obtained from the Swedish Cancer Society, the Swedish Medical Research Council, the Department of Agriculture, Konung Gustav V:s 80-årsfond, and the University of Uppsala.

[2] A Radioreceptor Assay for Platelet-Derived Growth Factor

By JAI PAL SINGH

Platelet-derived growth factor (PDGF) is a polypeptide released during platelet aggregation and formation of thrombus in response to tissue injury. *In vitro* PDGF is strongly mitogenic to smooth musle cells, skin fibroblasts, 3T3 cells, and glial cells.[1-4] It is believed that the growth-promoting properties of PDGF play an important role in tissue repair mechanisms *in vivo*. Under normal circumstances the circulating levels of PDGF are very low.[5] Measurable quantities of PDGF are found only when blood is allowed to clot,[1,2] suggesting that PDGF in blood appears only for a brief period and most probably remains localized to the area of injury. It has been speculated that continual presence of PDGF in the tissues may contribute to the development of certain proliferative diseases.[6-8] Ross and Glomset[6] have hypothesized that PDGF produced at the surface of repeatedly injured blood vessels may contribute to the development of fibrous plaque in arteriosclerosis. Recent studies show that a variety of transformed cells secrete PDGF-like molecules in their growth media (see Antoniades and Pantazis [3], this volume). Cell se-

---

[1] R. Ross, J. Glomset, B. Kariya, and L. Harker, *Proc. Natl. Acad. Sci. U.S.A.* **71**, 1207 (1974).

[2] N. Kohler and A. Lipton, *Exp. Cell Res.* **87**, 297 (1974).

[3] B. Westermark and Å. Wasteson, *Exp. Cell Res.* **98**, 170 (1976).

[4] C. D. Scher, R. C. Shepard, H. N. Antoniades, and C. D. Stiles, *Biochem. Biophys. Acta* **560**, 217 (1979).

[5] J. P. Singh, M. A. Chaikin, and C. D. Stiles, *J. Cell Biol.* **95**, 667 (1982).

[6] R. Ross and J. A. Glomset, *N. Engl. J. Med.* **295**, 369 (1976).

[7] M. D. Waterfield, G. T. Serace, N. Whittle, P. Stroobant, A. Johnsson, Å. Wasteson, B. Westermark, C.-H. Heldin, J. S. Huang, and T. F. Deuel, *Nature (London)* **304**, 35 (1983).

[8] R. F. Doolittle, M. W. Hunkapiller, L. E. Hood, S. G. Devare, K. C. Robbins, S. A. Aaronson, and H. N. Antoniades, *Science* **221**, 275 (1983).

creted activity can bind to the PDGF receptor and stimulate growth through an autocrine mechanism. Such autocrine production of growth stimulatory polypeptides may cause unrestricted growth of producer cells and other cells in the vicinity leading to pathogenesis.[9] Thus the availability of a quantitative assay for the determination of PDGF in biological systems is very important in establishing relationships, if any, between the disease states and the occurrence of PDGF.

Until now, biological activity, as measured by induction of DNA synthesis in cultured fibroblasts, has been the most widely used measurement of PDGF.[10] Unfortunately the biological assay suffers from a serious drawback of not being specific. Other growth factors such as insulin, epidermal growth factor (EGF), fibroblast growth factor (FGF), and insulin-like growth factor (IGF) occurring in serum or biological sources also stimulate DNA synthesis in fibroblasts. In addition, the biological assay is lengthy, normally taking 48–72 hr to complete. A radioimmunoassay for PDGF has been described,[11,12] but the titers of antibody produced against human PDGF have been relatively low. Also, PDGF antibodies are not widely available. The radioreceptor assay (RRA) described here utilizes competition between radiolabeled ligand (^{125}I-labeled PDGF) and unlabeled ligand (PDGF) for binding to a high affinity and specific cell surface receptor (BALB/c 3T3 in this case). The assay is sensitive enough to detect 1–2 ng of PDGF in a milliliter of sample. Most importantly the RRA is specific for polypeptides containing PDGF-like functional domain for receptor binding.

Cell Cultures

Stock cultures of BALB/c 3T3 cells were maintained in Dulbecco's modified Eagle's medium (DME) supplemented with high glucose and 10% calf serum as described before[13] and elsewhere in this volume. Cultures were passaged at weekly intervals before they became confluent. Only early passage cells were used. Cultures older than 8–10 passages were discarded and new cultures were started from frozen stock. Cultures of clone-2, a spontaneously transformed variant of BALB/c 3T3,[14]

[9] M. B. Sporn and G. J. Todaro, N. Engl. J. Med. 303, 878 (1980).
[10] R. W. Pierson, Jr., and H. M. Temin, J. Cell. Physiol. 79, 319 (1972).
[11] C.-H. Heldin, B. Westermark, and Å. Wasteson, Exp. Cell Res. 136, 225 (1981).
[12] J. S. Huang, S. S. Huang, and T. F. Deuel, J. Cell Biol. 97, 383 (1983).
[13] W. J. Pledger, C. D. Stiles, H. N. Antoniades, and C. D. Scher, Proc. Natl. Acad. Sci. U.S.A. 74, 4481 (1977).
[14] W. J. Pledger, C. A. Hart, K. L. Locatell, and C. D. Scher, Proc. Natl. Acad. Sci. U.S.A. 78, 4358 (1981).

benzo[a]pyrene transformed cells,[15] SV40 transformed human skin fibro-blasts (SV80),[16] and Kirstan sarcoma virus transformed BALB/c 3T3 (K_{234}) cells,[17] used for the preparation of conditioned media, were similarly maintained. In most cases the transformed cells grow up to a 4- to 6-fold higher density than BALB/c 3T3 cells. Pluripotent teratocarcinoma stem cells PSA-1 were adapted for growth on gelatinized tissue culture dishes.[18] These cells spontaneously differentiate into fibroblast-like cells when grown at high cell densities. Stem cells were plated at 1×10^5/dish on 35-mm tissue culture dishes pretreated at 4° for 15 min with 0.1% gelatin in DME containing 10% calf serum. To obtain PSA-1-derived fibroblast-like cells, a dense culture of stem cells was trypsinized and replated in the absence of gelatin. The attached cells were retrypsinized and plated at 1×10^5 cells per dish in DME containing 10% calf serum.

PDGF Bioassay

Mitogenic activity of PDGF and media conditioned by transformed cells was determined by the stimulation of DNA synthesis in confluent cultures of mouse embryo fibroblast BALB/c 3T3 cells.[19] BALB/c 3T3 cells were plated in flat bottom 96-well microtiter plates in DME, 10% calf serum and grown to confluency by incubation at 37° in 5% CO_2 and 95% air. Confluent monolayers were washed twice with phosphate-buffered saline (PBS) and spent medium was replaced with fresh DME containing 5% platelet poor plasma, 5 μCi [³H]thymidine/ml, and serial dilutions of PDGF samples. After 24 hr, plates were fixed with methanol and DNA synthesis was determined by autoradiography. PDGF eluting off the Waters I-125 column (Fig. 1), was assayed after removal of ammonium acetate either by lyophilization or dialysis.

Preparation of PDGF and Radiolabeled PDGF

An outline of the methodology used for the preparation of purified PDGF and radioiodinated PDGF is shown in Fig. 1. The first two chro-

[15] R. W. Holley, J. H. Baldwin, J. A. Kieman, and T. O. Messer, *Proc. Natl. Acad. Sci. U.S.A.* **73**, 3229 (1976).
[16] C. D. Scher, W. J. Pledger, P. Martin, H. N. Antoniades, and C. D. Stiles, *J. Cell. Physiol.* **97**, 371 (1978).
[17] W. H. Kirsten and L. A. Myer, *J. Natl. Cancer Inst. (U.S.)* **39**, 311 (1967).
[18] L. J. Gudas, J. P. Singh, and C. D. Stiles, *Cold Spring Harbor Conf. Cell Proliferation* **10**, 229 (1983).
[19] H. N. Antoniades, C. D. Scher, and C. D. Stiles, *Proc. Natl. Acad. Sci. U.S.A.* **76**, 1809 (1979).

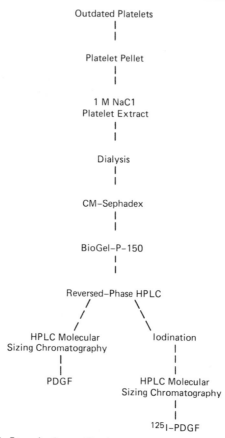

Outdated Platelets

Platelet Pellet

1 M NaCl
Platelet Extract

Dialysis

CM–Sephadex

BioGel–P–150

Reversed–Phase HPLC

HPLC Molecular   Iodination
Sizing Chromatography

PDGF   HPLC Molecular
Sizing Chromatography

^{125}I–PDGF

FIG. 1. Steps in the purification and radiiodination of PDGF.

matographic steps of this purification scheme were adopted from the previously described method of Antoniades et al.[19] Outdated platelets were
separated from plasma by centrifugation at 10,000 g for 30 min. Platelet
pellets were resuspended in 1 M NaCl, heated to 80° for 10 min, cooled to
4° by placing on ice, and then centrifuged (10,000 g for 30 min). The
supernatants were dialyzed at 4° against 10 mM sodium phosphate buffer,
pH 7.4, containing 0.1 M NaCl and chromatographed on a CM-Sephadex
followed by a BioGel P-150 column.[19] Active fractions from the BioGel P-
150 column were pooled, lyophilized, and resuspended in the initial solvent (A) prepared for reversed-phase, C_{18} HPLC column. Reversed-phase
HPLC chromatography was performed using a Waters Associates C_{18}
column (7.5 mm × 30 cm) and a Micromeritics HPLC system. Similar
results are obtained on a Dupont C_8 column. Initial solvent (A) was pre-

pared by mixing 5% 2-propanol, 20% acetonitrile, 0.1% trifluoroacetic acid, and H_2O. Solvent B was 100% 2-propanol containing 0.1% trifluoroacetic acid. Solvents A and B were degassed by bubbling helium. All solvents and H_2O were of HPLC grade. BioGel P-150 purified PDGF containing 2–4 mg protein was applied to the previously equilibrated reversed-phase column at a flow of 1.7 ml/min. After sample injection (if the volume is large, multiple injections are required), the column was washed with solvent A for 5 min and then eluted with a linear gradient formed between solvent A and solvent B over a period of 40 min. The final concentration of solvent A was 65% and solvent B was 35%. The column eluent was monitored at 280 nm using a variable UV detector. Reproducibly, the PDGF activity was eluted in the early part of the gradient corresponding to 17–23% 2-propanol. The reversed-phase chromatographic fractionation produced about 15- to 20-fold purification and 50–60% recovery of PDGF activity. Active fractions from the reversed-phase column were pooled, lyophilized, and resuspended in minimum volume (0.2–0.5 ml) of 1 M ammonium acetate. Half of the sample containing approximately 0.1–0.2 mg protein was fractionated on Waters Associates' I-125 molecular sizing columns. For optimum purification three ^{125}I columns (7.5 mm × 30 cm) were connected in series and equilibrated with 1 M ammonium acetate at a flow rate of 0.4 ml/min. Ammonium acetate (1 M) was found useful for dissociation of multimeric forms of a major low-molecular-weight contaminant present in the reversed-phase purified PDGF. Ammonium acetate is more suitable as compared to other salts because it can be lyophilized. PDGF from the sizing columns was eluted at a molecular weight corresponding to 30,000–35,000.

The other half of the reversed-phase purified samples was radioiodinated using chloramine-T by the method of Hunter and Greenwood.[20] For radioiodination 80 μl of the sample was mixed with 20 μl of 0.5 M sodium phosphate buffer, pH 7.4, 1–2 mCi of Na ^{125}I, and 10 μl of 10 mM chloramine-T solution prepared freshly in 0.5 M sodium phosphate buffer, pH 7.4. After 2–3 min on ice the sample was fractionated on the I-125 molecular sizing columns as described for unlabeled PDGF. The rapid fractionation of the iodination reaction mixture accomplishes termination of reaction, separation of labeled proteins from the free isotope, and purification of radioiodinated PDGF in a single step. It is important to perform the iodination and fractionation in a chemical hood equipped for radioiodination. The retention time for ^{125}I-labeled PDGF on the sizing column was similar to that of unlabeled PDGF. PDGF and ^{125}I-labeled PDGF obtained from the sizing column chromatography were lyophilized and resus-

[20] W. M. Hunter and F. C. Greenwood, *Nature (London)* **194**, 495 (1962).

pended in DME for binding studies. It is important to remove ammonium acetate from the PDGF sample because it may interfere with the biological activity and internalization of the cell-associated PDGF.

Preparation of Serum Samples and Platelet-Poor Plama

Human serum was prepared from freshly drawn blood in acid citrate–phosphate dextrose buffer by recalcification with stoichiometric equivalent of $CaCl_2$ and clotting by incubation at 37°. Blood from other animals was drawn by cardiac puncture or venipuncture and serums were prepared by allowing freshly drawn blood to clot at 37°. To prepare platelet-poor plasma (PPP) an aliquot of whole blood was centrifuged for 30 min at 28,000 g at 4° to assure removal of platelets. All serum and plasma samples were heated at 56° for 30 min and dialyzed against DME before use in the assay.

Preparation of Media Conditioned by Transformed Cells

Conditioned media from the cell lines listed in Table I were prepared by plating 2×10^5 cells per T-150 tissue culture flask in DME containing

TABLE I
DEMONSTRATION OF PDGF-LIKE ACTIVITY IN MEDIA CONDITIONED BY
TRANSFORMED CELLS

Source of conditioned medium	Inhibition of 125I-labeled PDGF binding[a] (%)	Calculated PDGF-like activity (ng/ml)[b]	Stimulation of DNA synthesis (% labeled cells)[c]
BALB/c 3T3	0.0	—	8
Clone-2	20.0	1.3	20
Benzo[a]pyrene transformed	40.0	2.7	70
K_{A-234}	33.0	2.2	80
SV80	35.0	2.4	80
PSA-1[d]	74.0	8.9	90

[a] 125I-Labeled PDGF binding assays were performed as in Fig. 2 except that 0.3 ml of 6-fold concentrated conditioned media was used in place of serum.

[b] Values estimated from the standard curve as in Fig. 2 are expressed as PDGF-like activity in unconcentrated media.

[c] Two-fold concentrated media were assayed for the stimulation of DNA synthesis as described under PDGF bioassay.

[d] Conditioned medium was concentrated only 3-fold.

10% calf serum. Plated cells were incubated at 37° in an atmosphere of 5% CO_2, 95% air. After 3–5 days when the cultures reached confluency, the spent medium was removed, and the cells were rinsed 2–3 times with phosphate-buffered saline. Washed cells were further incubated in DME containing 0.1% platelet-poor plasma. The exception to this was teratocarcinoma stem cells PSA-1, which were plated on gelatinized T-150 flasks at a density of 5×10^6 cells per flask. When the cells reached 2×10^7 per flask, they were washed and incubated with DME containing 0.1% platelet-poor plasma, 1 μg/ml transferrin, and 2 μg/ml insulin. After 48 hr medium was harvested by centrifugation (10,000 g, 15 min) to remove cellular debris. All conditioned media were concentrated 5- to 6-fold using a Diaflo filtration apparatus fitted with 5000 molecular weight cutoff membrane. The concentrated media were dialyzed against DME and kept frozen until use.

RRA Protocol and Results

BALB/c 3T3 cells were prepared by plating $1-2 \times 10^4$ cells in 35-mm tissue culture dishes in DME, 10% calf serum. The plates were incubated at 37° under 5% CO_2, 95% air for 4–5 days so that a confluent monolayer of cells was obtained. The final cell density was normally $4-5 \times 10^5$ cells/dish. Confluent cells were washed three times with cold PBS, covered with about 1 ml of washing solution, and transferred to the cold room at 4°. All plates were arranged on a flat surface. After 10 min of temperature equilibration, overlaying PBS was aspirated and replaced with the assay mixture. The assay mixture consisted of DME containing 1 mg/ml bovine serum albumin (BSA), ^{125}I-labeled PDGF, and the serum or conditioned media samples. For example, a typical assay for the determination of PDGF in serum contained 0.390 ml of DME/BSA (1 mg/ml), 10 μl of ^{125}I-labeled PDGF (60,000–80,000 cpm), and 0.3 ml of dialyzed serum sample. The total assay volume was adjusted to 0.7 ml with DME/BSA (1 mg/ml). All components of the assay mixture were precooled to 4° before addition to the cells. The plates were incubated for 2 hr with occasional shaking. Unbound ^{125}I-labeled PDGF was then removed by aspiration of the assay mixture and washing the monolayers 4–5 times with PBS containing 1 mg/ml BSA. The cells were then incubated with 1 ml of 1% Triton X-100 containing 10% glycerol at room temperature for 15 min. The contents of the dishes were solubilized and transferred to tubes for gamma counting. Nonspecific binding was determined by including a control containing 100-fold molar excess of unlabeled PDGF.

A standard curve generated using this protocol and known quantities of pure PDGF is shown in Fig. 2. A sample of freshly prepared human

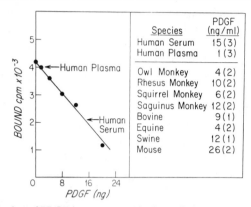

Species	PDGF (ng/ml)
Human Serum	15(3)
Human Plasma	1(3)
Owl Monkey	4(2)
Rhesus Monkey	10(2)
Squirrel Monkey	6(2)
Saguinus Monkey	12(2)
Bovine	9(1)
Equine	4(2)
Swine	12(1)
Mouse	26(2)

FIG. 2. Quantitation of PDGF in serum samples by radioreceptor assay. Between 300 and 400 μl of each sample was admixed with 8 ng of ^{125}I-labeled PDGF (specific activity 6000 cpm/ng). The final assay volume was 700 μl. The indicated PDGF content of the serum samples was read from the standard curve and normalized to 100% concentration. Human platelet-poor plasma which contains very little PDGF was actually assayed at 100% concentration. The numbers in parentheses indicate the number of determinations done for each sample. Experimental variations in these determinations were <5%.

serum produced 57% inhibition of PDGF binding, a competition equivalent to 15 ng/ml of pure PDGF. On the other hand platelet-poor plasma prepared from the same blood sample contained only 1 ng of PDGF per ml. These concentrations of PDGF in human serum and plasma are in close agreement with the concentrations previously determined by purification data and biological activity.[19]

Many transformed cells are known to exhibit lower serum requirements for growth.[16] We found that these cells also exhibit low binding of ^{125}I-labeled PDGF. We analyzed the media conditioned by the transformed cells (Table I) for PDGF-like activity using the radioreceptor assay. The transformed cells appear to produce greater than 1–2 ng PDGF like activity per ml of culture medium. These amounts are equivalent to PDGF found in 10% calf serum suggesting that the PDGF-like activity secreted by the cells may be sufficient to support their growth in low serum. The presence of competing activity determined by RRA also corresponds to the presence of mitogenic activity (Table I).

Sensitivity

BALB/c 3T3 cells exhibit about 150,000 high affinity ($K_d \sim 5 \times 10^{-10}$ M) receptors per cell.[5] These cells, therefore, provide a sensitive means of measuring PDGF binding activity. The standard curve presented in Fig. 2 was developed using ^{125}I-labeled PDGF with a specific activity of 6000–8000 cpm/ng. The detection limit of this assay was about 1 ng of PDGF present in 1 ml of sample. Higher specific activity ^{125}I-labeled PDGF is expected to provide better sensitivity. Assuming the molecular weight of PDGF to be 33,000, this assay can detect up to 0.03 pmol of PDGF/ml. This sensitivity is comparable to a typical ELISA or RIA.

Specificity

The most important advantage of RRA over a biological assay is its specificity as demonstrated in Fig. 3. In this experiment ^{125}I-labeled PDGF binding was determined in the presence of various concentrations of PDGF or other growth factors active on BALB/c 3T3 cells. ^{125}I-Labeled PDGF binding to BALB/c 3T3 was quantitatively competed out by approximately 10-fold excess of PDGF. On the other hand EGF, FGF, or insulin (not shown here) did not prevent ^{125}I-labeled PDGF binding even at a concentration in excess of 1000-fold. Similarly, other platelet-derived proteins (platelet factor IV, β-thromboglobulin) failed to prevent ^{125}I-labeled PDGF binding to BALB/c 3T3 cells. The results shown in the inset of Fig. 3 show similar kinetics of ^{125}I-labeled PDGF displacement by unlabeled PDGF or serum suggesting that the competition exhibited by serum is due to PDGF. However, it should be noted that the determinations by RRA depend upon the equilibrium dissociation constant (K_d) and

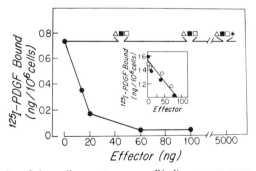

FIG. 3. Specificity of the radioreceptor assay. Binding assays were performed as described under protocol by mixing 4 ng of ^{125}I-labeled PDGF (specific activity 8000 cpm/ng) with the indicated quantities of unlabeled PDGF (●), EGF (■), pituitary FGF (□), platelet factor IV (▷), or β-thromboglobulin (∗). Inset: 8 ng of ^{125}I-labeled PDGF tracer was admixed with unlabeled PDGF (●) or human serum (○). The units on the abscissa are ng × 10⁻¹ for PDGF and percentage of serum.

the estimates may differ from the actual concentration if the K_d of the competing molecule for the receptor is significantly higher or lower than ^{125}I-labeled PDGF.

Acknowledgment

The author is thankful to Rose Erler for preparation of the manuscript.

[3] Structural and Functional Identification of Platelet-Derived Growth Factor-Like Proteins Produced by Mammalian Cells

By HARRY N. ANTONIADES and PANAYOTIS PANTAZIS

Introduction

The amino-terminal amino acid sequence of human platelet-derived growth factor (PDGF) revealed that it consists of two homologous polypeptide chains named PDGF-1 and PDGF-2.[1] The PDGF-2 chain has been shown to be near identical to a portion of the transforming gene product of the simian sarcoma virus (SSV),[2,3] an acute transforming retrovirus of primate origin. Characterization of its genome has localized its transforming gene to its cell-derived *onc* sequence, v-*sis*[4,5] which codes for a protein (p28[sis]) consisting of 226 amino acid residues.[6] The region of p28[sis] corresponding to PDGF starts at the serine residue in position 67, which follows a double basic (Lys-Arg) sequence at positions 65–66.[2] This appears to be the processing point yielding a polypeptide of 160 residues with a molecular size of 18,056 Da, essentially the same size estimated for the PDGF-2 chain on the basis of sodium dodecyl sulfate–polyacrylamide gel electrophoresis.[1] However, the SSV *onc* gene codes only for the PDGF-2 chain of the PDGF polypeptide. It was important, therefore, to establish whether the *onc* gene product was functioning as a single polypeptide chain or in a conformational manner similar to the dimeric formation of

[1] H. N. Antoniades and M. W. Hunkapiller, *Science* **220**, 963 (1983).
[2] R. F. Doolittle *et al., Science* **221**, 275 (1983).
[3] M. Waterfield *et al., Nature* (*London*) **304**, 35 (1983).
[4] K. C. Robbins *et al., Proc. Natl. Acad. Sci. U.S.A.* **78**, 2918 (1981).
[5] E. P. Gelman *et al., Proc. Natl. Acad. Sci. U.S.A.* **78**, 3373 (1981).
[6] S. G. Devare *et al., Proc. Natl. Acad. Sci. U.S.A.* **80**, 731 (1983).

biologically active PDGF. This important question has been resolved recently by the demonstration that, in SSV-transformed cells, p28^{sis} undergoes a series of discrete processing steps including dimer formation and proteolytic cleavage to yield molecules structurally and immunologically resembling the disulfide-linked dimeric forms of PDGF.[7] More recent studies have shown that this processed product is secreted by SSV-transformed fibroblasts into their culture media in a biologically active form which is recognized by PDGF antisera.[8] These studies established conclusively that the biologically active SSV *onc* gene product is a homodimer, consisting of two PDGF-2 chains linked together by disulfide bonds. Its immunologic reactivity and biologic properties were shown to be identical to those of human PDGF, derived from platelets.

The studies summarized above revealed the structural, functional, and immunologic identity between PDGF and the *onc* gene product of the simian sarcoma virus. This information was obtained from the application of powerful procedures, described below, which enabled the investigation of the synthesis, intracellular processing, and secretion of the *sis* products as well as their function. More recently, these procedures have been applied successfully to the investigations of the c-*sis*/PDGF-2 gene products synthesized by human malignant cells of mesenchymal origin, such as glioblastoma, fibrosarcoma, and osteosarcoma cells. *sis* transcripts have been demonstrated in these malignant cells,[9–11] a finding consistent with their ability to synthesize PDGF-like proteins. The human protooncogene, c-*sis*, a counterpart to viral oncogene v-*sis*, has also be shown to code for the PDGF-2 chain.[12]

Following is a description of the methods and techniques applicable to investigations of the synthesis, processing, secretion, and function of the *sis onc* gene products synthesized by SSV-transformed cells or by cultured human malignant cells exhibiting c-*sis* transcripts. The *sis* products synthesized by these cell lines represent only trace amounts. For this reason, their detection involves sensitive procedures including their metabolic labeling in cells in culture, followed by their immunoprecipitation with specific PDGF antisera and analysis of the products by SDS–PAGE. Biologic properties can be assessed by cell culture procedures, while additional specificity can be evaluated using PDGF receptor binding assays and induction of protein phosphorylation of specific cell membrane

---

[7] K. C. Robbins *et al.*, *Nature (London)* **305**, 605 (1983).
[8] A. J. Owen *et al.*, *Science* **225**, 54 (1984).
[9] A. Eva *et al.*, *Nature (London)* **295**, 116 (1982).
[10] D. T. Graves *et al.*, *Science* **226**, 972 (1984).
[11] P. Pantazis *et al.*, *Proc. Natl. Acad. Sci. U.S.A.* **82**, 2404 (1985).
[12] S. F. Josephs *et al.*, *Science* **223**, 487 (1984).

proteins. Most of the techniques described here have been applied for the investigation of the *sis*/PDGF-2 products synthesized by SSV-transformed cells and by human malignant cells of mesenchymal origin (see also Stroobant *et al.*,[4] this volume).

Procedures for the Study of the Synthesis, Processing, and Secretion of the *sis*/PDGF-2 Gene Products

Immunoprecipitation of Metabolically Labeled sis Products in Cell Culture with Specific Antisera

This procedure involves the introduction of labeled amino acids to cell culture. The labeled amino acids are incorporated into the *sis* precursor products synthesized by the cells. Continuous labeling (pulse-labeling) for a given period of time will provide information for the presence of *sis* precursors and processed products which appear during the labeling period. Labeling for defined periods (pulse-chase labeling) will provide information on the time course of processing events of the *sis* product.

The radiolabeled *sis* products which appear during pulse-labeling or pulse-chase labeling can be immunoprecipitated from the lysates and the conditioned media using specific PDGF antisera or antisera against synthetic polypeptides corresponding to various PDGF regions. The immunoprecipitates can be analyzed by SDS–PAGE and fluorography. These procedures include the following steps: metabolic labeling of cultured cells, preparation of cell lysates and conditioned media for immunoprecipitation with specific PDGF antisera, analysis of immunoprecipitated proteins by SDS–PAGE, and fluorography.

Step 1: Metabolic Labeling of Cells in Culture. Adherent cells in culture are grown in T75 flasks (Falcon) in Dulbecco's modified Eagle's minimal essential medium (DMEM) supplemented with 10% calf serum. The cultures are incubated in humidified 5% CO_2 atmosphere at 37°.

For metabolic labeling using [^{35}S]cysteine and [^{35}S]methionine, the media are aspirated from confluent cultures containing $4-5 \times 10^6$ cells per flask, and the cells are rinsed with 8–10 ml of serum-free/cystine-free/methionine-free DMEM (GIBCO; Grand Island, NY), prewarmed at 37°. The rinsed cells are then incubated in 5 ml of serum-free/cystine-free/methionine-free DMEM containing 250–300 μCi [^{35}S]cysteine and 100 μCi [^{35}S]methionine (~1200 Ci/mmol, Amersham) per ml.

For studies involving continuous labeling (pulse-labeling), the cultures are incubated in the presence of the radiolabeled amino acids for 2 to 4 hr. For prolonged pulse-labeling periods (14–16 hr) the cells are labeled in media containing 10% of the normal cysteine and methionine concentrations plus 0.5% fetal calf serum, to secure cell viability. For pulse-chase labeling, the cells are pulse-labeled for a short period (15 to 30 min) and then chased with DMEM supplemented with 250 μg of unlabeled cysteine and methionine per ml for the desired periods, usually 0 to 120 min. At the end of the labeling, or the chase period, the culture media are removed immediately and saved for immunoprecipitation of the secreted products, as described below. The cells are washed once with ice-cold phosphate-buffered saline (PBS) containing a mixture of protease inhibitors (10 μg/ml leupeptin, 10 μg/ml pepstatin A, and 1 mM phenylmethylsulfonyl fluoride). The washed cells are collected in 10 ml of PBS plus inhibitors, by scraping gently with a rubber cell remover (A. H. Thomas, Philadelphia, PA). The detached cells in PBS are pelleted by gentle centrifugation and they are washed once in ice-cold inhibitor-containing PBS. The washed cell pellet is prepared for immunoprecipitation as described in step 2.

The procedure described above is applicable to adherent cells in culture. For cells cultured in suspension, including human leukemic cell lines, the process is similar with the following modifications. The cells are grown in RPMI 1640 media (GIBCO, Grand Island, NY; Biofluids Inc., Rockville, MD) containing 10% heat-inactivated human serum[13] or 10% fetal calf serum, using Falcon sterile plasticware. For cell labeling, cultures of logarithmically grown cells (about 1 × 10⁶ cells per ml) are gently pelleted by centrifugation, and the cell pellet is resuspended in prewarmed cystine-free/methionine-free/serum-free RPMI 1640 medium containing 250–300 μCi [³⁵S]cysteine and 100 μCi [³⁵S]methionine per ml. The cell suspensions (about 5 ml) are placed in T25 flasks and incubated for 3 to 4 hr at 37° in a 5% CO_2 atmosphere. At the end of the pulse-labeling period, the cells are pelleted at 800 g for 10 min and washed once in cold PBS containing protein inhibitors as described above. The conditioned media and the washed cell pellets are prepared for immunoprecipitation. For pulse-chase labeling the protocol is the same as the one described above for the adherent cells in culture.

Step 2: Immunoprecipitation of Metabolically Labeled Proteins. The washed cell pellets obtained from step 1 are suspended in 1.0–1.5 ml of radioimmunoprecipitation (RIP) buffer consisting of 150 mM NaCl, 50 mM Tris, pH 7.5, 1% Triton X-100, 1% sodium deoxycholate, 0.1% SDS, 1 mM PMSF, 10 μg/ml leupeptin, 10 μg/ml pepstatin A, and lysed for 10 min under continuous, gentle agitation on a vertical rotator at 4°. The cell suspension is microcentrifuged (Fisher microcentrifuge) for 10 min to remove insoluble material. The clarified supernatant fluid is precleared

¹³ P. Pantazis *et al.*, "Cancer Cells 3: Growth Factors and Transformation" (J. Feramisco, B. Ozanne, and C. Stiles, eds.), p. 153. Cold Spring Harbor Lab., Cold Spring Harbor, New York, 1985.

with about 40 μl of protein A bound to Sepharose CL-4 beads (Pharmacia, Sweden). The precleared lysate is divided into three equal portions and incubated with normal rabbit serum (NRS), antiserum to PDGF or to synthetic polypeptides (anti-PDGF), and with anti-PDGF premixed with excess purified PDGF ($\sim$500 ng). Incubation is for 16–18 hr at 4° under continuous gentle agitation. Protein A beads are added to each incubation mixture and incubation continues for 60 min at 4° to allow binding of protein A to complexes of IgG/PDGF-like proteins. The beads are pelleted by microcentrifugation and washed three times in RIP buffer and once in 15 mM Tris, pH 7.5. Bead pellets are prepared for SDS–gel electrophoresis as described below.

For immunoprecipitation of conditioned media, the media are centrifuged at 5000 g (Sorvall) to remove insoluble material. The clarified fluid is lyophilized and reconstituted with 1 ml RIP buffer per 10 ml original media, and dialyzed against 200 volumes of the same buffer for 3 hr at 4°. The dialyzed material is divided into three equal portions and subjected to immunoprecipitation as described above.

Step 3: SDS–Gel Electrophoresis. Following immunoprecipitation, pellets of protein A beads are suspended in 100–150 μl of SDS sample buffer (70 mM Tris, pH 6.8, 2% SDS, 10% glycerol, 0.1% bromphenol blue) and placed in boiling water for 3 min. The beads are pelleted by microcentrifugation, and the supernatant fluid is divided into two equal volumes: one volume remains unreduced, and the other is reduced by addition of 2-mercaptoethanol (5% v/v) and incubated at room temperature for 1 hr. Unreduced and reduced samples are electrophoresed on 16% acrylamide slab gels.[14] Alternatively, equal volumes of reduced samples are analyzed on 8 and 15% acrylamide gels for resolution of high- and low-molecular-weight proteins, respectively. Molecular weight protein markers (Bio-Rad, Richmond, California; Pharmacia, Sweden) are used unreduced and reduced. Upon termination of electrophoresis, the gels are stained, destained, treated with Amplify (Amersham), and dried under heat and vacuum. Bands of protein markers on dried gels are marked with ^{14}C-labeled ink. The dried gels are exposed to Kodak XAR film at −80° for 1–3 weeks.

Electrophoretic Protein Transfer

The immunoprecipitation procedures described above require biosynthetic labeling of the *sis* products with radioactive amino acids and analysis of the radiolabeled immunoprecipitate by SDS–polyacrylamide gel. Other procedures in which unlabeled proteins are first separated by SDS–

[14] P. Pantazis and W. M. Bonner, *J. Biol. Chem.* **256**, 4669 (1981).

polyacrylamide gel slabs and then identified by a radioimmunoassay have been applied for the detection of the *sis* products.[15,16] They are based on the transfer of proteins from the polyacrylamide gel to commercially available nitrocellulose sheets. The transfer can be achieved by electrophoresis, or by diffusion, yielding replicas of the original gel pattern. The immobilized proteins on the replica are exposed to antisera resulting in the *in situ* localization of the antisera-reactive proteins. Following is a description of the method, introduced by Towbin *et al.*,[17] and applied by us and by Niman[15,16] for the identification of *sis* products.

Procedure. Cells are grown to confluence as described above. The cells are rinsed with serum-free media and incubation continues in the serum-free media. The cells are harvested at the desired time interval and they are lysed in a buffer similar to RIP buffer, containing protein inhibitors, but with the concentration of SDS made to 0.5% instead of 0.1%. The cell lysate is microcentrifuged for 10 min to remove insoluble material. The clarified supernatant fluid is pretreated with normal serum, homologous to antiserum, and protein A bound to Sepharose Cl-4 beads, as described. The clarified lysates are electrophoresed onto a 7.5–17% polyacrylamide gel. Electrotransfers to nitrocellulose sheets are performed as described[17] in a buffer of 25 mM Tris, pH 8.3, 200 mM glycine, containing 20% methanol (v/v) at 50 V, overnight at 4°. Nonspecific binding is blocked by saturating the sheets with 3% human or bovine serum albumin and 0.1% Triton X-100 in PBS. The sheets are incubated for 1 hr at 37° with PDGF antisera, or antisera to synthetic region of the *sis* product, made in 3% albumin, 0.1% Triton X-100 in PBS. After washing three times with 0.1% Triton X-100 in PBS the sheets are incubated with 10^7 cpm of ^{125}I-labeled protein A, or ^{125}I-labeled second antibody, washed with 0.1% Triton X-100 in PBS, and subjected to autoradiography using XRP-1 film (Kodak) and Cronex Hi-plus intensifying screen (Dupont) at −80° for 3.5 hr.

General Comments. The use of both methods, immunoprecipitation and protein transfer, is desirable. The immunoprecipitation procedure is more sensitive for proteins of molecular weight above 100K, and it is the only choice for the chase studies. The protein transfer is sensitive for proteins with molecular weight under 50K, and can be used for a simultaneous treatment of a large number of unlabeled samples. These procedures aim at the identification of trace amounts of PDGF-like proteins synthesized by cells in culture. Their success depends on the specificity of

[15] H. L. Niman, *Nature (London)* **307**, 180 (1984).
[16] H. L. Niman *et al.*, *Science* **226**, 701 (1984).
[17] H. Towbin *et al.*, *Proc. Natl. Acad. Sci. U.S.A.* **76**, 4350 (1979).

the PDGF antisera used in these studies. Antisera to contaminating proteins will confuse the results and will provide misleading information. Using an IgG fraction from the antiserum can help, providing cleaner backgrounds. The inclusion of appropriate controls is mandatory. They include parallel studies with nonimmune serum, or its IgG fraction, and competition studies with excess pure PDGF (500 ng) added to PDGF antiserum. If antisera to synthetic polypeptides are used, competition studies must include excess of the synthetic polypeptide. Contaminating proteins in the immunoprecipitates may be reduced by avoiding excess amount of antiserum and by washing the IgG protein A immunoprecipitate fraction, as recommended.

Application of the Procedures and Illustrative Data

The procedures described above have been applied successfully for pulse labeling and pulse-chase labeling studies of SSV-transformed cells and of human malignant cells of mesenchymal origin. These studies have revealed the production of the precursor protein p28[sis], by SSV-transformed cells, its rapid dimerization through disulfide bonds, and its processing to form disulfide-linked homodimers similar to those of human PDGF (Fig. 1).[7] These data, obtained from the pulse-chase labeling studies, established the important fact that in addition to sequence identity there is also conformational identity between the v-*sis*/PDGF-2 *onc* gene product and PDGF.

Figure 2 demonstrates the detection of a 34-kDa polypeptide in the conditioned media of pulse-labeled SSV/NRK cells.[8] Upon reduction, the molecular size of this secretory polypeptide was reduced to 17 kDa, a finding consistent with the disulfide-linked dimeric nature of PDGF-2 produced by the SSV-transformed cells. The media containing the 34-kDa polypeptide exhibited functions similar to those of PDGF.[8]

Similar studies, with pulse labeling techniques, have shown the synthesis and secretion of PDGF-like proteins by human glioblastoma, fibrosarcoma,[11] and osteosarcoma[10] cells in culture, which were shown to express the protooncogene (c-*sis*) homolog of v-*sis*. The data shown in Fig. 3 provide an example for the intracellular synthesis and secretion of PDGF-like polypeptides by human glioblastoma cells in culture. Under nonreducing conditions, the immunoprecipitates of the cell lysates (Fig. 3, lane b) were resolved into three distinct polypeptides of 36, 31, and 24 kDa. Control studies demonstrated that these polypeptides did not precipitate with preimmune rabbit sera (Fig. 3, lane a) and were competed by excess PDGF (500 ng) (Fig. 3, lane c). The 36- and 31-kDa polypeptides in lane b are within the molecular range reported for biologically active,

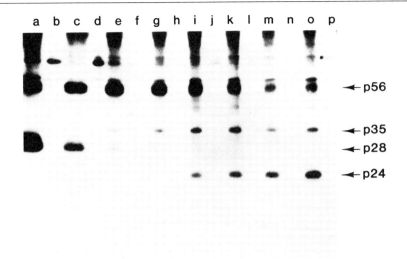

FIG. 1. Pulse-chase labeling analysis of posttranslational processing of p28^{sis} HF/SSV cells were pulse labeled for 15 min with [^{35}S]methionine and cysteine and chased for periods of 0 (lanes a and b), 15 (lanes c and d), 30 (lanes e and f), 45 (lanes g and h), 50 (lanes i and j), 75 (lanes k and l), 90 (lanes m and n), and 105 (lanes o and p) min. Immediately after the chase periods, cells were washed in ice cold PBS and disrupted. Cell lysates were immunoprecipitated with anti-PDGF (lanes a, c, e, g, i, k, and m) or normal rabbit serum (lanes b, d, f, h, j, l, and n) under nonreducing conditions and immunoprecipitates were analyzed by SDS–PAGE in the absence of reducing agent. (From Robbins *et al.*[7])

unreduced PDGF. The 24-kDa component represents an additionally processed product (Fig. 3, lane b) which is similar to the 24-kDa polypeptide shown in Fig. 1, and discussed elsewhere.[7] Under nonreducing conditions, the 31-kDa polypeptide was the primary PDGF-like product secreted in the conditioned media of the glioblastoma cells. Under reducing conditions, it was converted to the monomeric 16-kDa form. These values are similar to those of partially purified PDGF-like polypeptides derived from the conditioned media of human glioblastoma cells.[18] The media containing the 31-kDa polypeptide had biologic activity and properties similar to those of PDGF.[11] Similar results were obtained in pulse-labeled cultures of human fibrosarcoma[11] and osteosarcoma[8,19] cells in culture. Direct evidence for the nature of the PDGF-like polypeptides synthesized by the osteosarcoma cells was obtained from partial nucleotide sequence of the c-*sis* cDNA derived from the osteosarcoma cells. This partial sequence of the c-*sis* cDNA was shown to code for the carboxy-terminal

[18] N. Nister *et al., Proc. Natl. Acad. Sci. U.S.A.* **81**, 926 (1984).
[19] C. Betsholtz *et al., Biochem. Biophys. Res. Commun.* **117**, 176 (1983).

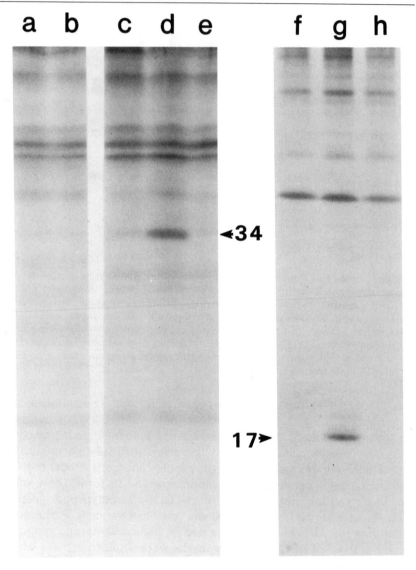

A. Unreduced   B. Reduced

FIG. 2. Antiserum to PDGF recognized the SSV oncogene product released into the conditioned medium of SSV-transformed NRK cells. In lanes a–e SDS–electrophoresis was performed under nonreducing conditions and in lanes f–h under reducing conditions. The PDGF antisera precipitated a 34-kDa protein from the SSV-transformed NRK cells (lane b). Nonimmune, control sera (lanes a and c) did not precipitate the 34-kDa protein. Excess of unlabeled PDGF (500 μg) prevented its precipitation with the antisera (lane e). Under

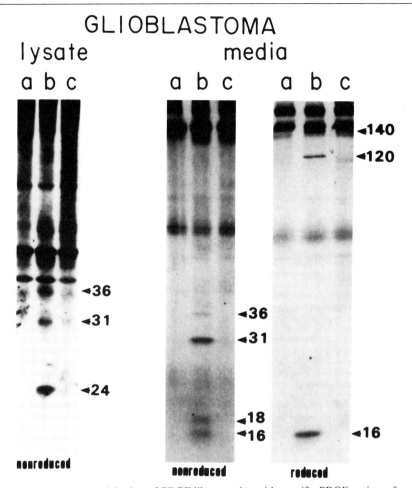

FIG. 3. Immunoprecipitation of PDGF-like proteins with specific PDGF antisera from lysates and media of human glioblastoma cells (A172) metabolically labeled with [³⁵S]cysteine. Precipitation was with normal, nonimmune serum (lane a), PDGF-antiserum (lane b), and PDGF antiserum plus excess PDGF (500 µg) (lane c). (From Pantazis *et al.*[11])

reducing conditions the 34-kDa was converted to its monomeric, 17-kDa component (lane g). Control, nonimmune sera did not precipitate the 17-kDa protein (lane f), and excess of unlabeled PDGF (500 µg) competed its binding to PDGF-antiserum (lane h). (From Owen *et al.*[8])

region of the PDGF-2 chain of the biologically active molecule of the dimeric PDGF.[20]

Procedures for Characterization of the Functions of the *sis onc* Gene Products

The procedures described above can be used effectively for the study of the synthesis, processing, and secretion of the *sis* products. They do not provide, however, information on the functional properties of these products. Such information can be obtained by other procedures, described below, which are suitable to yield information on the PDGF-like activity of the *sis* products, including their ability to stimulate DNA synthesis in target cells in culture, to induce protein phosphorylation at tyrosine residue of specific membrane proteins, and to compete with ^{125}I-labeled PDGF for binding to specific cell membrane PDGF receptors. The source material for these studies is usually serum-free conditioned media derived from cultured cells. Since the PDGF-like polypeptides are present in the media only in trace amounts, it is important to first concentrate them before attempting the functional studies.

Concentration of PDGF-Like Proteins in Serum-Free Conditioned Meda. Cells are grown to confluence in media supplemented with 10% calf serum, as described above. The serum-containing media are removed from the confluent cultures and the cells are rinsed twice with serum-free medium, prewarmed at 37°. The rinsed cells are incubated in serum-free media for 18 to 24 hr or for longer periods, if the cells remain viable. At the end of the incubation period the serum-free media are collected, they are brought to 0.2% human or bovine serum albumin, and clarified by centrifugation at 5000 *g* (Sorvall). Addition of albumin serves to minimize surface losses of the trace amounts of the PDGF-like components during the process of their concentration. Further handling of the clarified media depends on the scope of the experiment. For functional studies, the media can be concentrated by lyophilization, or they can be subjected first to partial purification before concentration.

For concentration without prior purification the media are lyophilized in plastic containers. For this purpose, aliquots (10 ml) of the clarified media are placed in plastic (polypropylene) 50-ml disposable centrifuge tubes (Corning or Falcon). The containers are covered with their plastic covers. Tiny holes are punctured in the covers and the containers are placed in an alcohol–dry ice bath for 30 min or at −80° overnight. The

[20] H. N. Antoniades *et al.,* "Cancer Cells 3: Growth Factors and Transformation" (J. Feramisco, B. Ozanne, and C. Stiles, eds.), p. 145. Cold Spring Harbor Lab., Cold Spring Harbor, New York, 1985.

frozen samples are lyophilized to dryness and reconstituted with cold 1 ml H_2O. Pooled, reconstituted fractions are dialyzed at 4° against 200 volumes of phosphate-buffered saline, or the desired buffer, with two buffer changes. Aliquots of the dialyzed, concentrated preparations are stored at $-20°$.

Partial purification of the clarified serum-free media can be achieved by cation-exchange chromatography. For this purpose, the media are mixed with CM-Sephadex C-50 (Pharmacia), preswollen in 0.08 M NaCl–0.01 M sodium phosphate buffer, pH 7.4. For small volumes of serum-free conditioned media ($\sim$100 ml), 10 ml of the CM-Sephadex is added to the media. The mixtures are incubated overnight at 4° under continuous, gentle shaking. The gel is allowed to settle, the supernatant fluid is removed, and the gel is placed in a column and washed with 30 ml of 0.08 M NaCl–0.01 M sodium phosphate buffer, pH 7.4. The column is eluted with 15 ml of 1 M NaCl. The eluate is dialyzed overnight at 4° against 200 volumes of 1 M acetic acid with one change. Aliquots of the dialyzed eluate are lyophilized. The lyophilized material is reconstituted in a minimum volume of phosphate-buffered saline, or other desired buffer, containing 0.2% human or bovine serum albumin.

Assays of PDGF-Like Activity by Cell Culture. In order to find out whether the media contain factors with PDGF-like activity, the concentrated, and/or partially purified preparations are assayed for their ability to stimulate DNA synthesis in cultures of 3T3 fibroblasts. Autoradiography or uptake of [^{3}H]thymidine by the cultured cells is used for the evaluation of DNA synthesis.[21,22]

For autoradiography, BALb/c 3T3 cells (clone A-31) are plated in 0.3-cm^2 microtiter wells (Falcon) in Dulbecco–Vogt modified Eagle's medium (DMEM) supplemented with 10% calf serum (Biofluids) and incubated at 37°. When confluent monolayers are formed, the spend medium is aspirated and replaced with 0.2 ml of fresh DMEM supplemented with 5% human platelet-poor plasma, [^{3}H]thymidine (5 μCi/ml; 6.7 Ci/mmol, New England Nuclear), and 10 μl of sample to be tested for PDGF activity. Platelet-poor plasma does not contain PDGF but it is required for the optimal replicative response. After the sample addition, the microtiter cell cultures are incubated at 37° for 24 hr and then fixed with methanol and processed for autoradiography. Kodak nuclear tract photographic emulsion NTB2 is added (about 100 μl) into each microtiter well and then removed immediately using a Titertek 12-channel pipettor (Flow Laboratories). The plates are then kept in the dark in the presence of calcium

[21] C. D. Scher *et al.*, *Nature (London)* **247**, 279 (1974).
[22] H. N. Antoniades *et al.*, *Proc. Natl. Acad. Sci. U.S.A.* **78**, 7314 (1981).

sulfate desiccant for 24 hr. The plates are developed by pipetting with the Titertek pipettor about 50 μl/well of Kodak Microdol-X developer and fixed with Kodak Rapid Fixer (100 μl). The cultures are then stained with Giemsa. Percentage labeled nuclei is determined by counting at least 200 cells with microscopic grid while scanning approximately 10,000 cells. On the basis of this assay, a unit of PDGF activity has been defined as the amount required to induce 50% (approximately 10^4) of the cells to synthesize DNA.

For estimation of the uptake of [^{3}H]thymidine, the 3T3 cells are incubated for 18 hr, as described above, and then they are pulse-labeled for 6 hr with [^{3}H]thymidine (5 μCi/ml). At the end of the 6 hr incubation the cells are rinsed with cold PBS, fixed in 200 μl of 10% trichloroacetic acid, and then rinsed with distilled water. The cells are solubilized in 200 μl of 1% SDS and the radioactivity is measured in a liquid scintillation counter.

Additional Criteria of PDGF-Like Activity. The ability of the conditioned media to stimulate DNA synthesis in cell culture is not sufficient to establish that the activity is due to PDGF-like polypeptides and not to other unrelated mitogen. Additional evidence is needed to establish this fact. Such evidence can be obtained from examination of the heat stability of the activity in the preparations, the sensitivity to reducing agents, and the neutralization of the activity by specific PDGF antisera. These criteria, described below, have been applied successfully for the investigation of PDGF activity in conditioned media derived from SSV-transformed cells[8] and from human fibrosarcoma,[11] glioblastoma,[11] and osteosarcoma[10] cell lines.

Heat Stability. The activity of human PDGF has been shown to be stable to heating at 100° for 10–20 min.[21,23] In order to test the heat stability of the activity present in the concentrates derived from the conditioned media, aliquots of these media are placed in borosilicated glass tubes (75 × 13 mm, Fisher) and the tubes are immersed in boiling water for 10 min. After cooling, the heated preparations along with the unheated, control, preparations are subjected to cell culture assay, as described above.

Sensitivity to Reducing Agents. The biologic activity of human PDGF is totally abolished by reducing agents.[23] This property has been used to establish whether the biologic activity of concentrates from conditioned media is also affected by reduction. For this purpose, aliquots of the preparations are incubated with 2-mercaptoethanol (5% v/v) for 2 hr at room temperature. Control, unreduced, preparations are also incubated for the same period of time. After incubation, both the reduced and the

[23] H. N. Antoniades *et al., Proc. Natl. Acad. Sci. U.S.A.* **76,** 1809 (1979).

unreduced sample are dialyzed, in separate containers for 24 hr at 4°, against 200 volumes of phosphate-buffered saline with three buffer changes. The dialyzed samples are assayed for PDGF-like activity as described.

Neutralization of the Activity by PDGF Antisera. Polyclonal antisera to human PDGF have been shown to neutralize the biologic effect of PDGF in cultures of untransformed mouse and human fibroblasts.[24,25] Similarly, the PDGF antisera were shown to be capable of neutralizing the biologic activity of the PDGF-like mitogen obtained from the conditioned media of human osteosarcoma cells.[25,26] This neutralizing effect of the PDGF antisera provides an additional criterion for determining whether the activity derived from the conditioned media of cultured cells is indeed due to PDGF-like polypeptides.

In practice, aliquots of the test preparations are incubated with and without PDGF antisera or, preferably, with their IgG fraction. In parallel studies, identical aliquots are incubated with and without the same amounts of preimmune serum or its IgG fractions. In a third series, a PDGF standard is also incubated with and without the PDGF antiserum or its IgG fraction. After the overnight incubation, the samples are brought to 37° and assayed for activity in cultures of 3T3 cells as described above.

General Comments. For assay for PDGF-like activity, we recommend that all samples are tested at least in triplicate, and in three different concentrations (10, 25, and 50 μl per well). Control samples must be always included in the assay along with the treated samples, and in the same microtiter plate. If the PDGF-like activity is compared to the activity of known PDGF standards, or other samples, again all samples, including controls, must be assessed in triplicate, in the same microtiter plate. Each microtiter plate should include in triplicate a negative control consisting of the serum-free medium used in the assay culture, and triplicates of standards containing 1, 5, and 10% fetal calf serum. These standards provide a good measure of the sensitivity of the assay system, and the responsiveness of the 3T3 cells used in the assay.

Binding of PDGF-Like Polypeptides to PDGF Cell Membrane Receptors

Competitive binding of PDGF-like mitogens for PDGF receptors has been used by several investigators as a criterion in order to establish a

---

[24] A. J. Owen *et al., Proc. Natl. Acad. Sci. U.S.A.* **79**, 3203 (1982).
[25] D. T. Graves *et al., Cancer Res.* **43**, 83 (1983).
[26] C.-H. Heldin *et al., J. Cell. Physiol.* **105**, 235 (1980).

similarity between these mitogens and PDGF.[18,27] This approach is justified by the fact that binding of PDGF to its receptor represents a highly specific event, requiring both structural integrity and biologic activity.[28-31] For example, reduced, inactivated PDGF preparations are not recognized by PDGF receptors.[31] Other growth factor polypeptides, such as epidermal growth factor (EGF), fibroblast growth factor (FGF), and insulin do not compete with PDGF for binding to its receptor.[28-30] Thus, competitive receptor binding between the *sis onc* gene products and PDGF is an important criterion for assessing the PDGF-like nature of these products.

An important requirement for these studies is the availability of pure ^{125}I-labeled PDGF which serves as the ligand for the receptor binding studies.

Preparation of Labeled PDGF. For radioiodination, 5 to 10 μg of pure PDGF is dissolved in 100 μl of 0.1 M sodium phosphate buffer, pH 7.0, in a borosilicate test tube (10 × 75 mm). One to two millicuries of ^{125}I (about 5 μl in 0.1 M NaOH) is added, followed by the addition of a single Iodo-Bead (Pierce Chemical), a chloramine-T derivatized polystyrene bead. After 40 min, the radioiodinated PDGF is removed and the test tube with the Iodo-Bead is rinsed twice with 200 μl of buffer and once with 200 μl of 0.02% SDS. The radioiodinated PDGF and washes are dialyzed against 0.05% SDS for 6 hr at room temperature, with frequent changes of dialysis fluid. After dialysis the solution is placed in an equal volume of 10% human serum albumin. Aliquots (100 μl) of this stock solution are kept frozen. The radioiodinated PDGF contains 15,000 to 20,000 cpm/ng of protein. Radioiodinated PDGF prepared with the procedure described above retains biologic activity as judged by its ability to stimulate incorporation of [^{3}H]thymidine into 3T3 fibroblasts. For these assays, ^{125}I-labeled PDGF is washed out prior to measuring [^{3}H]thymidine uptake, and the tritium channel of the scintillation counter is adjusted so that the small amounts of ^{125}I present in the cells do not significantly interfere with the measurement of [^{3}H]thymidine uptake. The purity of the final preparation can be checked by taking a small amount (before adding the albumin) and analyzing by SDS–PAGE and autoradiography. It is recommended that all subsequent dilutions of the stock ^{125}I-labeled PDGF are made in 1% serum albumin solution.

[27] D. F. Bowen-Pope *et al.*, *Proc. Natl. Acad. Sci. U.S.A.* **81**, 2396 (1984).
[28] C.-H. Heldin *et al.*, *Proc. Natl. Acad. Sci. U.S.A.* **78**, 3664 (1981).
[29] D. F. Bowen-Pope and R. Ross, *J. Biol. Chem.* **257**, 5161 (1982).
[30] J. S. Huang *et al.*, *J. Biol. Chem.* **275**, 8130 (1982).
[31] L. T. William *et al.*, *Proc. Natl. Acad. Sci. U.S.A.* **79**, 5876 (1982).

Assay for PDGF Receptor Competing Activity. PDGF-receptor competing activity can be measured on cultures of BALb/c 3T3 (clone A31) fibroblasts or human foreskin fibroblasts (see also chapter by Singh [2], this volume). The cells are cultured on 24-well plates (Flow Laboratories, McLean, VA), grown to confluence, changed to depletion medium containing 0.5% fetal calf serum, and incubated for 72 hr. ^{125}I-Labeled PDGF with unlabeled PDGF or concentrated conditioned medium is added to cells in cold binding medium (DPBS–1% HSA, 4°). Incubation is carried out for 3 hr at 4°, either with the culture plates at rest, or while they are being gently shaken. Binding medium is recovered and cells are rinsed 3 times with DPBS–0.1% HSA, or as recently recommended,[18] the cells are rinsed 6 times with DPBS–0.1% FCS, instead of 0.1% albumin. Cell-bound ^{125}I-labeled PDGF is extracted with 1% SDS, which was found more efficient than 1% Triton X-100. ^{125}I-Labeled PDGF is counted in a gamma scintillation counter. Nonspecific binding in these studies is determined as the amount of ^{125}I-labeled PDGF bound in the presence of excess (50 to 100 ng) of unlabeled PDGF. PDGF or competing activity of samples is converted to equivalent ng/ml of PDGF using for this purpose a standard curve derived from binding data of pure unlabeled PDGF.

Ability of the PDGF-Like Polypeptides to Stimulate Phosphorylation of Cell Membrane Proteins

Studies from various laboratories have shown that PDGF can stimulate tyrosine-specific kinases capable of phosphorylating cell membrane and cellular proteins at the tyrosine residue.[32–35] Protein kinases phosphorylate mostly either serine or threonine residues. The fact that PDGF, retroviral transforming proteins,[36] EGF,[37–41] and insulin[42] promote cell growth and share in their specificity of stimulating tyrosine-specific kinases has generated a strong interest for a possible connection between cellular protein tyrosine phosphorylation and cellular growth.

[32] B. Ek and C.-H. Heldin, *J. Biol. Chem.* **257,** 10486 (1982).
[33] J. Nishimura *et al., Proc. Natl. Acad. Sci. U.S.A.* **79,** 4303 (1982).
[34] J. A. Cooper *et al., Cell* **31,** 263 (1982).
[35] D. T. Graves *et al., Cancer Res.* **44,** 2966 (1984).
[36] T. Hunter and B. Sefton, *Proc. Natl. Acad. Sci. U.S.A.* **77,** 1131 (1980).
[37] G. Carpenter *et al., J. Biol. Chem.* **254,** 4884 (1979).
[38] S. Cohen *et al., J. Biol. Chem.* **257,** 1523 (1982).
[39] J. A. Fernandez-Pol, *J. Biol. Chem.* **256,** 9742 (1981).
[40] T. Hunter and J. A. Cooper, *Cell* **24,** 741 (1981).
[41] L. E. King *et al., Biochemistry* **19,** 1524 (1980).
[42] M. Kasuga *et al., J. Biol. Chem.* **257,** 9891 (1982).

The target of protein phosphorylation, by PDGF, in membranes of human fibroblasts appears to have a molecular weight of about 185K,[35,43] and 180K in Swiss mouse 3T3 cells.[33] This property of PDGF can serve to establish whether the *sis onc* gene products share a similar function and are capable of stimulating phosphorylation of the cell membrane proteins in target cells of PDGF action. Application of this procedure has shown that media derived from SSV/NRK cells were indeed capable of phosphorylating the same membrane protein like PDGF,[8] thus providing additional evidence that these *sis onc* gene products exert functions similar to those of PDGF.

Following is a description of the procedures used to study cell membrane protein phosphorylation (see also chapter by Pike [33], Vol. 146 of this series).

Membrane Preparation. Human diploid fibroblasts, GM-10 (Human Mutant Cell Repository, Camden, NJ) are grown to confluence in 850-cm^2 roller bottles. The cells are then depleted for 48 hr in MEM supplemented with 0.5% (v/v) platelet-poor plasma. The cells are harvested by treating them with Ca^{2+}/Mg^{2+}-free Earle's balanced salt solution containing 0.1% EDTA and 0.2 mM PMSF, and then pelleted at 1000 g for 10 min. All subsequent steps are done at 4°. Cells are resuspended in homogenization buffer (25 mM sucrose, 4 mM Tris, pH 8.4, 0.2 mM PMSF) and homogenized by 20 strokes of a ground glass Dounce homogenizer. The homogenate is pelleted at 4000 g for 10 min, and the supernatant is collected. The pellet is resuspended in homogenization buffer by four strokes of the homogenizer. The second homogenate is centrifuged again at 4000 g for 10 min. The supernatants of the two homogenizations are combined and then centrifuged at 30,000 g for 60 min. The high speed pellet is resuspended in 4 mM Tris, pH 7.2, it is layered over 2 volumes of 35% sucrose in 4 mM Tris, pH 7.2, and centrifuged at 100,000 g for 60 min. The material accumulating at the interface is stored frozen until used.

Phosphorylation Reactions. About 20 μg (10 μl) of purified membrane protein is added to 30 μl of 2× phosphorylation buffer (40 mM HEPES, pH 7.4, 20 mM MgCl$_2$, 0.2 mM MnCl$_2$). The membranes are preincubated with PDGF or with concentrated test media for 20 min at 4°. The kinase reaction is initiated by the addition of 0.6 nmol of [γ-^{32}P]ATP (6–10 Ci/mmol). The final incubation volume is 60 μl. Incubation proceeds for an appropriate time, typically 10–20 min, at 4°. It is then terminated by the addition of 40 μl of concentrated SDS sample buffer (175 mM Tris, pH 6.8, 5% SDS, 25% glycerol, 12.5% 2-mercaptoethanol, 0.25% bromphenol blue) followed by boiling for 3 min. Samples are then applied to SDS–

[43] B. Ek *et al.*, *Nature (London)* **295**, 419 (1982).

polyacrylamide gel electrophoresis (10% acrylamide). Gels containing duplicate samples are gently shaken in 1 N NaOH (1 hr at 50° for 0.75-mm-thick slab gels) to reveal specific tyrosine residue phosphorylation.[44] All gels are fixed and stained in Coomassie brilliant blue. The gels are then dried, the bands of protein standards are marked with radiolabeled ink, and the phosphoproteins are detected by autoradiography (−80° for 48 hr). The autoradiographs are quantitated by densitometry. Alternatively, using the autoradiographic images, gel pieces with the bands of interest are removed, digested overnight in digestion mixture (94 parts 30% H_2O_2: 6 parts 30% NH_4OH) as described,[14] and incorporated radioactivity is measured by scintillation counting using acidified scintillation cocktail.

Modification of the Phosphorylation Reaction as Applied to Conditioned Media Derived from Cultures of SSV/NRK Cells.[8] Approximately 30 μg of membrane protein in 1% human serum albumin–Dulbecco's phosphate-buffered saline (HSA–DPBS) is placed in centrifuge tubes containing 20 or 60 μl of concentrated (10-fold) conditioned media from cultures of SSV/NRK cells or untransformed NRK cells. Controls consist of 5 ng of PDGF in HSA–DPBS and HSA–DPBS alone. The total volume in all tubes is 1 ml. After incubation for 20 min at 4°, the tubes are centrifuged (30,000 g) for 30 min, and the pellet is resuspended in 30 μl of 20 mM HEPES, pH 7.4, 0.1 mM MnCl$_2$. Then 20 μCi of [γ-^{32}P]ATP is added to each tube, and the tubes are incubated for 20 min at 4°. The reaction is stopped, and the samples analyzed as described above.

General Discussion

This chapter describes procedures applicable for the identification and characterization of the structural and functional properties of the *sis onc* gene products derived from SSV-transformed cells, or from human malignant cells of mesenchymal origin expressing the c-*sis*/PDGF-2 gene. Recent information has shown that several leukemic cell lines also produce PDGF-like proteins.[45] However, these proteins do not appear to be coded by the c-*sis*/PDGF-2 locus since no *sis* mRNA was detectable in these cells (HL-60, ML-3, K562, HEL).[16,43] This underlines the importance of the diverse procedures described here which can yield significant information on structual and functional properties of the PDGF-like proteins produced by certain cells in culture. Since these PDGF-like products are present only in trace amounts in the cell cultures, extreme care must be

[44] J. A. Cooper and T. Hunter, *Mol. Cell. Biol.* **1**, 165 (1981).
[45] P. Pantazis, L. Lanfracone, P. G. Pelicci, R. Dalla-Ravera, and H. N. Antoniades, *Proc. Natl. Acad. Sci. U.S.A.* **83**, 5526 (1986).

exercised in the design of each study, including all appropriate controls, in order to obtain valid and meaningful data.

Acknowledgments

We thank Gertrude Easterly and Lori Teare for preparing the manuscript. Work conducted in our laboratories was supported by National Institutes of Health Grants CA-30301, HL-29583 (to H.N.A.) and CA-38784 (to P.P.) and grants from the American Cancer Society, and the Council for Tobacco Research, U.S.A. (to H.N.A.).

[4] Purification of Fibroblast-Derived Growth Factor

By P. STROOBANT, M. D. WATERFIELD, and E. ROZENGURT

Introduction

Fibroblast-derived growth factor (FDGF) is a potent polypeptide mitogen secreted into the growth medium by a simian virus 40 (SV40) transformed hamster fibroblast cell line (SV28).[1-3] Studies with partially and highly purified FDGF have established that its physical and biological properties are strikingly similar to those of human and porcine platelet-derived growth factor (PDGF) and that FDGF is immunologically related to human PDGF[1-3] (see also Heldin et al. [1], Singh [2], Antoniades and Pantazis [3], and Raines and Ross [5], this volume). Although the production of PDGF-like molecules by a number of cell types in culture has been reported,[4-8] including cells transformed by simian sarcoma virus whose sis oncogene encodes the B chain of human PDGF,[9-14] the purification of

[1] H. Bourne and E. Rozengurt, Proc. Natl. Acad. Sci. U.S.A. 73, 4555 (1976).

[2] P. Dicker, P. Pohjanpelto, P. Pettican, and E. Rozengurt, Exp. Cell Res. 135, 221 (1981).

[3] P. Stroobant, W. J. Gullick, M. D. Waterfield, and E. Rozengurt, EMBO J. 4, 1945 (1985).

[4] M. Nister, C.-H. Heldin, Å. Wasteson, and B. Westermark, Ann. N.Y. Acad. Sci. 397, 25 (1982).

[5] M. Nister, C.-H. Heldin, Å. Wasteson, and B. Westermark, Proc. Natl. Acad. Sci. U.S.A. 81, 926 (1984).

[6] D. F. Bowen-Pope, A. Vogel, and R. Ross, Proc. Natl. Acad. Sci. U.S.A. 81, 2396 (1984).

[7] R. A. Seifert, S. M. Schwartz, and D. F. Bowen-Pope, Nature (London) 311, 669 (1984).

[8] P. E. Dicorleto, Exp. Cell Res. 153, 167 (1984).

[9] K. C. Robbins, H. N. Antoniades, S. G. Devare, M. W. Hunkapiller, and S. A. Aaronson, Nature (London) 305, 605 (1983).

[10] T. F. Deuel, J. S. Huang, S. S. Huang, P. Stroobant, and M. D. Waterfield, Science 221, 1348 (1983).

these factors has proven difficult, probably due to their cationic and hydrophobic properties and their presence in conditioned media at low levels. However, the purification of FDGF to homogeneity from the conditioned medium of SV28 cells is possible using conventional column chromatography techniques followed by HPLC methods originally developed for PDGF[15] and then SDS–gel electrophoresis.

Properties of FDGF

FDGF is a protein of apparent molecular weight 31,000, whose hydrophobic and cationic properties make it necessary to take precautions against loss by adsorption to surfaces during purification. Plastic labware and test tubes should be used (polypropylene if possible and not "tissue-culture" treated plastics). Repeated freezing and thawing, and sterilization using membrane filters should be avoided. Since FDGF is a disulfide cross-linked protein, reducing agents will cleave it into subunits which are not mitogenically active. FDGF is extraordinarily stable to heat (1 hr at 100°), acids (1 N acetic, 0.1 N formic), denaturing agents (6 M guanidine, 2% sodium dodecyl sulfate), and organic solvents (acetonitrile) and some of these properties are exploited in its purification.

FDGF is a potent mitogen which synergizes with other growth factors and is able to elicit a wide variety of biological responses culminating in cell division.[1–3]

Assay of FDGF

FDGF is normally assayed for DNA synthesis activity using Swiss 3T3 cells and measuring the incorporation of [3H]thymidine into acid insoluble material after a 40 hr incubation in 2 ml of medium containing 0.9–1.25 μM thymidine.[16] The control for maximal stimulation is made using 10% fetal calf serum. When quantitative measurements are required, a dose–response analysis of the sample is made, and units of activity measured after a Probit analysis of the data. One unit of activity produces half-maximal incorporation of [3H]thymidine into DNA of that obtained with 10% fetal calf serum in 2 ml medium. FDGF can also be

[11] H. L. Niman, *Nature (London)* **307**, 180 (1984).
[12] A. J. Owen, P. Pantazis, and H. N. Antoniades, *Science* **225**, 54 (1984).
[13] H.-S. Thiel and R. Hafeurichter, *Virology* **136**, 414 (1984).
[14] H. L. Niman, R. A. Houghton, and D. F. Bowen-Pope, *Science* **226**, 701 (1984).
[15] P. Stroobant and M. D. Waterfield, *EMBO J.* **3**, 2963 (1984).
[16] P. Dicker and E. Rozengurt, *Nature (London)* **287**, 607 (1980).

assayed using an ^{125}I-labeled EGF binding assay at 37° with Swiss 3T3 cells, in which samples containing FDGF show diminished EGF binding.[17] In contrast to other growth factors that bind directly to the EGF receptor (e.g., TGFα/EGF), FDGF inhibits ^{125}I-labeled EGF binding more effectively at 37 than at 4°. This property can be used to give additional evidence for the presence of FDGF.

Purification Procedure

All steps should be carried out at 2–4°, and column fractions stored at $-20°$ as soon as possible. Protein is monitored during the purification using absorbance at 280 nm.

Preparation of Conditioned Medium

SV28 cells, which are stored, grown, and handled using standard techniques, are grown in 2-liter plastic pots containing a plastic spiral of surface 8000 cm² (Sterilin). Each pot is seeded with 1.5–2.0×10^8 cells in 2 liters Dulbecco's modified Eagle's medium (DME) containing 10% calf serum, penicillin (100 units/ml), and streptomycin (100 μg/ml) and bubbled slowly with 10% CO_2/90% air. After 2–3 days when the cells are confluent, the cell sheets in each pot are washed with $2\times$ 500 ml phosphate-buffered saline followed by 200 ml DME. The pots are then filled with DME/Waymouth (MB 752/1 containing 1.6 mg $FeSO_4$/liter) medium (1 : 1) and after 4–5 days incubation with aeration at 37°, the conditioned medium is harvested by centrifugation at 1200 g for 30 min. It should be emphasized that FDGF is synthesized by SV28 cells and secreted into serum-free medium in a time-dependent manner, and does not represent PDGF which could conceivably be adsorbed to the cells from serum and subsequently eluted.[2]

Concentration of Conditioned Medium

Ammonium sulfate is added to the conditioned medium (667 g/liter), which is then stirred overnight at 4° and centrifuged at 1200 g for 2 hr. The precipitate is dissolved in 20 ml phosphate-buffered saline (lacking added calcium and magnesium) and dialyzed in Spectrapor 1 dialysis tubing (Spectrum Medical Industries) against 5 $\times$ 10 liter changes of 10 mM formic acid. The dialyzate is stored at $-20°$ until 1250 ml batches of the concentrated medium (equivalent to 40 liters conditioned medium) are available.

---

[17] E. Rozengurt, M. Collins, K. D. Brown, and P. Pettican, *J. Biol. Chem.* **257**, 3680 (1982).

CM-Sephadex Chromatography

The concentrated medium (1250 ml) is thawed, centrifuged at 1200 g for 1 hr, and the supernatant adjusted to pH 7.4 with 100 mM NaOH. CM-Sephadex (C-50, 25 g dry beads) is added, and after 15 min the pH is readjusted to pH 7.4, and 7 liters 5 mM sodium phosphate, pH 7.4 is added. The mixture is stirred overnight at 4° using an overhead paddle stirrer.

A column is poured with the settled slurry which is then eluted with a 4 liter linear gradient of 0–1 M NaCl in 5 mM sodium phosphate, pH 7.4, at 1.7 ml/min collecting 9 ml fractions. The active fractions (Fig. 1A) are pooled (about 1100 ml), dialyzed against 5 × 10 liter changes of 10 mM formic acid using Spectrapor 4 dialysis tubing, lyophilized, and resuspended in 2 ml 100 mM formic acid.

Sephadex G-75 Chromatography

The active pool is applied to a column of Sephadex G-75 (30 g swollen and preequilibrated in 140 mM NaCl/20 mM HCl) and run in this solvent at 6 ml/hr collecting 3 ml fractions. Convenient molecular weight standards are RNase, chymotrypsinogen, ovalbumin, and bovine serum albumin (Pharmacia LMW gel filtration calibration kit). The active fractions (Fig. 1B), which run with an apparent molecular weight of about 31,000,

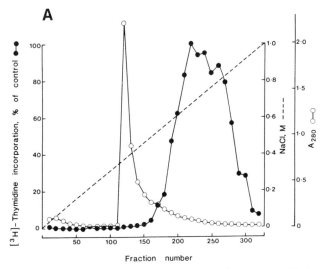

FIG. 1. Purification of FDGF. (A) CM-Sephadex chromatography; (B) Sephadex G-75 chromatography; (C) gel permeation HPLC; (D) Reversed-phase HPLC; (E) preparative gel electrophoresis.

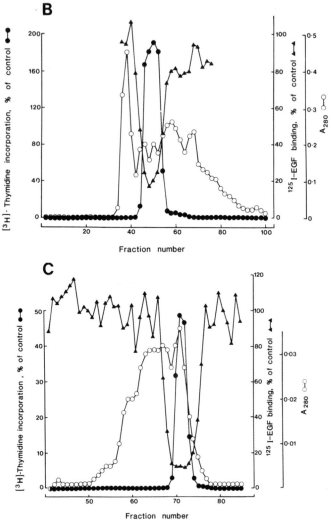

FIG. 1. (*continued*)

are pooled (about 33 ml), dialyzed against 10 mM formic acid as previously, lyophilized, and resuspended in 500 μl 6 M guanidine–HCl (Sigma)/0.1 M potassium phosphate, pH 4.5.

Gel Permeation HPLC

FDGF is then fractionated in two runs by gel permeation HPLC in 6 M guanidine–HCl/0.1 M potassium phosphate, pH 4.5, using a TSK SW

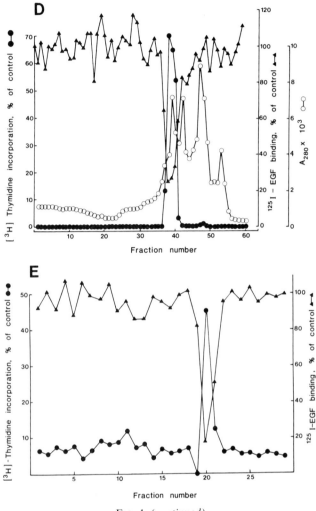

FIG. 1. (*continued*)

3000 column (0.7 × 60 cm; LKB) and run at a flow rate of 0.5 ml/min collecting 0.25 ml fractions. The active fractions (Fig. 1C) are then pooled.

Reversed-Phase HPLC

The active pool is fractionated in one run by reversed-phase HPLC in 0.1% trifluoroacetic acid on a Synchropac RPP column (0.46 × 7.5 cm; Synchrom, Linden, Indiana) using a linear gradient of acetonitrile from

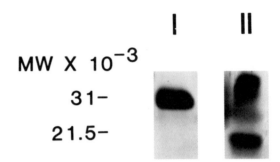

FIG. 2. SDS–polyacrylamide gel electrophoresis of nonreduced (I) and reduced (II) FDGF carried out using 12–22% gradient gels, as described by Laemmli,[18] and silver-stained using the method of Sammons *et al.*[19]

1 to 60% at 1 ml/min collecting 0.25 ml fractions over 35 min. The active fractions (Fig. 1D) are then pooled and lyophilized.

Preparative Gel Electrophoresis

The active fractions are resuspended in sample buffer without reducing reagent and further purified in two runs on a 1.5-mm-thick 12 slot 12–22% gradient SDS gel.[18] The gel is sliced horizontally into 2 mm pieces, which are then extracted at 4° overnight with shaking using 400 μl phosphate-buffered saline containing 0.02% (w/v) SDS. Aliquots (10 μl) are assayed for inhibition of EGF binding and DNA synthesis in the presence of 1 μg/ml insulin and 1 mg/ml fatty acid free bovine serum albumin (Sigma) (Fig. 1E). A single peak of activity is present which corresponds to purified FDGF.

Purified FDGF

When a sample of purified FDGF is rerun in the same SDS gel system it gives a single broad silver-staining band[19] with an apparent molecular weight of 31,000 (Fig. 2, track I). After reduction, the band is replaced by two major components with apparent molecular weights of about 19,000 and 34,000 (Fig. 2, track II). A summary of the purification of FDGF from 40 liters SV28 conditioned medium is shown in Table I. Difficulties in accurately measuring the protein concentration of the FDGF eluted from the final preparative SDS–polyacrylamide gel purification step necessitate an approximation for the final specific activity, based on the relative

[18] U. K. Laemmli, *Nature (London)* **227**, 680 (1970).
[19] D. W. Sammons, L. D. Adams, and E. E. Nishizawa, *Electrophoresis* **2**, 135 (1981).

TABLE I
FDGF PURIFICATION

Step	DNA synthesis activity (U)[a]	Total protein (mg)[b]	Specific activity (U/mg protein)
Ammonium sulfate dialyzate from 40 liters medium	25,000	1960	13
CM-Sephadex column	4,708	170	18
Sephadex G-75 column	2,357	34	69
Gel permeation HPLC	750	0.108	6,944
Reversed-phase HPLC	190	0.039[c]	4,872
Preparative gel[d]	190[d]	0.013	14,615

[a] One unit of activity produces half-maximal incorporation of [³H]thymidine into DNA of that obtained with 10% fetal calf serum.
[b] Determined as described by M. Bradford, *Anal. Biochem.* **72**, 248 (1976).
[c] Determined by amino acid analysis following hydrolysis in 6 N HCl *in vacuo* using a Beckman 6300 analyzer.
[d] Estimate from relative staining intensity of a silver-stained SDS gel of FDGF at reversed-phase HPLC step.

staining intensity of a silver-stained SDS gel of the reversed-phase step. This estimate suggests that 1 unit of DNA synthesis activity for pure FDGF is about 30 ng/ml (1 nM). This specific activity, which represents about a 1100-fold purification, is comparable to those observed for pure human PDGF (44 ng/ml) and porcine PDGF (14 ng/ml) assayed under identical conditions[15] using confluent and quiescent cultures of Swiss 3T3 cells. Highly purified FDGF gives positive Western blots using anti-PDGF antiserum, demonstrating the relatedness between FDGF, which is presumably derived from a hamster gene, and human PDGF.[3] These observations are reinforced by data indicating that anti-PDGF antibodies can inhibit FDGF-stimulated mitogenesis.[20]

[20] E. Rozengurt, J. Sinnett-Smith, and J. Taylor-Papadimitriou, *Int. J. Cancer* **36**, 247 (1985).

[5] Identification and Assay of Platelet-Derived Growth Factor-Binding Proteins

By ELAINE W. RAINES and RUSSELL ROSS

Introduction

Platelet-derived growth factor (PDGF) is a potent mitogen for connective tissue cells which was discovered as a result of the ability of platelets to reconstitute the mitogenic activity of whole blood serum lacking in serum derived from cell-free plasma.[1,2] Release of PDGF and other mitogens from platelet α-granules is dependent on platelet activation,[3-5] and it has been postulated that focal release of PDGF at sites of injury may play a physiological role in wound healing and tissue repair[6,7,7a] and a pathological role in the formation of lesions of atherosclerosis characterized by smooth muscle cell proliferation.[8,9] The possible importance of focal release of PDGF and local PDGF concentrations at the site of release is suggested by *in vivo* clearance studies with purified and ^{125}I-labeled PDGF. These studies demonstrated a $t_{1/2}$ for clearance of PDGF from blood of approximately 2 min.[10] One possible mediator of this rapid clearance, which would be consistent with the concept that PDGF acts as a local mitogen, is the presence in plasma of specific binding proteins for PDGF[10-13] which inhibit the binding of PDGF to its cell surface recep-

[1] R. Ross, J. Glomset, B. Kariya, and L. Harker, *Proc. Natl. Acad. Sci. U.S.A.* **71**, 1207 (1974).

[2] N. Kohler and A. Lipton, *Exp. Cell Res.* **87**, 297 (1974).

[3] K. L. Kaplan, M. J. Broekman, A. Chernoff, G. R. Lesznik, and M. Drillings, *Blood* **53**, 604 (1979).

[4] D. R. Kaplan, F. C. Chao, C. D. Stiles, H. N. Antoniades, and C. D. Scher, *Blood* **53**, 1043 (1979).

[5] L. D. Witte, K. L. Kaplan, H. L. Nossel, B. A. Lages, H. J. Weiss, and D. S. Goodman, *Circ. Res.* **42**, 402 (1978).

[6] R. Ross and A. Vogel, *Cell* **14**, 203 (1978).

[7] C. D. Scher, R. C. Shepard, H. N. Antoniades, and C. D. Stiles, *Biochim. Biophys. Acta* **560**, 217 (1979).

[7a] R. Ross, E. W. Raines, and D. F. Bowen-Pope, *Cell* **46**, 155 (1986).

[8] R. Ross and J. A. Glomset, *N. Engl. J. Med.* **295**, 369 and 420 (1976).

[9] R. Ross, *Arteriosclerosis* **1**, 293 (1981).

[10] D. F. Bowen-Pope, T. W. Malpass, D. M. Foster, and R. Ross, *Blood* **64**, 458 (1984).

[11] J. S. Huang, S. S. Huang, and T. F. Deuel, *J. Cell Biol.* **97**, 383 (1983).

[12] E. W. Raines, D. F. Bowen-Pope, and R. Ross, *Proc. Natl. Acad. Sci. U.S.A.* **81**, 3424 (1984).

[13] J. S. Huang, S. S. Huang, and T. F. Deuel, *Proc. Natl. Acad. Sci. U.S.A.* **81**, 342 (1984).

tor.[10,12] It has been shown that at low concentrations of PDGF and high concentrations of plasma, i.e., under conditions existing *in vivo* at sites peripheral to sites of platelet release, 80 to 85% of the PDGF would be bound to, and inhibited by, the plasma components. Since PDGF is an extremely potent mitogen ($ED_{50} \simeq 10^{-11}$) and is not normally present in plasma,[10] any modulation of the action of PDGF at the initial point of interaction between PDGF and a responsive cell is of great interest.

Amino acid sequence data of purified PDGF has demonstrated that one of the chains of PDGF is virtually identical to the transforming gene product of simian sarcoma virus[14,15] and cells transformed by this virus produce a PDGF-like molecule[16-20] (see also chapters by Antoniades and Pantazis [3] and Singh [2], this volume). However, cells transformed by agents which do not themselves encode a sequence homologous with PDGF have also been shown to produce a PDGF-like molecule,[17] suggesting activation of a cellular gene for PDGF as a result of transformation. In addition, *in vitro* studies have shown that PDGF-like molecules are produced by a number of normal cells: cultured vascular endothelial cells,[21] newborn rat aortic smooth muscle cells,[22] explants of first trimester human placenta,[23] rat arterial smooth muscle cells from intimal proliferative lesions,[24] and activated macrophages.[25] Although little is known about activation of the PDGF gene *in vivo*, the possibility of such broad expression of the PDGF molecule suggests that PDGF may play a role in several forms of growth regulation. Therefore, the possible modulation of these

[14] R. F. Doolittle, M. W. Hunkapiller, L. E. Hood, S. G. Devare, K. C. Robbins, S. A. Aaronson, and H. N. Antoniades, *Science* **221**, 275 (1983).

[15] M. D. Waterfield, G. T. Scrace, N. Whittle, P. Stroobant, A. Johnsson, Å. Wasteson, B. Westermark, C.-H. Heldin, J. S. Huang, and T. F. Deuel, *Nature (London)* **304**, 35 (1983).

[16] T. F. Deuel, J. S. Huang, S. S. Huang, P. Stroobant, and M. D. Waterfield, *Science* **221**, 1348 (1983).

[17] D. F. Bowen-Pope, A. Vogel, and R. Ross, *Proc. Natl. Acad. Sci. U.S.A.* **81**, 2396 (1984).

[18] J. S. Huang, S. S. Huang, and T. F. Deuel, *Cell* **39**, 79 (1984).

[19] A. J. Owen, P. Pantazis, and H. N. Antoniades, *Science* **225**, 54 (1984).

[20] A. Johnsson, C. Betsholtz, K. von der Helm, C.-H. Heldin, and B. Westermark, *Proc. Natl. Acad. Sci. U.S.A.* **82**, 1721 (1985).

[21] P. E. DiCorleto and D. F. Bowen-Pope, *Proc. Natl. Acad. Sci. U.S.A.* **80**, 1919 (1983).

[22] R. S. Seifert, S. M. Schwartz, and D. F. Bowen-Pope, *Nature (London)* **311**, 669 (1984).

[23] A. S. Goustin, C. Betsholtz, S. Pfeifer-Ohlsson, H. Persson, J. Rydnert, M. Bywater, G. Holmgren, C.-H. Heldin, B. Westermark, and R. Ohlsson, *Cell* **41**, 301 (1985).

[24] L. N. Walker, D. F. Bowen-Pope, R. Ross, and M. A. Reidy, *Proc. Natl. Acad. Sci. U.S.A.* **83**, 7311 (1986).

[25] K. Shimokado, E. W. Raines, D. K. Madtes, T. B. Barrett, E. P. Benditt, and R. Ross, *Cell* **43**, 277 (1985).

PDGF-like molecules by binding proteins has possible importance in both normal and abnormal growth of cells.

Among the known growth factors, a number of examples exist for the presence of specific binding proteins which modulate the activity of these mitogens when in the bound state. These include epidermal growth factor,[26] nerve growth factor,[27] the insulin-like growth factors (IGFs),[28,29] and the biologically active phorbol esters,[30] which mimic many of the actions of growth factors.[31] One of the most extensively studied systems is the IGFs, in which it has been shown that the binding proteins increase the circulating half-life of the IGFs from minutes to several hours[32] and inhibit the biological effects of IGF *in vitro*[28,29] (see also chapter by Smith *et al.* [26], Vol. 146 of this series). It has been suggested that plasma-binding proteins may serve as an inactive storage pool *in vivo* and also regulate the distribution of IGFs in the extracellular space.[33] Quantitative alterations in the levels of the IGF-binding proteins have been demonstrated during development,[34] suggesting that modulation of the IGF-binding proteins may be a mechanism for regulation of the circulating levels of IGF *in vivo*. It is conceivable that the levels of binding proteins for different growth factors may regulate the amount of unbound and active growth factor. By preventing binding of the factor to its receptor on responsive cells, binding proteins may alter their biological activity. By altering the molecular size of the growth factor by complex formation, the accessibility of the growth factor to target cells may be altered and ultimately clearance from the circulation may be modulated. It is also possible that complex formation between a growth factor and its binding protein may protect the growth factor against degradative enzymes or other mechanisms of inactivation and thus prolong its potential availability. A better definition of the role of these binding proteins will be important to our understanding of the action of growth factors *in vivo*.

This chapter will focus on known characteristics of PDGF-binding proteins and factors which affect PDGF-binding protein complexes, iden-

[26] K. J. Lembach, *Proc. Natl. Acad. Sci. U.S.A.* **73**, 183 (1976).

[27] E. A. Berger and E. M. Shooter, *Proc. Natl. Acad. Sci. U.S.A.* **74**, 3647 (1977).

[28] J. Zapf, E. Schoenle, G. Jagars, I. Sand, J. Grunwald, and E. R. Froesch, *J. Clin. Invest.* **63**, 1077 (1979).

[29] D. J. Knauer and G. L. Smith, *Proc. Natl. Acad. Sci. U.S.A.* **77**, 7252 (1980).

[30] M. Shoyab and G. J. Todaro, *J. Biol. Chem.* **257**, 439 (1982).

[31] P. Dicker and E. Rozengurt, *Nature (London)* **276**, 723 (1978).

[32] K. L. Cohen and S. P. Nissley, *Acta Endocrinol.* **83**, 243 (1976).

[33] J. Zapf, E. R. Froesch, and R. E. Humbel, *Curr. Top. Cell. Regul.* **19**, 257 (1981).

[34] R. M. White, S. P. Nissley, P. A. Short, M. M. Rechler, and I. Fennoy, *J. Clin. Invest.* **69**, 1239 (1982).

tification of specific binding proteins by various assays, and limitations in the analysis of the functional role of these binding proteins.

Characteristics of PDGF-Binding Proteins

Detection and General Properties

Binding proteins for PDGF were first identified in plasma because they intefered with measurement of PDGF levels in plasma and serum by radioreceptor assay[10,12] and by immunoassay.[11] In the case of immunoassay, it was found that ^{125}I-labeled PDGF interacted with plasma proteins and was therefore not precipitated with anti-PDGF IgG.[11] Further, specific associations of ^{125}I-labeled PDGF with higher molecular weight plasma fractions were demonstrated by gel filtration.[11,12] Figure 1A shows a gel filtration profile of ^{125}I-labeled PDGF in the presence and absence of plasma. The fractions which specifically associate with ^{125}I-labeled PDGF when chromatographed in the presence of plasma coincide with plasma fractions which specifically inhibit the binding of PDGF to its cell surface receptor[12] (Fig. 1B). These plasma fractions do not act by blocking or permanently altering the properties of the PDGF cell surface receptor, but rather appear to block the site on PDGF responsible for receptor binding or alter the PDGF molecule so that it is no longer able to bind to its receptor. Plasma fractions capable of binding PDGF with approximate molecular weights of 40,000, 150,000, and greater than 500,000 were further confirmed by gel filtration of whole blood serum and elution of PDGF from these same size fractions under dissociating conditions (Fig. 1C). Under physiologic conditions (Tris/saline), active PDGF was eluted only at the same position as ^{125}I-labeled PDGF run in the absence of plasma (Fig. 1A). However, if the higher molecular weight fractions obtained from whole blood serum were rechromatographed under dissociating conditions (4 M GuHCl), PDGF was eluted from the higher molecular weight peaks (Fig. 1C). Thus, the specific association of ^{125}I-labeled PDGF with plasma fractions separated on the basis of size is strongly correlated with fractions which inhibit the binding of PDGF to its cell surface receptor and with fractions which can be shown to contain dissociable PDGF when whole blood serum is chromatographed on the same column (Fig. 1).

α_2-Macroglobulin as a PDGF-Binding Protein

The fact that α_2-macroglobulin (α_2-M) is the major plasma protein >500,000 molecular weight[35] led to studies to determine whether α_2-M

[35] P. C. Harpel, this series, Vol. 45, p. 639.

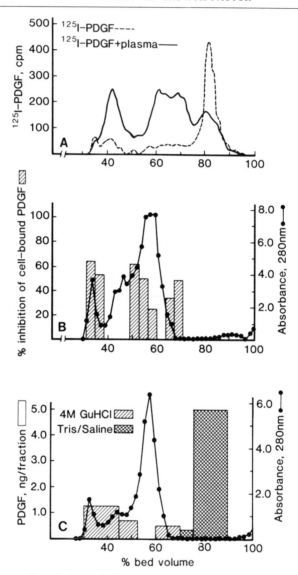

FIG. 1. Correlation of plasma ^{125}I-labeled PDGF interactions with plasma inhibition of PDGF cellular binding and with dissociation of PDGF from higher molecular weight fractions of whole blood serum. A BioGel A-0.5 m column (2.5 × 95 cm) equilibrated in 0.01 M Tris–HCl, 0.09 M NaCl, pH 7.4, with a flow rate of 15 ml/hr and 8.2 ml fractions, was used to make the following comparisons: association of ^{125}I-labeled PDGF with plasma fractions (A): ^{125}I-labeled PDGF at a final concentration of 2.5 ng/ml was incubated for 1 hr at 37° with column buffer containing 2.5 mg/ml BSA or with 500 μl plasma-derived serum (PDS, prepared from cell-free plasma by recalcification and centrifugation to remove the fibrin clot[12]

was one of the PDGF-binding proteins. α_2-M is present in plasma at a concentration of approximately 2 mg/ml and is composed of two noncovalently bound subunits of $\approx$320,000 which are in turn composed of two identical covalently bound subunits of 185,000. It has been most extensively studied as a protease inhibitor,[36] but has also been shown to bind nerve growth factor[37] and human growth hormone.[38] In addition, it has been shown that fibroblasts have specific receptors for α_2-M which have a high affinity for α_2-M–protease complexes.[39]

Utilizing commercially available purified α_2-M and antisera to α_2-M, several pieces of evidence summarized in Table I suggest that α_2-M is one of the PDGF-binding proteins. Huang et al.[13] were also able to show that α_2-M and a PDGF-binding protein activity copurified, and Libby et al.[40] found that fetuin contaminated with α_2-M contained PDGF. However, antiserum to α_2-M only removed 50% of PDGF-binding activity from plasma as determined by inhibition of PDGF binding to its receptor.[12] α_2-M, therefore, appears to account for only part of the plasma-binding activity for PDGF.

The Possibility of More Than One Population of PDGF-Binding Protein Complexes

Specific PDGF-binding proteins have been characterized in plasma by gel filtration and SDS gel electrophoresis. Table II outlines some of the properties of the complexes detected by these two approaches. Associa-

---

[36] A. J. Barrett and P. M. Starkey, *Biochem. J.* **133**, 709 (1973).
[37] H. Ronne, H. Anundi, L. Rask, and P. A. Peterson, *Biochem. Biophys. Res. Commun.* **87**, 330 (1979).
[38] N. F. Adham, Z. H. Chakmakjian, J. W. Mehl, and J. E. Bethune, *Arch. Biochem. Biophys.* **132**, 175 (1969).
[39] F. Van Leuven, J.-J. Cassiman, and H. Van Den Berghe, *J. Biol. Chem.* **254**, 5155 (1979).
[40] P. Libby, E. W. Raines, P. M. Cullinane, and R. Ross, *J. Cell. Physiol.* **125**, 357 (1985).

---

and then chromatographed and analyzed as described in the text; plasma inhibition of PDGF binding to its cell-surface receptor (B): 6 ml of human PDS was chromatographed and pooled fractions were assayed for inhibition of PDGF cellular binding as described in the text (adapted from Fig. 2 of Raines et al.[12]); dissociation of PDGF from higher molecular weight fractions of whole blood serum (C): 5 ml of human whole blood serum was chromatographed and the pooled fractions in the first 70% of the column bed volume (which contained no apparent PDGF by radioreceptor assay) were each rechromatographed on a BioGel column (1.5 × 90 cm) equilibrated in 4 M guanidine–HCl in Tris-buffered saline, pH 7.4. The elution position of free PDGF (75–90% of total column volume) was assayed for PDGF after each fraction was rechromatographed under dissociating conditions. The PDGF content of each fraction was determined by radioreceptor assay as previously described[43] (adapted from Fig. 5 of Raines et al.[12]).

TABLE I
FACTORS IMPLICATING α_2-MACROGLOBULIN IN THE BINDING OF PDGF IN PLASMA

Factors	References
α_2-M is the major plasma protein with a molecular weight >500,000	Harpel[35]
α_2-M inhibits the binding of PDGF to its cell surface receptor, $ED_{50} \approx 5.5 \times 10^{-5}$ g/ml	Bowen-Pope et al.,[10] Raines et al.[12]
Antisera to α_2-M	
Partially removed PDGF-binding activity from plasma	Raines et al.[12]
Immunoprecipitated ^{125}I-labeled PDGF-binding protein complexes	Huang et al.[13]
As determined by SDS–gel electrophoresis	
Complexes of ^{125}I-labeled PDGF and α_2-M and complexes of ^{125}I-labeled PDGF plasma-binding protein comigrated	Huang et al.[13]
α_2-M and PDGF-binding protein copurified	Huang et al.[13]
PDGF was found associated only with fetuin fractions contaminated with α_2-M	Libby et al.[40]

tion of PDGF with its binding proteins prevents recognition by anti-PDGF antibody[11] and inhibits binding of PDGF to its cell surface receptor,[10,12] but only partially inhibits the mitogenic activity of PDGF.[13] Use of purified α_2-M and antisera to α_2-M strongly suggests that α_2-M is one of the plasma binding proteins (Table I). However, the characteristics of the complexes detected by gel filtration and SDS gel electrophoresis (summarized in Table II) suggest that there may be more than one population of PDGF-binding protein complexes. Data obtained by gel filtration suggest PDGF can be dissociated from its binding proteins with 4 M guanidine–HCl or 1 N acetic acid. Active PDGF can be recovered from PDGF-binding protein complexes, including fractions consistent with the molecular weight of α_2-M. However, the complexes detected by SDS gel electrophoresis (only complexes consistent with the subunit structure α_2-M are detected) are not dissociable with 8 M urea, 1% SDS, 1 N acetic acid, or 0.1 M NaOH, but reduction with 2% mercaptoethanol dissociates these complexes. These data suggest that formation of covalent complexes with α_2-M have occurred, possibly similar to those described for α_2-M and proteases.[36] Unfortunately, lack of quantitative data from the SDS gel experiments makes it difficult to determine the extent of covalent complex formation. In the gel filtration experiments, the presence of covalent complexes cannot be excluded, but appear to represent an undetectable or extremely small percentage of the total detected by gel filtration as determined by rechromatography under dissociating conditions.[41] The possible presence

[41] E. W. Raines and R. Ross, unpublished observations (1983).

TABLE II
CHARACTERISTICS OF PDGF-BINDING PROTEIN COMPLEXES

Effect	Agent	References
Complexes detected by gel filtration or incubation with PDGF-Sepharose (three molecular weight classes: >500,000, 150,000, and 40,000)		
Dissociable	4 M guanidine–HCl 1 N acetic acid	Raines et al.[12]
Not recognized by	Anti-PDGF antibody	Huang et al.,[13] Raines and Ross[41]
310,000 molecular weight complex stable to SDS	SDS–gel electrophoresis	Huang et al.[11]
Formation >500,000 blocked	10^{-6} M unlabeled PDGF	Raines et al.[12]
Retained ≈50% mitogenic activity	On 3T3 cells	Huang et al.[13]
Complexes detected on SDS–polyacrylamide gel electrophoresis (apparent molecular weight: 310,000)		
Dissociable	Reduction with 2% mercaptoethanol	Huang et al.[13]
Formation blocked	10^{-6} M unlabeled PDGF N-Ethylmaleimide p-Chloromercuric benzoate	Huang et al.[13]
Formation not blocked	EGF, FGF, insulin Platelet factor 4 Protamine sulfate	Huang et al.[13]
Mobility not altered by	8 M urea, 1% SDS 1 N acetic acid, 0.1 M NaOH Methylamine	Huang et al.[13]
Mobility coincided with α_2-macroglobulin peak	Ultrogel AcA 34/AcA 22 (2 : 1, v/v)	Huang et al.[13]

of covalent complexes must be kept in mind in evaluation of assays for both PDGF and PDGF-binding proteins.

Physiologic Fluids Known to Contain PDGF-Binding Proteins

As detailed in Fig. 1 and Tables I and II, PDGF-binding proteins can be identified in human plasma by a number of different criteria. As assayed by inhibition of PDGF cellular binding, plasma and plasma-derived serum (PDS, prepared from plasma by recalcification and centrifugation to remove the fibrin clot) from a number of different species contain approximately equivalent amounts of PDGF-binding activity.[12] Using the same approach, PDGF-binding activity has been found in conditioned medium from human peritoneal macrophages[25] (Fig. 4) and in fractionated human urine.[41]

Methods Used to Study PDGF-Binding Proteins

General Considerations

The methods described here require highly purified reagents, some of which are not available commercially. A procedure for purifying PDGF[42] and radioiodinating PDGF,[43] as well as a description of one of the PDGF-binding proteins, α_2-M,[35] are found in earlier volumes of *Methods of Enzymology*.

Direct Detection of PDGF-Binding Activity

Gel Filtration. As shown in Fig. 1, gel filtration of ^{125}I-labeled PDGF in the presence and absence of plasma demonstrated association of PDGF with higher molecular weight plasma fractions. This approach is useful as a means of evaluating a mixture of different proteins. The specificity of this association of ^{125}I-labeled PDGF can be further characterized by preincubation with unlabeled PDGF. Figure 2 demonstrates that only association with the highest molecular weight plasma fraction (>500,000) was inhibited when 10^{-6} M unlabeled PDGF was preincubated with the plasma prior to the addition of ^{125}I-labeled PDGF. Because of the multiplicity of binding and the observed differences in competition, the separation is best performed using a gel filtration media such as BioGel A-0.5m (Figs. 1 and 2) or Sephacryl S-500 with a nominal exclusion of at least 500,000.

Materials

[125]I-Labeled PDGF: prepared as described by Bowen-Pope and Ross[43] with a specific activity of 25,000 to 35,000 cpm/ng.

Unlabeled PDGF: PDGF purified by the method of Raines and Ross[42] is used to determine the level of competable binding (15 μg/0.5 ml sample).

Column buffer: 0.01 M Tris–HCl, pH 7.4, containing 0.15 M NaCl, 1 mM MgSO$_4$, 1 mM ZnSO$_4$, and 1 mM CaCl$_2$.

Column: 50–100 ml BioGel A-0.5m (Bio-Rad Laboratories, Richmond, CA) equilibrated in column buffer and poured in, for example, a column 1.8 × 28 cm.

Incubation buffer: same as the column buffer but in addition contains 2.5 mg/ml bovine serum albumin (BSA). It is important that the BSA is pretested, as we have found many commercial BSA preparations contain PDGF-binding activity and PDGF.

[42] E. W. Raines and R. Ross, this series, Vol. 109, p. 749.
[43] D. F. Bowen-Pope and R. Ross, this series, Vol. 109, p. 69.

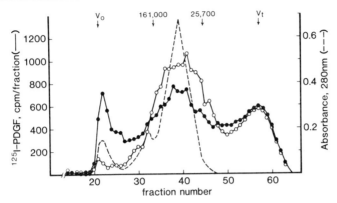

FIG. 2. Specific binding of ^{125}I-labeled PDGF to a high-molecular-weight plasma fraction. Human PDS (50 μl) was first incubated in buffer (●) or buffer containing 15 μg unlabeled PDGF (○), followed by incubation with ^{125}I-labeled PDGF to give a final concentration of 2.5 ng/ml and then chromatographed on a BioGel A-0.5 m column (2.0 × 28 cm) as described in the text. Thyroglobulin (void volume, V_0), aldolase (161,000), chymotrypsinogen (25,700), and ^3H_2O (total column volume, V_t) were used to calibrate the column (adapted from Fig. 4 of Raines et al.[12]).

Procedure. The test sample (0.05–0.5 ml, at ≃ pH 7.4) is incubated with 400 μl incubation buffer or 400 μl incubation buffer containing 15 μg/ ml PDGF (final concentration, 10^{-6} M) for 1 hr at 37°; this is followed by an additional 1 hr at 37° after the addition of ^{125}I-labeled PDGF to give a final concentration of 2.5 ng/ml. Total incubation volume is 0.5 to 1 ml. The incubated sample is loaded on BioGel A-0.5m column and run at a flow rate of 12 ml/hr at room temperature and 1-ml fractions are collected for direct counting in a gamma counter.

The results obtained with the basic protocol described above allow identification of size classes in a test sample interacting with PDGF and the degree of competability with unlabeled PDGF. Association which is not competed, for example, fractions 30–40 in Fig. 2, either represents lower affinity interactions or fractions which have a very high binding capacity which have not been saturated under these conditions (see below for differentiation between these two possibilities). This method of identification can also be used as a purification procedure.

Ligand Saturation Gel Filtration. The association of PDGF with specific fractions can be analyzed quantitatively using ligand saturation gel filtration originally described by Hummel and Dreyer[44] and modified by Cuatrecasas et al.[45] for quantitative analysis. This procedure is designed

---

[44] J. P. Hummel and W. J. Dreyer, *Biochim. Biophys. Acta* **63**, 530 (1962).
[45] P. Cuatrecasas, S. Fuchs, and C. B. Anfinsen, *J. Biol. Chem.* **242**, 3063 (1967).

to detect reversible interactions between macromolecules and allows quantitative evaluation of even quite weak interactions.

Materials

[125]I-Labeled PDGF: prepared as described above is added to the sample to be analyzed at 0.1 ng/ml (final concentration).

Column buffer: 0.01 M Tris–HCl, pH 7.4, containing 0.15 M NaCl, 1 mM MgSO$_4$, 1 mM ZnSO$_4$, 1 mM CaCl$_2$, and 0.1 ng/ml ^{125}I-labeled PDGF and 2.5 mg/ml gelatin.

Column: 10–100 ml Sephacryl S-200 (or Sephadex G-100) equilibrated in column buffer and packed in a column (preferably at least 1.0 cm diameter).

Procedure. The test sample is incubated with ^{125}I-labeled PDGF (0.1 ng/ml final concentration) for 1 hr at 37° and then loaded on the column equilibrated with the same concentration of PDGF. Fractions of ≈1% of the total column volume are collected and counted directly in the gamma counter.

Figure 3 shows the profile of ^{125}I-labeled PDGF obtained when 50 μl of human plasma was loaded on a column of Sephacryl S-200, both equilibrated with 0.1 ng/ml ^{125}I-labeled PDGF. The plasma proteins move down the column and the largest proteins (approximately 5% of the plasma proteins) are excluded from the gel. As they move down the column, they remove additional ^{125}I-labeled PDGF from solution until equilibrium is reached with the baseline concentration of ^{125}I-labeled PDGF. The ^{125}I-labeled PDGF bound to the plasma-binding proteins is therefore excluded from the gel and appears as a peak of radioactivity at the void volume of the column. The peak is followed by a trough which results from the extraction of ^{125}I-labeled PDGF from the column buffer by the PDGF-binding proteins. Thus, both the area of the peak and the trough represent independent determinations of the amount of bound PDGF. In the example shown in Fig. 3, 80% of the ^{125}I-labeled PDGF in the peak fraction was present as a complex. If the concentration of the protein in plasma responsible for the binding is known, the apparent dissociation constant can be determined from this data.[45]

Other Methods of Direct Detection. In an attempt to develop methods to evaluate the affinity of PDGF-binding protein interactions, we have bound varying amounts of PDGF-binding protein fractions to nitrocellulose and then incubated the nitrocellulose with ^{125}I-labeled PDGF. As little as 30 μg of plasma fraction I (30–40% of column volume, Fig. 1) gave significant binding above background.[41] It therefore should be possible to separate a mixture of proteins electrophoretically, transfer them to

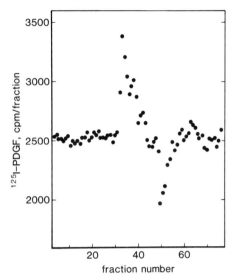

FIG. 3. Association of ^{125}I-labeled PDGF and plasma analyzed by ligand saturation gel filtration. A Sephacryl S-200 column (2.0 × 30 cm) was equilibrated in Tris-buffered saline containing 0.1 ng/ml ^{125}I-labeled PDGF and 2.5 mg/ml gelatin as described in the text. Human PDS (50 μl) was incubated with the same concentration of ^{125}I-labeled PDGF and then loaded on the column and 1 ml fractions collected and counted directly in the gamma counter.

nitrocellulose, and then detect PDGF-binding proteins by incubation with ^{125}I-labeled PDGF.

Indirect Effects on PDGF Assays

Importance of a Sequential Assay for PDGF. The importance of a sequential assay to determine the concentration of PDGF by radioimmunoassay or radioreceptor assay is depicted schematically in Fig. 4. When ^{125}I-labeled PDGF is incubated simultaneously (Fig. 4A) in a radioimmunoassay or radioreceptor assay with samples which contain large amounts of PDGF (left side) or PDGF-binding protein (right side), no ^{125}I-labeled PDGF is bound in either case. It is therefore impossible to know whether a sample contains binding protein, PDGF, or both. However, by doing the incubations sequentially (Fig. 4B), first (1) incubating the sample with the antibody (bound to Sepharose or other material) or with cells which have PDGF receptors, washing away unbound sample, and then (2) adding labeled PDGF, only competition for ^{125}I-labeled PDGF binding by the

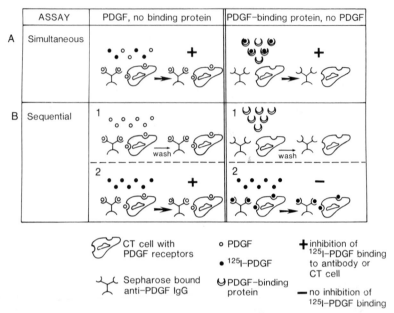

FIG. 4. Schematic of simultaneous and sequential assays for determination of levels of PDGF in test samples. A radioreceptor assay and a radioimmunoassay for PDGF are illustrated which evaluate the ability of test samples containing PDGF or PDGF-binding proteins to compete with ^{125}I-labeled PDGF for receptor binding on connective tissue (CT) cells and for binding to anti-PDGF bound to Sepharose. The left side of the diagram represents a sample which contains PDGF but no binding protein, and the right side represents a sample with PDGF-binding protein but no PDGF. (A) depicts a simultaneous assay in which the test sample and ^{125}I-labeled PDGF are incubated simultaneously with receptor-bearing cells or Sepharose-bound antibody, washed, and the amount of receptor- or antibody-bound ^{125}I-labeled PDGF is determined. Since large amounts of PDGF in a sample (left side) compete with ^{125}I-labeled PDGF for receptor/antibody binding, and large amounts of PDGF-binding protein in a sample (right side) prevent ^{125}I-labeled PDGF from binding to its receptor or antibody, it is impossible to determine whether a sample contains PDGF, binding protein, or both. (B) illustrates the sequential assay in which (1) test samples are first incubated with receptor-bearing cells or Sepharose-bound antibody, which are then washed to remove any of the test sample not specifically bound to receptors or antibody, followed by (2) incubation with ^{125}I-labeled PDGF, wash, and determination of receptor- or antibody-bound ^{125}I-labeled PDGF. In the sequential assay, if a sample contains PDGF (left side, B), PDGF bound to its receptor or antibody during incubation (1) will remain bound and inhibit ^{125}I-labeled PDGF binding in incubation (2). In samples containing PDGF-binding proteins (right side, B), the binding proteins will be washed away after the first incubation in the sequential assay and will therefore have no effect on ^{125}I-labeled PDGF binding to antibody or receptor in incubation (2).

antibody-bound or receptor-bound PDGF in the sample will be detected. In samples containing binding protein (Fig. 4B, right side), the binding protein is washed away in the wash after the first incubation and therefore does not compete with ^{125}I-labeled PDGF for binding to antibody or to cell surface receptors. We have also found that our quantitative ELISA, in which 1 ng/ml PDGF is coated on microtiter wells, is not susceptible to false positives seen with sequential radioimmunoassays and therefore provides an alternative to a sequential radioimmunoassay.[41]

If both PDGF and binding proteins are present in the same sample, internal standards can be added to samples to partially correct for the presence of binding proteins. However, even this does not allow detection of irreversible complexes of PDGF and binding protein or PDGF bound to a low-capacity binding component. Therefore, if both PDGF and binding protein are present in the same sample, ideally the binding protein should be purified away from the PDGF.

Inhibition of Cellular Binding. The PDGF-binding proteins in plasma were first described as a result of their interference with the PDGF radioreceptor assay[10,12] and radioimmunoassay,[11] two assays commonly used to determine the concentration of PDGF in fluids. In the case of the radioreceptor assay, this interference with the ability of PDGF to bind to its cell surface receptor suggests a possible modulatory role for the binding proteins. Therefore, to study the possible role of these proteins in modulating PDGF activity, an assay which determines the ability of a particular PDGF-binding activity (determined from one of the assays described above) to functionally inhibit PDGF receptor binding is described. Since the assay is a modification of the PDGF radioreceptor assay, an extensive discussion of binding conditions and receptor characteristics can be found in Bowen-Pope and Ross.[43]

Materials

^{125}I-Labeled PDGF and unlabeled PDGF: as described above.
Monolayer cultures of human fibroblasts: normal human fibroblasts (for example, American Type Culture Collection, CRL 1564) are seeded at $1–2 \times 10^4$ cells per culture well in 24-well cluster trays (Costar, Cambridge, MA). The cells are plated in DMEM containing 1% human plasma-derived serum using 1 ml per well and are normally used for assay 2–7 days after plating.
Binding medium: Ham's medium F12 buffered at pH 7.4 with HEPES and supplemented with 0.25% BSA (pretested and known to be free of PDGF and PDGF-binding activity).
Binding rinse: phosphate-buffered saline containing 1 mg/ml BSA.
Solubilization buffer: 1% Triton X-100 in water with 0.1% BSA.

Procedure

1. Samples to be tested must be prescreened for the presence of PDGF by use of a sequential assay (Fig. 4). Samples with undetectable levels of PDGF can be tested directly. If PDGF is present, it should be removed before assay by gel filtration or by incubation with anti-PDGF antibody bound to Sepharose.

2. Samples to be tested are diluted in binding medium and ^{125}I-labeled PDGF added to give a final concentration of 0.5 ng/ml. Samples are normally tested on three wells (0.5 ml/well). Binding medium alone, plasma [at 20, 40, 60, and 80% (v/v)], and 20 ng/ml unlabeled PDGF are normally run to determine total binding, plasma inhibition as a reference, and nonspecific binding, respectively. All samples are preincubated for 1 hr at 37° and then cooled to 4°.

3. The subconfluent cultures of normal human fibroblasts are rinsed once with cold binding rinse, drained, and 0.5 ml of sample added per well. The trays are incubated for 2 hr at 4° on an oscillating shaking platform at 2.5 cycles per second.

4. At the end of the incubation time, the trays are placed on ice, aspirated, rinsed three times with cold binding rinse, and drained.

5. Solubilization buffer (1 ml per well) is added to one tray at a time and placed on an oscillating table (2.5 cycles per second) for 4 min. After 4 min shaking, the solubilized monolayer is transferred to tubes for counting. Alternatively, if a shaking table is not available, the solubilization buffer can be squirted up and down several times over the culture surface with a Pasteur pipet. It is important to avoid prolonged incubation of solubilization buffer with the culture wells since ^{125}I-labeled PDGF bound to the culture dish can also be slowly eluted as described by Bowen-Pope and Ross.[43]

A characteristic inhibition curve is shown in Fig. 5 and the assay is depicted schematically in Fig. 4A (right panel). The results are expressed as specific binding in which nonspecific binding (^{125}I-labeled PDGF in the presence of 20 ng/ml unlabeled PDGF, depicted in Fig. 4A, left panel) is subtracted from ^{125}I-labeled PDGF bound in the presence of the test samples. This specific binding is then expressed as a percentage of the specific binding in control wells containing ^{125}I-labeled PDGF in binding medium without test samples. Figure 5 demonstrates that the inhibition of PDGF cellular binding is specific to plasma or the high-molecular-weight fraction of plasma and not a general property of concentrated protein solutions such as BSA. It also demonstrates that a commercial preparation of α_2-M and conditioned medium from a cell known to produce α_2-M, the macrophage,[46] each inhibits PDGF cellular binding.

[46] T. Hovi, D. Mosher, and A. Vaheri, *J. Exp. Med.* **145**, 1580 (1977).

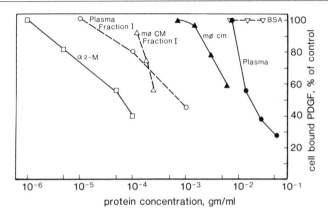

FIG. 5. Inhibition of PDGF binding to its cell surface receptor. The ability of increasing concentrations of various test substances to inhibit the binding of ^{125}I-labeled PDGF to its cell surface receptor was determined as described in the text. Test substances were BSA, plasma, macrophage-conditioned medium (mϕ CM), plasma fraction I (20–40% bed volume) of PDS chromatographed on BioGel A-0.5m (Fig. 1B), the highest molecular weight fraction from chromatography of macrophage-conditioned medium (mϕ CM fraction I) on G-100 in 1.0 N acetic acid, and purified α_2-macroglobulin (α_2-M) (Calbiochem). The human macrophage-conditioned medium was prepared from cells collected from patients on chronic peritoneal ambulatory dialysis and fractionated as described by Shimokado et al.[25]

Inhibition of Antibody Binding. As described above for the radioreceptor assay and illustrated in Fig. 4 for both the radioreceptor and radioimmunoassay, the ability of binding proteins to form a complex with ^{125}I-labeled PDGF and therefore prevent binding, to a cellular receptor or to antibody bound to Sepharose, in a simultaneous assay can result in the false identification of a test sample as containing PDGF (Fig. 4A, right panel). Although we have not attempted to standardize the inhibition of the radioimmunoassay, an assay as described above for cell receptor binding could be utilized to detect PDGF binding proteins. Huang et al.[13] found that ^{125}I-labeled PDGF plasma-binding complexes and ^{125}I-labeled PDGF–α_2-M complexes lost 97 and 98% of their immunoreactivity as compared with uncomplexed ^{125}I-labeled PDGF. Our individual analysis of each of the three peaks of ^{125}I-labeled PDGF plasma–protein complexes depicted in Fig. 1 also established a total loss of immunoreactivity for all three.[41] In addition, the mitogenic activity of PDGF-like molecules in peritoneal macrophage-conditioned medium containing binding protein activity (Fig. 5) was not neutralized by anti-PDGF IgG, while the PDGF-like mitogenic activity of alveolar macrophage-conditioned medium which contained no detectable binding protein activity was neutralized by anti-PDGF.[25] Therefore, to date, all reported PDGF-binding protein complexes lose their immunoreactivity with antibody to PDGF.

Assays for α_2-Macroglobulin

The availability of purified α_2-M and antibody to α_2-M facilitates quantitation of the levels of α_2-M in various test samples. An ELISA developed in our laboratory for α_2-M using commercially available reagents has an ED_{50} of approximately 1 μg/ml and is currently being utilized to screen possible sources of PDGF-binding protein activity in addition to the inhibition of cellular binding described above. α_2-M is also a potent enzyme inhibitor and can therefore be detected by assaying trypsin binding activity using N-α-benzoyl-DL-arginine p-nitroanilide (BAPA).[47] However, it is not at present known how the enzyme binding activity of α_2-M correlates with the PDGF-binding activity. In addition, since α_2-M constitutes only approximately 50% of the binding protein activity (Table I and text), any assay for α_2-M is limited by the fact that other PDGF-binding proteins exist.

Acknowledgments

We thank Katie Sprugel for helpful comments, Arnie Hestness for drafting the figures, and Mary Hillman for assistance in preparing the manuscript. This work was partially supported by grants from the National Institutes of Health (HL-18645) and RJR Nabisco, Inc.

[47] P.-O. Ganrot, *Clin. Chim. Acta* **14**, 493 (1966).

[6] Differential Colony Hybridization: Molecular Cloning from a Zero Data Base

By BRENT H. COCHRAN, PETER ZUMSTEIN, JOHN ZULLO, BARRETT ROLLINS, MARK MERCOLA, and CHARLES D. STILES

Problems in structure and function of animal cell growth factors are readily amenable to recombinant DNA technology. Powerful cloning strategies allow isolation of growth factors, their receptors, or growth factor-regulated gene sequences provided that a data base of knowledge already exists for the gene of interest. The data base may consist of no more than a partial amino acid sequence. Such sequence data can be used to generate oligonucleotide probes to the corresponding structural genes. This strategy was successfully applied to the isolation of gene sequences encoding epidermal growth factor,[1,2] nerve growth factor,[3,4] the insulin-like growth factors,[5-8] granulocyte–macrophage colony-stimulating factor,[9] and erythropoietin.[10] This strategy has also been applied to the isolation of structural genes encoding the insulin receptor[11,12] and interleukin-2 receptor.[13,14]

Polyclonal or monoclonal antibodies may form the data base. By reacting with partially translated nascent peptides on polysomal RNA, antibodies can be used to enrich for a specific mRNA of interest. This tactic was employed in the cloning of the low-density lipoprotein (LDL) receptor for example.[15] Antibodies may also be used to screen bacterial expression libraries. This tactic was applied with success in molecular cloning of

[1] A. Gray, T. J. Dull, and A. Ullrich, *Nature (London)* **303**, 722 (1983).

[2] J. Scott, M. Urdea, M. Quiroga, R. Sanchez-Pescador, N. Fong, M. Selby, W. J. Rutter, and G. I. Bell, *Science* **221**, 236 (1983).

[3] J. Scott, M. Selby, M. Urdea, M. Quiroga, G. I. Bell, and W. J. Rutter, *Nature (London)* **302**, 538 (1983).

[4] A. Ullrich, A. Gray, C. Berman, and T. J. Dull, *Nature (London)* **303**, 825 (1983).

[5] M. Jansen, F. M. A. van Schaik, A. T. Ricker, B. Bullock, D. E. Woods, K. H. Gabbay, A. L. Nussbaum, J. S. Sussenbach, and J. L. Van den Brande, *Nature (London)* **306**, 609 (1983).

[6] G. I. Bell, J. P. Merryweather, R. Sanchez-Pescador, M. M. Stempien, L. Priestley, J. Scott, and L. B. Rall, *Nature (London)* **310**, 775 (1984).

[7] T. J. Dull, A. Gray, J. S. Hayflick, and A. Ullrich, *Nature (London)* **310**, 777 (1984).

[8] H. J. Whitfield, C. B. Bruni, R. Frunzio, J. E. Terrell, S. P. Nissley, and M. M. Rechler, *Nature (London)* **312**, 277 (1984).

[9] N. M. Gough, J. Gough, D. Metcalf, A. Kelso, D. Grail, N. A. Nicola, A. W. Burgess, and A. R. Dunn, *Nature (London)* **309**, 763 (1984).

[10] K. Jacobs, C. Shoemaker, R. Rudersdorf, S. D. Neill, R. J. Kaufman, A. Mufson, J. Seehra, S. S. Jones, R. Hewick, E. F. Fritsch, M. Kawakita, T. Shimizu, and T. Miyake, *Nature (London)* **313**, 806 (1985).

[11] A. Ullrich, J. R. Bell, E. Y. Chen, R. Herrera, L. M. Petruzzelli, T. J. Dull, A. Gray, L. Coussens, Y.-C. Liao, M. Tsubokawa, A. Mason, P. H. Seeburg, C. Grunfeld, O. M. Rosen, and J. Ramachandran, *Nature (London)* **313**, 756 (1985).

[12] Y. Ebina, L. Ellis, K. Jarnagin, M. Edery, L. Graf, E. Clauser, J.-H. Ou, F. Masiarz, Y. W. Kan, I. D. Goldfine, R. A. Roth, and W. J. Rutter, *Cell* **40**, 747 (1985).

[13] W. J. Leonard, J. M. Depper, G. R. Crabtree, S. Rudikoff, J. Pumphrey, R. J. Robb, M. Kronke, P. B. Svetlik, N. J. Peffer, T. A. Waldmann, and W. C. Greene, *Nature (London)* **311**, 626 (1984).

[14] T. Nikaido, A. Shimizu, N. Ishida, H. Sabe, K. Teshigawara, M. Maeda, T. Uchiyama, J. Yodoi, and T. Honjo, *Nature (London)* **311**, 631 (1984).

[15] D. W. Russell, T. Yamamoto, W. J. Schneider, C. J. Slaughter, M. S. Brown, and J. L. Goldstein, *Proc. Natl. Acad. Sci. U.S.A.* **80**, 7501 (1983).

the EGF receptor.[16] If antibodies are directed to a determinant located on the cell surface, they may be used as optical tags for gene cloning by cell sorting technology. This tactic was used to isolate molecular clones of the transferrin receptor gene.[17,18]

The data base may consist of nothing more than an effective bioassay. The genes encoding interleukin-1,[19,20] interleukin-2,[21] interleukin-3,[22] and mast cell growth factor[23] have been cloned with variations of the following protocol: (1) fractionate total mRNA by sucrose density gradient, (2) translate mRNA fractions (in frog oocytes) and screen the translation products by bioassay, (3) prepare a cDNA library from the mRNA fraction which preferentially directs translation of the material, (4) fractionate the recombinant DNA library into pools of individual bacterial clones (usually 10–20 clones per pool), and (5) prepare plasmid DNA from the pools and screen for pools which contain the gene of interest by hybrid selection and *in vitro* translation of mRNA.

Interesting variations of this Poissonian screening system based on bioassay have been applied to the isolation of the genes encoding interleukin-2[24] and granulocyte–macrophage colony-stimulating factor 1.[25]

This review deals with a cloning strategy which requires no preexisting data base of knowledge at all. The strategy of "differential colony hybridization" or "+/− screening" requires nothing more than the existence of two closely matched tissues. Genes are selected on the basis of unequal levels of expression in the two cell populations. Using this technique, it has been possible to clone genes whose expression is tissue specific, stage specific, cell cycle regulated, or hormone inducible. Differential colony hybridization has been used to isolate molecular clones of

[16] C. R. Lin, W. S. Chen, W. Kruiger, L. S. Stolarsky, W. Weber, R. M. Evans, I. M. Verma, G. N. Gill, and M. G. Rosenfeld, *Science* **224**, 843 (1984).

[17] R. Newman, D. Domingo, J. Trotter, and I. Trowbridge, *Nature (London)* **304**, 643 (1983).

[18] L. C. Kuhn, A. McClelland, and F. H. Ruddle, *Cell* **37**, 95 (1984).

[19] P. T. Lomedico, U. Gubler, C. P. Hellmann, M. Dukovich, J. G. Giri, Y.-C. E. Pan, K. Collier, R. Seminow, A. O. Chua, and S. B. Mizel, *Nature (London)* **313**, 458 (1984).

[20] P. E. Auron, A. C. Webb, L. J. Rosenwasser, S. F. Mucci, A. Rich, S. M. Wolff, and C. A. Dinarello, *Proc. Natl. Acad. Sci. U.S.A.* **81**, 7907 (1984).

[21] T. Taniguchi, H. Matsui, T. Fujita, C. Takeoka, N. Kashima, R. Yoshimoto, and J. Hamuro, *Nature (London)* **302**, 305 (1983).

[22] M. C. Fung, A. J. Hapel, S. Ymer, D. R. Cohen, R. M. Johnson, H. D. Campbell, and I. G. Young, *Nature (London)* **307**, 232 (1984).

[23] T. Yokota, F. Lee, D. Rennick, C. Hall, N. Arai, T. Mosmann, G. Nabel, H. Cantor, and K.-I. Arai, *Proc. Natl. Acad. Sci. U.S.A.* **81**, 1070 (1984).

[24] T. Yokota, N. Arai, F. Lee, D. Rennick, T. Mosmann, and K.-I. Arai, *Proc. Natl. Acad. Sci. U.S.A.* **82**, 68 (1985).

[25] G. G. Wong, J. S. Witek, P. A. Temple, K. M. Wilkens, A. C. Leary, D. P. Luxenberg, S. S. Jones, E. L. Brown, R. M. Kay, E. C. Orr, C. Shoemaker, D. W. Golde, R. J. Kaufman, R. M. Hewick, E. A. Wang, and S. C. Clark, *Science* **228**, 810 (1985).

gene sequences which are induced by serum,[26,27] epidermal growth factor,[28] retinoic acid,[29] and platelet-derived growth factor.[30] Other examples include the isolation of genes inducible by galactose in yeast,[31] heat shock in *Drosophila*,[32] and testosterone in mice.[33] Differential colony hybridization has been used to study developmental programs of adipocyte[34] and slime mold differentiation.[35] The potential applications of this technique are virtually endless and the number of laboratories using it is rapidly expanding.

In this chapter we will attempt to provide a guide to differential colony hybridization for the cDNA cloning novice. Since there are many good articles and manuals available to the cDNA cloning novice,[36–39] some of the procedures outlined below will not be described in exhaustive detail.

In applying the differential colony hybridization technique, the reader would be wise to consult these other sources and keep in mind that molecular cloning protocols are in constant flux. Procedures can become obsolete rapidly. The protocols detailed below are adaptations of procedures developed by many others and are the ones that are presently being used by our laboratory. We will emphasize practical as well as theoretical considerations in the choices of alternative procedures and try to point out commonly encountered problems with the procedures.

Rapid Cloning of Full Length cDNA

To clone full length cDNAs it is of utmost importance to have intact mRNA. There are three general rules for successfully isolating full length mRNAs.

[26] D. I. H. Linzer and D. Nathans, *Proc. Natl. Acad. Sci. U.S.A.* **80**, 4271 (1983).

[27] R. R. Hirschhorn, P. Aller, Z.-A. Yuan, C. W. Gibson, and R. Baserga, *Proc. Natl. Acad. Sci. U.S.A.* **81**, 6004 (1984).

[28] D. N. Foster, L. J. Schmidt, C. P. Hodgson, H. L. Moses, and M. J. Getz, *Proc. Natl. Acad. Sci. U.S.A.* **79**, 7317 (1982).

[29] S.-Y. Wang and L. J. Gudas, *Proc. Natl. Acad. Sci. U.S.A.* **80**, 5880 (1983).

[30] B. H. Cochran, A. C. Reffel, and C. D. Stiles, *Cell* **33**, 939 (1983).

[31] T. P. St. John and R. W. Davis, *Cell* **16**, 443 (1979).

[32] J. T. Lis, W. Neckameyer, R. Dubensky, and N. Costlow, *Gene* **15**, 67 (1981).

[33] F. G. Berger, K. W. Gross, and G. Watson, *J. Biol. Chem.* **256**, 7006 (1981).

[34] B. M. Spiegelman, M. Frank, and H. Green, *J. Biol. Chem.* **258**, 10083 (1983).

[35] J. G. Williams and M. M. Lloyd, *J. Mol. Biol.* **129**, 19 (1979).

[36] T. Maniatis, E. F. Fritsch, and J. Sambrook, "Molecular Cloning: A Laboratory Manual." Cold Spring Harbor Lab., Cold Spring Harbor, New York, 1982.

[37] H. M. Goodman and R. J. MacDonald, this series, Vol. 68, p. 75.

[38] A. Efstratiadis and L. Villa-Kamaroff, *in* "Genetic Engineering," Vol. 1, p. 15. Plenum, New York, 1979.

[39] J. G. Williams, *in* "Genetic Engineering" (R. Williamson, ed.), Vol. 1, p. 1. Academic Press, New York, 1981.

1. Destroy any ribonucleases which may come into contact with the RNA. This can be done by autoclaving for 90 min any material such as Eppendorf tubes and salt solutions which may come into contact with the RNA. (Gas-sterilized plasticware is essentially RNase free.)

2. Wear gloves to prevent contamination of RNA with ribonucleases found on the hands. Use RNase inhibitors when possible. We usually include SDS in the oligo(dT) column buffer and placental ribonuclease inhibitor (Promega Biotec) in the first-strand cDNA synthesis reaction.

3. Isolate RNA by the guanidium isothiocyanate procedure.[40,41] Guanidium isothiocyanate is a very powerful protein denaturant and when used in combination with reducing agents, rapidly inactivates cellular ribonucleases.

RNA Isolation

1. Prepare lysis buffer consisting of 4 M guanidium isothiocyanate, 25 mM sodium citrate, pH 7.0, and 0.1 M 2-mercaptoethanol. Do not autoclave this solution. Filter sterilize instead. Use this solution directly to lyse cells or tissue of interest. Lyse from 5×10^6 to 1×10^8 cells per 3 ml of lysis buffer.

2. Force the lysate through a 20-gauge needle several times to break up the DNA and reduce the viscosity of the solution. This improves the yield.

3. Carefully layer onto a cushion of autoclaved 5.7 M CsCl, 25 mM sodium acetate, pH 5.0. Use 3 ml of lysis buffer to 2 ml of 5.7 M CsCl solution in a Beckman SW 50.1 polyallomer tube. Centrifuge at 40,000 rpm for 18 hr at 20°.

4. Aspirate the top of the gradient and then pour off the remainder of the CsCl taking care not to contaminate the pellet with the upper buffer.

5. Cut off the bottom of the tube with a pair of scissors or a razor blade.

6. Resuspend the RNA pellet in 300 μl autoclaved H_2O. The pellet usually has to be scraped off the bottom of the tube with the pipet tip. Extract once with an equal volume of phenol/chloroform solution.

7. Add 30 μl of 3 M sodium acetate, pH 5.5 (autoclaved) and 1 ml of 95% ethanol. Leave at $-20°$ overnight.

8. Pellet RNA in microfuge for 10 min. Remove the ethanol supernatant and dry pellet *in vacuo*. Resuspend RNA in autoclaved H_2O. Use directly for selection of poly(A)$^+$ RNA.

[40] J. Chirgwin, A. Aeyble, R. MacDonald, and W. Rutter, *Biochemistry* **18**, 5294 (1979).
[41] K. Kelly, B. H. Cochran, C. D. Stiles, and P. Leder, *Cell* **35**, 603 (1983).

Isolation of Poly(A)$^+$ RNA

Poly(A)$^+$ RNA can be selected by binding to either poly(U)-Sepharose[42,43] or oligo(dT)-cellulose[44] columns. One cycle of binding yields poly(A)$^+$ RNA of sufficient purity to synthesize cDNA. Two cycles of binding, however, will increase the purity of the poly(A)$^+$ RNA by about 50%. The following protocol was adapted from that of Aviv and Leder.[44]

1. Suspend oligo(dT)-cellulose (Collaborative Research) according to manufacturer's instructions. Prepare a column of 0.2 g of oligo(dT)-cellulose in a 3-ml sterile plastic syringe. Use autoclaved, siliconized, glass wool to plug the column.

2. Prepare binding buffer (0.5 M NaCl, 10 mM Tris, pH 7.5, 0.5% SDS, 1 mM EDTA), wash buffer (100 mM NaCl, 10 mM Tris, pH 7.5, 0.5% SDS, 1 mM EDTA) and elution buffer (10 mM Tris, pH 7.5, 1 mM EDTA, 0.05% SDS). Autoclave to inactivate ribonucleases.

3. Preelute the column with 10 ml of elution buffer at 60°. Equilibrate with binding buffer at room temperature.

4. Suspend RNA in 3–5 ml of binding buffer. Pass the RNA over the column 4 times at room temperature.

5. Wash column with 20 ml of binding buffer. Follow with 10 ml of washing buffer.

6. Elute poly(A)$^+$ RNA with 60° elution buffer. Collect 400 μl fractions in autoclaved 1.5-ml tubes.

7. Read the OD$_{260}$s of the fractions to determine the location of the poly(A)$^+$ RNA. Most of the RNA is usually found in the first two fractions. Pool the RNA-containing fractions and determine the quantity of RNA recovered on the basis that 1 OD$_{260}$ unit equals approximately 40 μg/ml RNA. The yield of poly(A)$^+$ RNA after one purification cycle is usually 1–5% of the total RNA input.

8. Add autoclaved 3.0 M sodium acetate, pH 5.5, to the RNA solution to a final sodium acetate concentration of 0.3 M. Precipitate the poly(A)$^+$ RNA overnight at −20° by the addition of 2.5 volumes of 95% ethanol.

9. Pellet the RNA by centrifugation at 10,000 g for 10 min at 0°.

10. Wash the pellet with 70% ethanol at −20° and repellet as above.

11. Remove supernatant and lyophilize pellet to dryness. (Pellet may not be visible.) Resuspend pellet in autoclaved H$_2$O at approximately 1 mg/ml. It may be useful to add 10 units of RNasin to the RNA at this point. Freeze at −70° if not used immediately.

[42] J. M. Taylor and T. P. H. Tse, *J. Biol. Chem.* **251**, 7461 (1976).
[43] D. J. Shapiro and R. T. Schimke, *J. Biol. Chem.* **250**, 1759 (1975).
[44] H. Aviv and P. Leder, *Proc. Natl. Acad. Sci. U.S.A.* **69**, 1408 (1972).

First-Strand Synthesis

The following procedures are modified from Gubler and Hoffman,[45] Okayama and Berg,[46,47] and Mullins.[48] The basic protocol for cDNA synthesis is outlined directly below. For further discussion of methods for analyzing the reactions, see the "Controls and Problems with cDNA Synthesis" section which follows.

1. In an autoclaved 1.5-ml tube, mix 10 μl of poly(A)$^+$ RNA (1 mg/ml in H_2O) with 1 μl of 100 mM CH_3HgOH (Alfa Chemicals). Leave 3–5 min at room temperature. The methylmercury denatures the RNA to increase the chances of getting full length transcripts.

2. Add 2 μl of 700 mM 2-mercaptoethanol. Leave 2 min at room temperature. This high concentration of 2-mercaptoethanol reverses the reaction of CH_3HgOH with RNA.[49]

3. Add 1 μl or at least 10 units of placental RNase inhibitor (Promega Biotec).

4. Add 40 μl of 2× salt buffer containing 100 mM Tris (pH 8.3), 20 mM $MgCl_2$, and 60 mM KCl. This solution should be autoclaved.

5. Add 10 μl or a mass equivalent to the amount of input RNA of oligo(dT)$_{12-18}$ (1 μg/μl) (Collaborative Research).

6. Add 8 μl of 10× dNTP solution (10 mM each + 25 μCi [^{3}H]CTP).

7. Add 2 μl autoclaved H_2O. Warm the mixture to 37°.

8. Add 2 μl of autoclaved 160 mM sodium pyrophosphate.

9. Add 5 μl or 50–100 units of reverse transcriptase. Mix and incubate at 42° for 30 min. A 2.5-μl aliquot can be taken at this point and added to 10 μCi of [^{32}P]GTP for the purpose of analyzing the length of the first-strand reaction products. See discussion below. Reverse transcriptase can be obtained from many sources and varies in quality from supplier to supplier and batch to batch. The major problem with reverse transcriptase seems to be variable contamination with an endoribonuclease. Ideally, each batch of transcriptase should be tested for its ability to produce full length transcripts from a defined RNA species of greater than 2 kb in length. (Such RNA templates can be prepared from RNA viruses or by *in vitro* transcription from bacterial RNA polymerase promoters.) Currently we are obtaining the enzyme from Seikagaku America, St. Petersburg, Florida. It is important not to let the reaction run longer than 30 min.

[45] V. Gubler and B. J. Hoffman, *Gene* **25**, 263 (1983).
[46] H. Okayama and P. Berg, *Mol. Cell. Biol.* **2**, 151 (1982).
[47] H. Okayama and P. Berg, *Mol. Cell. Biol.* **3**, 280 (1983).
[48] J. Mullins, personal communication.
[49] J. M. Bailey and N. Davidson, *Anal. Biochem.* **70**, 75 (1976).

10. Add EDTA to 20 mM to stop the reaction.
11. Extract with phenol/chloroform (optional).
12. Precipitate RNA–DNA hybrids by adding ammonium acetate to 2 M and adding 2 volumes of 95% ethanol. Mix. Leave on dry ice for 30 min. Centrifuge for 10 min in microfuge. Wash pellet with 70% ethanol at −20°. For second-strand synthesis, resuspend in 300 μl H_2O. Note: The RNA–DNA hybrids at the end of the first strand synthesis often stick avidly to the bottoms of 1.5 ml plastic tubes. To avoid this problem, add 20 μg of glycogen or tRNA as carrier for the precipitation. If tRNA is used as the carrier, it should be removed by ribonuclease treatment and by gel filtration chromatography (as described below) after the second strand synthesis, but before the tailing reaction.

Second-Strand Synthesis

1. To the 300 μl of the first strand, add 80 μl of 5× salt buffer [100 mM Tris, pH 7.5, 25 mM $MgCl_2$, 50 mM $(NH_4)_2SO_4$, 50 mM KCl, 250 μg/ml BSA, 50 mM DTT, 200 μM each of dNTPs and 20 μCi of [α-^{32}P]dGTP].
2. Add 4 μl RNase H (1300 units/ml stock).
3. Add 4 μl 15 mM B-NAD$^+$ (fresh).
4. Add E. coli DNA polymerase I to a concentration of 230 units/ml (approximately 180 units total). Be sure that the enzyme is diluted so that there is less than a 10% final concentration of glycerol in the reaction.
5. Add 0.2 μg E. coli DNA ligase and incubate overnight at 14°.
6. Stop the reaction by adding EDTA to 20 mM.
7. Phenol/$CHCl_3$ extract (optional).
8. Precipitate twice from 2 M ammonium acetate and ethanol as with step 12 above.

Controls and Problems with cDNA Synthesis

As a control, the cDNA cloning reaction should be monitored by measuring the efficiency of conversion of RNA to DNA and by determining the size of the reaction products. These measurements can be made by the inclusion of radiolabeled nucleotides in the first- and second-strand reactions.

There are a variety of ways of following the reaction and here we will suggest but one. In the first strand reaction include 25 μCi of [^{3}H]CTP and after adding reverse transcriptase remove a 3-μl aliquot to which is added 10 μCi of [^{32}P]CTP. Both reactions are then incubated in parallel.

From the percent of [^{3}H]CTP incorporated into TCA precipitable counts, an estimate of the percentage mass conversion of RNA into cDNA can be made as follows: [(fraction ^{3}H incorporated into TCA precipitate counts) $\times$ (4 mM dNTP $\times$ 0.080 ml $\times$ 350 g/mol dNTP)/(μg of input RNA)]. For cDNA made against a heterogeneous population of poly(A)$^+$ RNA, the reaction efficiency usually runs in the range of 10–30%. If the efficiency is less than 10%, then controls should be run on the reaction components. A good control reaction is to make cDNA from globin RNA, which can be purchased commercially (BRL, Bethesda). Portions of the ^{32}P-labeled side reaction can be used to estimate the length of the first-strand reaction products by electrophoresis through 1.4% alkaline agarose gels and exposure to X-ray film. (Before loading onto the gel, unincorporated nucleotide triphosphates should be removed by chromatography over a 1–2 ml Sephadex G-75 column or equivalent.)

For monitoring the second-strand reaction, 20 μCi of a ^{32}P-labeled dNTP can be included directly in the reaction mixture. The small amount of label incorporated does not interfere with quantitating the tailing reaction later. TCA precipitable counts can be taken as a measure of the efficiency of second-strand synthesis. Efficiency is calculated as [(fraction ^{32}P incorporated into TCA precipitable counts) $\times$ (0.8 mM dNTP $\times$ 0.4 ml $\times$ 350 g/mol dNTP)/(μg input of first-strand template)]. The efficiency of the second-strand reaction is usually on the order of 50%. After removal of unincorporated nucleotides, the second-strand reaction products can be sized by electrophoresis through denaturing or nondenaturing agarose gels. The reaction products of both strands typically run as a smear from 500 to 7000 base pairs. Optimally, the size of the second strand should equal, but not exceed that of the first. In addition, the size should be the same on denaturing or nondenaturing gels. If, under denaturing conditions, the size of the reaction products increases, then there are probably a large number of hairpins at one end.[45] This will reduce the cloning efficiency. Using the original Gubler–Hoffman protocol, we have sometimes noticed second-strand products that were several times larger than the first-strand products. We believe these large products could be due to the ligation of several cDNAs together caused by excessive ligase concentrations. The protocol given here uses much less ligase in the second-strand reaction to avoid this problem. Recently, it has been suggested that it may be preferable to leave the ligase and the RNase H entirely out of the reaction.[49a]

Sometimes there are problems with the first-strand synthesis reaction caused by pyrophosphate precipitates which inhibit the reaction. To alleviate this problem it is best to use the highest purity grade of NaPP$_i$

[49a] J. M. D'Alessio, M. C. Noon, H. L. Loy III, and G. F. Gerard, *BRL Focus* **9:1**, 1 (1987).

available and add it to the prewarmed reaction just before reverse transcriptase. If precipitates are a problem, then it may be helpful to reduce the concentration of Tris in the buffer. Also, avoid freezing solutions containing pyrophosphate as this promotes precipitation. The purpose of the NaPP$_i$ is to suppress DNA-dependent DNA synthesis by reverse transcriptase by inhibiting the endogenous RNase H activity of the enzyme.

Other details that should be paid attention to are the 30 min incubation time of the first-strand reaction and the pH of the buffer at 43°. Artifactually long first-strand reaction products are sometimes observed for incubation periods longer than 30 min. Inefficient first-strand synthesis may sometimes be due to changes in the Tris buffer pH with temperature. The pH should be 8.3 at 43°.

Often the cDNA synthesis reactions will generate many very short products. It is a good idea to eliminate them by running the second-strand reaction products over a 5 ml Sepharose 4B column and collecting the excluded volume. The size of the reaction products eluted from the column can be monitored by electrophoresing aliquots through 1% agarose gels.

Choice of Vector

There are two basic choices of vectors into which cDNA can be cloned—plasmids or λ phage. Each has its own advantages and disadvantages, and either will serve the same purpose. Lambda vectors have the advantage of the easy storage and propagation of large numbers of clones. In addition, our experience is that phage plaques give cleaner signals after filter hybridization. The disadvantages are that the handling of λ phage requires more manipulation in the cloning procedures (although the commercial availability of *in vitro* packaging kits and purified vector arms is changing this situation) and the fact that the cDNA insert is such a small percentage of the total vector. The latter factor usually necessitates the subcloning of a cDNA of interest into a plasmid vector for further characterization. Plasmid vectors are relatively easy to handle and the availability of vectors with promoters for bacterial RNA polymerases for cloning give plasmids a slight edge for most cDNA cloning applications. Currently our laboratory is using the Gemini vector of Promega Biotec which has the T7 and SP6 RNA polymerase promoters oriented in opposite directions across multiple cloning sites so that either strand of a given DNA insert can be transcribed into RNA without any further manipulations of the vector. This vector allows the easy production of high-sensitivity single-stranded RNA probes of either orientation. In addition the vector replicates quite well so that milligram quantities of DNA can be obtained readily.

Tailing versus Linkers

After precipitation of double-stranded cDNA, one must decide on a method of recombining the cDNA into the vector. The two principal methods to be considered are (1) homopolymer tailing of cDNA insert and vector followed by annealing and (2) addition of defined sequence linkers to the cDNA and ligating to the vector. The homopolymer tailing method entails the fewest experimental manipulations and is therefore the easiest method to use. However, this method does have the drawback that comes from the difficulties of not being able to precisely control the length of the dG:dC tails. Long dG:dC tails interfere with sequencing by the primer extension method, with transcription by SP6 RNA polymerase off the SP6 promoter, and possibly with plasmid replication. The use of linkers avoids these problems entirely, but requires several more steps including the enzymatic modification of the cDNA to protect from cutting by the restriction enzyme after linker addition, 2 ligation reactions, and kinasing of the linkers. In our experience, linkers, while giving a preferable outcome to the tailing method, are experimentally more problematic and not recommended to the cloning novice.

In addition, the problems of tailing are largely from having tail lengths of larger than 30 bp. By arranging the tailing conditions appropriately, the long dG:dC stretches can be minimized or avoided entirely.

Homopolymer Tailing Reaction

The most widely used method to introduce cDNA into the vector is based on the addition of oligo(dG) tails to the plasmid and of complementary oligo(dC) tails to the cDNA.[50,51] The stepwise addition of deoxynucleotide triphosphate to the 3′ termini of DNA is catalyzed by terminal deoxynucleotidyltransferase (TdT). Single-stranded DNA or duplex DNA with protruding 3′ ends are efficiently used as substrates for terminal transferase in the presence of Co^{2+} or Mg^{2+} ions. The resulting polymerized tails have a relatively uniform length distribution.[52,53] In contrast, a recent study by Michelson and Orkin[54] demonstrated that blunt ended

[50] L. A. Villa-Komaroff, A. Efstratiadis, S. Broome, P. Lomedico, R. Tizard, S. P. Naker, W. L. Chick, and W. Gilbert, *Proc. Natl. Acad. Sci. U.S.A.* **75,** 3727 (1978).
[51] W. Rowekamp and R. A. Firtel, *Dev. Biol.* **79,** 409 (1980).
[52] P. E. Lobban and A. D. Kaiser, *J. Mol. Biol.* **78,** 453 (1973).
[53] R. Roychoudhury and R. Wu, this series, Vol. 65, p. 43.
[54] A. M. Michelson and S. H. Orkin, *J. Biol. Chem.* **257,** 14773 (1982).

DNA is less efficiently used as a substrate even by high concentrations of terminal transferase and in the presence of Co^{2+} ions. Thus, the initiation of the polymerization reaction on blunt ends is asynchronous even under the most favorable conditions, and the tailed products are more heterogeneous in length.[54] The efficiency of the blunt-end tailing reaction cannot be substantially increased by varying the enzyme or deoxynucleotide triphosphate concentration or the incubation temperature.[54] This is a potentially limiting factor for cloning of cDNA by this method.

Taking these facts into account, we have worked out and successfully used the following protocol which is adapted from the work of Roychoudhury and Wu[53] and Michelson and Orkin.[54]

Oligo(dG) Tailing of the Plasmid

Cloning of cDNA into the *Pst*I site of plasmid vectors by dG : dC tailing has the advantage of regenerating the recognition site for *Pst*I.[50] Thus, cloned inserts can be recovered from the plasmid by digestion with *Pst*I. The *Sac*I site and dC : dG tailing may be used for a similar strategy (unpublished results). Both restriction enzymes create 3' protruding ends, a condition suitable to obtain a product with uniform tail length.[54] Many plasmids vectors, including the Gemini vector of Promega Biotec, have both of these restriction sites in the polylinker region.

In our hands, the conditions for reproducibly achieving a uniform tail length of about 10 residues per end are as follows.

1. Add plasmid DNA such that final DNA concentration of the reaction will be 30–50 pmol 3' termini/ml.
2. Add dGTP to 10 μM, from a 10× stock solution in water, pH 7, stored frozen. Add 5–10 μCi of [α-^{32}P]dGTP (400 Ci/mmol).
3. Add reaction buffer from a 5× stock solution stored in frozen aliquots. The final reaction buffer composition should be 125 mM, K^+ (or Na^+) cacodylate, pH 7.2, 10 mM $MgCl_2$, 0.2 mM dithiothreitol (DTT), and 0.5 mg/ml bovine serum albumin.
4. Initiate the reaction by the addition of 15 units terminal transferase per pmol of 3' DNA termini.
5. Incubate for 10 min at 20°.
6. Stop the reaction by the addition of EDTA to 25 mM and heating to 65° for 10 min.

For dC tailing the vector at a restriction site with 3' overhangs (e.g., *Sac*I to regenerate the recognition site) use 10 μM dCTP and lower the incubation temperature to 12°.

Individual lots of terminal transferase can vary widely in activity.

Therefore, for best results we recommend setting up a series of pilot reactions, with 1 μg of plasmid vector in 20–30 μl, to test the enzyme activity and to establish the reaction time for the desired tail length. The tail length is calculated from the amount of oligonucletide incorporated into the plasmid as determined from trichloroacetic acid precipitable radioactivity (pmol NTP incorporated/pmol 3′ termini).[36] In determining the amount of specific TCA-precipitable counts incorporated, be sure to control for nonspecific trapping of the ^{32}P-labeled nucleotide. This number is determined by precipitating an aliquot of the reaction mixture before adding the enzyme. Alternatively, the tail length on vectors with a polylinker can be visualized directly.[54] After digestion at a restriction site close to the tailed ends, the small tailed and therefore labeled fragments can be analyzed on a DNA sequencing gel.[54] Reproducibly, we find that the results of both assays give very similar results. These results also demonstrate that most of the 3′ ends of the vector become tailed.

To prepare a larger amount of tailed vector, e.g., 10–15 μg, the reaction volume can be easily scaled up, typically to 200–300 μl under otherwise identical conditions (30–50 pmol 3′ ends/ml, TdT 15 units/pmol 3′ end, dGTP 10 μM).

The tailed product is desalted on a column (5 ml, in a plastic tissue culture pipet) of Sephadex G-75 in 10 mM Tris–HCl, pH 8, 1 mM EDTA, and can be used directly for annealing and transformation (see below). If the background of transformation is higher than 1–2 colonies/ng, the tailed plasmid can be purified over oligo(dC)-cellulose.[55] For long-term storage of the tailed vector, we recommend omitting the radioactive label. An aliquot of the reaction mixture containing the radioactive nucleotide can be run in parallel to monitor the tail length. Tailed vector can be stored at −20° for several weeks, or for long-term storage, precipitated in ethanol.[36]

Oligo(dC) Tailing of Double-Stranded cDNA

The homopolymer tailing of duplex cDNA presents inherent difficulties in achieving the desired average tail length. A uniform size distribution cannot be expected for several reasons. First, the cDNA is blunt ended and, thus, the diminished ability of flushed ends to act as a substrate for terminal transferase results in only about 40% of available termini being tailed.[35] This means that only about 16% of the cDNA will be cloned due to the predicted absence of sufficiently long dC tails on both

[55] H. Land, M. Grez, H. Hauser, W. Lindenmaier, and G. Schutz, *Nucleic Acids Res.* **9,** 2251 (1981).

ends to form a stable hybrid with the dG tailed vector. Even this low efficiency is likely to be an overestimate since these are the tailing efficiencies of unique sequence blunt end fragments. Tailing of heterogeneous sequence cDNA populations is even less efficient. (In fact, in general we have found that enzymatic reaction conditions that have been optimized for RNA and cDNA populations of defined sequence complexity cannot be extrapolated to heterogeneous populations.) Another difficulty faced in controlling the tail length of the cDNA is that the concentration of cDNA ends is based on an estimate of the average length of cDNA as determined by analytical gel electrophoresis. As a rule, average length is deliberately estimated to be somewhat shorter than that revealed by gel electrophoresis. Alternatively and more accurately, the average length of the cDNA can be estimated from the median of a normalized size distribution. The gel used to analyze the cDNA is cut into several slices of equal size and counted for radioactivity. The cpm per gel slice divided by the mean base pair length for each slice is plotted as a histogram against the base pair length. The mean value taken from this graph is then used to calculate the concentration of 3' ends.

We have used the following protocol to achieve an average cDNA length of approximately 10 residues:

1. Add cDNA to approximately 30 pmol 3' ends/ml.
2. Add dCTP to 10 μM from a 10× stock solution in water (stored in frozen aliquots). Then 5–10 μCi of [α-^{32}P]dCTP (specific activity 400 Ci/mmol) should be added to monitor the reaction.
3. Add buffer from 5× stock solution stored in frozen aliquots to a final concentration of 125 mM K$^+$ (or Na$^+$) cacodylate, pH 7.2, 1 mM CoCl$_2$, 0.2 mM dithiothreitol (DTT), and 0.5 mg/ml bovine serum albumin.
4. Add terminal transferase to 15 units/pmol 3' end.
5. Incubate at 20° for 30 min.
6. Stop the reaction by adding EDTA to 25 mM and heating to 65° for 10 min. Remove nucleotide triphosphates by running through a column of Sephadex G-75, as described before, or if additional size fractionation is desired, on Sepharose CL-6B. (Note: Removing dNTPs by ethanol precipitation from ammonium acetate may result in the formation of insoluble Co^{2+} DNA complexes.) Freezing the DNA/Co^{2+} mixture can also result in precipitation and is to be avoided.
7. Monitor the reaction by determining the picomoles of dCTP incorporated by measuring TCA precipitable counts. The functionality of the tailed cDNA can also be determined by measuring the percentage of cDNA which is able to bind to an oligo(dG)-cellulose column. If the

reaction was very inefficient, it may be helpful to increase the dCTP concentration in the reaction to 900 μM.[45] If you are willing to sacrifice control of tail length for more efficient tailing, a 10 mM concentration of dCTP can be used.

Annealing and Transformation

Only circular DNA molecules transform bacteria with an acceptable efficiency. Therefore, the objective of annealing the cDNA to the plasmid vector is not only to link the two DNAs, but also to produce a recircularized plasmid. Optimal conditions for the annealing reaction cannot be predicted since the fraction of functionally tailed cDNA is unknown. Therefore, conditions may vary with different cDNA preparations. It may be desirable to test for the presence of functional dC tails on the cDNA by binding the tailed cDNA to oligo(dG)-cellulose in a manner analogous to that used to select for poly(A)$^+$ RNA on oligo(dT)-cellulose columns. (Omit the SDS, however.) This step serves several purposes. It concentrates the cDNA eluted from the sizing column, enriches for functionally tailed cDNA, and gives a measure of the efficiency of the tailing reaction. For these reasons, we recommend this simple step before using the cDNA for annealing.

The annealing reaction should first be optimized in a series of small-scale pilot reactions. Varying amounts of dC tailed cDNA are annealed to a constant amount of dG tailed vector in order to find the cDNA to vector ratio which gives the maximum number of transformants per nanogram of cDNA. Controls are included with supercoiled vector to test the competence of the host bacteria and with tailed vector alone to assess the background transformation activity. (If the number of background transformants obtained from tailed vector alone is more than 2 per nanogram vector, it would be wise to clean up the vector by binding and elution from an oligo(dC)-cellulose column, if this has not been done already.)

1. In a series of tubes, set up annealing reactions each containing 10 ng tailed vector and from 0.1 to 10 ng tailed cDNA in 0.1 M NaCl, 1 mM EDTA, 10 mM Tris, pH 7.5, in a total volume of 50 μl per tube.
2. Heat for 3 min at 65°.
3. Incubate for 2 hr at 37°.
4. Leave at room temperature overnight. (The long overnight incubation may be critical for DNAs with short tails.)

Low concentrations of cDNA (500 ng/ml or less) should favor intramolecular circularization, but suboptimal concentrations reduce the number of clones obtained per nanogram cDNA. Thus, with increasing concentrations of cDNA, a peak ratio of cDNA to vector will be reached for transformation efficiency. At greater cDNA concentrations, the number of transformants obtained per nanogram of cDNA should drop. DNA from the annealing reaction is used directly to transform competent host bacteria. A $recA^+$ strain of $E.$ $coli$ such as C600 gives a higher transformation efficiency than $recA^-$ strains. The procedure we use to produce competent bacteria is essentially that of Dagert and Ehrlich.[56]

1. Grow C600 at 37° in L broth (10 g NaCl, 5 g yeast extract, 10 g bactotryptone per liter) until the cultures reach an OD_{550} of 0.2.
2. Pellet the cells by low-speed centrifugation and resuspend in 1/2 volume of 0.1 M $CaCl_2$ (sterile).
3. Leave on ice for 20 min and pellet again.
4. Resuspend in 1/10 the original volume of 0.1 M $CaCl_2$ and leave on ice for 20–24 hr.
5. Add 50 μl (or a volume containing 10 ng of vector) of the annealed cDNA and vector to 0.2 ml of competent cells.
6. Leave on ice for 10 min.
7. Heat shock at 37° for 5 min.
8. Add 1 ml of L broth and incubate with shaking at 37° for 1 hr to allow time for the drug resistance gene on the plasmid vector to be expressed.
9. Plate out 100–500 μl aliquots onto L agar plates containing selective drug (usually 50–100 μg/ml ampicillin or 15 μg/ml tetracycline).
10. Incubate 24 hr at 37°.

Optimal molar ratios of cDNA to vector range from 1 : 40 to 1 : 3 depending on the "quality" of cDNA preparation used. By this procedure, 200–1000 colonies/ng of cDNA may be obtained for optimal ratios. Background transformation is usually below 1 colony/ng of vector. It may be possible to increase the efficiency of transformation (if desired) by using the procedure of preparing competent bacteria recently described by Hanahan.[57] Competent cells prepared by this method may be stored at $-70°$ and are commercially available from BRL and Stratagene.

The results from the pilot experiments can be used to scale up the annealing reaction. Typically, 200–400 ng of vector and the amount of cDNA required for the optimal ratio of cDNA : vector are annealed in a reaction volume of 1 ml under otherwise identical conditions. An aliquot from this solution containing 10 ng of vector is finally tested for transformation efficiency. The total number of transformants expected in the

[56] M. Dagert and S. D. Ehrlich, $Gene$ **6,** 23 (1979).
[57] D. Hanahan, $J.$ $Mol.$ $Biol.$ **166,** 557 (1983).

cDNA library can be estimated from this result. After annealing, this solution can be stored at −20° or precipitated in ethanol for long-term storage. Transformation cannot be scaled up directly to produce a linear increase in the number of transformants (see Dagert and Ehrlich[56] for explanation). Large scale transformation should be performed exactly as described before, in multiple tubes, each containing 10 ng of annealed vector (in 20–50 μl) and 0.2 ml of competent cells. The number of colonies that can reasonably be screened by the technique of differential colony hybridization is about 10,000–15,000. This means that usually only part of the total annealed cDNA–vector hybrid is used. The rest can be stored as described before. Alternatively this DNA can be transfected into bacteria which are then stored in 15% glycerol at −70°.[36]

Differential Colony Hybridization

The general scheme for the differential colony hybridization protocol is depicted in Fig. 1. A cDNA library is constructed which contains the gene sequences of interest. Clones from this library are then transferred to bacterial plates in an ordered array and replica plated onto duplicate nitrocellulose filters. After growth of the colonies on the filters, followed

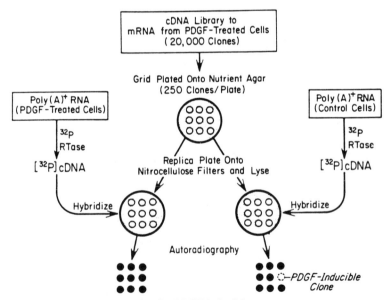

FIG. 1. Screening for PDGF-inducible gene sequences.

by lysis and baking, each filter is then hybridized to two different ^{32}P-labeled cDNA probes made from poly(A)[+] RNA. The RNA from which the cDNA probes are made can be prepared from any two cell populations that can be expected to have differences in gene expression. This could be, as in our case, cells before and after treatment with platelet-derived growth factor.[30] Any other biological response modifier which could be expected to have effects on gene expression could be studied in the same way. Alternatively, RNA could be extracted from two different cell types, such as B and T lymphocytes, for example, in order to look for tissue-specific gene expression. Thus, the technique of differential colony hybridization can be applied to a wide variety of biological problems. After the replica filters are hybridized to the two cDNA probes and autoradiographed, each colony can be scored for the intensity of its hybridization to both probes. Clones which show a stronger signal intensity after hybridization with one of the probes as compared to the other are picked for further analysis.

The other strategy which has been used for differential screening of libraries is subtractive hybridization. This approach was used in the isolation of the T cell receptor.[58] Subtractive hybridization, rather than using 2 different probes to screen replica filters, uses one probe which has been enriched for the sequences of interest to screen a unique filter set. An enriched probe is made by making ^{32}P-labeled cDNA from the RNA of interest and hybridizing this probe to unlabeled control RNA. The resulting products are fractionated over hydroxylapatite columns to separate the single-stranded cDNA from the double-stranded hybrids. The single-stranded cDNA which is eluted from the column is therefore enriched for sequences not present in the control RNA.

Each of these approaches has unique advantages and disadvantages. Subtractive hybridization is the more sensitive of the two. This method allows the detection of low abundance genes such as the T cell receptor.[58] The increased sensitivity is due to the fact that the signal-to-noise ratio is higher because fewer dpm/ml of a less heterogeneous probe is used in screening. Thus, only a low percentage of positive signals has to be detected against a background of mostly negative colonies. Differential colony hybridization uses more dpm/ml of probe and generates a much higher percentage of positive signals. In addition, the background of non-specific hybridization to clones which should be negative is greater. Thus, the signal-to-noise ratio is lower. We estimate that the differential colony

[58] S. M. Hedrick, D. I. Cohen, E. A. Nielsen, and M. M. Davis, *Nature (London)* **308**, 149 (1984).

hybridization method can detect genes down to 0.05% of the total poly(A)$^+$ RNA population.[30] This is approximately 150 mRNAs/cell. It may be possible to increase this sensitivity further by using cDNA probes with higher specific activity and/or concentration combined with hybridization and washing conditions designed to further reduce background. Even if differential hybridization were made more sensitive, however, significant problems would still remain from the fact that more clones would need to be screened to find a very low abundance gene—perhaps more than 100,000. Since every clone has to be individually scored, this would be a formidable task. In addition, screening this many more clones would require larger amounts of cDNA probe which could present further problems if the sources of poly(A)$^+$ RNA are limited.

Subtractive hybridization avoids these problems by preparing a probe which is specific for genes present in one population that are not represented in the control RNA populations. Since the probe will primarily detect genes of interest, it is thus possible to screen colonies or plaques at much higher densities. Thus, the signal-to-noise ratio is greater and more clones can be screened at once. However, this approach also has drawbacks. In principle, the major disadvantage to the subtractive hybridization strategy is that genes whose expression is regulated in a quantitative manner rather than a qualitative one may be missed. Most subtractive hybridization protocols call for multiple rounds of hybridization of the ^{32}P-labeled cDNA against an excess quantity of control RNA. If a particular mRNA is present in the control RNA preparation but simply at a lower level than in the RNA sample of interest, then it is possible to deplete the ^{32}P-labeled cDNA probe of a potentially interesting sequence. This problem can be avoided, in theory, by controlling the extent of hybridization by a careful R_0t analysis. In practice, however, limiting the extent of subtractive probe hybridization prior to selection over HAP columns may increase the number of false positive sequences in the probe—which is perhaps the major problem with the procedure anyway. Other problems with this procedure are the requirements for relatively large amounts of control RNA needed for hybridization and difficulties in obtaining enough ^{32}P-labeled cDNA subtracted probe to screen a large number of clones. Often, it is necessary to prepare a subtracted cDNA library for successful screening by this procedure.

We recommend that if the gene or genes of interest are expressed at or below the 0.05% level of the poly(A)$^+$ RNA, then use the subtractive method. Otherwise, the "differential" or "+/−" screening method presented below is recommended. It is easier, allows for direct detection of quantitative changes, and has been reproduced in several laboratories.

cDNA Probe Synthesis

To make ^{32}P-labeled cDNA probes for differential colony screening, a first-strand cDNA reaction is run with a few modifications. Poly(A)$^+$ RNA is prepared from the two cell population of interest as previously described. Of each RNA 0.5 μg is then incubated in a reaction mixture of 50 μCi [^{32}P]dCTP (400–800 Ci/mmol), 0.5 mM each of dATP, dGTP, dTTP, 50 mM Tris, pH 8.3, 30 mM 2-mercaptoethanol, 10 mM MgCl$_2$, 0.5 μg oligo(dT)$_{12-18}$ primer, and 5 U of AMV reverse transcriptase in a 25 μl volume. The reaction is incubated at 37° for 1 hr. It is terminated by the addition of 175 μl of 0.3 N NaOH, 0.2 mM EDTA, and boiled for 3 min to hydrolyze the RNA strand. The mixture is then neutralized by the addition of 100 μl of 4.5 M sodium acetate, pH 5.5, and run over a Sephadex G-50 column to remove free nucleotides. A typical reaction of this type produces 3 × 10^7 cpm of labeled probe.

Colony Growth

For differential colony hybridization, bacterial clones from the cDNA library must be replica plated onto nitrocellulose filters for screening. This may be done from plates with either random colony patterns or colonies transferred onto plates into a regular array. We recommend that the colonies from the library be gridded into arrays. This facilitates scoring the clones after autoradiography. If the colonies are randomly spaced, then a strong signal from one colony can overshadow the signal from nearby neighbors. In addition, since each and every colony must be scored, the random grids tend to produce eyestrain. Therefore, using sterile toothpicks grid out the cDNA library onto LB + selective drug plates at a density of 100 colonies/100 mm dish or 250 colonies/150 mm dish. These colonies are then grown up overnight. These master plates are then used to make replica filters and can be stored at 4° for several months if not allowed to dry out. (It is useful to arrange the colony grid asymmetrically so that the orientation of the filters can be readily determined.)

Replicas are made onto nitrocellulose filters from the master plate either by using standard microbiological techniques[59,60] or disposable replicate colony transfer pads (FMC Corp., Rockland, ME). Nitrocellulose filters (Triton free filters give better results in our hands) are placed on LB agar plates plus selective drug. The colony grids are then transferred to

[59] J. Lederberg and E. M. Lederberg, *J. Bacteriol.* **63**, 399 (1952).
[60] D. Hanahan and M. Meselson, *Gene* **10**, 63 (1980).

the nitrocellulose filters via the replica plating device used. The colonies are grown up on the nitrocellulose overnight at 37°. Each filter is then transferred to an agar plate containing 170 μg/ml chloramphenicol to amplify the plasmid. After an overnight incubation on chloramphenicol plates at 37°, the colonies are ready to be lysed.

Complete and uniform colony lysis is critically important for the success of the differential screening method. Incomplete lysis of an individual colony could result in the clone being falsely picked as a differentially expressed gene. We have found that the colony lysis procedure of Thayer[61] consistently gives reproducible results in the differential colony hybridization procedure.

Colonies are lysed by blotting filters onto a series of five buffers. The basic procedure involves soaking a piece of Whatman 3MM paper with one of the lysis solutions, and then placing the filter with the colonies face up on the paper. After 1 min, the filter is blotted on a piece of dry 3MM or equivalent. This process is then repeated twice more for each filter before applying the next buffer.

The five solutions in order of application are (1) 25% sucrose, 50 mM Tris, pH 8.0, +1.5 mg/ml lysozyme (fresh). (Repeat twice.) This step is performed in the cold room. All subsequent steps are performed at room temperature; (2) 0.2% Triton X-100, 0.5 N NaOH (repeat twice); (3) 0.5 N NaOH (repeat twice); (4) 1 M Tris, pH 7.5 (repeat twice); and (5) 0.15 M NaCl, 0.1 M Tris, pH 7.5 (repeat twice). After the lysis steps, the filters are air dried and baked in a vacuum oven for 2 hr at 80°.

Filters are prehybridized for 4 hr at 65° in 6× SSC (1× SSC = 0.15 M NaCl, 0.015 M sodium citrate, pH 7.0) − 1× Denhardt's solution [0.02% BSA, 0.02% Ficoll 400, 0.02% poly(vinylpyrollidone) (w/v)[62]]. Filters are then hybridized in the same solution plus 100 μg/ml denatured salmon sperm DNA and 0.5×10^6 cpm ^{32}P-labeled cDNA probe per ml. We use 5 ml of hybridization solution per 150 mm nitrocellulose filter containing 250 colonies. Hybridization is for 3 days at 65°. The concentrations and specific activities of probe given here have not been optimized, but have been used to isolate an inducible gene that is 0.2% of the poly(A)$^+$ RNA.[30]

After hybridization is complete, the filters are washed four times for 1 hr per wash in 6× SSC–1× Denhardt's solution at 65°. The filters are then blotted dry and exposed to preflashed X-ray film with intensifying screens for 5 days at −70°.[63]

To read the films for differential expression, lay a piece of transparent

[61] R. E. Thayer, *Anal. Biochem.* **98**, 60 (1979).
[62] D. T. Denhardt, *Biochem. Biophys. Res. Commun.* **23**, 641 (1966).
[63] R. A. Laskey, this series, Vol. 65, p. 363.

plastic over the developed X-ray film from the filter hybridized to the "+" probe and mark the positions of all the colonies giving signals. Then align the transparency over the autoradiograph of the corresponding filter that was hybridized to the "−" probe. Compare each mark on the transparency with each signal on the filter and pick colonies which show differences. Although not essential, it may be preferable to screen each plate in duplicate to give a higher reliability of interpretation. All putative clones of interest should be picked, arrayed onto a new plate, and screened differentially twice more. All clones which still appear to be differentially expressed should be individually characterized for quantitative differences in expression levels by Northern gel or other similar analysis.

Acknowledgments

We thank Greg LaRosa and Belinda Wagner for useful criticisms of the manuscript and Carol Barnstead and Eileen Feldman for help in its preparation.

Work from the author's laboratory is supported by the National Institute of Health and the Anjinomoto Company. C.D.S. is supported by a faculty research award from the American Cancer Society. B.H.C. is a Rita Allen Foundation Scholar.

[7] Identification of Platelet-Derived Growth Factor-Modulated Proteins

By CHARLES D. SCHER and W. JACKSON PLEDGER

Treatment of responsive cells with one or more polypeptide growth factors stimulates a pleotypic response, which results in an increase in the synthesis of all proteins.[1] This leads to an increase in cell mass required for cell replication. In addition certain proteins are selectively synthesized, i.e., synthesized at a greater rate than the bulk of cellular proteins.[2] The preferential synthesis of some of these proteins occurs long before resting cells become committed to synthesize DNA and is growth factor specific.[3] This chapter will describe methodology for detection and quantitation of these growth factor-modulated proteins. The system that has

[1] A. Hershko, P. Mamon, R. Shields, and G. Tomkins, *Nature (London) New Biol.* **232**, 206 (1971).

[2] G. Thomas, G. Thomas, and H. Luther, *Proc. Natl. Acad. Sci. U.S.A.* **78**, 5712 (1981).

[3] W. J. Pledger, C. A. Hart, K. L. Locatell, and C. D. Scher, *Proc. Natl. Acad. Sci. U.S.A.* **78**, 4358 (1981).

been extensively studied is platelet-derived growth factor (PDGF)-modulated protein synthesis by quiescent BALB/c 3T3 cells. The principles utilized should be applicable to a variety of growth factor-modulated proteins, whether synthesized during growth and/or differentiation[4] (see Nilsen-Hamilton and Hamilton [39], this volume).

Cell Synchronization

It is best to start with a "synchronized" cohort of cells. Various methods can be utilized including a double-thymidine block (early S phase) and mitotic shake.[5] Perhaps the most generally applicable method is the use of density-arrested monolayers. Quiescent monolayers of BALB/c 3T3 cells are arrested in G_0/G_1 phase. They display a replicative response to a variety of growth factors and selectively synthesize certain mRNAs and proteins. In practice, cells are grown to confluence in Dulbecco's modified Eagle's medium (high glucose formulation) containing 10% bovine serum and are used 3 days after medium replenishment. At this time flow microfluorimetry indicates that more than 97% of the cells have a G_1 DNA content. Such density-arrested BALB/c 3T3 cells can be utilized as is to study protein synthesis. Alternatively, the cultures can be treated with a variety of growth factors in the presence or absence of various nutrients or reversible inhibitors of DNA synthesis to obtain populations of cells arrested at various points in G_0/G_1 or early S phase as is detailed in this volume.[6] Because 3T3 cells readily become transformed spontaneously, it is imperative to use cells that have a low saturation density in 10% serum (5–8 × 10^4 cell/cm^2). Of course, proteins that are modulated by growth factors in density arrested cells may not be produced in growth factor treated cells which are arrested at other growth arrest points or are growing exponentially. In such cases other growth factor-modulated proteins may be found.

Specificity

BALB/c 3T3 cells synthesize DNA in response to several growth factors including PDGF, EGF, and somatomedin C (or insulin). As yet, however, only PDGF has been found to stimulate the selective protein synthesis.[3] These selectively synthesized proteins include pI, a nuclear

[4] S. P. Hann, C. B. Thompson, and R. N. Eisenman, *Nature* (*London*) **314**, 366 (1985).
[5] D. M. Prescott, "Reproduction of Eukaryotic Cells," pp. 19–35. Academic Press, New York, 1976.
[6] M. Harrington and W. J. Pledger, this volume [36].

protein, and pII, which has been identified as the lysosomal protein MEP.[6a] In addition, these proteins are optimally synthesized in the absence of PDGF by cells which do not require PDGF for growth.[3] Thus the synthesis of these proteins is specifically regulated by PDGF and is not a result of cell cycle traverse. In contrast, proteins such as dihydrofolate reductase, thymidine kinase, and histone H2b are synthesized as cells enter the S phase; their synthesis is not growth factor specific.[4]

Growth Factor Treatment

To study selective protein synthesis one should use electrophoretically homogeneous growth factors. Purified growth factors such as PDGF, a highly charged hydrophobic molecule, are quickly lost in the absence of added protein. It is best to add the growth factors in the presence of a carrier. PDGF may be added to a small amount (0.3%) of platelet-poor plasma, a blood fraction which lacks PDGF. This concentration of plasma is sufficient to prevent loss of PDGF and to help maintain cellular viability. Methods for preparing human platelet-poor plasma are given in this volume.[6] Alternatively, 1% bovine serum albumin may be used. In either case controls should be treated with the carrier without addition of the mitogen.

For PDGF the preferential synthesis of proteins (as well as DNA synthesis) is regulated by both the growth factor concentration and the duration of treatment.[7] Furthermore, the time until the peak of protein synthesis varies from one protein to another even when cells are continuously exposed to the mitogen. For example, the peak synthesis of pI occurs 1–2 hr after PDGF addition and is undetectable at 6 hr. Peak synthesis of MEP occurs at 6–24 hr, a time when pI synthesis is undetectable (Fig. 1). Thus one should sample a number of time points in order to detect a variety of growth factor-modulated proteins.

Labeling

Growth factor-modulated proteins are detected by fluorography after one- or two-dimensional gel electrophoresis. In order to obtain sufficient counts to be detectable by this technique, a radioactive isotope with a relatively short half-life (and high specific activity) is used. ^{35}S meets these criteria; both [^{35}S]methionine and [^{35}S]cysteine are available commercially.

[6a] C. D. Scher, R. L. Dick, A. P. Whipple, and K. L. Locatell, *Mol. Cell. Biol.* **3**, 70 (1983).
[7] W. J. Pledger, C. D. Stiles, H. N. Antoniades, and C. D. Scher, *Proc. Natl. Acad. Sci. U.S.A.* **74**, 4481 (1977).

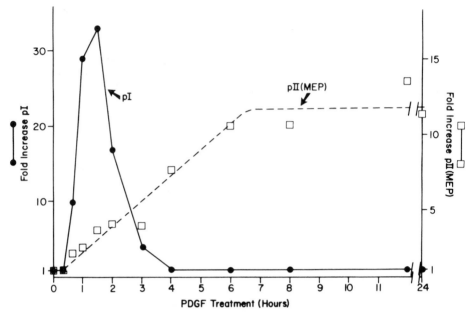

Fig. 1. The sequential synthesis of PDGF-modulated proteins during continuous treatment with PDGF.

The amount of radioactivity added and the specific activity of the ^{35}S-labeled amino acid will vary from experiment to experiment. To compare the rates of synthesis of a protein under different conditions (e.g., in response to various growth factors), two features should be kept in mind: (1) the incorporation of the labeled amino acid should increase linearly with time during treatment with the labeled precursor, and (2) sufficient counts should be incorporated by the "coldest" samples to allow fluorography.

For BALB/c 3T3 cells and their transformed derivatives, addition of Dulbecco's medium containing 2.5% (0.75 mg/liter) of the usual methionine concentration (30 mg/liter) allows linear incorporation of [35S]methionine for several hours. Methionine-free medium can be obtained commercially (Irvine Scientific, Santa Ana, CA) or can be made from individual components. Addition of 2.5 ml of standard medium to 97.5 ml of methionine-free medium will give the correct concentration of all components.

In order to study rapidly induced proteins, it is often necessary to incubate the cultures for several hours in low-methionine medium before adding growth factors and [35S]methionine. Such pretreatment depletes intracellular methionine pools and allows the rapid incorporation of exogenous label into protein. Such cultures can be successfully processed within 10 min of growth factor and [^{35}S]methionine addition. The growth factor-treated cultures usually have an increased synthesis of the bulk of the cellular proteins as compared to the untreated controls. Thus it may be necessary to increase the amount of [^{35}S]methionine in the control cultures (while leaving the cold methionine constant) in order to obtain sufficient counts for fluorography. For example, 25 μCi/ml may suffice for the growth factor-treated cultures, whereas 100 μCi/ml may be required for the untreated ones. Of course, if cost is not a factor, 100 μCi/ml may be used for all cultures.

The length of labeling may affect the form of the protein detected. Short labeling periods (10–30 min) allow the detection of newly synthesized proteins usually within the cytoplasm. Longer labeling periods (2–3 hr) or pulse-chase experiments allow the detection of products of further processing, and allow subcellular localization. For example, one PDGF-modulated protein, pI, is detected within the cytoplasm after a short labeling period[3] but is found in the nucleus when longer labeling times are used.[8] MEP, another "cytoplasmic" protein, is secreted,[9] or transported into lysosomes where it is processed into two smaller forms.[10] Thus longer labeling periods are useful for detecting transport and processing, but can complicate interpretation of rates of synthesis.

Harvesting Cultures

The method of harvesting depends upon both the cell fraction under study and the method of analysis. To analyze secreted proteins, the medium is collected, clarified by centrifugation, and dialyzed extensively against 1 M acetic acid or ammonium acetate. The secreted material is concentrated by lyophilization, which also removes the acetic acid or ammonium acetate.

If whole cell extracts are to be resolved on two-dimensional gels, the cells (in 35 or 60 mm dishes) are washed once with serum-free medium and solubilized in 100 μl of isoelectric focusing buffer [9.6 M urea, 2% Nonidet P-40, 5% 2-mercaptoethanol, 1.6% ampholines (pH 5 to 7), 0.5% ampholines (pH 3.5 to 10)] and clarified by low-speed centrifugation. Aliquots can be stored at −70° until used.

For analysis on one-dimensional SDS gels whole cells can be solubi-

[8] N. E. Olashaw and W. J. Pledger, *Nature (London)* **306**, 272 (1983).
[9] M. M. Gottesman, *Proc. Natl. Acad. Sci. U.S.A.* **75**, 2767 (1978).
[10] S. Gal, M. C. Willingham, and M. M. Gottesman, *J. Cell Biol.* **100**, 535 (1985).

lized in the following buffer: 150 mM NaCl, 20 mM Tris–HCl (pH 7.5), 1 mM EDTA, 0.5% Triton X-100, and 0.1% SDS. Many investigators add a serine-protease inhibitor such as TPCK, but we have found that to be unnecessary provided that extracts are kept at 0° and processed rapidly. Aliquots can be directly applied to gels or can be immunoprecipitated. They are stored at −70°.

Most procedures for isolating cell nuclei utilize 1% Nonidet P-40 to solubilize cytoplasmic proteins. However, Nonidet P-40 has been shown to leach pI, a growth factor-modulated protein.[8] Thus it is best to start without Nonidet P-40. If one shows that Nonidet P-40 does not leach out the protein of interest, the detergent can be utilized in other experiments. Labeled cells are suspended in reticulocyte saline buffer: 10 mM NaCl, 15 mM MgCl$_2$, 10 mM Tris–HCl (pH 7.4), and the nuclei pelleted by low-speed centrifugation. The nuclei are washed three times with buffer, suspended in buffer containing 1 M sucrose, overlaid on 1.8 M sucrose, and centrifuged at 35,000 rpm for 30 min in an SW 50.1 rotor. The pellet of purified nuclei can be solubilized in an appropriate buffer for immunoprecipitation or electrophoresis.

Immunoprecipitation

The method for immunoprecipitation given below uses rabbit heterosera directed against MEP, a mouse protein. Whole, fixed *Staphylococcus aureus* A are utilized to precipitate the antigen–antibody complex. With antisera directed against MEP, there is little nonspecific precipitation of unrelated proteins. In principle the method employed can be used with a large variety of proteins provided that specific heterosera are available. Monoclonal antibodies can be used, provided that they bind to protein A. If they do not bind, a bridge technique employing a heteroserum directed against the monoclonal antibody followed by *Staphylococcus aureus* A can be used. Alternatively one might use a Sepharose-coupled antiserum directed against mouse immunoglobulin (Cooper Biomedical, Malvern, PA).

The labeled cell lysate is thawed (on ice) and clarified by centrifugation. The number of acid-insoluble counts for each lysate is determined so that equal amounts of acid-insoluble counts can be immunoprecipitated. Under the conditions of labeling described herein, it is usually unnecessary to treat with 5% TCA at 80° to hydrolyze methionine-charged tRNA because at least 95% of the acid-insoluble counts are in protein. Immunoprecipitation is begun by adding the antisera to the clarified extract.

For each antiserum and antigen, optimization of conditions is necessary. Specifically one determines the amount of antiserum needed and the number of acid-insoluble counts required. For unknown samples, it is best to immunoprecipitate a large number (i.e., about 10^6) of counts. By adding increasing quantities of antiserum, one can determine the optimal amount of antibody needed by electrophoresis and fluorography. The amount of *Staphlococcus aureus* A must be adjusted according to the volume of antiserum. In general we use 50 μl of a 10% suspension of *Staphlococcus aureus* A (IgGsorb, New England Enzyme Center, Boston, MA) for 3 μl of rabbit antiserum.

In practice 0.25–1.0 × 10^6 acid-insoluble counts of cell sample are diluted to 1.0 ml with immunoprecipitation buffer A [150 mM NaCl, 50 mM Tris–HCl (pH 7.4), 0.5% Nonidet P-40, and 0.05% SDS] and incubated with specific antiserum at 4° for 30 min. The IgGsorb is then added and incubation continues at 4° for another 30 min. The bacteria are then pelleted in a microfuge, washed three times with cold immune precipitation buffer B [150 mM NaCl, 50 mM Tris–HCl (pH 7.4), 0.5% Nonidet P-40, and 2.5 M KCl], once with buffer A, and once with deionized water. An appropriate buffer is then added for one- or two-dimensional gel electrophoresis.

The clarification of the sample before addition of antisera, the short periods of incubation, and extensive washing eliminate most nonspecific adsorption. However, it may be useful to eliminate proteins that bind to antigen–antibody–*Staphlococcus aureus* A protein complexes by first adding a known unlabeled antigen (e.g., BSA) and antisera (anti-BSA) followed by *Staphlococcus aureus* A. After clarification by centrifugation, specific antiserum to the labeled protein can be added followed by *Staphlococcus aureus* A.

Electrophoresis and Fluorography

Methods for one-[11,12] and two-dimensional[13,14] gels have been described in detail. For one-dimensional gels, it is useful to start with a 5–20% gradient of polyacrylamide in order to establish the optimal acrylamide concentration needed for detection of the band of interest. Such precautions are not necessary if immune precipitation techniques are used because the M_r is then known. For two-dimensional gels, nonequilibrium pH gradient gel electrophoresis (NEPHGE) is used for cationic proteins, and isoelectric focusing for anionic proteins. A duplicate gel is run in the

[11] U. K. Laemmli, *Nature (London)* **227**, 680 (1970).

[12] R. F. Schleif and P. C. Wensink, "Practical Methods in Molecular Biology." Springer-Verlag, New York, 1981.

[13] P. H. O'Farrell, *J. Biol. Chem.* **250**, 4007 (1975).

[14] P. H. O'Farrell, H. M. Goodman, and P. H. O'Farrell, *Cell* **12**, 1133 (1977).

first dimension and frozen after the run. This gel is cut into 0.5–1.0 cm slices and the ampholines eluted with water to allow a determination of the pH gradient.

Fluorography is performed after fixation of the gels. In general, impregnation of gels with PPO-DMSO gives sharper bands than other methods. The gels are dried under vacuum, preflashed film is applied, and the gels incubated at −70°. Kodak XAR-5 film is very sensitive. Preflashing is used to assure a linear relationship between the quantity of radioactivity and the density of a "light" band or spot.[15] An additional problem is that "heavy" densities may result from overexposure and cause an underestimate of the quantity of protein synthesized. In questionable cases one should sequentially overlay the gel with several films using different exposure periods; the density of the band should be proportional to the exposure time. The bands of interest can be scanned with a densitometer to obtain the relative rate of synthesis. Alternatively, the band of interest can be cut out of the original gel and counted with an appropriate fluor in a scintillation counter.

[15] R. A. Laskey and A. D. Mills, *Eur. J. Biochem.* **56**, 335 (1975).

Section II

Angiogenesis, Endothelial and Fibroblast Growth Factors

[8] Purification of Endothelial Cell Growth Factors by Heparin Affinity Chromatography

By MICHAEL KLAGSBRUN, ROBERT SULLIVAN, SANDRA SMITH, RUSSELL RYBKA, and YUEN SHING

Introduction

Growth factors that stimulate the proliferation of endothelial cells appear to have a strong affinity for heparin. These growth factors bind tightly to columns of heparin immobilized to Sepharose and are eluted with 0.9–1.8 M NaCl. Heparin-binding growth factors have been purified from tumors[1] as well as normal tissue such as brain,[2-4] pituitary,[2] hypothalamus,[5] retina,[6] and cartilage.[7] We have analyzed most of these endothelial cell growth factors by heparin affinity chromatography. The concentrations of NaCl required to elute these growth factors from heparin-Sepharose columns and structural properties such as approximate size and charge are summarized in Table I. Both cationic (pI of about 8–10) and anionic (pI of about 5) endothelial cell growth factors bind to heparin-Sepharose. Anionic endothelial cell growth factors are eluted with 0.9–1.1 M NaCl and cationic endothelial cell growth factors are eluted with 1.3–1.8 M NaCl. The molecular weights of these growth factors are between 16,000 and 19,000.

Heparin-Sepharose affinity chromatography has greatly facilitated the purification of heparin-binding endothelial cell growth factors. We have used a two-step procedure combining BioRex 70 cation-exchange and heparin-Sepharose affinity chromatography to purify to homogeneity a cationic chondrosarcoma-derived endothelial cell growth factor with a molecular weight of 18,000.[1] This two-step purification scheme can be used for the purification of other cationic endothelial cell growth factors as well. In this report we describe the large-scale purification to homogeneity of cationic endothelial cell growth factors from three sources: a

[1] Y. Shing, J. Folkman, R. Sullivan, C. Butterfield, J. Murray, and M. Klagsbrun, *Science* **223**, 1296 (1984).
[2] D. Gospodarowicz, J. Cheng, G.-M. Lui, A. Baird, and P. Bohlen, *Proc. Natl. Acad. Sci. U.S.A.* **81**, 6963 (1984).
[3] T. Maciag, T. Mehlman, R. Friesel, and A. B. Schreiber, *Science* **222**, 932 (1985).
[4] R. R. Lobb and J. W. Fett, *Biochemistry* **23**, 6295 (1985).
[5] M. Klagsbrun and Y. Shing, *Proc. Natl. Acad. Sci. U.S.A.* **82**, 805 (1985).
[6] P. D'Amore and M. Klagsbrun, *J. Cell Biol.* **99**, 1543 (1984).
[7] R. Sullivan and M. Klagsbrun, *J. Cell Chem.* **260**, 2399 (1985).

TABLE I
PROPERTIES OF HEPARIN-BINDING ENDOTHELIAL CELL GROWTH FACTORS

Source	Elutes at NaCl concentration (M)	Isoelectric point	Molecular weight
Chondrosarcoma	1.3–1.5	9.8	18,000
Hepatoma	1.4–1.6	Cationic	19,000
Cartilage	1.5–1.8	9.8	19,000
Pituitary (cationic FGF)	1.4–1.6	9.6	18,000
Brain (cationic FGF)	1.4–1.6	9.6	18,000
Hypothalamus	1.3–1.5	8	18,000
Brain (anionic FGF)	0.9–1.1	5	16,000
Hypothalamus (ECGF)	0.9–1.1	5	16,000
Retina	0.9–1.1	5	16,000

transplantable tumor (rat chondrosarcoma), a cultured tumor cell line (SK HEP-1 human hepatoma), and a normal tissue (bovine hypothalamus). (For additional applications of heparin-Sepharose affinity chromatography, see the chapter in this volume by Gospodarowicz [9] for purification of acidic and basic forms of FGF.)

General Purification Procedures

Overall Strategy

Endothelial cell growth factors can be purified from both tissue and from cultured cells. Growth factor can be extracted from tissue either by collagenase digestion or by NaCl extraction. In cell culture, we find that there is usually more growth factor activity associated with the cells than with the conditioned medium. The cell-associated growth factor is extracted from cells with NaCl. Purification of cationic endothelial cell growth factors to homogeneity can be accomplished in two biochemical steps. BioRex 70 cation-exchange chromatography is used as a first step to process gram amounts of extract. This procedure produces milligram amounts of partially purified growth factor which can then be purified to homogeneity by gradient heparin-Sepharose affinity chromatography.

Extraction of Growth Factor

Tissue such as hypothalamus and chondrosarcoma are homogenized and digested with collagenase (*Clostridium histolyticum* clostridiopeptidase A (EC 3.4.24.3, CLS 11, Cooper Biomedical) at a concentration of 2 mg/ml in phosphate-buffered saline (PBS, Gibco) at 37° for 2–24 hr using a rotatory shaker. When the tissue is completely digested, the digest is centrifuged and the clarified supernatant fraction is used for growth factor purification. Cells such as SK-HEP-1 human hepatoma cells are frozen, thawed, resuspended in 1 M NaCl, and lysed by sonication or homogenization. The lysate is centrifuged and the clarified supernatant fraction is used for growth factor purification.

BioRex 70 Cation-Exchange Chromatography

BioRex 70 cation-exchange chromatography with batch elution is used to process large amounts of crude extract. BioRex 70 (Bio-Rad; either 100–200 or 200–400 mesh; one bottle contains 1 pound and 700 ml of resin) is prepared for chromatography as follows. Two pounds of BioRex 70 are washed with 2000 ml of 0.5 M NaCl, 0.5 M Tris–HCl, pH 7.5, once and with 4000 ml of equilibration buffer (0.1 M NaCl, 0.01 M Tris–HCl, pH 7.5) 3 times. Conductivity and pH are monitored to make sure that the BioRex 70 is properly equilibrated. For purification, growth factor extracts that have been clarified by centrifugation are mixed with BioRex 70 (extract volume to BioRex volume ranging from 1 : 1 to 4 : 1, up to 100 mg protein/ml BioRex) for 4 to 20 hr at 4° using a rotatory shaker or an overhead stirring device. The BioRex 70 is allowed to settle, is washed once with equilibration buffer, and is poured into a column. The column is washed with 2–5 volumes of equilibration buffer and growth factor is eluted batchwise with 1–2 column volumes of 0.6 M NaCl, 0.01 M Tris–HCl, pH 7.5, at a flow rate of 60–90 ml/hr. Fractions are collected and monitored for growth factor activity. Details of this procedure are described below for the various growth factors. After chromatography, BioRex 70 is regenerated for reuse by washing 2 pounds of used resin with 4000 ml of 3 M NaCl, 0.01 M Tris–HCl, pH 7.5, twice followed by 4000 ml 0.1 M NaCl, 0.01 M Tris–HCl, pH 7.5, once. BioRex 70 is stored at 4° in the equilibration buffer supplemented with 0.4% sodium azide.

Heparin-Sepharose Affinity Chromatography

Growth factors partially purified by BioRex 70 cation-exchange chromatography are purified to homogeneity by heparin-Sepharose affinity chromatography. Heparin-Sepharose is obtained from Pharmacia in lots of 40 ml. Each 40 ml lot contains 10 g of which 136 mg is heparin. Heparin-Sepharose is prepared for chromatography by washing 40 ml of heparin-Sepharose with 500 ml equilibration buffer (from 0.1 to 0.6 M NaCl in 0.01 M Tris–HCl, pH 7.5) three times. For growth factor purification, active fractions obtained by BioRex chromatography are applied directly (up to

10 mg protein/ml heparin-Sepharose) to 7–40 ml columns of heparin-Sepharose at 4°. After a wash with 5–15 column volumes, growth factor is eluted with a 300–400 ml gradient of NaCl. The gradients are produced by a gradient maker, either of our own design (2 cylindrical chambers holding up to 170 ml each) or one purchased from Bethesda Research Laboratories (2 concentric chambers of 400 ml each) and are monitored by measuring conductivity (Radiometer conductivity meter). The flow rates are about 40–60 ml/hr and fractions are collected and monitored for growth factor activity. Details of the heparin-Sepharose chromatography procedure are outlined for the various growth factors below. After chromatography, heparin-Sepharose is regenerated by washing with 500 ml of 6 M urea three times, with H_2O twice, and with equilibration buffer twice. Heparin-Sepharose is stored at 4° either in H_2O or in equilibration buffer. Heparin-Sepharose can be reused at least 10 times.

Storage of Growth Factor

For short-term storage (up to 4 weeks), growth factor prepared by heparin-Sepharose chromatography is stored at 4° at the concentration of NaCl in which it is eluted. The growth factor preparations are usually concentrated enough so that they can be diluted into physiological salt solution for use in cell culture. For long-term storage growth factor is desalted, lyophilized, and stored at −20°. One method of desalting is dialysis. Extensive dialysis leads to a loss of growth factor. Therefore, growth factor preparations are dialyzed for a minimum of 6 hr up to a maximum of 18 hr against H_2O at 4°. The volume of H_2O is about 100 times that of the growth factor solutions in the dialysis bags. An alternative method for desalting growth factor solutions is reverse phase chromatography on HPLC C_3 columns. Fractions of growth factor produced by heparin-Sepharose chromatography are pooled and pumped onto a C_3 RPLC column. Growth factor adheres to the column while the salt elutes in the void volume. Growth factor is eluted from the column with a gradient of 0–60% organics (acetonitrile and 2-propanol 50:50 v/v) in 0.1% TFA. However, under these chromatographic conditions growth factor activity is inactivated. Thus this desalting technique is useful only for structural studies such as amino acid and sequence analysis.

Growth Factor Assays

Two assays are used to monitor growth factor activity. The growth factors described in this chapter stimulate both DNA synthesis in 3T3 cells and the proliferation of capillary endothelial (CE) cells. Measurement of the stimulation of DNA synthesis in 3T3 cells is a relatively nonspecific assay that will detect many growth factors such as fibroblast growth factor (FGF), platelet-derived growth factor (PDGF), and epidermal growth factor (EGF). Nonetheless, this assay is relatively simple and rapid and allows screening of many samples such as are generated by column chromatography. The CE cell proliferation assay is much more specific. For example, PDGF and EGF do not stimulate CE cell proliferation. However this bioassay is relatively time consuming compared to the 3T3 cell assay and is limited by the difficulty of obtaining large numbers of CE cells. The strategy used in purification is to assay for 3T3 growth factor activity first in order to rapidly screen large numbers of fractions and then to test fractions of interest for the ability to stimulate CE cell proliferation.

Measurement of DNA Synthesis in 3T3 Cells. Mouse BALB/c 3T3 cells (clone A 31) are used. These cells are grown in water-jacketed incubators at 37° and 10% CO_2 in Dulbecco's modified Eagle's medium (DMEM) containing 4.5 g/liter glucose (Gibco), 10% calf serum (Colorado Serum Company), 50 U/ml penicillin (Gibco), and 50 µg/ml streptomycin (Gibco). In the course of passaging cells, care is taken not to allow the 3T3 cells to become more than 50% confluent. Otherwise they lose their property of being density inhibited. The bioassay is carried out as follows: BALB/c 3T3 are plated into 96-well microtiter plates (Costar, 5000–10,000 cells/200 µl/well). The cells become confluent in about 3 days. About a week after plating the 3T3 cells have become quiescent because the growth factor activity of the medium has become depleted. For bioassay, microtiter plates containing quiescent 3T3 cells can be used in the 7- to 14-day period after initial plating. In the bioassay, 10–50 µl of sample and 10 µl of tritiated thymidine (ICN, catalog #24066, 6.7 Ci/mmol, 1 µCi/well) are added to the 200 µl of medium atop the 3T3 cells. After a minimum incubation period of 36 hr (and up to 48 hr), cells are fixed, unincorporated thymidine is removed, and radioactive DNA is precipitated. This is done by washing the microtiter plates consecutively with 0.15 M NaCl, methanol (twice for 5 min each), H_2O (4 rinses), 5% cold trichloroacetic acid (twice for 10 min each), and H_2O (4 rinses). The cells are lysed by addition of 200 µl 0.3 M NaOH for 1 min and the lysates are transferred to scintillation vials. Scintillation fluid (Beckman Ready Solv EP) is added and the vials are counted in a Beckman model LS1800 scintillation counter. Background incorporation in this assay is about 1000–2000 cpm. Maximal stimulation, when all the 3T3 cells are fully synthesizing DNA, is about 150,000–200,000 cpm. One unit of 3T3 growth factor activity is defined as the amount of growth factor in 250 µl that is required to stimulate half-maximal DNA synthesis. The heparin-binding endothelial cell growth factors described in this chapter contain about 5 units/ng and are active at about 1 ng/ml.

Measurement of CE Cell Proliferation. Capillary endothelial (CE) cells are prepared from bovine adrenal glands and grown on gelatin-coated dishes as described by Folkman *et al.*[8] The CE cells are maintained in sarcoma 180 conditioned medium. To measure growth factor activity, CE cells are resuspended in DMEM/10% calf serum and plated sparsely into 24-well microtiter plates (Costar, 10,000 cells/0.5 ml/well). On the following day, unattached CE cells are removed and samples in 0.5 ml DMEM/10% calf serum are added to the attached cells. After approximately 72 hr at 37°, CE cells are detached from the wells with 0.05% trypsin, 0.02% EDTA (Gibco) and counted in a Coulter model zf electron particle counter. Under these conditions after 72 hr the background is about 20,000 CE cells and maximal proliferation results in about 75,000–90,000 CE cells. To facilitate the CE proliferation assay, the number of 3T3 growth factor activity units in a fraction is determined first. Then concentrations of growth factor containing 1–20 3T3 units are assayed for CE cell stimulation.

After column chromatography where gradients of NaCl are used to elute growth factor, fractions can be assayed either directly or after dialysis. If fractions are to be assayed directly without dialysis care must be taken not to add too large a fraction volume to the cells, otherwise toxicity due to excessive NaCl concentration will occur. As a rule, no more than 10 μl of a fraction in 1.5 M NaCl can be added to 200 μl of medium atop cells.

Large-Scale Purification of Chondrosarcoma-Derived Growth Factor

Transfer and Extraction of Chondrosarcoma

Chondrosarcoma is grown in Sprague–Dawley CD male rats (100–125 g) purchased from Charles River. The tumors are grown subcutaneously on both hips. When the tumors become about 5 cm in diameter, the rats are anesthetized by exposure to carbon dioxide and the tumors are excised with a pair of scissors and placed in a beaker of 0.15 M NaCl kept on ice. The capsule surrounding each tumor is removed using a scalpel and a forceps with teeth. The tumors are cut into quarters in preparation for growth factor extraction or for tumor transfer. Typically, for growth factor extraction, 70–80 tumors (15–20 g per tumor) are ground up by one

---

[8] J. Folkman, C. Haudenschild, and B. R. Zetter, *Proc. Natl. Acad. Sci. U.S.A.* **76,** 5217 (1979).

pass through an Oster high output meat/food grinder that has been fitted with 2 types of screen, an inner 0.4-mm mesh screen held in place with an outer 2.5-mm mesh screen. The ground-up tumor is collected into petri dishes and either stored frozen at $-40°$ or digested with collagenase. For collagenase digestion, filter-sterilized clostridial collagenase (2 mg/ml in PBS, pH 7.4, supplemented with 20 units/ml penicillin and 200 μg/ml streptomycin) is added to tumor extract (2 ml collagenase solution/g of tumor). Typically, a suspension of 2 liters of collagenase and 1 kg of tumor extract is incubated for 2 hr at 37° while shaking on a rotary platform shaker. Afterward, the suspension is centrifuged at 11,000 g for 1 hr at 4°. The supernatant fraction, about 3000 ml, and containing about 10% of the original tumor protein, is decanted and the pellet is discarded. The soluble supernatant fraction is used as the starting material for purification of chondrosarcoma-derived growth factor.

A slightly modified procedure is used to prepare tumor for transfer. About four decapsulated, quartered tumors are forced through a stainless steel cylinder with a 0.4-mm screen using a plunger. A tumor suspension is prepared by mixing 15 ml of the strained tumor with 35 ml of lactated Ringer's solution. Rats are injected with about 0.5 ml (about 2 million cells) per hip. It takes about 4 weeks to produce a tumor that is 5 cm in diameter.

BioRex 70 Ion-Exchange Chromatography

Chondrosarcoma extract (about 40 g, 3000 ml) and 1700 ml of BioRex 70 equilibrated with 0.1 M NaCl, 0.01 M Tris–HCl, pH 7.5, are mixed for 4 hr at 4° with occasional shaking. The BioRex 70 is allowed to settle for 2 hr and then is washed with 8000 ml of equilibration buffer. A slurry of BioRex 70 is poured into a glass column (5 × 100 cm) and washed with 4000 ml of equilibration buffer or until no more protein is being eluted as monitored by measuring absorbance at 280 nm using an LKB Uvicord-S spectrophotometer. Chondrosarcoma-derived growth factor (ChDGF) is eluted from the column with 2000 ml of 0.6 M NaCl, 0.01 M Tris–HCl, pH 7.5. Fractions (20 ml) are collected at a flow rate of 90 ml/hr and monitored for growth factor activity.

Heparin-Sepharose Affinity Chromatography

The active fractions obtained by BioRex 70 chromatography are pooled (about 150 mg, 400 ml) and applied to a column (1 × 9 cm, 7 ml) of heparin-Sepharose equilibrated with 0.6 M NaCl, pH 7.5. After a wash with 100 ml of equilibration buffer, ChDGF is eluted with 300 ml of a gradient of 0.6–2 M NaCl in 0.01 M Tris–HCl, pH 7.5, at a flow rate of 40

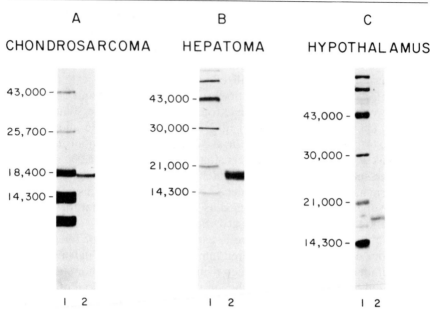

FIG. 1. Sodium dodecyl sulfate–polyacrylamide gel electrophoresis (SDS–PAGE) of purified growth factors. Cationic endothelial cell growth factors purified by a combination of BioRex 70 and heparin-Sepharose chromatography were analyzed by SDS–PAGE [U. K. Laemmli, *Nature (London)* **277**, 680 (1970)] and silver stain [B. R. Oakley, D. R. Kirsch, and N. R. Morris, *Anal. Biochem.* **105**, 362 (1975)]. (A) Chondrosarcoma-derived growth factor; (B) hepatoma-derived growth factor; (C) hypothalamus-derived growth factor. In each set, molecular weight markers are shown in lane 1 and the growth factor in lane 2.

ml/hr. Fractions (8 ml) are collected and monitored for growth factor activity. ChDGF is eluted at about 1.5 M NaCl and appears as a single band on SDS–PAGE (Fig. 1A). In this purification procedure, the yield of ChDGF starting with 1200 g of tumor (70 tumors) is about 20 μg.

Large-Scale Purification of Hepatoma-Derived Growth Factor

Extraction of Hepatoma Cells

The endothelial cell growth factor produced by the human hepatoma cell line, SK HEP-1, in culture is found to be associated predominantly with the cells rather than with the conditioned medium. SK HEP-1 cells[9]

[9] J. Fogh, J. M. Fogh, and R. Orfeo, *J. Natl. Cancer Inst.* **59**, 221 (1977).

obtained originally from Dr. J. Fogh of the Sloan-Kettering Institute (New York, NY) are growth either in monolayer or in suspension with Dulbecco's modified Eagle's medium supplemented with 10% calf serum. Using 10 roller bottles, it is possible to produce about 2×10^9 cells/week. For large scale culture, SK HEP-1 cells are grown as follows: SK HEP-1 cells are grown in monolayer in T-75 cm^2 flasks. At confluence, cells from four of these flasks are transferred into a 1-liter spinner flask. When the cells reach maximum density (about 10^6 cells/ml, they are transferred and grown first in 3-liter and then in 12-liter spinner flasks. The cells from two to four 12-liter spinner flasks (at about $1–5 \times 10^5$ cells/ml) are introduced into a 100-liter Vibromixer reactor.[10] When the cell density is about 10^6/ml, the cells are collected by centrifugation. About 10^{11} SK HEP-1 cells are produced in a 100-liter reactor and when centrifuged produce a 300 ml pellet.

For growth factor purification, a 150 ml pellet containing about 5×10^{10} SK HEP-1 cells is frozen, thawed, and resuspended in 1500 ml of 1.0 M NaCl, 0.01 M Tris–HCl, pH 7.5. The cells are disrupted by homogenization in a Waring blendor for 1 min at room temperature. The homogenate is stirred overnight at 4° with a magnetic stirring bar and is centrifuged at 25,000 g for 30 min. The supernatant fraction (1200 ml) is used for further purification.

BioRex 70 Cation-Exchange Chromatography

About 1200 ml of clarified SK HEP-1 extract is dialyzed against about 100 volumes 0.01 M Tris–HCl, pH 7.5, to obtain a sample that is 0.15 M NaCl, 0.01 M Tris–HCl, pH 7.5. The sample (about 3 million units of growth factor activity) is mixed with 1750 ml of BioRex 70, equilibrated in 0.15 M NaCl, 0.01 M Tris–HCl, pH 7.5, overnight at 4° with constant stirring from overhead using a motor-driven stirrer (Heller 4 blade propeller, model GT-21, Thomas Scientific). The BioRex 70 is allowed to settle for 45 min and the supernatant is discarded. A slurry of BioRex 70 is poured into a glass column (5 × 100 cm) and is washed with equilibration buffer until protein is no longer eluted as monitored with an LKB Uvicord-S spectrophotometer that measures absorbance at 280 nm. The wash is usually about 2000 ml. Hepatoma-derived growth factor is eluted by washing the column with 1800 ml of 0.6 M NaCl, 0.01 M Tris, pH 7.5, at a flow rate of 60 ml/hr. Fractions (20 ml) are collected and monitored for growth factor activity.

[10] W. R. Tolbert, R. A. Schoenfeld, C. Lewis, and J. Feder, *Biotechnol. Bioeng.* **24**, 1671 (1982).

Heparin-Sepharose Affinity Chromatography

The active fractions containing hepatoma-derived growth factor are pooled (about 1–2 million units, 300 ml) and applied to a column of heparin-Sepharose (2.0 × 13 cm, 40 ml) equilibrated with 0.6 M NaCl, 0.01 M Tris, pH 7.5. After a wash of about 200 ml, hepatoma-derived growth factor is eluted with 400 ml of a gradient of 0.6–2.5 M NaCl in 0.01 M Tris–HCl, pH 7.5, at a flow rate of 40 ml/hr. Fractions (7 ml) are collected and monitored for growth factor activity. Hepatoma-derived growth factor elutes at about 1.6 M NaCl and appears as a single band on SDS–PAGE (Fig. 1B). In this procedure the yield of hepatoma-derived growth factor starting with 5 × 10^{10} cells is about 100–150 μg.

Large-Scale Purification of Cationic Hypothalamus-Derived
    Growth Factor

Extraction of Hypothalamus

Bovine hypothalami are purchased from Pel Freez (Rogers, Arkansas) and kept frozen at −20°. For growth factor preparation 25 hypothalami, weighing about 15 g each, are partially thawed and homogenized in a food processor. A metal blade with a repeat pulse action is used until the tissue is totally homogenized. The homogenate is resuspended in 500 ml of phosphate-buffered saline (PBS), pH 7.4, and digested with collagenase (2 mg/ml) at 37° for 3 hr. The digest is centrifuged at 20,000 g for 70 min. The supernatant fraction is recovered and the pellet is washed twice with 500 ml PBS. The supernatant fractions which contain about 10% of the original hypothalamus protein are pooled and stored frozen at −20°.

BioRex 70 Chromatography

The supernatant fraction (about 50 g, 1300 ml) and BioRex 70 (400 ml), which has been equilibrated with 0.15 M NaCl, 0.01 M Tris–HCl, pH 7.5, are combined and incubated overnight at 4° with continuous shaking. The BioRex 70 is allowed to settle for about 45 min. The supernatant is decanted and discarded. The BioRex 70 is resuspended in 300 ml of the equilibration buffer and poured into a glass column (5 × 20 cm). The BioRex is washed with 1700 ml of equilibration buffer or until no more protein is eluted from the column as monitored with an LKB Uvicord-S spectrophotometer. The flow rate is 60 ml/hr. Growth factor is eluted by washing the column with 500 ml of 0.6 M NaCl, 0.01 M Tris–HCl, pH 7.5. Fractions (8 ml) are collected and monitored for growth factor activity.

Heparin-Sepharose Affinity Chromatography

The active fractions are pooled (about 110 mg, 160 ml) and applied to a column of heparin-Sepharose (1 × 13 cm, 10 ml) equilibrated with 0.6 M NaCl, 0.01 M Tris–HCl, pH 7.5. After a rinse of about 100 ml, a gradient (320 ml) of 0.6 to 2.5 M NaCl in 0.01 M Tris–HCl, pH 7.5, is applied. Fractions (8 ml) are collected at a flow rate of 45 ml/hr and monitored for growth factor activity. Cationic hypothalamus-derived growth factor elutes from heparin-Sepharose at about 1.4–1.6 M NaCl and appears as a single band on SDS–PAGE (Fig. 1C). The recovery of cationic hypothalamus-derived growth factor is about 5–10 μg starting with 25 hypothalami (400 g).

Summary and Perspectives

Cationic endothelial cell growth factors can be purified to homogeneity on a large scale using a two-step procedure of BioRex 70 cation-exchange chromatography followed by heparin-Sepharose affinity chromatography. The same procedure can be used to purify other growth of this class including a cartilage-derived growth factor (CDGF) (see Klagsbrun *et al.* [3a], Vol. 146, this series) and cationic FGF found in brain and pituitary (see Gospodarowicz [9] and Thomas [10], this volume, on acidic and basic FGFs). An alternative method for purification is to recycle growth factor on successive columns of heparin-Sepharose. For example CDGF has been purified to homogeneity using 3 cycles of heparin-Sepharose affinity chromatography without any BioRex chromatography.[7] The availability of large amounts of homogeneous growth factor will facilitate biological and structural studies. [Editors' note: Recent studies have shown cationic endothelial cell growth factors to be identical to, or multiple molecular weight forms of basic FGF.[11]]

[11] M. Klagsbrun, S. Smith, R. Sullivan, Y. Shing, and S. Davidson, *Proc. Natl. Acad. Sci. U.S.A.* **84**, 1839 (1987).

[9] Isolation and Characterization of Acidic and Basic Fibroblast Growth Factor

By D. GOSPODAROWICZ

Introduction

Basic fibroblast growth factor (FGF) is a single chain polypeptide (MW of 16K and 14.5K) originally isolated from bovine brain and pituitary.[1,2] It is also present in a wide range of other organs (Table I). It has been shown to be mitogenic for endothelial cells and a wide variety of mesoderm and neuroectoderm-derived cells (Table II; Ref. 3). In addition, it is also a differentiation factor preventing the dedifferentiation of cultured vascular and corneal endothelial cells, chondrocytes, and myoblast, and inducing differentiation in nerve cells (PC-12) and conversion of fibroblasts into adipocytes (reviewed in Ref. 4). At the cellular level, FGF has been shown to bind to specific and high affinity cell surface receptor,[5] and to trigger the pleiotypic cell response,[6] as well as increase expression of mRNA transcripts of the cellular oncogenes c-*myc* and c-*fos*.[7] *In vivo* basic FGF is an angiogenic factor.[8] Related to basic FGF is acidic FGF, which has until now only been reported present in neural tissue. It is mitogenic for the same cell types as basic FGF.[9]

Although methods for the isolation of basic and acidic FGF have been described,[10–12] the yields are low, and the multistep procedures are time consuming. We are here describing our present procedure for the isolation of acidic and basic FGF. The method is based on the observation by

[1] D. Gospodarowicz, *Nature (London)* **249**, 123 (1974).

[2] D. Gospodarowicz, *J. Biol. Chem.* **250**, 2515 (1975).

[3] D. Gospodarowicz, G. Greenburg, and H. Bialecki, *In Vitro* **14**, 85 (1978).

[4] D. Gospodarowicz, in "Mediators in Cell Growth and Differentiation" (R. J. Ford and A. L. Maizel, eds.), p. 109. Raven Press, New York, 1984.

[5] G. Neufeld and D. Gospodarowicz, *J. Biol. Chem.* **260**, 13860 (1985).

[6] P. S. Rudland, D. Gospodarowicz, and W. E. Seifert, *Proc. Natl. Acad. Sci. U.S.A.* **71**, 2600 (1974).

[7] R. Muller, R. Bravo, J. Burckhardt, and T. Curran, *Nature (London)* **312**, 716 (1984).

[8] D. Gospodarowicz, H. Bialecki, and T. K. Thakral, *Exp. Eye Res.* **28**, 501 (1979).

[9] P. Bohlen, F. Esch, A. Baird, and D. Gospodarowicz, *EMBO J.* **4**, 1951 (1985).

[10] P. Bohlen, A. Baird, F. Esch, N. Ling, and D. Gospodarowicz, *Proc. Natl. Acad. Sci. U.S.A.* **81**, 6963 (1984).

[11] D. Gospodarowicz, S. Massoglia, J. Cheng, G.-M. Lui, and P. Bohlen, *J. Cell. Physiol.* **122**, 323 (1985).

[12] K. A. Thomas, M. Rios Candelore, and S. Fitzpatrick, *Proc. Natl. Acad. Sci. U.S.A.* **81**, 357 (1984).

METHODS IN ENZYMOLOGY, VOL. 147

TABLE I
CELL TYPES FOR WHICH BASIC OR ACIDIC FGF IS
MITOGENIC

Cell type	Basic FGF	Acidic FGF
Normal diploid cells		
Glial and astroglial cells	+ (D)[a]	+
Trabecular meshwork cells	+	?
Capillary, large vessel, and endocardium endothelial cells	+ (D)	+ (D)
Corneal endothelial cells	+ (D)	+ (D)
Fibroblast	+	+
Myoblast	+ (D)	? (D)
Vascular smooth muscle	+	+
Chondrocytes	+ (D)	+ (D)
Blastema cells	+	?
Adrenal cortex cells	+	+
Granulosa cells	+	+
Established cell lines		
Rat fibroblast-1	+	?
BALB/c 3T3	+	+
Swiss 3T3	+	+
BHK-21	+	+
PC-12	(D)	(D)

[a] (D), Induce differentiation.

Shing *et al.* that angiogenic factors have a strong affinity for heparin.[13] Using heparin-Sepharose affinity chromatography, one can, in 32 hr, purify to homogeneity both basic and acidic FGF, with yields as high as 500 μg/kg of wet tissue.[9,14] In view of the potency of basic FGF, which saturates at concentrations of 10 to 20 pM, it makes it the most abundant growth factor found in mammalian tissue.[4]

FGF Bioassay and Radioimmunoassay

Bioassay

The mitogenic activity of column fractions can be determined using bovine vascular endothelial cells derived from adult aortic arch as de-

[13] Y. Shing, F. Folkman, R. Sullivan, C. Butterfield, J. Murray, and M. Klagsbrun, *Science* **223**, 1296 (1984).
[14] D. Gospodarowicz, J. Cheng, G.-M. Lui, A. Baird, and P. Bohlen, *Proc. Natl. Acad. Sci. U.S.A.* **81**, 6963 (1984).

TABLE II
FGF FROM VARIOUS SOURCES

Tissue	Species	Biological activity[a]	Radio-immuno-assay[b]	Purified to homo-geneity	AA residue	Yield of basic FGF (μg/kg)	Acidi FGF
Brain	Bovine	+	+	+	1–146	35–50	+
	Human	+	+	+	1–146	ND[c]	+
Pituitary	Bovine	+	+	+	1–146	350–600	–
Kidney	Bovine	+	–	+	16–146	90–150	–
Adrenal glands	Bovine	+	+	+	1–146 and 16–146	200	–
Corpus luteum	Bovine	+	–	+	16–146	50–70	–
Placenta	Human	+	+	+	1–146 and 16–146	30–40	–
Retina	Bovine	+	+	+	1–146 and 16–146	ND	+
Macrophage	Murine	+	+	–	?	ND	?
Chondrosarcoma	Rodent	+	+	?	?	ND	?
Thymus	Bovine	+	+	+	1–146 and 16–146	35–70	–

[a] Biological activity detected on the various cell types shown in Fig. 1.
[b] Radioimmunoassay using either monoclonal antibodies directed against pituitary basic FGF or a described in Ref. 18.
[c] Not determined.

scribed[15,16] or capillary endothelial cells derived either from adrenal or brain cortex and corpus luteum.[17] Briefly, cells are seeded at an initial density of 2×10^4 cells per 35-mm dish containing 2 ml of DMEM H-16 supplemented with 10% calf serum and antibiotics.[15–17] Six hours later, a set of triplicate plates is trypsinized and cells are counted to determine the plating efficiency. Ten-microliter aliquots of the appropriate dilution of each fraction (with DME medium/0.5% bovine serum albumin) are then added to the dishes every other day. After 4 to 5 days in culture, triplicate plates are trypsinized, and final cell densities are determined by counting cells in a Coulter counter (Fig. 1).

The FGF mitogenic activity can also be tested on bovine vascular smooth muscle cells, human umbilical endothelial cells, bovine granulosa cells, adrenal cortex cells, and rabbit costal chondrocyte cultures maintained as previously described.[4,11] Cells are seeded at an initial density of 2 or 4×10^4 cells per 35-mm dish. Assays are conducted as described for bovine vascular endothelial cells (Fig. 1).

[15] D. Gospodarowicz, G.-M. Lui, and J. Cheng, *J. Biol. Chem.* **257**, 12266 (1982).
[16] D. Gospodarowicz, J. Moran, D. Braun, and C. R. Birdwell, *Proc. Natl. Acad. Sci. U.S.A.* **73**, 4120 (1976).
[17] D. Gospodarowicz, S. Massoglia, J. Cheng, and D. K. Fujii, *J. Cell. Physiol.* **127**, 121 (1986).

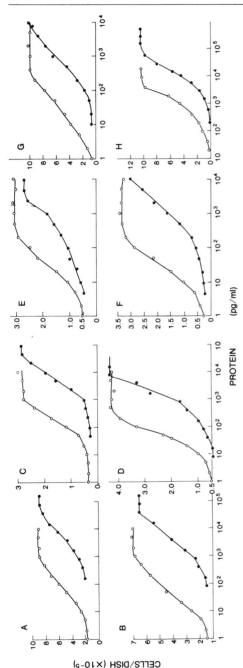

FIG. 1. Effects of acidic (○) and basic (●) FGF on the proliferation of various mesoderm-derived cell types *in vitro*. Mitogenic activity of purified basic and acidic FGF was determined using cultures of bovine brain (A) and adrenal cortex-derived capillary endothelial cells (B), human umbilical vein endothelial (HUE) cells (C), bovine adrenal cortex cells (D), granulosa cells (E), vascular smooth muscle cells (F), rabbit coastal chondrocytes (G), and BHK-21 cells (H). Culture and assay conditions were similar to those described earlier [Refs. 11, 15, and 17; see also G. Neufeld, S. Massoglia, and D. Gospodarowicz, *Regul. Peptides* **13**, 253 (1986)]. Brain and adrenal cortex-derived capillary endothelial cells were maintained in DMEM supplemented with 10% calf serum, and HUE cells were maintained in Medium 199 supplemented with 20% fetal calf serum. Adrenal cortex cells were maintained in F12 medium supplemented with 10% calf serum, granulosa cells were maintained in DMEM supplemented with 1% calf serum, vascular smooth muscle cells were maintained in DMEM supplemented with 5% defibrinated bovine plasma, chondrocytes and BHK-21 cells were maintained in a DMEM/F12 medium, (v/v) supplemented with 10% calf serum for chondrocytes or with 250 μg HDL protein/ml and 1 μg transferrin/ml for BHK-21 cells.[30] All cells were seeded at an initial density of 2 × 10⁴ cells/35-mm dish (except for HUE cells, where the density was 4 × 10⁴ cells/35-mm dish) and maintained in the presence of various concentrations of either acidic or basic FGF (10-μl aliquots added on days 2 and 4) for 5–6 days after which they were trypsinized and counted in a Coulter counter.

Radioimmunoassay (RIA)

An RIA using amino-terminally directed antibodies against a synthetic decapeptide representing the first 9 terminal amino acids of pituitary FGF coupled to tyrosine has been developed.[10,18] These antibodies recognize both synthetic antigens and native FGF on an equimolar basis. However, since in some tissue FGF exists under an amino-terminally truncated form (Des. 1–15 FGF), this assay may not detect it in all tissues. RIA using antibodies directed against synthetic antigens representing the central portion of the native molecule are actually being developed. Monoclonal antibodies raised against basic pituitary FGF and which recognize either the N-terminal or central portion of the molecule have also been developed and can be used for RIA.[19] One can also use monoclonal antibodies (HIS and H-9) directed against endothelial cell growth factor (ECGF). Although these antibodies have been claimed to be highly specific for ECGF,[20-22] by dot blot analysis, or by Western blotting they strongly react with basic FGF. No reaction of these monoclonal antibodies can be observed with either acidic FGF or ECGF.

Isolation Procedure

For extraction of FGF, 2.3 kg of pituitaries or 5 kg of brain is homogenized at 4° in 8 liters of 0.15 M $(NH_4)_2SO_4$. The pH of the suspension is adjusted to 4.5 and the suspension stirred for 2 hr at 4°. It is then centrifuged and the pH of the supernatant is adjusted to 6.0 $(NH_4)_2SO_4$ (230 g/liter, final concentration 1.9 M) is added and the precipitate removed by centrifugation. Further addition of $(NH_4)_2SO_4$ (250 g/liter final concentration 3.8 M) to the supernatant gives a precipitate that is collected by centrifugation, redissolved in water, and dialyzed against cold water overnight. Most of the bioactivity present in the original extract (90%) is recovered in the 1.9 to 3.8 M $(NH_4)_2SO_4$ fraction. This fraction is called crude extract.

Pituitary or brain FGF is purified by the successive application of two chromatographic steps consisting of cation-exchange chromatography

[18] A. Baird, P. Bohlen, N. Ling, and R. Guillemin, *Regul. Peptides* **10**, 309 (1985).
[19] S. Massoglia, J. Kenney, and D. Gospodarowicz, *J. Cell. Physiol.*, in press.
[20] T. Maciag, T. Mehlman, R. Friesel, and A. B. Schreiber, *Science* **225**, 932 (1984).
[21] A. B. Schreiber, J. S. Kenney, W. J. Kowalski, R. Friesel, T. Mehlman, and T. Maciag, *Proc. Natl. Acad. Sci. U.S.A.* **82**, 6138 (1985).
[22] A. B. Schreiber, J. Kenney, J. Kowalski, K. A. Thomas, G. Gimenez-Gallego, M. Rios-Candelore, J. Di Salvo, D. Barritault, J. Courty, Y. Courtois, M. Moenner, C. Loret, W. H. Burgess, T. Mehlman, R. Friesel, W. Johnson, and T. Maciag, *J. Cell Biol.* **101**, 1623 (1985).

and heparin-Sepharose (HS) affinity chromatography. All chromatography is carried out at room temperature. Crude extracts prepared as above are applied to a column of carboxymethyl (CM)-Sephadex C-50 (3 × 20 cm bed volume 140 ml) that has been equilibrated with 0.1 M sodium phosphate, pH 6.0. The column was extensively washed with 0.1 M sodium phosphate (pH 6.0) and eluted sequentially by the stepwise addition of 0.15 and 0.6 M NaCl in 0.1 M sodium phosphate (pH 6.0) as described.[2,11,14] The unadsorbed material that elutes with 0.15 M NaCl contains less than 10% of the initial activity.[2,14] Most of the bioactive material (94%) is recovered in the 0.6 M NaCl fraction, resulting in an 58- to 75-fold purification factor.[2,14]

Mitogenic activity eluted with 0.6 M NaCl is either dialyzed and lyophilized, or directly loaded onto a HS column (1.6 × 8 cm, bed volume 15 ml) that had been equilibrated at room temperature with 10 mM Tris (pH 7.0), 0.6 M NaCl. To ensure a constant flow rate the column is regulated at 70 ml an hour with a peristaltic pump. The column is washed (flow rate 70 ml/hr) with 10 mM Tris–HCl (pH 7.0), 0.6 M NaCl, until the absorbancy of the eluate at 280 nm became negligible. Most of the protein (99%) does not bind to the column, and the unadsorbed material has little biological activity (less than 0.1% that of the input).

In the case of brain FGF, stepwise elution of the HS column *at a flow rate which should not exceed 20 ml/hr*, with 1 M NaCl and 2 M NaCl solutions, yields two fractions with mitogenic activity for ABAE cells (Fig. 2A). The mitogenic activity present in the 1 M NaCl fraction was identified as acidic FGF. The active material in this fraction was purified 2500-fold from the crude brain FGF preparation and stimulates half-maximal proliferation of ABAE cells at a dose of 6 ng/ml (Fig. 2B). Mitogenic activity in the 2 M salt fraction was purified 300,000-fold as compared to the crude FGF preparation and stimulated half-maximal ABAE cell proliferation at 60 pg/ml. It was ascertained that this material corresponds to basic FGF. The yields of brain acidic and basic FGFs from three different purifications ranged from 620 to 710 and 37 to 45 μg/kg, respectively. Due however to the higher biological specific activity of basic versus acidic FGF only 8.7% of the activity originally present in the crude extract is recovered in the acidic FGF fractions, versus 72% in basic FGF fractions[9] (Table III).

In the case of pituitary FGF (Fig. 2C and D), stepwise elution of the HS column with 1 M NaCl yields only a fraction which had a mitogenic activity similar to that of the input. Elution with 2 M NaCl yields a mitogenic fraction which was purified 140,000-fold, as compared to the crude FGF preparation. It was ascertained (see below) that this material corresponds to basic FGF. The yields of pituitary basic FGF from 3

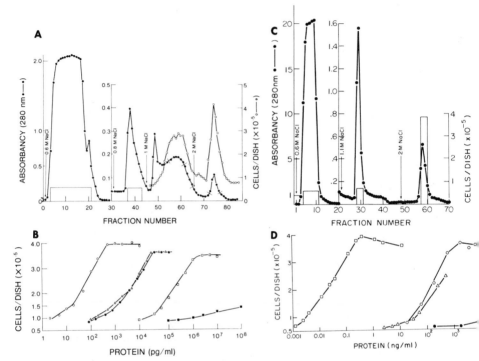

FIG. 2. HS affinity chromatography of pituitary and brain FGFs. (A) HS affinity chromatography of bovine brain FGF. The 0.6 M NaCl fractions containing mitogenic activity from
the CM-Sephadex C-50 chromatography (110 ml, 7.6 mg/ml) were loaded onto a heparin-
Sepharose column as described in the text. Fractions of 9 ml were collected during loading
and washing of the column; afterward the fraction size was 2 ml. For fractions 3–21, protein
concentration was determined by weighing a 2-ml aliquot after dialysis and lyophilization
and bioassays were performed by diluting aliquots with DMEM/0.5% BSA to a concentration of 0.7 mg/ml and adding 10-μl aliquots of the diluted samples to low density ABAE cell
cultures. For the fractions containing 0.8–2.0 M NaCl, mitogenic activity was determined by
adding 10-μl aliquots (after dilution 1 : 10, 1 : 50, or 1 : 500 with DMEM/0.5% BSA for fractions containing 0.8, 1.0, and 2.0 M NaCl, respectively) to low density ABAE cultures. The
histogram shows final cell densities of pooled fractions 3–21 and 37–43 after 4 days in
culture. Densities of control cultures after 4 days were 1.65 × 10^5 cells/35-mm dish. After
use, the heparin-Sepharose column was stripped with 3 M NaCl/10 mM Tris–HCl, pH 7.0,
and reequilibrated with 0.6 M NaCl/10 mM Tris–HCl, pH 7.0. (B) Mitogenic activities of
fractions taken from various purification steps. Carboxymethyl-Sephadex 0.6 M NaCl fractions (□), heparin-Sepharose fractions: unabsorbed (■), 1 M NaCl (pooled fractions 55–61,
△, and 62–65, ●), and 2 M NaCl (pooled fractions 73–75, ○). (C) HS affinity chromatography of bovine pituitary FGF. The lyophilized 0.6 M NaCl fractions containing mitogenic
activity from the CM-Sephadex C-50 chromatography step corresponding to the purification
of 4 kg of pituitaries were dissolved in 0.6 M NaCl, 10 mM Tris–HCl, pH 7.0 (50 ml, 24 mg/
ml) and loaded onto a HS column as described in the text. Fractions of 9 ml were collected
during loading and washing; afterward the fraction size was 1.8 ml. For fractions 4–11,

TABLE III

PURIFICATION OF ACIDIC AND BASIC FGF FROM BOVINE BRAIN

Purification step	Protein recovered (mg)	Maximal mitogenic effect (ng/ml)	$ED_{50}{}^{a}$ (ng/ml)	Total activity[b] (units × 10^5)	Recovery of biological activity (%)	Purification factor
Brain (1 kg)						
Crude extract	18,000	150×10^3	15×10^3	12.0	100	1
Carboxymethyl-Sephadex C-50	190	2×10^3	200	9.5	80	75
HS, 1 M NaCl (fract. 55–65)[c]	0.630	30–40	6	1.05	8.7	2500
HS, 2 M NaCl (fract. 73–77)[d]	0.043	0.5	0.05	8.6	72	300,000

[a] Concentration of FGF preparation required to give a 50% maximal response in the assay system.
[b] One unit of activity is defined as the quantity of FGF required to give half-maximal stimulation of cell proliferation in the assay system described.
[c] Acidic FGF.
[d] Basic FGF. HS, Heparin-Sepharose.

different purifications ranged from 350 μg to 600 μg/kg of pituitaries (Table I). The whole purification process of either brain or pituitary FGF should take no longer than 32 hr, provided that the 0.6 M NaCl, CM-Sephadex C-50 fraction is directly chromatographed on HS.

The heparin-Sepharose step alone is responsible for 2400- and 5300-fold purification of pituitary and brain basic FGF, respectively, thus constituting an extremely powerful technique. It also has the advantage, in the case of tissues containing both basic and acidic FGF, that it permits purification of both mitogens simultaneously. This reflects the different affinity of acidic versus basic FGF for HS with basic FGF having the highest affinity. The use of HS chromatography is, however, not without disadvantages. Since it is primarily based on the presence of an affinity

protein concentration was determined by weighing a 2-ml aliquot after dialysis and lyophilization and bioassays were performed by diluting aliquots with DMEM/0.5% BSA to a concentration of 0.5 mg/ml and adding 10-μl aliquots. For fraction 9–12 (1.1 M NaCl) and 57–60 (2 M NaCl) protein concentration was determined by the dye fixation assay as described[14] and 10-μl aliquots of the pooled fractions, after dilution 1 : 10 or 1 : 500 with DMEM/0.5% BSA for fractions containing 1.1 and 2.0 M NaCl, respectively, were added to low density ABAE cultures. The histogram shows final cell densities of pooled fractions 9–12 and 57–60 after 4 days in culture. Densities of control cultures after 4 days were 0.7×10^5 cells/35-mm dish. After use, the heparin-Sepharose column was stripped with 3 M NaCl/10 mM Tris–HCl, pH 7.0, and reequilibrated with 0.6 NaCl/10 mM Tris–HCl, pH 7.0. (D) Mitogenic activities of fractions taken from various purification steps. CM-Sephadex C-50, 0.6 M NaCl fractions (○), heparin-Sepharose fractions: unabsorbed (●), 1.1 M NaCl (pooled fractions 9–12, △), and 2 M NaCl (fractions 57–60, □).

site not yet defined, any degradation product of FGF, either present in the tissue or generated during the extraction, which will contain those affinity sites will copurify with the native molecules. It is therefore recommended to start with as freshly collected tissue as possible and to store them if needed at $-80°$. In our experience, most commercial sources of frozen pituitary are unsuitable, with the exception of J.R. Scientific (Woodland, CA). One should also be aware that there can be considerable variation in the HS batches marketed by various companies. Even within the same company (Pharmacia), the retention of acidic and basic FGF on HS will vary depending on the batch. It is therefore necessary to first screen various batches for their retention behavior by running a salt gradient (0.6 to 2 M NaCl, 10 mM Tris–Cl, pH 7.0) in order to calibrate the column.[14] Due to the extreme affinity of basic FGF for HS, the first time one uses the column, the yield of basic FGF will be relatively low. However, once the sites where basic FGF binds irreversibly are saturated, the yield increases dramatically. In the case of pituitary basic FGF, the yield of 3 consecutive chromatographies on the same HS column was 266, 422, and 55 μg/kg of pituitaries.

Characterization of Acidic and Basic FGF

The homogeneity of the HS affinity purified acidic and basic FGF can be assessed by SDS–polyacrylamide gel electrophoresis and/or by HPLC. Pituitary and brain basic FGF migrate as a single band with apparent MW of 16,000. Brain acidic FGF migrate as a doublet with apparent MW of 15,000, and a quantitatively minor component, which is biologically active, but with a somewhat higher molecular weight is also present (Fig. 3). Isoelectric focusing essentially confirmed the basic and acidic natures of the brain-derived mitogens. While basic brain FGF sharply focused with a pI of 9.6, acidic FGF gave 3 incompletely resolved bioactive peaks with pI of 6.0–5.9, 5.4, and 5.0–4.8 (Fig. 4). When analyzed by HPLC (Vydac C$_4$ column 0.46 × 25 cm, 300 Å pore size, 5 μm particle size, the separation group, Hesperia, CA) using a linear 2 hr linear gradient of 30 to 45% (v/v) acetonitrile in 0.1% (v/v) trifluoroacetic acid as the mobile phase, brain acidic FGF gave a major biologically active peak together with 2 minor peaks.[10] For both pituitary and brain basic FGF, the column is developed with a 90 min acetonitrile gradient (26–36% v/v) in 0.1% trifluoroacetic acid.[14] In the case of pituitary basic FGF, a single major biologically active peak is observed. In the case of brain basic FGF, distinct peaks are observed which all possess reduced biological activity and are immunoreactive.[14] Amino-terminal sequence analysis of the various forms of brain FGF have shown that they are identical to pituitary

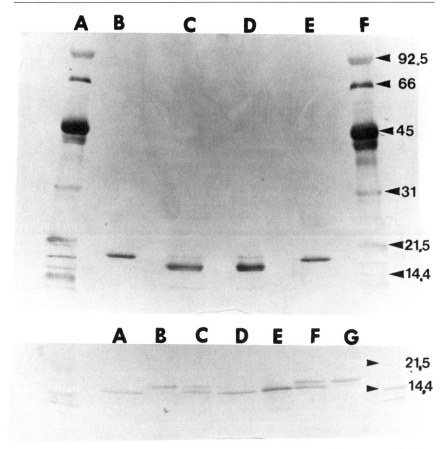

Fig. 3. SDS–polyacrylamide gel electrophoresis of HS-purified FGFs. Top panel: Samples of 10 μl (0.9 μg protein) were applied to the gel. Lanes A and F: protein standard mixture, including phosphorylase *b* (92.5 kDa), BSA (66 kDa), ovalbumin (45 kDa), carbonate dehydratase (31 kDa), soybean trypsin inhibitor (21.5 kDa), and lysozyme (14.4 kDa). Lane B: bovine pituitary FGF; 2 *M* NaCl eluate pool of fractions 57–60 (Fig. 2C). Lanes C and D: bovine brain acidic FGF; 1 *M* NaCl eluate, pool of fractions 55–61 and 62–65, respectively (Fig. 2A). Lane E: bovine brain basic FGF, 2 *M* NaCl eluate, pool of fractions 73–77 (Fig. 2A). Gels were run and stained with silver nitrate as described in Refs. 11 and 14. Bottom panel: Comparison of the electrophoretic behavior of HS purified corpus luteum (A), pituitary (B), adrenal (C, F), corpus luteum (D), kidney (E), and brain (G) basic FGF. Protein (1 to 1.2 μg) was applied to the gels and gels were run and stained as described in Refs. 11 and 14.

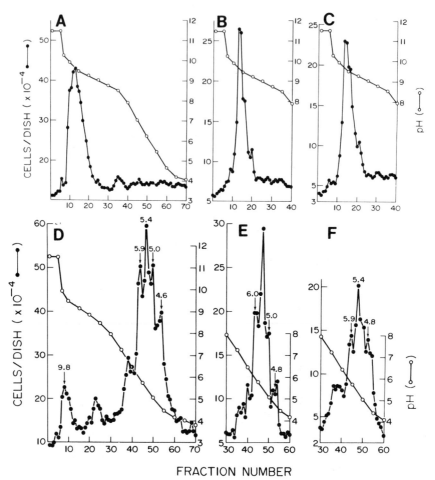

FRACTION NUMBER

FIG. 4. Isoelectric focusing of basic and acidic FGF. (A–C) Heparin-Sepharose affinity purified basic pituitary FGF (100 μg) was dialyzed against 0.1 M NH_4 carbonate overnight, lyophilized, and redissolved together with 1 mg of an inactive side fraction in 2 ml gradient solution. This sample was mixed in the mid position of a sucrose density gradient (5–50% w/ v) using 1 ml of LKB ampholytes pH 3 to 10 and 3 ml of pH 9 to 11. Isoelectric focusing was performed using a 110 ml LKB column.[11,15] The separation was carried out at 4° for 48 hr at 1000 V. The gradient was then collected in 1.4 ml fractions. The pH of every fifth tube was determined (○) and aliquots of each fraction diluted 1 to 1000 in DMEM, 0.5% BSA before being tested for their ability to stimulate (●) the proliferation of bovine brain-derived capillary endothelial (A), bovine adrenal cortex cells (B), and human umbilical endothelial cells (C). Bioassay was conducted as described in Fig. 2. (D–F) Heparin-Sepharose affinity purified acidic FGF (1 mg) was dialyzed and lyophilized as described above. It was then mixed in the mid position of a sucrose density gradient (5–50% w/v) using 2 ml of LKB ampholytes pH 3 to 10 and 2 ml of pH 9 to 11. Isoelectric focusing was performed as described above. Aliquots of each fraction diluted 1 to 50 in DMEM, 0.5% BSA were tested for their ability to stimulate the proliferation of bovine brain-derived capillary endothelial cells (D), bovine adrenal cortex cells (E), and human umbilical endothelial cells (F).

basic FGF and all forms are immunoreactive in an RIA using specific antisera raised against a synthetic replicate of the amino-terminal sequence of pituitary basic FGF.[14] The nature of the structural differences between the various forms of brain basic FGF is not yet clear, and it is likely that they are reflections of protein side chain modifications or microheterogeneity of FGF.

It should be noted that in contrast to acidic FGF that can withstand acidic conditions, basic FGF, when exposed to a pH lower than 4.0, loses 95 to 97.5% of its potency.[11] Therefore, basic FGF submitted to HPLC can only be used to assess the homogeneity of the preparation or for structural studies,[10,23] but not for biological studies.

Tissue Distribution of Basic and Acidic FGF: Similarities and Differences between These Two Mitogens

Basic FGF has been shown to be present in a wide variety of organs (Table II). In contrast, acidic FGF seems to be only present in detectable quantities in neural tissue (brain and retina). This does not necessarily mean that it is totally absent elsewhere, rather it indicates that its contribution to the mitogenic potential of other organs will be minimal as compared to that of basic FGF. Even so, it is already minimal in the case of brain and retina.

The presence of basic FGF in adrenals,[24] corpus luteum,[25] kidney,[26] placenta,[27] retina,[28] and macrophage,[29] and its ability to stimulate the proliferation *in vitro* of capillary endothelial cells correlates with the well recognized angiogenic properties of those organs (reviewed in Refs. 24 and 25). The yield of basic FGF from these organs routinely exceeds the apparent mitogenic content of the crude extracts. This is particularly true in the case of kidney extracts where the yield of basic FGF is 30-fold higher than that apparently present in kidney crude extract. It reflects the elimination during the purification procedure of various growth inhibitors

[23] F. Esch, A. Baird, N. Ling, N. Ueno, F. Hill, L. Deneroy, R. Klepper, D. Gospodarowicz, P. Bohlen, and R. Guillemin, *Proc. Natl. Acad. Sci. U.S.A.* **85,** 6507 (1985).

[24] D. Gospodarowicz, J. Cheng, G. M. Lui, A. Baird, F. Esch, and P. Bohlen, *Endocrinology* **117,** 2283 (1985).

[25] D. Gospodarowicz, A. Baird, J. Cheng, G. M. Lui, F. Esch, and P. Bohlen, *Endocrinology* **12,** 201 (1986).

[26] A. Baird, F. Esch, N. Ling, and D. Gospodarowicz, *Regul. Peptides* **12,** 201 (1985).

[27] D. Gospodarowicz, J. Cheng, G.-M. Lui, D. K. Fujii, A. Baird, and P. Bohlen, *Commun. Biochem. Biophys. Res. Commun.* **30,** 554 (1985).

[28] A. Baird, F. Esch, D. Gospodarowicz, and R. Guillemin, *Biochemistry* **24,** 7855 (1985).

[29] A. Baird, P. Mormede, and P. Bohlen, *Biochem. Biophys. Res. Commun.* **126,** 358 (1986).

and cytotoxic factors which would otherwise interfere with the activity of basic FGF.[26]

Basic FGF purified from different sources has been shown to be present in two forms, one of which is identical with brain and pituitary FGF, while the other form is an amino-terminally truncated form missing the first 15 AA residues (Des. 1–15 FGF). One or the other forms are ordinarily predominant within a given organ. Brain and pituitary contain the native molecule as a major form,[10,11,14] while in kidney and corpus luteum Des. 1–15 FGF is the major form.[25,26] In some organs, such as the adrenal gland,[24] both forms are present in roughly equivalent amounts. The truncated form of FGF is as potent as native FGF in stimulating cell proliferation indicating a lack of involvement of the NH_2-terminal region in the active sites of basic FGF,[25,26] or in its binding domain.[5] It remains to be seen whether Des. 1–15 FGF occurs in tissue under physiological conditions or is generated during tissue extraction or FGF purification, as a result of enzymatic degradation.

The available data have until now clearly shown that basic and acidic FGF are two different molecular entities. Acidic FGF has pI ranging from 5.0 to 5.9 and a MW of 14K to 15K. Basic FGF has a pI of 9.6 and a MW of 16K for the native molecule and 14.5K for Des. 1–15 FGF. Their amino acid composition[9–12,14] and their N-terminal amino acid sequence also apparently differed (Fig. 5). Those differences, however, may not be as significant as one would have initially believed. Both mitogens bind strongly and very selectively to heparin,[9,14] and they also display a remarkable similarity in their biological activity toward endothelial and other mesoderm-derived cells.[9,11,14,23] Recent studies[30] have shown that they interact with the same cell surface receptor and can displace each other, suggesting that acidic FGF contains within its sequence a region that is similar or even identical to a sequence segment of basic FGF that carries biological activity (receptor binding site). This possibility is supported by a striking sequence homology between residues 4 and 29 of acidic FGF versus residues 13 and 38 of basic FGF.[23] The fact that antisera raised against a synthetic peptide from basic FGF, (Tyr[69]) FGF (69–87) NH_2 cross-react with acidic FGF strongly suggests that additional sequence homologies exist between the midportions of these two proteins.[23]

The primary structure of acidic FGF shows a 53% absolute homology with basic FGF and up to 64% of the remaining structure may involve nucleotide substitution where a single base change could result in amino

[30] G. Neufeld and D. Gospodarowicz, *J. Biol. Chem.* **261**, 5631 (1986).

```
                           10              20            30      35
BRAIN ACIDIC FGF       F N L P L, G N Y K K, P K L L Y, C S N G G, Y F L R I,L P D G T, V D G T . . . . .

PITUITARY BASIC FGF    P A L P E, D G G S G, A F P P G, H F K D P, K R L Y C, K N G G F, F L R I L, P D G . . . . .


ADRENAL (a) BASIC FGF  P A L P E, D G G S G, A F P P . . . . .

ADRENAL (b) BASIC FGF                           H F K D P, K R L Y X, K N G G . . . .


CORPUS LUTEUM, BASIC FGF                         H F K D P, K R L Y X, K N G G X, F L .. . . . .
```

Fig. 5. Amino-terminal sequences of bovine FGFs from various tissues. X, Residue not identified; amino acids are abbreviated by one-letter symbols: D, Asp; N, Asn; S, Ser; E, Glu; P, Pro; G, Gly; A, Ala; C, Cys; L, Leu; Y, Tyr; F, Phe; H, His; K, Lys; R, Arg.

acid replacement and concomitant homology.[31] Since acidic FGF is highly homologous to basic FGF and since both bind to the same receptor, acidic FGF can be considered a weak agonist of basic FGF, based on its 30- to 100-fold lower potency.

Due to the wide distribution of basic FGF in various tissue and due to the wide range of cell types on which both basic and acidic FGF are acting, numerous pseudonyms exist for those two closely related factors. Cartilage-derived growth factor, chondrosarcoma-derived growth factor, hepatoma growth factor, endothelial cell growth factor II, eye-derived growth factor I, astroglial growth factor 2, human prostatic growth factor, and heparin binding growth factor β, although they have not yet been characterized structurally, are likely to be identical to basic FGF. Endothelial cell growth factor, retina-derived growth factor, eye-derived growth factor II, heparin binding growth factor α, astroglial growth factor 1, and anionic endothelial cell growth factor I, also structurally uncharacterized, are likely to be identical to acidic FGF.

Note added in proof. For more recent developments in the biology of FGF the following review can be consulted: D. Gospodarowicz, N. Ferrara, L. Schweigerer, and G. Neufeld, *Endocrine Rev.* **8,** 1 (1987).

---

[31] F. Esch, N. Ueno, A. Baird, F. Hill, L. Denoroy, N. Ling, D. Gospodarowicz, and R. Guillemin, *Biochem. Biophys. Res. Commun.* **133,** 554 (1985).

[10] Purification and Characterization of Acidic Fibroblast Growth Factor

By KENNETH A. THOMAS

Introduction

Whole brain, acting principally through its "transducer" organ the pituitary, has long been recognized to control the levels of many hormones. Less well recognized is that brain itself is a potent source of hormone-like growth factors. Mitogenic activity for fibroblasts was identified in brain extracts nearly 50 years ago.[1,2] More recently, whole brain has been observed to be a plentiful source of mitogenic activity *in vitro* for a wide variety of types of cells.[3] After approximately a 1000-fold purification, one such brain-derived mitogen, fibroblast growth factor,[4] was proposed to be a family of proteolytic degradation fragments of myelin basic protein,[5] an abundant constituent of the myelin sheath surrounding many neurons of the central and peripheral nervous systems. We refuted this claim based on (1) the retention of most of the protein but not the active mitogen(s) on an antimyelin basic protein immunoaffinity column and (2) the lack of mitogenic activity of proteolytic fragments of myelin basic protein that closely resembled those proposed to be active.[6] Therefore, although the principal polypeptides in this preparation were confirmed to be fragments of myelin basic protein, the active mitogens were still minor contaminants. Furthermore, both acidic[6] and basic mitogens[7] for BALB/c 3T3 fibroblasts were identified by isoelectric focusing of these partially purified samples.

The purification and characterization[8] of the brain acidic fibroblast

[1] O. A. Trowell, B. Chir, and E. N. Willmer, *J. Exp. Biol.* **16,** 60 (1939).

[2] R. S. Hoffman, *Growth* **4,** 361 (1940).

[3] D. Gospodarowicz and J. S. Moran, *Annu. Rev. Biochem.* **45,** 531 (1976).

[4] D. Gospodarowicz, J. S. Moran, and H. Bialecki, *in* "Third International Symposium on Growth Hormone and Related Peptides" (A. Pecile and E. Muller, eds.), p. 141. Excerpta Medica, New York, 1976.

[5] F. C. Westall, V. A. Lennon, and D. Gospodarowicz, *Proc. Natl. Acad. Sci. U.S.A.* **75,** 4675 (1978).

[6] K. A. Thomas, M. C. Riley, S. K. Lemmon, N. C. Baglan, and R. A. Bradshaw, *J. Biol. Chem.* **255,** 5517 (1980).

[7] S. K. Lemmon, M. C. Riley, K. A. Thomas, G. A. Hoover, T. Maciag, and R. A. Bradshaw, *J. Cell Biol.* **95,** 162 (1982).

[8] K. A. Thomas, M. Rios-Candelore, and S. Fitzpatrick, *Proc. Natl. Acad. Sci. U.S.A.* **81,** 357 (1984).

growth factor (aFGF) is reviewed and discussed. The pure growth factor is mitogenic for a variety of types of normal cells in culture including fibroblasts, glial cells, osteoblasts, and vascular endothelial cells and induces blood vessel growth *in vivo*.[9] Complete amino acid sequence determinations of both this pure acidic[10-12] and a basic[13] mitogen from brain have confirmed that each is unique and that neither molecule is derived from myelin basic protein. Pure aFGF has been shown to have amino acid sequence homology with human interleukin-1s (IL-1s)[9,10,14] thereby establishing a new family of homologous growth factors.

Method of Assay

The cell line used to monitor purification of aFGF is the BALB/c 3T3 fibroblast line (clone A31 from American Type Culture Collection, Rockville, MD). If stock cultures are passaged just prior to reaching confluence then they can be used indefinitely for assays. Mitogenic activity is monitored by measuring the incorporation of [*methyl*-³H]thymidine into the DNA of subconfluent cell cultures in the presence of dexamethasone and low concentrations of serum. The stock culture standard "growth media" is composed of Dulbecco's modified Eagle's medium (DMEM) containing 4.5 g/liter glucose, 10% (v/v) heat-inactivated bovine calf serum, 10^5 units of penicillin, 100 mg of streptomycin, and 100 mg of glutamine (Gibco) per liter. Stock cultures are trypsinized, resuspended in growth media, and plated into 35-mm-diameter wells (6-well assay dishes, Costar) at 2×10^4 cells in 2 ml/well. We have increased the plating volume from 1 to 2 ml/well because the miniscus formed by the 1 ml solution caused more cells to settle around the edges than the center of the 35-mm assay wells. The assay plates are placed at 37° in a humidified tissue culture incubator with the percentage CO_2 calibrated to give a pH of from about 7.3 to 7.4 in the low concentration of calf serum used to later assay the mitogen. A setting of 7–10% CO_2 usually gives a pH in the desired range. If the incubator is maintained at less than 100% relative humidity then evaporation can take

[9] K. A. Thomas, M. Rios-Candelore, G. Gimenez-Gallago, J. DiSalvo, C. Bennett, J. Rodkey, and S. Fitzpatrick, *Proc. Natl. Acad. Sci. U.S.A.* **82,** 6409 (1985).

[10] G. Gimenez-Gallego, J. Rodkey, C. Bennett, M. Rios-Candelore, J. DiSalvo, and K. Thomas, *Science* **230,** 1385 (1985).

[11] F. Esch, N. Ueno, A. Baird, F. Hill, L. Denoroy, N. Ling, D. Gospodarowicz, and R. Guillemin, *Biochem. Biophys. Res. Commun.* **133,** 554 (1985).

[12] D. J. Strydom, J. W. Harper, and R. R. Lobb, *Biochemistry* **25,** 945 (1986).

[13] F. Esch, A. Baird, N. Ling, N. Ueno, F. Hill, L. Denoroy, R. Klepper, D. Gospodarowicz, P. Bohlen, and R. Guillemin, *Proc. Natl. Acad. Sci. U.S.A.* **82,** 6507 (1985).

[14] K. A. Thomas and G. Gimenez-Gallego, *Trends Biochem. Sci.* **11,** 81 (1986).

place during the week-long assay resulting in a change in osmolality of the media and altered cell proliferation. The relative humidity can vary throughout an incubator leading to various degrees of evaporation in different assay wells as determined by measurement of the liquid volumes in the wells at the end of the assay. We routinely change the positions and orientations of the assay plates each time media changes or additions are made thereby minimizing the effect of the variable humidity within the incubator.

About 6 to 8 hr after plating the cells are attached and beginning to spread out. The DMEM medium (containing all of the above nonserum supplements) is changed to 1 ml/well of a low maintenance level of serum that allows almost all of the cells to survive the week-long assay but does not cause appreciable DNA synthesis. The amount of serum required to attain a stable but quiescent state varies from lot to lot but is typically 0.3 to 0.5% by volume. The low serum medium is changed once more about 18 hr later.

The protein samples are diluted into a solution of 1 mg of bovine serum albumin (BSA, Sigma, A7511) per ml of DMEM in polypropylene tubes, filter sterilized through a 0.22-μm Millex syringe filter (Millipore), and serially diluted into filter sterilized BSA/DMEM. Following addition of 20 μl of a stock solution of 56 μg of dexamethasone (Sigma) per ml of 25% ethanol (the diluted ethanol solution is filtered prior to the addition of dexamethasone) to each assay well the samples are added in 100-μl aliquots. Control solutions of either 100 μl of BSA/DMEM or 100% serum are also included in each assay.

Between 12 and 16 hr after sample addition, a filter sterilized solution of 2 μCi of [$methyl$-^{3}H]thymidine (1 mCi/ml, 20.0 Ci/mmol, New England Nuclear) and 3 μg of unlabeled thymidine (Sigma) in 50 μl of DMEM is added to each well. After 25 hr of radiolabel incorporation the cells are rinsed twice with 1 ml of 1 mg BSA per ml of 0.05 M HEPES buffered (pH adjusted to 7.3 at 37°) Hanks' balanced salt solution (Gibco). To each well is added 1 ml of ice cold 5% trichloroacetic acid (TCA) (w/v) and the 6-well plates are floated in an ice bath for 30 min. After 2 additional washes with cold 5% TCA a 1 ml solution of 2% Na_2CO_3 (w/v) in 0.1 M NaOH is added to each well and the assay plates placed at 37° for 1 hr. From each well 750 μl is transferred to a scintillation vial containing enough 50% TCA (w/v) to neutralize the basic sample solution. The volume of 50% TCA required is calibrated before each assay and is typically found to range from 70 to 100 μl.

Protein samples are assayed as a function of dose from no, to optimal, stimulation over a minimum of 4 log orders of concentration. Each dose of a given sample is assayed in triplicate and the response averaged. One unit of activity is defined as the amount of protein required to elicit a half-maximal mitogenic response. The optimal stimulation induced by the mitogenic samples is typically 50 to 100% that of 10% serum, varying to some degree with the lot of serum. Nearly full inhibition of DNA synthesis by BALB/c 3T3 cells can be observed at concentrations of aFGF that are 100- to 1000-fold greater than is required for optimal stimulation of mitogenesis. Inhibition is correlated with the formation of long processes associated with cell chemokinesis as established by time-lapse cinematography and chemotaxis monitored by checkerboard analysis using Boyden chambers. As a result of the high concentration inhibition, single concentration assays can give a misleading measure of activity resulting from the ambiguity of which side of the dose–response peak the readings lie.

Purification

The initial steps in the purification of aFGF are based on the protocol for partial purification of brain fibroblast growth factor described by Gospodarowicz et al.[4] with minor modifications.[8] To achieve homogeneous aFGF, over a 100-fold purification incorporating isoelectric focusing and HPLC reversed-phase chromatography is required.

Step 1. Extraction and Salt Precipitation. Fresh bovine brains (about 1/3 kg each) are obtained from a local slaughter house and transported to the laboratory on ice. Both visible blood clots and the outer meningeal membranes are removed, the brains diced into cubes of about 2 cm on a side, quick frozen in liquid nitrogen, and stored at $-70°$. All subsequent steps are performed at $4°$ unless otherwise noted. Distilled water is used to make all solutions and the pHs are adjusted against pH standards at the same temperature.

Four kilograms of diced brains is thawed in 0.15 M $(NH_4)_2SO_4$, initially at about $20°$, and homogenized in a 6-liter model CB-6 Waring blender. The pH of the viscous homogenate is adjusted to 4.5 with 6 M HCl while stirring vigorously with a motor-driven 6-in.-diameter propeller stirrer. After 1 hr the homogenate is centrifuged at 13,800 g for 40 min. The supernatant is adjusted to pH 6.75 with 1 M NaOH, 200 g of $(NH_4)_2SO_4$ slowly added per liter (1.52 M), and the mixture centrifuged at 13,800 g for 30 min. An additional 250 g of $(NH_4)_2SO_4$ is added per liter of supernatant (3.41 M final concentration), the mixture is centrifuged as before for 30 min, the resulting pellet dissolved in 200 ml of water, dialyzed for 18 hr in M_r 6000–8000 cutoff bags (Spectrum Medical Industries, Los Angeles) against two 14-liter volumes of water, and lyophilized.

Step 2. CM-Sephadex Chromatography. Lyophilized protein, about 72 g, from the salt precipitate of 16 kg of brains is dissolved in 900 ml of

0.05 M sodium phosphate, pH 6.0, and the mixture readjusted to pH 6.0 with 1 M NaOH prior to clarification by centrifugation at 23,300 g for 30 min. The supernatant is stirred for 15 min with 800 ml of hydrated CM-Sephadex C-50 (40 g dry weight, Pharmacia) preequilibrated with 0.1 M phosphate buffer, the unadsorbed protein sucked out on a coarse sintered glass filter, the resin washed with 3 liters of 0.1 M buffer, packed into an 8.3-cm-diameter column, and sequentially eluted at 30 ml/min with 0.1 M buffer containing 0.15 and 0.6 M NaCl collecting 22.5 ml fractions. The protein peak fractions eluted by the high salt buffer are pooled as shown in Fig. 1, dialyzed in M_r 6000–8000 cutoff bags for 18 hr against two 14-liter volumes of water, and lyophilized.

Step 3. Sephadex G-75 Chromatography. One quarter of the lyophi-

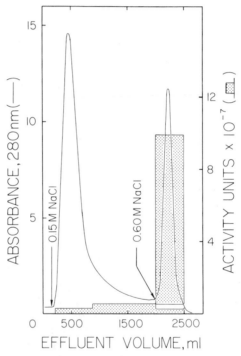

Fig. 1. Chromatography on CM-Sephadex C-50. Lyophilized protein (72 g) from the dialyzed product of 3.41 M (NH$_4$)$_2$SO$_4$-precipitated crude extract from 16 kg of bovine brains is batch adsorbed to 40 g of Sephadex C-50 equilibrated with 0.1 M sodium phosphate (pH 6.0). The resin is rinsed with the same buffer, packed into an 8.3-cm-diameter column, washed with 0.1 M sodium phosphate, pH 6.0, 0.15 M NaCl, and eluted with buffer containing 0.6 M NaCl. The flow rate is 30 ml/min and 22.5 ml fractions are collected. The material pooled for subsequent purification is indicated by the open horizontal bar. Taken from Thomas *et al.*[8]

lized protein, eluted by 0.6 M NaCl from the CM-50 column (about 590 mg), is dissolved in 20 ml of 0.1 M ammonium bicarbonate (pH 8.5) and clarified by centrifugation for 15 min at 27,000 g. The sample is loaded on a 5 × 90 cm Sephadex G-75 (particle size, 40–120 μm, Pharmacia) column equilibrated in the same buffer and eluted at 74 ml/hr collecting 17.5 ml fractions. Most of the mitogenic activity elutes in the peak from about 885 to 1075 ml as shown in Fig. 2. This region is pooled and directly lyophilized.

Step 4. CM-Cellulose Chromatography. Protein from the Sephadex G-75 column (about 210 mg) is dissolved in 10 ml of 0.1 M ammonium formate (pH 6.0), the pH readjusted to 6.0 with 0.1 M formic acid, the solution clarified by centrifugation at 27,000 g for 15 min, and the supernatant is loaded on a 1.5 × 6.5 cm CM-cellulose 52 (Whatman) column preequilibrated with 0.2 M ammonium formate pH 6.0 buffer. The resin is washed with 0.2 M, and eluted with 0.6 M, ammonium formate, pH 6.0, at a flow rate of 60 ml/hr collecting 4.25 ml fractions as shown in Fig. 3. The active pool eluted by 0.6 M salt is lyophilized directly.

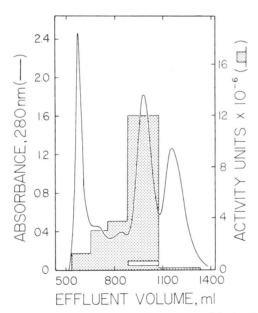

EFFLUENT VOLUME, ml

Fɪɢ. 2. Chromatography on Sephadex G-75. One-quarter of the lyophilized protein (590 mg) in the dialyzed 0.6 M NaCl-eluted pool from the CM-Sephadex G-50 column is loaded on a 5 × 90 cm Sephadex G-75 (particle size, 40–120 μm) column equilibrated with the same buffer. The column is eluted at a flow rate of 74 ml/hr and 17.5 ml fractions are collected. The material pooled for subsequent purification is indicated by the open bar. Taken from Thomas *et al.*[8]

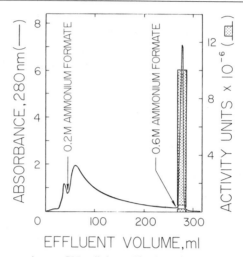

FIG. 3. Chromatography on CM-cellulose. The lyophilized protein (210 mg) from the highest activity pool (885 to 1075 ml elution volume) of the G-75 column is loaded on a 1.5 × 6.5 cm CM-52 column equilibrated with 0.2 M ammonium formate (pH 6.0). The resin is washed with 0.2 M, and eluted with 0.6 M, ammonium formate (pH 6.0). The column is eluted at a flow rate of 60 ml/hr and 4.25 ml fractions are collected. The material pooled for subsequent purification is indicated by the open horizontal bar. Taken from Thomas et al.[8]

Step 5. Isoelectric Focusing. Samples are focused in a miniaturized LKB Multiphor flatbed isoelectric focusing apparatus to optimize recoveries of μg amounts of protein. A 0.5 × 10 cm focusing lane is formed on a 12.5 × 26 cm glass plate out of General Electric clear silicone (GE 361) set around a plastic form wrapped in Parafilm. After the silicone sets overnight the form is removed and the lane can be used repeatedly. Ultrodex (75 mg, LKB) is hydrated in 1.9 ml of water containing 126 μl of pH 3–10 Pharmalyte (Pharmacia) and 47 μl of pH 9–11 Ampholine (LKB). The suspension is poured into the form, placed on the LKB Multiphor cooling plate set to 37°, containing cellulose paper wicks at each end hydrated with either 1 M NaOH (cathode) or 1 M H_3PO_4 (anode), and evaporated 32% by weight under a gentle stream of air from an overhead fan. After the temperature of the cooling block is lowered to 7° the sample is dissolved in 100 μl of the diluted ampholyte solution and loaded by spotting over the length of the resin. The protein (about 1.25 mg) is focused at 7° for 2800 V × hr at a maximum of 0.12 W for 24 hr under a N_2 atmosphere. After the proteins have reached equilibrium, as monitored by focusing of colored standard proteins in a control run, the resin is divided into 10 1-cm slices and each segment is eluted by 3 5-min centrifugations with 333 μl of 0.6 M NaCl at 800 g in a MF-1 microfiltration tube containing a 1 μm pore size regenerated cellulose RC-60 filter (Bioanalytic Systems, West Lafayette, IN). The pH of each sample is measured against pH standards at 0–5°. As shown in Fig. 4, two peaks of activity are observed, one in the basic and one in the acidic region. The basic mitogen is still heavily contaminated with myelin basic protein fragments but the acidic mitogen is very nearly pure. The acidic mitogen focuses from approximately pH 5 to nearly 7, varying somewhat from preparation to preparation. The acidic pools are injected on the C_4 HPLC column as soon as possible, usually within 1–2 hr after elution from the focusing resin. Recently, the size of the minifocusing lane has been increased to 0.8 × 24.2 cm and 10 mg of protein loaded onto the resin and focused using only the pH 3–10 ampholytes. In the larger lane, the protein is typically focused for 7,000 to 10,000 V × hr at a maximum of about 0.6 W.

Step 6. Reversed-Phase HPLC. After isoelectric focusing, the protein is usually greater than 90% pure but loses most of its mitogenic activity within about a day if stored in the ampholyte solution at 4°. Furthermore, the ampholytes interfere with characterization by polyacrylamide gel electrophoresis, amino acid analysis, or sequence determinations. At-

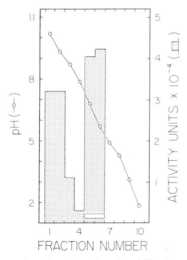

FIG. 4. Flatbed isoelectric focusing. A 1.25 mg lyophilized sample from the 0.6 M ammonium formate-eluted pool of the CM-52 column is loaded dropwise over the length of a 0.5 × 10 cm flatbed miniaturized focusing lane. The sample is focused at a maximum of 0.12 W for 2800 V × hr over a 24-hr period at 7° in a N_2 atmosphere. After focusing, the resin is sliced into 1 cm segments and these are eluted with 0.6 M NaCl. The acidic fractions typically pooled for subsequent purification are indicated by the open horizontal bar. Taken from Thomas *et al.*[8]

tempts to quantitatively remove the ampholytes by either dialysis, electrodialysis, gel filtration, or ion-exchange chromatography were unsuccessful. Poor recoveries of protein were observed on chromatography in 330 Å pore size, C_8 reversed-phase HPLC columns. To increase protein recovery less hydrophobic C_6 and C_4 custom HPLC columns were made. The C_4 column gave excellent resolution leading to complete purification and a 90% or greater recovery of activity with no binding of A_{210}-absorbing compounds from samples or ampholytes. The manufacturer (The Separations Group, Hesperia, CA) has subsequently marketed the C_4 column for protein purification.

This final HPLC purification step is performed on a 4.6 mm × 5 cm Vydac C_4 reversed-phase column (particle size, 5 μm; pore size, 330 Å) equilibrated in a 10 mM trifluoroacetic acid solution that is prepurified over a C_{18} preparative reversed-phase HPLC column to remove UV-absorbing contaminants. Elution solvents are vacuum degassed and stored under argon during use. The pure mitogen is eluted at 0.5 ml/min

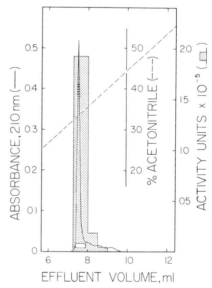

FIG. 5. Chromatography by reversed-phase C_4 HPLC. The acidic pool (about pH 5–7) from the isoelectric focusing of 4 mg of partially purified aFGF from the CM-52 column is loaded on a 4.6 mm × 5 cm C_4 Vydac HPLC column (particle size, 5 μm; pore size 330 Å) equilibrated with 10 mM trifluoroacetic acid and eluted with a 0–67% linear gradient of acetonitrile over 30 min at 20–22°. The flow rate is 0.5 ml/min and either the protein peak is collected manually or 0.25 ml fractions are collected in polypropylene tubes. The material pooled for subsequent characterization is indicated by the open horizontal bar. Taken from Thomas et al.[8]

with a 0–67% linear gradient of acetonitrile over 30 min at 20–22°. Nonprotein contaminants, including the ampholytes, elute in either the injection breakthrough or the early portion of the acetonitrile gradient. The peak of protein, shown in Fig. 5, is manually collected into a polypropylene tube, covered with argon, capped, and stored as a liquid (the acetonitrile prevents freezing) at −20°. Under these conditions the mitogenic activity is stable. We experience a 50–90% loss in mitogenic activity after lyophilization.

Properties

Purity. As shown in Table I, the growth factor is purified a minimum of 35,000-fold by the above protocol. The actual degree of purification very likely exceeds this value since the total mitogenic activity for BALB/c 3T3 cells in the crude brain homogenate probably contains other mitogens. If, in the other extreme, aFGF is recovered in 100% yield then it is purified about 850,000-fold. The two additional steps of isoelectric focusing and reversed-phase HPLC lead to an increased specific activity of 135-fold. Since aFGF accounts for only about one-half of the mitogenic activity observed on isoelectric focusing, the actual increase in level of purity achieved by these two steps is probably about 270-fold. Although we have

TABLE I
PURIFICATION OF aFGF[a]

Purification step	Protein recovery (mg)	Activity recovery		Specific activity (units/mg)	Purification factor
		Units	%		
Brain homogenization	1.1×10^{5b}	7.9×10^7	100	7.2×10^2	1.0
Salt fractionation	1.8×10^{4b}	5.2×10^7	66	2.9×10^3	4.0
Chromatography					
Sephadex C-50	5.9×10^{2c}	2.5×10^7	32	4.2×10^4	58.0
Sephadex G-75	2.1×10^{2c}	1.2×10^7	15	5.7×10^4	79.0
CM-52	54^d	1.0×10^7	13	1.85×10^5	2.6×10^2
Isoelectric focusing	NQ^e	3.6×10^6	4.6	—	—
C$_4$ HPLC	0.144^f	3.2×10^6	4.1	2.5×10^7	3.5×10^4

[a] Values are based on 4 kg of bovine brain. Except for the value for the homogenate, purification factors are minimum values estimated by assuming that all of the initial mitogenic activity is aFGF. Taken from Thomas *et al.*[8]
[b] Protein was estimated by the A_{260}/A_{280} ratio method.
[c] Protein was estimated by using $A_{280}^{1\%} = 10$.
[d] Protein was estimated by using $A_{280}^{1\%} = 7.9$.
[e] Not quantitated.
[f] Protein was quantitated by amino acid analysis.

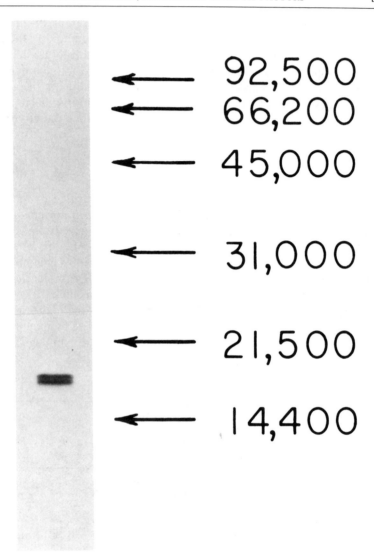

FIG. 6. SDS–polyacrylamide gel electrophoresis of reduced aFGF. About 100 ng of HPLC-purified aFGF is heat denatured and reduced with sodium dodecyl sulfate/2-mercaptoethanol, electrophoresed at 10 mA through a 0.75-mm-thick 15% polyacrylamide gel with a 4.5% stacking gel, and silver stained. The protein molecular weight standards are phosphorylase *b* (92,500), bovine serum albumin (66,200), ovalbumin (45,000), carbonate dehydratase (31,000), soybean trypsin inhibitor (21,500), and lysozyme (14,400). The lower limit of detection is about 1 ng of protein. Taken from Thomas *et al.*[8]

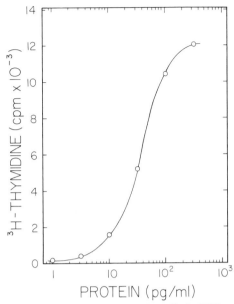

PROTEIN (pg/ml)

FIG. 7. DNA synthesis dose–response assay of purified aFGF on BALB/c 3T3 cells. Incorporation of [³H]thymidine into trichloroacetic acid-insoluble DNA of quiescent BALB/c 3T3 cells is measured as a function of the concentration of HPLC-purified aFGF. Background values with no sample added are 125 cpm. Full stimulation by 10% heat-inactivated calf serum is 13,750 cpm. Taken from Thomas et al.[8]

not characterized the basic mitogen(s), it may be similar or identical to the pituitary basic FGF described by others.[15,16]

The principal active peak that elutes from the HPLC column is very sharp. Electrophoresis of 100 ng of reduced (or nonreduced) protein in a sodium dodecyl sulfate–polyacrylamide slab gel followed by high sensitivity silver staining reveals two very close bands centered at about 16,000 Da (Fig. 6). This system is able to detect about 1 ng of protein per band, so the mitogen is very pure. The two bands have been determined to be microheterogeneous forms of the same protein (see below).

The absorbance value of a 10 mg/ml solution of pure protein in the HPLC elution solvent, as calculated from the amino acid composition of a solution of measured absorbance, is $A_{210\,nm}^{1\%,\,1\,cm} = 250$. The absorbances 1% solutions at 230 and 280 nm are 71 and 13, respectively.

Activity. The pure protein is a very potent mitogen for cells of the BALB/c 3T3 line. As shown in Fig. 7, typical half-maximal stimulation of

[15] S. K. Lemmon and R. A. Bradshaw, *J. Cell. Biochem.* **21**, 195 (1983).
[16] P. Bohlen, A. Baird, F. Esch, N. Ling, and D. Gospodarowicz, *Proc. Natl. Acad. Sci. U.S.A.* **81**, 5364 (1984).

incorporation of [³H]thymidine into DNA is about 40 pg/ml (2.4 pM) with a range from 10 to 100 pg/ml, depending on the particular lot of serum. The pure protein is also an excellent mitogen for both large vessel and microvascular endothelial cells in culture with half-maximal stimulation of [³H]thymidine incorporation at less than 1 ng/ml.[9] This activity appears to be enhanced to various amounts with different cultures of vascular endothelial cells by heparin,[17,18] a polysaccharide to which the mitogen has been shown to avidly bind *in vitro,* the basis for a recently described alternative purification protocol.[19] In the presence of heparin, pure aFGF causes blood vessels to grow *in vivo* in the chick egg chorioallantoic membrane angiogenesis assay.[9,20]

Structure. The complete amino acid sequences of the 140-residue bovine[10-12] and human[21] aFGFs are shown in Fig. 8. Two microheterogeneous forms of the bovine protein are present, one 6-residues longer at the N-terminus than the other. Although the relative amounts of the two electrophoretic bands vary among different preparations, the percentage of the higher mass band correlates well with the fraction of the molecules having the full-length amino-terminus. Two forms of human aFGF have also been found that either have, or are missing, the amino-terminal Phe residue. The microheterogeneity of the aFGFs, common among growth factors, could be generated either *in vivo* or during purification.

In Fig. 9 the complete amino acid sequences of bovine acidic and basic FGFs are aligned with human interleukins-1α and -1β. Although the IL-1s are lymphokines defined by their thymocyte mitogenic activity, IL-1β[22] and probably IL-1α are also mitogens for fibroblasts. Acidic and basic FGFs are very homologous having 55% amino acid sequence identity as aligned. A distant sequence homology is also seen between the FGFs and interleukin-1s. Based on a hydropathicity analysis, the FGFs and IL-1s are related to each other with a significance greater than 6 standard devia-

[17] A. B. Schreiber, J. Kenney, J. Kowalski, K. A. Thomas, G. Gimenez-Gallego, M. Rios-Candelore, J. DiSalvo, D. Barritault, J. Courty, Y. Courtois, M. Moenner, C. Loret, W. H. Burgess, T. Mehlman, R. Friesel, W. Johnson, and T. Maciag, *J. Cell Biol.* **101,** 1623 (1985).

[18] G. Gimenez-Gallego, G. Conn, V. B. Hatcher, and K. A. Thomas, *Biochem. Biophys. Res. Commun.* **135,** 541 (1986).

[19] R. R. Lobb and J. W. Fett, *Biochemistry* **23,** 6295 (1984).

[20] R. R. Lobb, E. M. Alderman, and J. W. Fett, *Biochemistry* **24,** 4969 (1985).

[21] G. Gimenez-Gallego, G. Conn, V. B. Hatcher, and K. A. Thomas, *Biochem. Biophys. Res. Commun.* **138,** 611 (1986).

[22] J. Schmidt, G. Limjuco, P. Cameron, J. Chin, E. Bayne, E. Rupp, S. Galuska, C. Bennett, J. Rodkey, J. Shapiro, J. Boger, and M. Tocci, *in* "The Physiological, Metabolic, and Immunologic Actions of Interleukin-1" (M. J. Kluger, J. J. Oppenheim, and M. C. Powanda, eds.), pp. 501–509. Liss, New York, 1985.

```
                  1                                            10
HaFGF:  PHE-ASN-LEU-PRO-PRO-GLY-ASN-TYR-LYS-LYS-PRO-LYS-LEU-LEU-TYR-
BaFGF:                                  -LEU-
                  20                                           30
        CYS-SER-ASN-GLY-GLY-HIS-PHE-LEU-ARG-ILE-LEU-PRO-ASP-GLY-THR-
                                   -TYR-
                  40
        VAL-ASP-GLY-THR-ARG-ASP-ARG-SER-ASP-GLN-HIS-ILE-GLN-LEU-GLN-
                                                     -LYS-
                  50                                          60
        LEU-SER-ALA-GLU-SER-VAL-GLY-GLU-VAL-TYR-ILE-LYS-SER-THR-GLU-
             -CYS-                  -ILE-
                                                     70
        THR-GLY-GLN-TYR-LEU-ALA-MET-ASP-THR-ASP-GLY-LEU-LEU-TYR-GLY-
                                    -PHE-
                  80                                          90
        SER-GLN-THR-PRO-ASN-GLU-GLU-CYS-LEU-PHE-LEU-GLU-ARG-LEU-GLU-
                                      100
        GLU-ASN-HIS-TYR-ASN-THR-TYR-ILE-SER-LYS-LYS-HIS-ALA-GLU-LYS-
                  110                                        120
        ASN-TRP-PHE-VAL-GLY-LEU-LYS-LYS-ASN-GLY-SER-CYS-LYS-ARG-GLY-
        HIS-                                   -ARG-SER-   -LEU-
                                   130
        PRO-ARG-THR-HIS-TYR-GLY-GLN-LYS-ALA-ILE-LEU-PHE-LEU-PRO-LEU-
                                   -PHE-
                  140
        PRO-VAL-SER-SER-ASP
```

FIG. 8. Complete amino acid sequences of human and bovine aFGFs. The human aFGF (HaFGF) sequence is written in full. Only the differences in the bovine (BaFGF) sequence are listed below the corresponding residues of the human sequence. Taken from Gimenez-Gallego et al.[21]

tions from random. The sequence similarities among these growth factors and lymphokines define a new family of homologous proteins. Moreover, the FGFs and IL-1s appear to have diverged from a common ancestral protein, followed by a split between the IL-1s and finally by the emergence of the two FGFs.[10,14]

Either aFGF or very similar molecules have been purified in different laboratories[17,19,23–27] and given names based on tissue source, heparin-binding capacity, or target cell as listed in Table II. These proteins have been proven, or are very likely, to be aFGF as assessed by mass, p*I*,

[23] P. A. D'Amore and M. Klagsbrun, *J. Cell Biol.* **99**, 1545 (1984).
[24] A. Baird, F. Esch, D. Gospodarowicz, and R. Guillemin, *Biochemistry* **24**, 7855 (1985).
[25] B. Pettmann, M. Weibel, M. Sensenbrenner, and G. Labourdette, *FEBS Lett.* **189**, 102 (1985).
[26] J. S. Huang, S. S. Huang, and M.-D. Kuo, *J. Biol. Chem.* **261**, 11600 (1986).
[27] J. W. Crabb, L. G. Armes, C. M. Johnson, and W. L. McKeehan, *Biochem. Biophys. Res. Commun.* **136**, 1155 (1986).

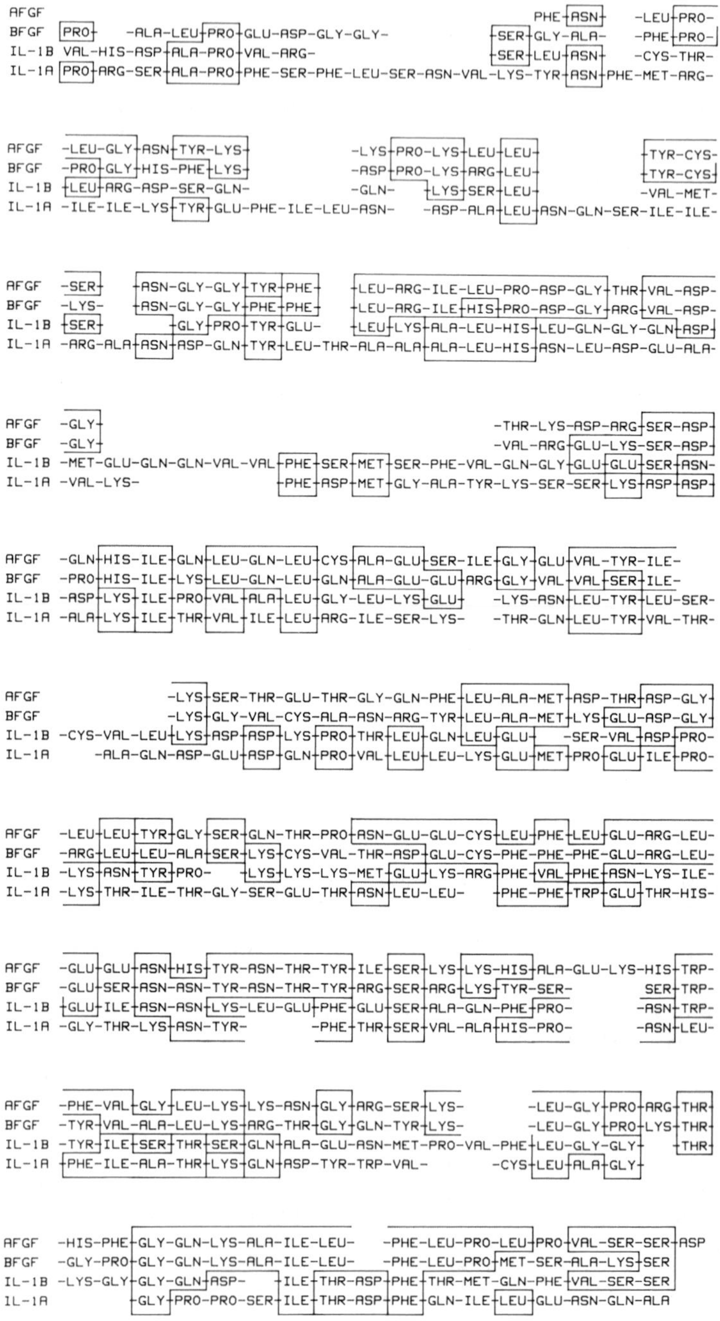

```
AFGF                                                      PHE-ASN-    -LEU-PRO-
BFGF   PRO-   -ALA-LEU-PRO-GLU-ASP-GLY-GLY-               SER-GLY-ALA-    -PHE-PRO-
IL-1B  VAL-HIS-ASP-ALA-PRO-VAL-ARG-                       SER-LEU-ASN-    -CYS-THR-
IL-1A  PRO-ARG-SER-ALA-PRO-PHE-SER-PHE-LEU-SER-ASN-VAL-LYS-TYR-ASN-PHE-MET-ARG-


AFGF   -LEU-GLY-ASN-TYR-LYS-           -LYS-PRO-LYS-LEU-LEU-       -TYR-CYS-
BFGF   -PRO-GLY-HIS-PHE-LYS-           -ASP-PRO-LYS-ARG-LEU-       -TYR-CYS-
IL-1B  LEU-ARG-ASP-SER-GLN-            -GLN-   LYS-SER-LEU-         -VAL-MET-
IL-1A  -ILE-ILE-LYS-TYR-GLU-PHE-ILE-LEU-ASN-  -ASP-ALA-LEU-ASN-GLN-SER-ILE-ILE-


AFGF   -SER-   ASN-GLY-GLY-TYR-PHE-    LEU-ARG-ILE-LEU-PRO-ASP-GLY-THR-VAL-ASP-
BFGF   -LYS-   ASN-GLY-GLY-PHE-PHE-    LEU-ARG-ILE-HIS-PRO-ASP-GLY-ARG-VAL-ASP-
IL-1B  SER-    GLY-PRO-TYR-GLU-        LEU-LYS-ALA-LEU-HIS-LEU-GLN-GLY-GLN-ASP-
IL-1A  -ARG-ALA-ASN-ASP-GLN-TYR-LEU-THR-ALA-ALA-ALA-LEU-HIS-ASN-LEU-ASP-GLU-ALA-


AFGF   -GLY-                           -THR-LYS-ASP-ARG-SER-ASP-
BFGF   -GLY-                           -VAL-ARG-GLU-LYS-SER-ASP-
IL-1B  -MET-GLU-GLN-GLN-VAL-VAL-PHE-SER-MET-SER-PHE-VAL-GLN-GLY-GLU-GLU-SER-ASN-
IL-1A  -VAL-LYS-                       PHE-ASP-MET-GLY-ALA-TYR-LYS-SER-SER-LYS-ASP-ASP


AFGF   -GLN-HIS-ILE-GLN-LEU-GLN-LEU-CYS-ALA-GLU-SER-ILE-GLY-GLU-VAL-TYR-ILE-
BFGF   -PRO-HIS-ILE-LYS-LEU-GLN-LEU-GLN-ALA-GLU-GLU-ARG-GLY-VAL-VAL-SER-ILE-
IL-1B  -ASP-LYS-ILE-PRO-VAL-ALA-LEU-GLY-LEU-LYS-GLU-   -LYS-ASN-LEU-TYR-LEU-SER-
IL-1A  -ALA-LYS-ILE-THR-VAL-ILE-LEU-ARG-ILE-SER-LYS-    -THR-GLN-LEU-TYR-VAL-THR-


AFGF            -LYS-SER-THR-GLU-THR-GLY-GLN-PHE-LEU-ALA-MET-ASP-THR-ASP-GLY-
BFGF            -LYS-GLY-VAL-CYS-ALA-ASN-ARG-TYR-LEU-ALA-MET-LYS-GLU-ASP-GLY-
IL-1B  -CYS-VAL-LEU-LYS-ASP-ASP-LYS-PRO-THR-LEU-GLN-LEU-GLU-   -SER-VAL-ASP-PRO-
IL-1A  -ALA-GLN-ASP-GLU-ASP-GLN-PRO-VAL-LEU-LEU-LYS-GLU-MET-PRO-GLU-ILE-PRO-


AFGF   -LEU-LEU-TYR-GLY-SER-GLN-THR-PRO-ASN-GLU-GLU-CYS-LEU-PHE-LEU-GLU-ARG-LEU-
BFGF   -ARG-LEU-LEU-ALA-SER-LYS-CYS-VAL-THR-ASP-GLU-CYS-PHE-PHE-PHE-GLU-ARG-LEU-
IL-1B  -LYS-ASN-TYR-PRO-    LYS-LYS-LYS-MET-GLU-LYS-ARG-PHE-VAL-PHE-ASN-LYS-ILE-
IL-1A  -LYS-THR-ILE-THR-GLY-SER-GLU-THR-ASN-LEU-LEU-    PHE-PHE-TRP-GLU-THR-HIS-


AFGF   -GLU-GLU-ASN-HIS-TYR-ASN-THR-TYR-ILE-SER-LYS-LYS-HIS-ALA-GLU-LYS-HIS-TRP-
BFGF   -GLU-SER-ASN-ASN-TYR-ASN-THR-TYR-ARG-SER-ARG-LYS-TYR-SER-       SER-TRP-
IL-1B  -GLU-ILE-ASN-ASN-LYS-LEU-GLU-PHE-GLU-SER-ALA-GLN-PHE-PRO-       -ASN-TRP-
IL-1A  -GLY-THR-LYS-ASN-TYR-    -PHE-THR-SER-VAL-ALA-HIS-PRO-          -ASN-LEU-


AFGF   -PHE-VAL-GLY-LEU-LYS-LYS-ASN-GLY-ARG-SER-LYS-       -LEU-GLY-PRO-ARG-THR-
BFGF   -TYR-VAL-ALA-LEU-LYS-ARG-THR-GLY-GLN-TYR-LYS-       -LEU-GLY-PRO-LYS-THR-
IL-1B  -TYR-ILE-SER-THR-SER-GLN-ALA-GLU-ASN-MET-PRO-VAL-PHE-LEU-GLY-GLY-   -THR-
IL-1A  -PHE-ILE-ALA-THR-LYS-GLN-ASP-TYR-TRP-VAL-       -CYS-LEU-ALA-GLY-


AFGF   -HIS-PHE-GLY-GLN-LYS-ALA-ILE-LEU-    -PHE-LEU-PRO-LEU-PRO-VAL-SER-SER-ASP
BFGF   -GLY-PRO-GLY-GLN-LYS-ALA-ILE-LEU-    -PHE-LEU-PRO-MET-SER-ALA-LYS-SER
IL-1B  -LYS-GLY-GLY-GLN-ASP-    ILE-THR-ASP-PHE-THR-MET-GLN-PHE-VAL-SER-SER
IL-1A          GLY-PRO-PRO-SER-ILE-THR-ASP-PHE-GLN-ILE-LEU-GLU-ASN-GLN-ALA
```

FIG. 9. See legend on p. 135.

TABLE II
GROWTH FACTORS THAT ARE EQUIVALENT TO aFGF

Growth factor	Abbreviation	Reference
Endothelial cell growth factor	ECGF	17
Heparin-binding growth factor α	HGF-α, HBGF-α	19
or class 1 heparin-binding growth factor		
Retinal-derived growth factor	RDGF	23
or α-retina-derived growth factor	αRDGF	24
Eye-derived growth factor II	EDGF II	17
Astroglial growth factor 1	AGF1	25
Brain-derived growth factor	BDGF	26
Prostatropin		27

heparin binding, antibody cross-reactivity, competitive receptor binding, amino acid composition, or amino acid sequence analysis. The recognition of identity among these growth factors should serve to facilitate the biological and physiological characterization of aFGF.

FIG. 9. Amino acid sequences of FGFs and homologies with interleukin-1s. The complete amino acid sequences of bovine brain-derived aFGF (AFGF) and pituitary-derived bFGF (BFGF) are aligned with the carboxyl-terminal halves of the precursors of both human IL-1β and IL-1α beginning at residues 114 and 110, respectively. The amino-terminus of IL-1β starts at Ala-117. Mature mouse IL-1α starts at a position equivalent to Ser-112 in the human protein. Common aligned residues are enclosed in boxes. Taken from Thomas and Gimenez-Gallego.[14]

[11] Assay of Capillary Endothelial Cell Migration

By BRUCE R. ZETTER

The growth and development of new blood vessels, referred to as angiogenesis or neovascularization, consists primarily of the movement and proliferation of a single cell type: the capillary endothelial cell.[1] New vascular outgrowth occurs not only in the developing embryo but also in wound healing, in immune reactions, in the inflammatory response, and in the maturation of the mammalian ovarian follicle.[2] Neovascularization is

[1] D. H. Ausprunk and J. Folkman, *Microvasc. Res.* **14**, 53 (1977).
[2] R. Auerbach, *Lymphokines* **4**, 69 (1981).

also a critical component of a number of pathological conditions including diabetes, atherosclerosis, arthritis, psoriasis, and neoplasia.[3]

Recent advances in the culture of capillary endothelial cells[4] have made it possible to assay directly for each of the cellular components of angiogenesis. These include (1) the secretion of hydrolytic enzymes necessary to degrade the basal lamina around the preexisting venules from which capillary sprouts arise,[5] (2) the migration of capillary endothelial cells,[6] (3) the subsequent proliferation of capillary endothelial cells,[7] and (4) the formation of a three-dimensional network of vascular tubes and loops necessary to transport plasma and cells.[8]

To date, the ability to stimulate the migration of capillary endothelial cells has been a property of every angiogenic factor studied.[6,9-11] Most but not all[10] of these factors have mitogenic activity also. Thus, while both cell migration and proliferation are normal components of the angiogenic process, only cell migration appears to be obligatory. This makes the *in vitro* assay of capillary endothelial cell migration an extremely useful indicator of angiogenic potential.

Assay

The most widely used method for the study of endothelial cell movements in response to angiogenic stimuli is the phagokinetic assay originally developed for the study of 3T3 cell motility by Albrecht-Buehler.[12] In this assay, cells are plated onto coverslips that are evenly coated with fine particles such as colloidal gold[12] or small plastic spheres.[13] The endothelial cells ingest the particles by phagocytosis. If the cells move, they leave clear tracks that reflect the path of their movements. In the original application of this technique, Albrecht-Buehler analyzed the pattern of

[3] J. Folkman, *Adv. Cancer Res.* **43**, 175 (1985).
[4] B. Zetter, *in* "Biology of Endothelial Cells" (E. A. Jaffe, ed.), p. 14. Nijhoff, Boston, 1984.
[5] D. Moscatelli, J. L. Gross, E. A. Jaffe, and D. B. Rifkin, *in* "Biology of Endothelial Cells" (E. A. Jaffe, ed.), p. 429. Nijhoff, Boston, 1984.
[6] B. R. Zetter, *Nature (London)* **285**, 41 (1985).
[7] J. Folkman, C. C. Haudenschild, and B. R. Zetter, *Proc. Natl. Acad. Sci. U.S.A.* **76**, 5217 (1979).
[8] J. Folkman and C. C. Haudenschild, *Nature (London)* **288**, 551 (1980).
[9] D. E. Mullins and D. B. Rifkin, *J. Cell. Physiol.* **119**, 247 (1984).
[10] M. J. Banda, D. R. Knighton, T. K. Hunt, and Z. Werb, *Proc. Natl. Acad. Sci. U.S.A.* **79**, 7773 (1982).
[11] B. R. McAuslan, W. G. Reilly, G. N. Hannan, and G. A. Gole, *Microvasc. Res.* **26**, 323 (1983).
[12] G. Albrect-Buehler, *Cell* **11**, 395 (1977).
[13] J. L. Obeso and R. Auerbach, *J. Immunol. Methods* **70**, 141 (1984).

directional movements of postmitotic cells. We and others[6,14] have modified this assay to yield quantitative analysis of cell motility by measuring the dimensions of the phagokinetic tracks using computerized image analysis.

Preparation of Protein-Coated Coverslips. Using a jewellers forceps, hold one corner of a 22 × 22 mm glass coverslip and immerse it in a solution of 1% bovine serum albumin (Boehringer-Mannheim, Indianapolis, IN). Blot any excess by pressing the bottom edge of the coverslip against a paper towel but do not allow it to dry completely. Quickly immerse the coverslip into 100% ethanol, blot as before, and allow to dry under a stream of warm air from a hot air dryer. Discard any coverslips that have dried unevenly and place the rest into individual 35 mm Falcon petri dishes.

It is possible to coat the dishes with molecules other than albumin. We have used other agents such as fibronectin, laminin, and polylysine. In using these agents, it is not necessary to dry the coverslips after coating. Instead, the coverslips are placed in 35 mm petri dishes containing 2 ml of a 20 μg/ml solution of any of the above reagents. After a minimum incubation of 30 min at room temperature, the coating solution is aspirated out of the dish and immediately replaced with a solution of colloidal gold or latex spheres before the coverslips have dried completely.

Preparation of a Colloidal Gold Solution

Reagents

14.5 × 10^{-3} M AuCl$_3$
36.5 × 10^{-3} M Na$_2$CO$_3$
0.1% HCHO

Mix the following reagents sequentially in a 250-ml Ehrlenmeyer flask with constant swirling: (1) 11 ml glass distilled water, (2) 1.8 ml gold chloride, and (3) 6.0 ml Na$_2$CO$_3$. At this step, the color of the solution will change from yellow to clear. Quickly place the flask on a preheated hot plate or over a bunsen burner. Five to ten seconds after the solution begins to boil, remove the flask from the heat and immediately pour in 1.8 ml of freshly diluted formaldehyde. The colloidal gold solution should be a transluscent blue-gold color. Discard and start again if the solution is dark purple or muddy brown. Pipet 2.0 ml of the colloidal gold solution over each precoated coverslip and leave at room temperature for 45 min. Then, use the jewellers forceps to transfer the coverslip to a fresh dish containing 2.0 ml of serum-free culture medium.

---

[14] B. R. McAuslan and W. Reilly, *Exp. Cell Res.* **130**, 147 (1980).

Preparation of Polystyrene Bead Monolayers. An alternative particulate coating of polystyrene beads can be prepared according to the method of Obeso and Auerbach.[13] In order to increase the number of samples that could be assayed at one time, these authors used Immulon II 96-well culture plates (Dynatech Laboratories, Alexandria, VA) that were precoated with fibrin, collagen, gelatin, or albumin. They then prepared a suspension of 1.4×10^8 beads (MX Covasphere beads, Covalent Technology Corp., Ann Arbor, MI) in cold serum-free Dulbecco's modified Eagle's medium (DMEM, Grand Island Biological Co., Grand Island, NY). The covasphere suspension was sonicated for 30 sec at 40 W using a W185 Sonifer cell disrupter (Heat System-Ultrasonics, Inc.); 200 μl of bead suspension was added to each well immediately after removal of the protein-coating solution from the wells. The plates were then centrifuged at 557 g for 15 min at 4° and subsequently incubated for at least 1 hr at 37° in an incubator aerated with 5% CO_2.

Experimental Protocols

Capillary endothelial cells can be cultured in gelatin-coated dishes with medium that has been preconditioned by incubation with tumor cells[7] or in medium supplemented with tumor-derived growth factors such as that recently purified by Shing *et al.*[15] (see also chapter by Klagsbrun *et al.* [8], this volume). Methods for the culture of capillary endothelial cells have been reviewed previously[4,16] and are not discussed in detail here.

Twenty-four to forty-eight hours before performing an experiment, incubate a confluent 25 cm^2 culture of capillary endothelial cells in nonconditioned medium containing 10% calf serum but without any added growth factors. Preincubation of cells in the absence of growth factors helps to lower the background motility of unstimulated endothelial cells.

Plating of Capillary Endothelial Cells. The cells are removed from their culture flasks by a 1–2 min incubation in 0.25% trypsin (Gibco). After the cells are in suspension, they are diluted 1 : 1 in medium containing 1.0% calf serum. Cell number is determined using a Coulter Zf particle counter (Coulter Electronics, Hialeah, FL). The cells are then centrifuged at 800 g in a table-top centrifuge and resuspended at a concentration of 1.5 $\times 10^3$ cells/ml in DMEM containing 1.0% calf serum. The cells are then taken up in a 5-ml syringe and expelled through the aperture of a 20-gauge needle in order to disperse any aggregates that may form during trypsini-

---

[15] Y. Shing, J. Folkman, R. Sullivan, C. Butterfield, J. Murray, and M. Klagsbrun, *Science* **223**, 1296 (1984).

[16] J. C. Voyta, D. P. Via, C. Butterfield, and B. R. Zetter, *J. Cell Biol.* **99**, 2034 (1984).

zation or centrifugation. This is extremely important for this assay because only tracks made by single cells can be adequately analyzed. Each coverslip-containing dish receives 2 ml of cell suspension (3×10^3 cells) and is then placed in a 37° incubator aerated with 10% CO_2.

After 90 min to allow attachment of the cells to the coverslips, the test samples are added. They should be added in no more than 200 μl of additional medium or buffer. The plates are then returned to the incubator for an additional 18 hr after which the experiment is terminated by addition of 2.5 ml phosphate-buffered 10% formalin (Fisher Scientific, Fair Lawn, NJ). Continuation of the experiment for longer periods of time can result in significant cell mitoses and a consequent increase of tracks with multiple cells.

Analysis of Cell Movement

Figure 1 shows representative phagokinetic tracks made by capillary endothelial cells incubated for 18 hr in the presence and absence of a tumor-derived migration-stimulating factor. Although the cells ingest gold particles under both conditions, the stimulated cells' tracks are significantly larger. In addition, the stimulated cells do not move in a straight line; their tracks are irregular in shape and vary in both length and width.

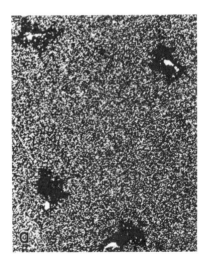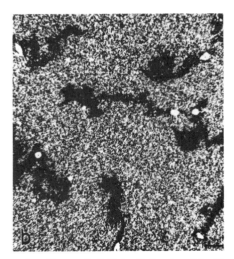

FIG. 1. Phagokinetic tracks of bovine adrenal capillary endothelial cells incubated for 18 hr in Dulbecco's modified Eagle's medium containing 10% calf serum (a) or in sarcoma-conditioned medium containing 10% calf serum (b). Bar = 10 μm. Reprinted with permission from Zetter et al.[17]

Two central questions emerge from these observations. (1) Are the tracks in the unstimulated cultures small because the cells are moving more slowly or could the cells be moving rapidly back and forth within a small area? (2) What is the best measurement to ascertain the size of phago-kinetic tracks made by cells moving in an irregular pattern?

To address the first question, we have performed time-lapse cinemato-graphic analysis of the stimulated and unstimulated cells. The films reveal that the unstimulated cells move slowly or not at all. Thus, for these cells, small tracks are indicative of a low rate of cell locomotion. The stimulated cells, on the other hand, move steadily throughout the 18-hr course of the assay. The path of the stimulated cells is extremely tortuous and irregular. The moving cells change direction frequently, sometimes traversing the same area twice. The films also show that after mitosis, daughter cells move away from each other. Cells that collide with each other also appear to move away from the site of the collision. These observations suggest that cell division and cell collisions can influence the rate of capillary endothelial cell motility independent of other migration stimuli. Conse-quently, it is essential to keep the length of the assay shorter than the generation time of the cells and to keep the initial cell density low in order to avoid unnecessary cell contacts.

Measurement of Phagokinetic Track Size. When viewed under inci-dent light, phagokinetic tracks appear black against the white background of the gold or plastic particles. This contrast enables digital image analysis equipment to easily distinguish and measure track size. We currently use a Bausch and Lomb Omnicon system although several other manufactur-ers make comparable equipment.

Because of the irregular nature of the tracks made by capillary endo-thelial cells, the degree of cell migration is not adequately represented by the end to end length of the tracks. The complete dimensions of the tracks are more accurately determined by measurement of track *area*. Thus, for each coverslip, we measure the areas of 64–100 tracks and calculate the mean for each experimental point.

To determine the validity of using track area as a measurement of cell motility, we have designed a computer program that simulates cell move-ment and track formation.[17] The parameters of cell movement that can be varied include cell speed (μm/hr), linearity of cell movement, and angle of directional change. Linearity of movement is defined as the distance

[17] B. R. Zetter, R. G. Azizkhan, J. C. Azizkhan, D. Brouty-Boye, J. Folkman, C. C. Haudenschild, M. Klagsbrun, R. Potash, and C. J. Scheiner, *in* "Plasma and Cellular Modulatory Proteins" (D. H. Bing and R. A. Rosenbaum, eds.), p. 59. Center for Blood Research Press, Boston, 1981.

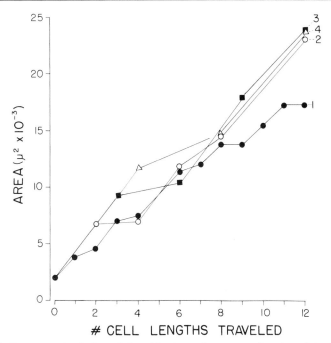

Fig. 2. Computer simulation of phagokinetic track area as a function of the directional linearity of cell movement. Tracks were constructed for computer-simulated capillary endothelial cells moving at constant speed and turning at random angles. The graph shows the increase in track area per distance traveled for individual cells. The numbers at the end of each line represent the linearity of movement (L) which is defined as the number of cell lengths traveled between each required change in direction. Reprinted with permission from Zetter et al.[17]

(measured in cell diameters) that a cell will move before it changes direction. The angle of directional change can be set as a constant, as a random distribution, or as a weighted set of possible angles such as that observed by Albrecht-Buehler for the movement of 3T3 cells.[18,19] The question we wished to ask was how different patterns of movement would be reflected in the areas of the phagokinetic tracks made as cells moved across a particle-coated surface.

In Fig. 2, we have plotted the increase in track area for 4 hypothetical cells moving at constant speed with the angle of directional change determined by the computer's random generator. We have varied only the degree of linearity (L) which corresponds to the frequency of directional

[18] G. Albrecht-Buehler, J. Cell Biol. **72**, 595 (1979).
[19] G. W. Chodak, C. J. Scheiner, and B. R. Zetter, N. Engl. J. Med. **305**, 869 (1981).

change. A value of $L = 2$, for example, would indicate that a cell changed direction every time it traveled a distance equivalent to 2 cell diameters. The rate of capillary endothelial cell migration is such that the maximum distance traveled over an 18-hr period would represent approximately 8 cell diameters. Therefore, a linear increase in cell area can be expected from any moving cell, regardless of how often and at what angle it changes direction. These simulations confirm that an increase in phagokinetic track area over an 18-hr period is primarily a function of the *rate* and not the pattern of cell movement.

We have therefore used the measurement of phagokinetic track area as an indicator of capillary endothelial cell motility. With this method we have been able to detect activity in tumor lysates[6] and in biological fluids of patients with known ocular or urologic tumors.[19,20] In addition, we have found that this assay has sufficient range and sensitivity to assess the activity of column fractions and guide a purification scheme for isolation of migration factors.[21] A representative fractionation of a rat tumor lysate on a Sephadex G-75 column is shown in Fig. 3.

Other Assays for the Analysis of Cell Migration

The major advantages of the assay described above are its sensitivity and reproducibility. Also, the phagokinetic assay requires relatively low numbers of cells and is particularly useful for cells such as human capillary endothelial cells that are difficult to grow in large numbers. The major drawback of the assay is that it is difficult to use to assess directional motion in response to a gradient of a chemical agent (chemotaxis). Thus the motility observed in this assay is generally not under any directional constraints (chemokinesis). Since certain types of neovascularization— such as tumor angiogenesis—have a distinct directional component, it would be useful to also have an assay that measures this type of directional movement.

Perhaps, the best known assay for directional cell motility is the Boyden chamber.[22] This assay has been extensively used for studies of leukocyte migration and excellent reviews have been written concerning its proper use.[23,24] Briefly, the device consists of two chambers separated

[20] D. Tapper, D. M. Albert, N. L. Robonson, and B. R. Zetter, *J. Natl. Cancer Inst. (U.S.)* **71**, 501 (1983).

[21] J. Azizkhan, R. Sullivan, R. Azizkhan, B. R. Zetter, and M. Klagsbrun, *Cancer Res.* **43**, 3281 (1983).

[22] S. Boyden, *J. Exp. Med.* **115**, 453 (1962).

[23] P. C. Wilkinson, *J. Immunol. Methods* **51**, 133 (1982).

[24] B. M. Czarnetski, E. Kownatzki, M. Dierich, and P. C. Frei, *Arch. Dermatol. Res.* **275**, 359 (1983).

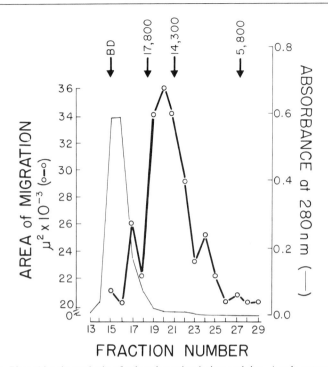

FIG. 3. Phagokinetic analysis of migration-stimulating activity. A salt extract prepared from a rat chondrosarcoma was prepared as previously described (see Ref. 21) and further fractionated on a Sephadex G-75 column equilibrated with 4 M guanidine–HCl and 5 mM dithiothreitol. The results demonstrate a peak of capillary endothelial cell migration-stimulating activity with an approximate molecular weight of 16,000. Reprinted with permission from Azizkhan et al.[21]

by a porous filter through which actively motile cells can pass. A test substance is placed in one chamber and a control substance in the other. The movement of cells from the control side of the apparatus to the side containing the test substance can be caused by chemokinesis, chemotaxis, or haptotaxis (movement along an adhesive gradient). Consequently, the inclusion of appropriate controls to distinguish among these possibilities is absolutely essential to the proper use of this technique.[23] The advantages of the modified Boyden chamber lie in its ease of use and its ability to provide quantitative data. Its primary disadvantage lies in the fact that gradients formed in this apparatus break down extremely rapidly.[25] Consequently this assay is more suited to the study of rapidly moving leukocytes than to more slowly moving endothelial cells or fibroblasts.

[25] S. H. Zigmond, Ciba Found. Symp. 71, 299 (1980).

The Boyden chamber assay has been used occasionally in the study of angiogenesis. In an early study, D'Amore *et al.* showed that tissue extracts contain a substance that is chemotactic for fetal bovine aortic endothelial cells.[26] More recently, Castellot *et al.* found a chemotactic factor for bovine adrenal capillary endothelial cells in angiogenic extracts prepared from the Swiss mouse 3T3 cell-derived adipocyte cell line.[27] These results suggest that certain angiogenic factors do have chemotactic as well as chemokinetic activity for endothelial cells. For general analysis of capillary endothelial cell migration, the advantages of this assay may be mitigated by its relative lack of sensitivity, the rapidity with which the gradients break down and the difficulty in determining the contribution of the various types of cell movement that the assay detects. Consequently, for routine analysis of capillary endothelial cell motility for the detection or purification of angiogenic factors the phagokinetic assay is the method of choice.

[26] P. A. D'Amore, B. M. Glaser, S. K. Brunson, and A. H. Fenselau, *Proc. Natl. Acad. Sci. U.S.A.* **78**, 3068 (1981).

[27] J. J. Castellot, M. J. Karnovsky, and B. M. Spiegelman, *Proc. Natl. Acad. Sci. U.S.A.* **79**, 3068 (1982).

[12] Spectrophotometric Assay for the Quantitation of Cell Migration in the Boyden Chamber Chemotaxis Assay

By GARY R. GROTENDORST

Introduction

Cell migration plays a key role in many biological processes as well as disease states, and there is considerable interest in identifying the factors which regulate these migrations and understanding how these factors function to control cell movement. In order to isolate and study the functional activity of migration factors, a wide variety of assays has been developed to quantitate the migratory activity of the target cells. These include migration of cells under an agarose layer,[1] phagokinetic tract assay,[2] cell orientation assays,[3] time-lapse cinematography,[4] and migra-

[1] J. Cutler, *Proc. Soc. Exp. Biol. Med.* **147**, 471 (1974).

[2] B. R. Zetter, *Nature (London)* **285**, 41 (1980).

[3] S. H. Zigmund, *J. Cell. Biol.* **75**, 606 (1977).

[4] M. Bessis and M. Burte, *Tex. Rep. Biol. Med.* **23**, 204 (1965).

tion of cells through a porous filter (Boyden chamber).[5] Each of these assays has its advantages and disadvantages and all are time consuming, either in the duration of the assay, the quantitation, or both. Some, such as the phagokinetic tract assay, require sophisticated equipment and computer analysis of the raw data.

The most widely used chemotaxis assay is the Boyden chamber assay where target cells migrate through a porous filter. This assay has been used with neutrophils,[6] monocytes,[7] fibroblasts,[8] smooth muscle cells,[9] and endothelial cells.[10] It is quantitated by either measuring the distance traveled by the leading front of cells when Millipore filters are used (100 μm thick), or by counting the number of cells per unit area on the lower surface of the filter when polycarbonate filters are employed (10 μm thick). Computer assisted image analysis has been utilized to quantitate the cell migration. This method does not work well for the author with the Nuclepore filters (polycarbonate) since the pore diameter is similar to that of the stained nucleus.

Because the Boyden chamber assay is applicable to a wide variety of cell types, is simple to perform, and can distinguish between random and directed cell migration, several methods which would allow for a rapid and quantitative assessment of cell migration in this system were investigated. Here a spectrophotometric method is described for the quantitation of the cell migration based on the extraction of stain from the nuclei of cells which have migrated through the filter. The method is rapid, accurate, and does not require any specialized equipment. The method can also be adapted for use with many cell types and circumvents the subjectiveness of counting cells by eye.

Materials and Methods

Cells

Fetal bovine aortic smooth muscle cells were prepared as described previously[9] and were cultured in Dulbecco's modified Eagle's medium

---

[5] S. V. Boyden, *J. Exp. Med.* **115**, 453 (1962).
[6] S. H. Zigmond and J. G. Hirsch, *J. Exp. Med.* **137**, 387 (1973).
[7] R. Snyderman, H. S. Shin, and M. S. Huasman, *Proc. Soc. Exp. Biol. Med.* **138**, 382 (1971).
[8] A. E. Postlethwaite, R. Snyderman, and A. E. Kang, *J. Exp. Med.* **144**, 1188 (1976).
[9] G. R. Grotendorst, H. E. J. Seppa, H. K. Kleinman, and G. R. Martin, *Proc. Natl. Acad. Sci. U.S.A.* **78**, 3669 (1981).
[10] B. M. Glaser, P. A. D'Amore, H. Seppa, S. Seppa, and E. Schiffmann, *Nature (London)* **288**, 483 (1980).

(DMEM) containing 10% fetal calf serum (FCS) and 50 μg/ml gentamycin. Human skin fibroblasts (CRL 1475) were obtained from ATCC (Rockville, MD). Tumor macrophages (U937) were obtained from Ralph Snyderman (Duke University, Durham, NC) and maintained in RPMI 1640 containing 10% FCS and 50 μg/ml gentamycin. These cells were stimulated to differentiate by adding 1 mM dibutyl-cAMP to the medium 8 hr prior to the chemotaxis assay. All cell cultures were maintained at 37° in humidified chambers containing 5% CO_2 (M199 and RPMI 1640) or 10% CO_2 (DMEM).

Preparation of Collagens, Attachment Factors, and Chemoattractants

Type I collagen was prepared by acid extraction of rat tail tendons as described by Bornstein and Piez.[11] Type V collagen was isolated from bovine placental membranes using the method of Bentz *et al.*[12] Fibronectin was purified from human plasma using gelatin affinity chromatography.[13] Laminin was purified from a murine tumor which secretes basement membrane components.[14] The platelet-derived growth factor was isolated from human platelets as described previously[15] using a modification of the methods described by Raines and Ross[16] and Heldin *et al.*[17] (see also [1] and [5], this volume). EGF and FGF were purchased from Biomedical Technologies, Inc. (Cambridge, MA). Synthetic *N*-formylmethionyl peptide (FMLP) was purchased from Sigma (St. Louis, MO).

Chemotaxis Assay

The chemotactic response of all cell types was assayed using the modified Boyden chamber.[6–10] The test substance was diluted in DMEM containing bovine serum albumin (BSA) (0.2 mg/ml) and added to the lower well of the blind well chamber. The lower well was then covered with a collagen-coated polycarbonate filter (Nuclepore, 8 μm diameter pores) ensuring a continuous contact between the lower surface of the filter and the solution. The upper well was then fixed in place and cells (3 $\times$ 10⁵ cells/assay) freshly released from tissue culture flasks were added to the upper well in DMEM containing BSA (2 mg/ml). After the chambers were

[11] P. Bornstein and K. A. Piez, *Biochemistry* **5**, 3460 (1966).
[12] H. Bentz, H. P. Bachinger, R. Glanville, and K. Kuhn, *Eur. J. Biochem.* **92**, 563 (1978).
[13] E. Engvall and E. Ruoslahti, *Int. J. Cancer* **20**, 1 (1977).
[14] R. Timpl, H. Rhode, P. Gehron-Robey, S. I. Rennard, J. M. Fordart, and G. R. Martin, *J. Biol. Chem.* **254**, 9933 (1979).
[15] G. R. Grotendorst, *Cell* **36**, 279 (1984).
[16] E. W. Raines and R. Ross, *J. Biol. Chem.* **257**, 5154 (1982).
[17] C. A. Heldin, B. Westermark, and Å. Wasteson, *Biochem. J.* **193**, 907 (1981).

incubated 4–6 hr at 37° in an atmosphere of 90% air/10% CO_2, the filters were removed from the chambers and the cells were fixed and stained for light microscopy using Diff-Quick stain (Harleco). The filters were placed on glass microscope slides with the lower surface of the filter adjacent to the glass slide. The upper surface cells were removed by gently scraping with a rubber policeman, and the chemotactic response was analyzed by counting the number of cell nuclei/microscopic field (400×) on the lower surface. The chemotactic response of the U937 cells was performed as above except that noncoated filters were used and the assays were incubated for only 1 hr. Each assay was performed in duplicate and the cell numbers varied by less than 10%. All data are expressed as the mean of duplicate experiments, each of which is the average cell number in 10 microscopic fields on two filters.

Spectrophotometric Analysis of Chemotaxis Assay

The chemotaxis assays were performed as described above up to the point of placing the stained filters on glass microscope slides. The upper surface cells were then carefully removed by scraping with a rubber policeman being certain to remove all cellular debris which contained stained material. The filters on the glass slides were then immersed in a petri dish containing distilled water and allowed to hydrate for 0.5–2.0 min. This allowed the filters to be removed from the slide and minimized the number of cells which remained attached to the glass slide. The filters were then dried by blotting on absorbant paper and placed in either a small test tube (13 × 75) or in the well of a 96-well ELISA plate. Nuclear stain was then extracted for 15 min with 0.1 N HCl (500 μl/tube or 300 μl/well). The solution was then thoroughly mixed and the absorbance at 600 nm is determined. In the 96-well plate, the solutions (250 μl) were transferred to another plate and the absorbance measured with Titertek ELISA spectrophotometer (Flow Labs, McLean, VA).

Results and Discussion

Comparison of the Spectrophotometric Analysis with Cell Counts

The standard method for quantitating cell migration in the Boyden chamber chemotaxis assay is to count the number of cells on the lower surface of the filter in 10 high-power fields. While this technique directly measures the number of nuclei, it is nonetheless a subjective determination, since it requires the individual to determine what contributes a nucleus and to remember which ones have already been counted. Because of its subjective nature as well as other problems such as nonhomogeneity of the filter and the fact that less than 0.1% of the total area of the filters is used for the quantitation of the response, the results are subject to a high degree of variability. It seemed that a different method which was objective and would utilize the entire area of the filter for the quantitation would be useful. Because the positive responses of stained filters were so easily picked out by visual examination, it seemed reasonable that the amount of stain would be proportional to the cell numbers. Also, since the amount of nuclear material per cell is constant, the amount of dye which reacts with this material should also be constant from cell to cell.

Initially, several different stains were employed including hematoxylin–eosin and a commercially available Wright–Giemsa (Diff-Quik) stain. The Diff-Quik stain gave the most consistent and most intense staining with any of the cell types tested and was used in all the studies described. The quantitation of the stain was performed by extraction of the dye with 0.1 N HCl. The amount of dye extracted could then be quantitated by measuring the absorbance of the extract at 600 nm. Figure 1 compares the migratory response of smooth muscle cells to different concentrations of PDGF as judged by cells/field or the A_{600} of extracted dye from the re-

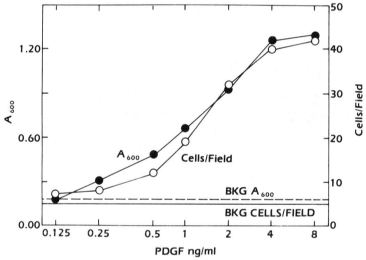

FIG. 1. Comparison of the quantitation of smooth muscle cell migration by cell counts and dye extraction. The chemotactic response of smooth muscle cells to different concentrations of PDGF was determined. After scraping the cells from the upper surface of the filter, the response was initially quantitated by counting the number of cells/field as described under Materials and Methods. The filters were rehydrated with distilled water transferred to test tubes (13 × 75) and the dye extracted with 0.1 N HCl (500 μl/tube). The absorbance at 600 nm was determined with a standard 1 cm cuvette in a Beckman Model 25 spectrophotometer.

sponding cells. The relationship between cells/field and A_{600} is linear with an elevated intercept of the y axis (Fig. 2). This is due to the amount of stain which is nonspecifically taken up by the filter. Staining filters without cells as controls and subtracting the value of dye extracted from these filters from all experimental filters results in a standard curve which intersects near the origin (Fig. 2).

The blind well chambers used above had a surface area of 50 mm² for cell migration and required larger amounts of both attractant (220 μl) and target cells (3.5 × 10⁵ cells/assay). A smaller chamber which is commercially available has a surface area of 18 mm² and requires only 30 μl of attractant solution and 6.0 × 10⁴ cells/assay. The spectrophotometric method can be adapted to the smaller chambers but the data are more variable due to the higher background stain relative to specific staining of nuclei.

Application to Quantitation of Chemotactic Response of Other Cell Types

The Boyden chamber chemotactic assay has been used to quantitate the migration of neutrophils, macrophages, fibroblasts, smooth muscle cells, and endothelial cells. We next determined whether the migration of these cell types other than smooth muscle could be quantitated with the

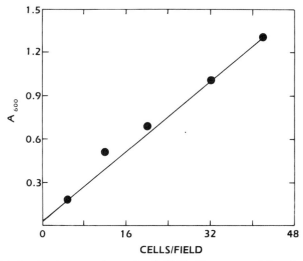

FIG. 2. Relationship between A_{600} and cells/field. The corrected data from Fig. 1 are plotted showing the linear relationship between A_{600} and cells/field. Correction for nonspecific staining of the filter was performed by subtraction of the absorbance value of filters without any cells.

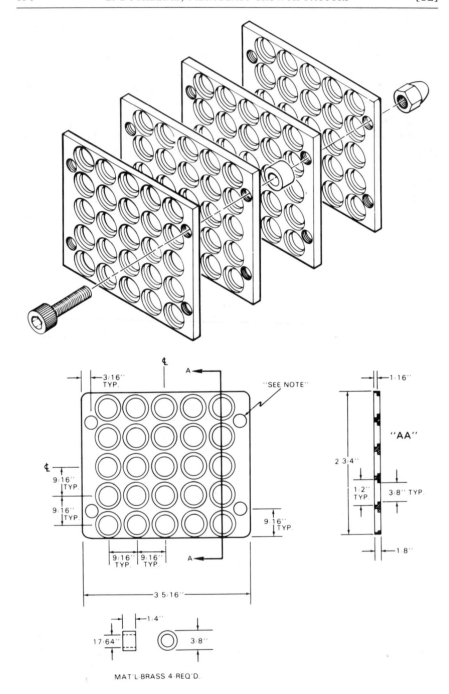

TABLE I
QUANTITATION OF THE CHEMOTACTIC RESPONSE OF VARIOUS CELLS TO
DIFFERENT CHEMOATTRACTANTS

	Chemotactic response[a] (corrected A_{600}) to chemoattractant				
Cell type	None	PDGF	FN[b]	FMLP[c]	Sm. cond. med.[d]
CRL 1475 fibroblasts	0.05	0.73	0.50	0.07	0.12
U937 monocytes	0.02	0.03	0.01	0.21	0.03
Bovine aortic endothelial cells	0.04	0.04	0.10	0.05	0.68

[a] The chemotactic response was determined as described under Materials and Methods. All A_{600} values have been corrected by subtracting the A_{600} values of filters without cells.
[b] Fibronectin (10 μg/ml).
[c] N-Formylmethionylleucylphenylalanine (10^{-9} M).
[d] Smooth muscle cell conditioned media.

spectrophotometric method. Neutrophils were too motile for quantitation with our Nucleopore filters and little difference could be detected between the control with its high level of random migration and stimulated migration. However, other cell types worked well in this system (Table I). Fibroblasts were responsive to both PDGF and fibronectin but not to N-formylmethionyl peptides, which are attractants for neutrophils and monocytes. Endothelial cells exhibited migration to smooth muscle cell conditioned media but were not responsive to PDGF or fibronectin. The U937 cells, a monocyte tumor line, were responsive to N-formylmethionyl peptides but not to any of the other attractants tested. These data indicated that this assay can be utilized with a wide variety of cell types.

Construction of a Filter Holder for Staining of Nuclepore Filters

This method quantitates the number of cells present on the lower surface of the filters from the amount of stain which can be extracted from those cells. Therefore, it is critical that each filter be uniformly exposed to the staining solution. A specialized filter holder was constructed to accomplish this (Fig. 3). The holder was machined from brass plates and had

FIG. 3. Schematic diagram of staining chamber used for holding Nucleopore filters. "Note" holes in the bottom plate are threaded to accommodate bolt. All materials are machined from solid brass.

25 holes/plate which were lined with rubber O rings. Filters were sandwiched between two plates and were held in place by the O rings. Our holder consisted of 2 pair of plates so that 50 filters at a time could be stained. In this way, each filter was in contact with the staining solution for equal lengths of time which removed any variation in the staining aspects of the assay.

Summary

The method described here for the quantitation of cell migration in Boyden chambers is applicable to many cell types, especially connective tissue cells and endothelial cells. Currently, there are many groups working toward the isolation of chemotactic factors for these cells and this method would be useful for the large numbers of repetitive assays done during a purification. In our hands the assay is very reproducible and can be learned within a short time. The major problem with the assay is that the upper surface cells must be completely removed as well as any cellular debris containing dye. However, if the proper care is used in removing these cells, the results vary less than those obtained by counting cells directly.

[13] Isolation of Two Proteinase Inhibitors from Human Hepatoma Cells That Stimulate Human Endothelial Cell Growth

By W. L. McKeehan and K. A. McKeehan

Rationale

In addition to an optimum nutrient medium (MCDB 107), epidermal growth factor, and high- or low-density lipoprotein, human endothelial cells require two additional synergistically acting classes of growth factors to proliferate in serum-free medium.[1] One class of factor is concentrated in neural tissue and can be distinguished by its heparin-binding property.[2] The second class of factor is concentrated in the medium of well-differentiated hepatoma cells.[1] We identified the second class of growth factors as low-molecular-weight, acid-stable proteinase inhibi-

---

[1] H. Hoshi and W. L. McKeehan, *Proc. Natl. Acad. Sci. U.S.A.* **81**, 6413 (1984).
[2] T. Maciag, T. Mehlman, R. Friesel, and A. B. Schreiber, *Science* **225**, 932 (1984).

tors.[3] These type proteinase inhibitors have several chemical properties similar to hormone-like growth factors.[3] They exhibit potent activity per unit protein,[3] have high affinity for specific receptors,[3,4] sequence homology,[5,6] sometimes compete for the same binding site,[7] and are tumor associated.[3,8,9] These striking similarities suggest a common evolutionary origin of such proteinase inhibitors and some growth factors. Here we describe the purification and characterization of two human hepatoma cell-associated proteinase inhibitors with apparent endothelial cell growth factor activity. Rapid, efficient purification was achieved by employment of high-performance (pressure) liquid chromatography (HPLC) from start to finish.

Bioassay

Activity was measured by growth of human endothelial cells from umbilical vein tissue[3,10] and quantitated by computerized videometry.

Hepatoma Cell Medium

Human hepatoma cells were grown to monolayer in plastic roller bottles with 850 cm[2] surface area. After confluency, the medium was replaced with factor-free, serum-free nutrient medium[3] and collected every 2 or 3 days. The medium was frozen until about 50 liters was collected for processing below.

Purification of Factors

Stepwise Reversed Phase. Fifty liters of medium from hepatoma cells was concentrated to 1 to 2 liters by hollow-fiber ultrafiltration (DC10L unit, Amicon Corp., Danvers, MA). The crude concentrated medium was fractionated by stepwise elution from reversed phase media. The concen-

[3] W. L. McKeehan, Y. Sakagami, H. Hoshi, and K. A. McKeehan, *J. Biol. Chem.* **261**, 5378 (1986).
[4] J. P. Vincent and M. Lazdunski, *Biochemistry* **11**, 2967 (1972).
[5] L. T. Hunt, W. C. Barker, and M. O. Dayhoff, *Biochem. Biophys. Res. Commun.* **60**, 1020 (1974).
[6] T. Yamamoto, Y. Nakamura, T. Niskide, M. Emi, M. Ogawa, T. Mori, and K. Matsuhara, *Biochem. Biophys. Res. Commun.* **132**, 605 (1985).
[7] D. A. Green and J. B. Moore, Jr., *Arch. Biochem. Biophys.* **202**, 201 (1980).
[8] D. F. Bowen-Pope, A. Vogel, and R. Ross, *Proc. Natl. Acad. Sci. U.S.A.* **81**, 2396 (1984).
[9] G. J. Todaro, C. M. Fryling, and J. E. De Larco, *Proc. Natl. Acad. Sci. U.S.A.* **77**, 5258 (1980).
[10] W. L. McKeehan and K. A. McKeehan, this volume [35].

trated medium was made 0.1% trifluoroacetic acid (TFA) and then pH 2.5 with concentrated HCl. The yellow solution (phenol red) was centrifuged at 10,000 g for 10 min, passed through a 1.2-μm filter, and then degassed under vacuum. The particulate-free solution was pumped onto a 22.5 × 150 mm column that was packed dry and by hand with Vydac C_{18}-silica (218TPB2030 Separations Group, Hesperia, CA). The column was washed with methanol and then water and then equilibrated with 0.1% TFA. New-packed columns were loaded first with 1 g human serum albumin followed by elution at 70% acetonitrile (AN) and 0.1% TFA and then again equilibrated with 0.1% TFA. About 1 liter of concentrated hepatoma cell medium containing up to 2500 A_{280} units was pumped onto the column at 50 ml/min with a Rainin HPX pump (Rainin Instrument Co., Woburn, MA). The eluent was monitored at A_{280} and the column was eluted stepwise with 0.1% TFA, 20% AN/0.1% TFA, 40% AN/0.1% TFA, and 80% AN/0.1% TFA. Elution was carried out at a rate of 50 ml/min at each step until the A_{280} dropped to the baseline. About 80% of activity eluted at 40% AN.

Preparative Gradient RP-HPLC. Activity was resolved into two discrete areas when the 40% AN elution was subjected to gradient elution on a preparative reversed-phase HPLC column. The material which eluted above at 40% AN/0.1% TFA containing about 250 A_{280} units was diluted to 20% AN with 0.1% TFA and pumped directly at 30 ml/min onto a 22.5 × 250 mm column guarded by a 1 × 150 mm precolumn containing Vydac C_{18}-silica (218TPB1520 Separations Group). The columns were packed for high performance (Custom LC, Houston, TX) and equilibrated with 20% AN/0.1% TFA before loading. New columns were loaded with serum albumin as described earlier to prevent irreversible absorption of medium components in the first run. The column was eluted at 10 ml/min with a gradient from 20% AN to 50% AN over 60 min using a Gilson HPLC unit and controller (Gilson Electronics, Middleton, WI). Fractions eluting between 26 and 28% AN were pooled and designated endothelial cell growth factor-2a (ECGF-2a).[11] Fractions eluting between 31 and 34% AN were designated ECGF-2b. A preliminary analysis of crude hepatoma cell-derived medium using flat-bed, preparative isoelectric focusing between pH 3 and 10 indicated the presence of an activity that ran against the acid electrode and another near pH 6. Subsequent analysis indicated that ECGF-2b was the acidic activity and exploitation of this property was necessary for its purification.

Ion-Exchange Separation of ECGF-2b. The pool from preparative RP-

---

[11] We reserved the term ECGF-1 for the heparin-binding class of endothelial cell growth factors.

HPLC containing ECGF-2b activity was made 25 mM triethylamine–acetate (pH 4.2) and pumped directly onto a Mono P HR/520 column (Pharmacia Fine Chemicals, Piscataway, NJ) at a rate of 2.8 ml/min (1000 psi). The column was previously equilibrated with 25 mM triethylamine–acetate (pH 4.2) and 10% AN. The loaded column was washed with the same solution until A_{280} dropped to baseline. The active material was then eluted with a solution containing 0.1% TFA and 2 M NaCl. Of the applied A_{280} absorbing material, 83% was unretained by the column and 14% was recovered in the 2 M NaCl elution which contained 90% of the activity. Both ECGF-2a and ECGF-2b were further purified by RP-HPLC using heptafluorobutyric acid as the ion pairing agent instead of trifluoroacetic acid.

Semipreparative RP-HPLC in HFBA. Pools of ECGF-2a containing 3 to 5 A_{280} units were diluted to 20% AN, made 0.13% HFBA, and pumped at 10 ml/min onto a 1 × 150 mm column packed with Vydac C$_4$-silica (214TP1010 Separations Group) and previously equilibrated with 20% AN/0.13% HFBA. The column was then eluted for 10 min with a gradient between 20% AN and 28% AN, then 35 min to 35% AN and then 10 min to 50% AN at a flow rate of 3 ml/min. Fractions were collected every minute. The absorption peak eluting at 31% AN was dried on a Speed-Vac concentrator (Savant Instruments, Hicksville, NY) for fractionation by molecular filtration.

The fraction that eluted at 2 M NaCl from the Mono P column containing ECGF-2b was similarly fractionated. Pools containing 5 to 15 A_{280} units were made 0.13% HFBA, pumped on the C$_4$-silica column, and eluted with a gradient between 20% AN and 50% AN over 70 min. Elution was carried out for 10 min to 30% AN, then to 40% AN for 50 min and to 50% for 10 more min. Fractions that eluted between 34% AN and 36% AN were pooled and dried on the Speed-Vac. Both ECGF-2a and ECGF-2b were then further purified according to size by molecular filtration HPLC.

Molecular Filtration. Both dried ECGF-2a and ECGF-2b were suspended in 100 μl of solution containing 10% AN, 0.1% TFA, and 0.10 M NaCl. Two 7.5 × 600 mm TSK 62000SW (Phenomenex, Rancho Palos Verdes, CA) columns in tandem with a 100 mm precolumn were equilibrated with the above solution. Solutions were injected, elution was carried out at flow rate of 1 ml/min, and fractions of 0.5 ml were collected. About 0.15 A_{280} units of ECGF-2a was applied and 1 unit of ECGF-2b was applied. Eluents were monitored at A_{215} and A_{280}, respectively. Columns were calibrated using human albumin, cytochrome c, and insulin under the same conditions.

ECGF-2a activity eluted as a sharp peak at a relative molecular weight of 6500. ECGF-2b activity migrated coincident with a peak at apparent molecular weight of 21,000. Separate analysis of molecular weight of ECGF-2b on polyacrylamide gels in sodium dodecyl sulfate under reducing conditions indicated a molecular weight of 27,000. Both factors were then subjected to a final desalting on RP-HPLC in 0.1% TFA.

Final Desalting on RP-HPLC. The ECGF-2a peak, which eluted at about M_r = 6500 in the TSK column, was injected directly onto a 4.6 × 250 mm Vydac C_4-silica column (214TP54 Separations Group) previously equilibrated with 20% AN and 0.1% TFA. Multiple sample injections were made where necessary and elution with the starting solution was carried out until absorption at A_{215} dropped to baseline. Gradient elution was carried out between 20% AN and 25% AN for 10 min then to 30% AN for 25 min and to 50% AN for 10 min at 0.5 ml/min. Fractions over the active peak eluting at 28% AN were pooled, dried, and used for sequence and activity analysis.

The peak of ECGF-2b, which eluted from the TSK column at apparent M_r = 21,000, was similarly applied directly to the C_4-silica Vydac column. The column was eluted between 25% AN and 30% AN for 5 min and then to 35% AN for another 25 min. Fractions across the peak eluting at 32% AN were pooled, dried, and used for sequence activity analysis.

Fifty liters of hepatoma cell medium yielded about 100 to 200 μg of ECGF-2a and 1 to 2 mg of ECGF-2b by the above procedure.

Properties of ECGF-2a and ECGF-2b

ECGF-2a migrated as a single, sharp band at 6500 molecular weight on silver-stained SDS–polyacrylamide gel electrophoresis. ECGF-2b migrated as a diffuse, single band at 27,000. NH$_2$-terminal sequence of both factors was determined by automated gas phase sequencing.[3] Both exhibited a single NH$_2$-terminal sequence. The clearly assignable residues of NH$_2$-terminal sequence and the total amino acid composition of ECGF-2a and ECGF-2b were identical to human pancreatic secretory trypsin inhibitor (HPSTI)[12] and the 27- to 30-kDa glycopeptide proteinase inhibitor (HI-30/ EDC1).[13,14] Both hepatoma cell-derived ECGF-2a and ECGF-2b inhibited pancreatic trypsin.[3] ECGF-2a and ECGF-2b stimulated endothelial cell growth at a half-maximal rate at 50 and 80 to 130 ng/ml, respectively.[3] When assayed under identical conditions, no effect of either factor on human smooth muscle cells, human hepatoma cells (the cell of origin), or human, rat, or mouse fibroblasts could be detected.[3]

[12] D. C. Bartelt, R. Shapanka, and L. J. Greene, *Arch. Biochem. Biophys.* **179**, 189 (1977).
[13] K. Hochstrasser, O. L. Schonberger, I. Rossmanith, and E. Wachter, *Hoppe-Seyler's Z. Physiol. Chem.* **362**, 1357 (1981).
[14] R. K. Chawla, A. D. Wadsworth, and D. Rudman, *J. Immunol.* **121**, 1636 (1978).

[14] Identification of Cell Surface Proteins Sensitive to Proteolysis by Thrombin

By JAMES A. THOMPSON and DENNIS D. CUNNINGHAM

Several lines of evidence have shown that the primary biochemical event in thrombin-stimulated cell division is cleavage of one or more cell surface proteins. First, cellular internalization of thrombin is not necessary for cell activation.[1] Second, thrombin is rendered nonmitogenic by derivatizating its catalytic site serine with a diisopropylphospho group or a methylsulfonyl group.[2] Third, although some cells possess thrombin-binding sites, mitogenic stimulation does not require binding of thrombin to these cell surface sites.[3]

These considerations prompted a search for cell surface proteins which are substrates for thrombin. Unfortunately, techniques such as two-dimensional gel electrophoresis, which very effectively resolve soluble proteins, usually provide much poorer resolution of membrane proteins. The latter proteins frequently show streaking on these gels which make quantitation of electrophoretograms difficult. The procedures described herein permit the separation of membrane proteins with better resolution.

Briefly, membrane proteins are first fractionated into four groups as depicted in Fig. 1. Fraction 1 is obtained by treating cells with a weak detergent and a calcium chelator (saponin and EDTA) under conditions that the cells retain their ability to exclude trypan blue. Purified plasma membranes are then prepared from these cells and extracted with a nonionic detergent (Triton X-114). After warming to 22° and centrifugation to produce a two-phase system, the detergent-rich phase contains Fraction 2 and the aqueous phase contains Fraction 3. The proteins that are insoluble in Triton X-114 are solubilized in a denaturing detergent in combination with a reducing agent [sodium dodecyl sulfate (SDS) and 2-mercaptoethanol] to produce Fraction 4. Subsequent one-dimensional gel electrophoresis of these four fractions produces resolution of membrane proteins that is more amenable to quantitation by scanning densitometry

[1] D. Carney and D. Cunningham, *Cell* **14**, 811 (1978).
[2] K. Glenn, D. Carney, J. Fenton II, and D. Cunningham, *J. Biol. Chem.* **255**, 6609 (1980).
[3] D. A. Low, H. S. Wiley, and D. D. Cunningham, *in* "Growth Factors and Transformation" (J. Feramisco, B. Ozanne, and C. Stiles, eds.), p. 401. Cold Spring Harbor Lab., Cold Spring Harbor, New York, 1985.

METHODS IN ENZYMOLOGY, VOL. 147

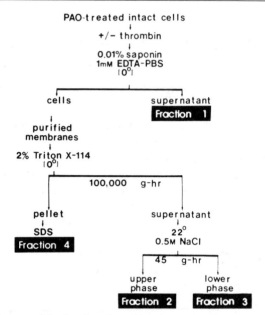

FIG. 1. Flow diagram of the fractionation of membrane proteins by selective extraction.

than two-dimensional gel electrophoresis of detergent-solubilized whole membranes.

Fractionation of Membrane Proteins

IIC9 cells, a subclone of Chinese hamster embryonic fibroblasts,[4] are used because of their high level of mitogenic responsiveness to thrombin and their virtual lack of thrombin-binding sites.[5] These cells were grown and maintained as previously described.[5] When cells grown in 150 mm dishes reach 80% confluency, they are arrested in G_0 of the cell cycle by deprivation of serum for 2 days. At this point membrane turnover is inhibited by rinsing cells 3 times with 10 ml phosphate-buffered saline (PBS) at 0° and incubation with 10^{-5} M phenylarsineoxide (PAO) in PBS containing 0.1 mM MgCl$_2$ for 30 min at 0°.[6] PAO is diluted from a stock solution of 0.1 M in dimethyl sulfoxide. (It is important to inhibit membrane turnover to detect cleavage of membrane proteins that have a very

[4] R. Sager and P. E. Kovac, *Somatic Cell Genet.* **4**, 375 (1978).
[5] D. A. Low, R. W. Scott, J. B. Baker, and D. D. Cunningham, *Nature (London)* **298**, 476 (1982).
[6] H. S. Wiley and D. D. Cunningham, *J. Biol. Chem.* **257**, 4222 (1982).

short turnover time.) Cells are then rinsed with 10 ml PBS to remove the PAO and incubated in the presence or absence of 10 μg/ml highly purified thrombin for 30 min at 37° in 10 ml 0.1 mM MgCl$_2$ in PBS. Once cells are treated with thrombin, all subsequent buffers contain the following protease inhibitors: 1 mM benzamidine–HCl, 10 μM chymostatin, 10 μM pepstatin, 10 μM leupeptin, 10 μM antipain, 10 μM aprotinin, and 1 mM phenylmethanesulfonyl fluoride. Unless noted otherwise, all steps described below are carried out at 0°.

After treatment of cells with thrombin, the first group of membrane-associated proteins is extracted by treatment of the intact cells with EDTA and a low concentration of saponin, a plant glycoside. Cells in 150-mm dishes are rinsed 3 times with 10 ml of PBS and extracted with 5 ml of isotonic 0.01% saponin, 1 mM EDTA, 0.15 M NaCl, 10 mM HEPES, pH 7.4, for 1 min. Cells extracted in this way retain their ability to exclude trypan blue. The extract is centrifuged at 100,000 g for 30 min to remove any insoluble material. These solubilized proteins are lyophilized to dryness, suspended in 0.4 ml of water, and desalted on a 1.5 ml column of Sephadex G-15 equilibrated with 0.01% lithium dodecyl sulfate (LDS), 0.06 mM Tris–HCl, pH 6.8.[7] The eluate of this column is lyophilized and stored until electrophoresis.

For further fractionation of membrane proteins, highly purified membranes are then prepared from cells that were extracted with saponin and EDTA. In order to enrich for surface membranes and avoid internal pools of membranes, the procedure of Gruber et al.[8] is followed exactly with two exceptions. (1) Cells are lysed in a saturated fluorescein mercuric acetate solution prepared as described by Warren et al.[9] (2) The large membrane sheets obtained are initially sedimented by centrifugation at 650 g for 10 min.

The next two groups of membrane proteins are obtained by dissolving the purified membranes in 100 μl 4% Triton X-114 (prepared as described by Bordier[10]) containing 10 mM HEPES, pH 7.4, by trituration with a Hamilton syringe (100 μl) at room temperature for 5 min. Insoluble material is removed by centrifugation at 100,000 g for 1 hr and saved. Enough 4 M NaCl is added to the supernatant to bring the solution to 0.5 M NaCl. This facilitates the following separation of the mixture into a detergent and an aqueous phase. At this salt concentration 15 min at 22° is enough to

---

[7] G. P. Tuszynski, L. Knight, J. R. Piperno, and P. N. Walsh, Anal. Biochem. **106,** 118 (1980).
[8] M. Y. Gruber, K. H. Cheng, J. R. Lepock, and J. E. Thompson, Anal. Biochem. **138,** 112 (1984).
[9] L. Warren, M. C. Glick, and M. K. Nass, J. Cell. Physiol. **68,** 269 (1967).
[10] C. Bordier, J. Biol. Chem. **256,** 1604 (1981).

precipitate the Triton. Centrifugation in 6 × 45 mm polypropylene tubes at 9000 g for 30 sec in the Beckman microfuge at 22° results in separation of a detergent-free upper phase and a detergent-rich lower phase. With these long slender tubes, only one precipitation is necessary if the upper phase is removed carefully. Membrane proteins fractionate according to hydrophobicity with the more hydrophobic proteins in the Triton-rich phase and the more hydrophilic proteins in the aqueous phase. Removal of excess Triton in the detergent-rich phase is accomplished by desalting using a Sephadex G-15 column equilibrated with 0.01% LDS, 0.06 mM Tris–HCl, pH 6.8. This step must be conducted 0° to prevent precipitation of Triton. To prepare both samples for electrophoresis and to remove tightly bound Triton, 0.9 ml of cold acetone is added to precipitate the proteins. After collection of proteins by centrifugation at 9000 g for 15 min, the pellet is dried *in vacuo* to remove residual acetone.

Solubilization of the above Triton-insoluble material in 5% LDS, 10% glycerol, 2% 2-mercaptoethanol, 62.5 mM Tris–HCl, pH 6.8, at 100° for 2 min yields Fraction 4 and completes the fractionation of membrane proteins into four groups (Fig. 1).

Gel Electrophoresis

Electrophoresis of each of the above four fractions is carried out on 7.5 to 15% acrylamide (w/v) gradient gels (16 × 21 × 0.075 cm) using the discontinuous buffer system of Laemmli.[11] The gradient is constructed by using 12.5 ml of 7.5% acrylamide with 12.5 ml 15% acrylamide, 30% glycerol in a gradient maker. In order to increase the sharpness of stained bands, high wattage is used on gels held at 10° with a Brinkman cold water circulator. LDS replaces SDS due to its increased solubility at 10° as described previously by Perdew *et al.*[12] However, because of the cation distribution during electrophoresis it is necessary to use LDS only in the sample buffer. SDS may be used in the upper reservoir buffer as long as detergent is omitted from the lower reservoir and gel buffers.

Specifically, lyophilized or acetone-precipitated samples are suspended in 100 μl of LDS sample buffer. After a 2 min incubation at 100° samples are centrifuged at 100,000 g for 1 hr to remove insoluble aggregates. Protein (2.5 μg) (quantitated by the method of Schaffner and Weissmann[13]) is added per lane and stacked at 100 V (constant voltage); pro-

---

[11] U. K. Laemmli, *Nature (London)* **227**, 680 (1970).
[12] G. H. Perdew, H. W. Schaup, and D. P. Selivonchick, *Anal. Biochem.* **135**, 453 (1983).
[13] W. Schaffner and C. Weissmann, *Anal. Biochem.* **56**, 502 (1973).

teins are separated at 5 W (constant power) for up to 6 hr. The bromphenol blue dye front is run off the gel; completion of the run is monitored using prestained molecular weight markers prepared by the method of Bosshard and Datyner.[14] The run is complete when the 20K molecular weight marker migrates within 1 in. of the bottom of the gel. Gels are prestained with Coomassie Blue R in methanol:acetic acid: water (5:1:5) for 10 min, thoroughly destained with ethanol:acetic acid:water (5:7:88), and subsequently silver stained using the method of Ohsawa and Ebata.[15] This sensitive protocol prevents negative staining while keeping background staining low. In Fig. 2, the composite silver-stained gel compares total membrane proteins (lane 1) and proteins in each of the four membrane fraction (lanes 2-5).

Quantitation of Silver Stained Gels

Protein bands on individual lanes of a silver-stained gel are digitized and analyzed using the general computer-assisted procedures described by Mariash et al.[16] This method employs a high-resolution video camera connected to an Apple II+ microcomputer via a digitizing board (Micro-works). For the present studies, the program of Mariash et al.[16] was modified; copies of this modified program will be sent to investigators who send a standard 5-1/4 in. flexible disc. Either stained gels or contact duplications on X-OMAT duplicating film (Kodak) can be scanned and analyzed. Gels from both control and thrombin-treated cells can be scanned and presented in graphic display.

The following points are important to consider when quantitating bands on gels by this computer-assisted procedure. First, since some bands can be wider than others, scanning the middle of a lane can lead to inaccurate quantitation. Therefore, the program delineates and scans the entire lane. Intensity values are compiled across the width of the gel to give an accurate representation of each band on the one-dimensional gel lane. Also, complex patterns of overlapping bands make differences between control and thrombin-treated fractions difficult to visualize; these can be resolved by digital subtraction. Peak positions are aligned and experimental intensity values are subtracted from control values. Graphs of these differences are displayed as a series of positively and negatively displaced peaks; positive peaks represent proteins that disappear or decrease in intensity and negative peaks represent proteins that appear after

[14] H. F. Bosshard and A. Datyner, Anal. Biochem. **82**, 112 (1984).
[15] K. Ohsawa and N. Ebata, Anal. Biochem. **135**, 409 (1983).
[16] C. N. Mariash, S. Seelig, and J. H. Oppenheimer, Anal. Biochem. **121**, 388 (1983).

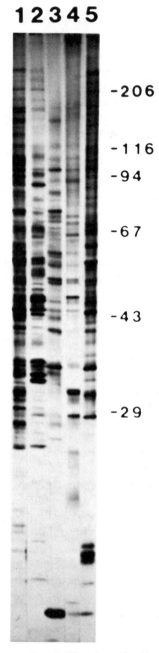

FIG. 2. Composite of silver-stained LDS–polyacrylamide gel of total membranes (lane 1) and Fractions 1 through 4 (lane 2 through 5, respectively). Protein (2.5 μg) was loaded per lane and electrophoresed as described in the text. The composite was made from a duplication on X-Omat duplicating film exposed for 3.5 sec on a standard fluorescent light box.

thrombin treatment. Critical to this type of analysis is the correct alignment of the positions of invariant protein peaks. Misalignment can occur when subtracting long stretches of gels. Therefore, short stretches must be subtracted to eliminate artificially produced peaks. Figure 3 represents this type of analysis on short stretches of two electrophoretograms. Figure 3A–C shows the appearance of bands (arrows) upon treatment with thrombin in Fraction 1. Figure 3D–F shows the disappearance of bands upon treatment with thrombin in Fraction 2. The difference can be quantitated by integration of peaks after digital subtraction.

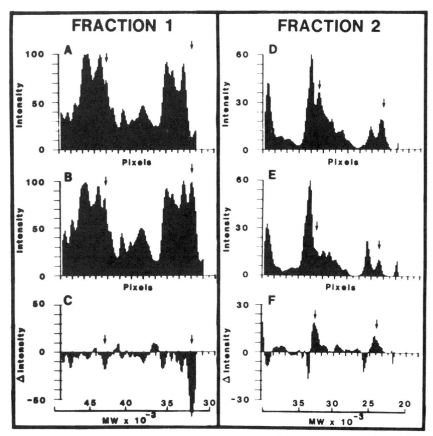

FIG. 3. Densitometer tracings and digital subtraction of short segments of electrophoretogram of Fraction 1 (A–C) and Fraction 2 (D–F). All segments were scanned as described in the text. (A) and (D) are scans derived from control cells; (B) and (E) are scans derived from thrombin-treated cells. (C) and (F) are graphs representing differences between scans from control and thrombin-treated cells after alignment of invariant peaks. The arrows represent examples of differences found after treatment of cells with thrombin as explained in the text.

Discussion

The advantages of this technique for analyzing membrane proteins are its ease, reproducibility, and ability to analyze large numbers of experimental points on the same gel. The present work demonstrates its utility in searching for cell surface proteins sensitive to the proteolytic mitogen, thrombin. This study can easily be expanded to the use of other mitogenic proteases or analysis of thrombin unresponsive mutants derived from a responsive cell line.

Use of these techniques in combination with cell surface labeling techniques can obviate the need for purification of membranes. Cell surface proteins can be labeled using the ^{125}I-labeled lactoperoxidase technique, a procedure which labels only those proteins that have exposed tyrosines. Then, the extraction procedure described herein can be applied to whole cells; the resulting four fractions are electrophoresed on one-dimensional gels followed by autoradiography. This results in four groups of labeled membrane proteins with little overlap between each group (unpublished results).

There are two features of the present techniques which increase the sensitivity of detecting proteolytic cleavages of membrane proteins. First, the membrane proteins are fractionated into four different groups with limited overlap between groups; this simplifies the one-dimensional gel patterns. Second, membrane turnover is inhibited, permitting the identification of cleavages of membrane proteins that turnover rapidly. Indeed, most past attempts to identify cleavages of membrane proteins using intact fibroblasts[17] or platelets[18] treated with thrombin have revealed only several sensitive proteins. Many more cleavages are identified when membranes are first purified and subsequently treated with thrombin; this is true both with platelets[19] and fibroblasts (unpublished results). By using the techniques described herein, similar patterns of thrombin-sensitive proteins are observed using thrombin treatment of intact cells or of isolated membranes.

Acknowledgments

This research was supported by Grant CA 12306 from the National Institutes of Health. We thank L. Cho and S. Sternberg for excellent technical assistance.

[17] M. Moss and D. D. Cunningham, *J. Supramol. Struct.* **15**, 49 (1981).
[18] D. R. Phillips and P. P. Agin, *Biochim. Biophys. Acta* **352**, 218 (1974).

Section III

Nerve and Glial Growth Factors

[15] Two-Site Enzyme Immunoassay for Nerve Growth Factor

By SIGRUN KORSCHING and HANS THOENEN

Introduction

Conflicting results have been obtained in the past using various assay methods for the quantitation of nerve growth factor (NGF). For example, serum NGF levels ranging between the undetectable (i.e., less than 0.05 ng/ml[1]) and 500 ng NGF/ml have been reported.[1–6] The difficulties encountered in NGF determinations are due to the presence of NGF-binding proteins and the strong nonspecific adsorptivity of NGF. These difficulties become significant at the extremely low NGF levels in tissues in which NGF has a physiological role.[1] Problems arising from the presence of binding proteins are not unique for NGF, but have also been reported for other rare proteins such as platelet-derived growth factor.[7] In this chapter we will deal only with immunological determinations of NGF, although it should be noted that probably as many difficulties occur in the determination of its biological activity (cf. Ref. 8).

Comparison of Competition and Two-Site Assay Concepts

Many false-positive NGF determinations have been obtained using competition assays (one-site assays).[2,4,5,9,10] Such results could not be confirmed with both reliable two-site immunoassays and biological assay methods.[1,6,11–14] Whenever the presence of binding proteins for the antigen

[1] S. Korsching and H. Thoenen, *Proc. Natl. Acad. Sci. U.S.A.* **80**, 3513 (1983).
[2] D. G. Johnson, P. Gorden, and I. J. Kopin, *J. Neurochem.* **18**, 2355 (1971).
[3] I. A. Hendry, *Biochem. J.* **128**, 1265 (1972).
[4] R. N. Fabricant, J. E. De Larco, and G. J. Todaro, *Proc. Natl. Acad. Sci. U.S.A.* **74**, 565 (1977).
[5] S. Furukawa and K. Hayashi, *Life Sci.* **26**, 837 (1980).
[6] K. Suda, Y.-A. Barde, and H. Thoenen, *Proc. Natl. Acad. Sci. U.S.A.* **75**, 4042 (1978).
[7] J. S. Huang, S. S. Huang, and T. F. Deuel, *J. Cell Biol.* **97**, 383 (1983).
[8] H. Thoenen and Y.-A. Barde, *Physiol. Rev.* **60**, 1284 (1980).
[9] J. P. Schwartz and X. O. Breakefield, *Proc. Natl. Acad. Sci. U.S.A.* **77**, 1154 (1980).
[10] G. A. Bray, Y. Shimomura, M. Ohtake, and P. Walker, *Endocrinology* **110**, 47 (1982).
[11] G. P. Harper, F. L. Pearce, and C. A. Vernon, *Dev. Biol.* **77**, 391 (1980).
[12] S. D. Skaper and S. Varon, *Exp. Neurol.* **76**, 655 (1982).
[13] S. Korsching, G. Auburger, R. Heumann, J. Scott, and H. Thoenen, *EMBO J.* **4**, 1389 (1985).
[14] S. Furukawa, I. Kamo, Y. Furukawa, S. Akazawa, E. Satoyoshi, K. Itoh, and K. Hayashi, *J. Neurochem.* **40**, 734 (1983).

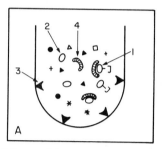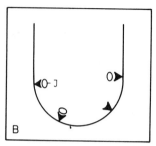

FIG. 1. Scheme of the competitive (one-site) radioimmunoassay. (A) ^{125}I-labeled NGF (1) and unlabeled NGF (2) compete for a small number of antibody molecules (3). Both unlabeled NGF and NGF-binding molecules have removed labeled NGF from the solid phase antibodies. NGF-binding molecule (4). (B) The radioactivity remaining after washing steps is quantified.

of interest cannot be excluded, a competition assay will run the risk of false-positive results. In this type of assay, radioactively labeled NGF added to the sample and endogenous NGF in the sample compete for a limited number of binding sites, which are either antibodies or receptors. Figure 1 illustrates such a competition assay. The decrease of bound labeled NGF can be interpreted as due to the presence of unlabeled NGF, but may equally well be caused by the presence of NGF-binding macromolecules in the sample. Such molecules, e.g., α_2-macroglobulin,[15] have been demonstrated in serum.[6] The effect of these binding proteins is especially serious, since they are present in far greater numbers than the antibody molecules (a small, limiting number of antibody molecules is pertinent to achieve a sensitive assay). Furthermore there is no possibility of correcting for errors resulting from the presence of such binding molecules.

In contrast, the presence of binding proteins does not interfere with the two-site immunoassay shown in Fig. 2. NGF present in the sample is bound to a large excess of antibodies linked to the solid phase. The bound NGF is then detected by a second, labeled antibody. In this type of assay, the presence of binding proteins leads to an underestimation of the true NGF content that can be corrected for by the determination of recovery of NGF added to the sample. However, in the two-site assay the purity of the antibody preparations used is much more critical than in the competition assay, as will be discussed below. Erroneously high NGF levels have been found in a variety of tissues, when analyzed by a two-site radioimmunoassay using impure antibody preparations.[3]

[15] H. Ronne, H. Anundi, L. Rask, and P. A. Peterson, *Biochem. Biophys. Res. Commun.* **87**, 330 (1979).

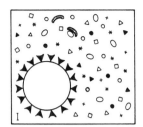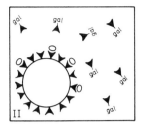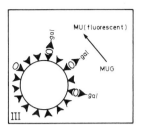

FIG. 2. Scheme of the two-site enzyme immunoassay (symbols as in Fig. 1). (I) Unlabeled NGF in the sample binds to an excess of solid phase antibodies and to NGF-binding molecule. (II) The galactosidase-labeled second antibody binds to NGF. (III) The bound second antibody is quantified by production of intensely fluorescent 4-methylumbelliferone (MU).

Several two-site immunoassays for NGF have been developed.[1,6,14,16,16a] However, we will restrict our description to the enzyme immunoassay (EIA) developed in our laboratory[1,17] because of its high sensitivity.

Antibody Preparation

Polyclonal sheep anti-NGF antibodies were obtained by intracutaneous immunization with a dose of 1 mg NGF[18] in Freund's complete adjuvant followed by a booster with half the dose 4 weeks later.[19] The antibodies were purified by affinity chromatography on a Sepharose 4B column to which NGF had been coupled via cyanogen bromide activation.[19] Similar results were obtained by subcutaneous immunization of rabbits with 5–100 μg NGF and repeating the same dose every 2 weeks.[17] The antibodies from the low-dose immunized rabbits did not possess higher than average affinity.[17] The affinity-purified antibodies from both rabbit and sheep were monospecific in an Ouchterlony double immunodiffusion[20] against NGF.[17] However, antibodies against possible contaminants of the NGF preparation will necessarily copurify during affinity purification. Since the usual procedures for NGF purification (Bocchini and Angeletti,[21] Mobley *et*

[16] T. Ebendal, L. Olson, and Å. Seiger, *Exp. Cell Res.* **148**, 311 (1983).

[16a] G. Weskamp and U. Otten, *J. Neurochem.*, in press.

[17] S. Korsching, Ph.D. thesis, Ludwigs-Maximilians-University, Munich, 1984.

[18] Unless otherwise specified, NGF refers to the 2.5 S form of mouse β-NGF (molecular weight 26K). The biological activity of the 2.5 S form is indistinguishable from that of β-NGF, which is the biologically active, neutrotrophic subunit of the 7 S NGF complex (cf. Ref. 8).

[19] K. Stöckel, C. Gagnon, G. Guroff, and H. Thoenen, *J. Neurochem.* **26**, 1207 (1976).

[20] Ö. Ouchterlony and L. Å. Nilsson, *in* "Handbook of Experimental Immunology" (D. M. Weir, ed.), pp. 19.1–19.39. Blackwell Scientific, Oxford, England, 1973.

[21] V. Bocchini and P. U. Angeletti, *Proc. Natl. Acad. Sci. U.S.A.* **64**, 787 (1969).

al.,[22] Suda *et al.*[6]) result in not more than 95% purity (a common contaminant of both β-NGF and 7 S NGF are mouse immunoglobulins[23,24]), one has to assume that affinity-purified polyclonal anti-NGF antibodies in general contain some contaminating antibodies, which may not be detected in standard procedures for purity control.

Monoclonal anti-NGF antibodies were prepared according to the procedure of Köhler and Milstein[25] (see also Neet *et al.* [16], this volume). BALB/c mice were immunized with 2.5 S mouse NGF by the method of Stähli *et al.*[26] The mouse myeloma line Ag8.653[27] was used for fusion using polyethylene glycol. Positive clones are selected by dot immunoassay[28] with NGF absorbed to nitrocellulose and detection of the bound antibodies by anti-mouse immunoglobulin–horseradish peroxidase conjugate.

The antibodies were purified from culture supernatants by affinity chromatography on protein A-Sepharose[29] at pH 8.5 (some mouse IgG subtypes do not bind to protein A below pH 8[29]). The mouse Ig preparations obtained contained between 30 and 70% fetal calf Ig originating from the fetal calf serum used for the culture (cf. Ref. 30). Should this constitute a problem (which was not the case with the batches of fetal calf serum we used), the monoclonal antibody 27/21 can also be produced in low-dose serum (stepwise reduction to 0.1% serum). This antibody was used in most of the experiments; its subtype is IgG_1, as determined by immunodiffusion[20] against subtype-specific antibodies (Meloy) in a 1% agarose gel containing 1% polyethylene glycol 4000. The cross-reactivity of 27/21 with bovine NGF[31] was determined to be 70% by EIA.

Unspecific polyclonal sheep antibodies were obtained from nonimmune serum by 40% ammonium sulfate precipitation at pH 7.7, followed by ion-exchange chromatography on DEAE-cellulose (DE-52, What-

[22] W. C. Mobley, A. Schenker, and E. M. Shooter, *Biochemistry* **15**, 5543 (1976).

[23] J. T. Tomita and S. Varon, *Neurobiology* **1**, 176 (1971).

[24] J. R. Carstairs, D. C. Edwards, F. L. Pearce, C. A. Vernon, and S. J. Walter, *Eur. J. Biochem.* **77**, 311 (1977).

[25] G. Koehler and C. Milstein, *Nature (London)* **256**, 495 (1975).

[26] C. Stähli, T. Staehelin, V. Miggiano, J. Schmidt, and P. Häring, *J. Immunol. Methods* **32**, 297 (1980).

[27] J. F. Kearney, A. Radbruch, B. Liesegang, and K. Rajewsky, *J. Immunol.* **123**, 1548 (1979).

[28] R. Hawkes, E. Niday, and J. Gordon, *Anal. Biochem.* **119**, 142 (1982).

[29] P. L. Ey, S. J. Prowse, and C. R. Jenkin, *Immunochemistry* **15**, 429 (1978).

[30] P. A. Underwood, J. F. Kelly, D. F. Harman, and H. M. MacMillan, *J. Immunol. Methods* **60**, 33 (1983).

[31] G. P. Harper, R. W. Glanville, and H. Thoenen, *J. Biol. Chem.* **257**, 8541 (1982).

man).[32] Unspecific mouse IgG was obtained by affinity chromatography of mouse serum on protein A-Sepharose at pH 8.5.[29]

Selection of the Marker Enzyme

Use of an enzyme as label introduces a theoretically unlimited amplification factor in the immunoassay. Three enzymes are mainly used as labels in the EIA: horseradish peroxidase, alkaline phosphatase, and β-galactosidase. We use β-galactosidase which has the following advantages: the turnover number is relatively high (6000 sec^{-1} at 25°[33]), unlike alkaline phosphatase and horseradish peroxidase, a stable fluorogenic substrate, 4-methylumbelliferyl-β-D-galactoside (MUG, Sigma), is available, for which the enzyme activity is 35% of that for the chromogenic substrate o-nitrophenyl-β-D-galactoside (the detection limit for 4-methylumbelliferone is 0.1 ng/ml using a Perkin Elmer 650-40 fluorometer), the enzyme reaction is linear for up to 30 hr (data not shown), the enzyme tolerates high salt (up to 5 M NaCl) and nonionic detergent concentrations (1% Triton X-100), and finally, it has free SH groups[33] that are not necessary for activity and therefore can be used for derivatization.

Coupling of the First Antibody to the Solid Phase

Most solid phase immunoassays described in the literature (cf. Refs. 34–37) use physical adsorption, usually to polystyrene, for the immobilization of the first antibody. We tested glass, silicone rubber, and polystyrene in the EIA for their suitability as solid phase and found no significant difference in the signal-to-noise ratios obtained (data not shown). The stability of the adsorbed antibodies was investigated by performing the EIA with radioactively labeled sheep anti-NGF antibodies. The antibodies were iodinated according to the lactoperoxidase technique of Marchalonis[38] as modified by Rohrer and Barde.[39] After overnight incubation at

[32] J. L. Fahey, in "Methods in Immunology and Immunochemistry" (C. A. Williams and M. W. Chase, eds.), Vol. 1, pp. 307–339. Academic Press, New York, 1967.

[33] K. Wallenfels and R. Weil, in "The Enzymes" (P. D. Boyer, ed.), Vol. 7, 3rd Ed., pp. 617–663. Academic Press, New York, 1972.

[34] A. H. W. M. Schuurs and B. K. van Weemen, Clin. Chim. Acta 81, 1 (1977).

[35] S. Avrameas, T. Ternynck, and J.-L. Guesdon, Scand. J. Immunol. 8 (Suppl. 7), 7 (1978).

[36] E. Ishikawa and K. Kato, Scand. J. Immunol. 8 (Suppl. 7), 43 (1978).

[37] J. E. Herrmann, this series, Vol. 73, p. 239.

[38] J. J. Marchalonis, Biochem. J. 113, 299 (1969).

[39] H. Rohrer and Y.-A. Barde, Dev. Biol. 89, 309 (1982).

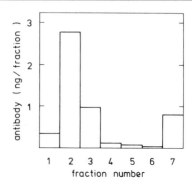

FIG. 3. Loss of adsorbed antibody during EIA. The amount of sheep anti-NGF antibodies recovered in the different steps of the EIA is shown. The coating solution contained no detergent, whereas both in EIA buffer (fractions 3 and 5) and W buffer (fractions 2, 4, and 6) 0.1% Triton X-100 is present. (1) Supernatant after coating; (2) 2× washing buffer; (3) supernatant of step I (see Fig. 2); (4) 2× washing buffer; (5) supernatant of step II (see Fig. 2); (6) 2× washing buffer; (7) solid phase (96-well microtiter plate, Greiner) after performance of steps 1–6. Taken from Korsching.[17]

room temperature with 1 μg antibody/ml 50 mM Tris–HCl, pH 9.0, 90% of the added antibodies initially bind to the polystyrene wells (Fig. 3), i.e., 130 ng/cm^2. For a monomolecular layer of antibody molecules packed at maximal density on the solid phase, 110 to 270 ng antibody per cm^2 were calculated, assuming a smooth well surface and an antibody molecule of diameter derived from the crystallographic (8 nm[40]) and the hydrodynamic radius (5.2 nm[17,41–43]), respectively.

Although initially nearly all the antibodies were coated, the majority (80–85% of initially adsorbed radioactivity) is lost during the EIA procedure (Fig. 3). This is probably due to the presence of nonionic detergents in the EIA buffer, which disturbs the hydrophobic binding interaction between antibody and polystyrene. However, the presence of detergent is essential to reduce nonspecific adsorption to the solid phase and thereby the background signal in the EIA. Interestingly, protein concentrations higher than 1 μg/ml in the coating solution led to smaller specific signals and therefore to lower signal-to-noise ratios (S/N) (Fig. 4). At these higher concentrations protein to protein adsorption should occur besides protein

[40] R. E. Cathou, in "Comprehensive Immunology" (G. W. Litman and R. A. Good, eds.), Vol. 5, pp. 37–83. Plenum, New York, 1978.
[41] E. Marler, C. A. Nelson, and C. Tanford, Biochemistry 3, 279 (1964).
[42] M. E. Noelken, C. A. Nelson, C. E. Buckley, and C. Tanford, J. Biol. Chem. 240, 218 (1965).
[43] J. T. Edsall, in "The Proteins: Chemistry, Biological Activity and Methods" (H. Neurath and K. Bailey, eds.), Vol. 1, pp. 549–726. Academic Press, New York, 1953.

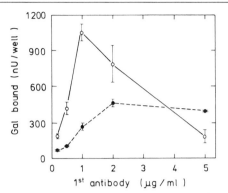

FIG. 4. Influence of antibody concentration in the coating solution. The concentration of sheep anti-NGF antibody in the coating solution was varied between 0.2 and 5 μg/ml. The EIA was performed in microtiter plate wells essentially under standard conditions. (○) Specific signal at 0.5 ng NGF/ml (mean ± SEM); (●) nonspecific background signal at 0 ng NGF/ml (mean ± SEM). Taken from Korsching.[17]

to plastic adsorption, the former being presumably less stable than the protein/polystyrene interaction.

Because of the drastic losses of adsorbed antibodies, we decided to use antibodies covalently coupled to glass beads as the solid phase. Antibodies were coupled to beads instead of wells since there the ratio of reactive surface to reaction volume is freely variable. A practical advantage of using beads is that the coupling reaction can easily be performed on a much larger scale than feasible for wells, i.e., several thousand beads can be derivatized at once.

Coupling Reaction

Amino groups were introduced to glass beads by condensation of surface hydroxyl groups to a silyl ester carrying an alkylamine chain.[44] The amino groups were derivatized with glutaraldehyde. After removal of the excess glutaraldehyde, the antibody was coupled via its amino groups. Standard reaction conditions are shown in the following flow diagram:

30 g glass beads (1 mm diameter, Sigma, pretreated with concentrated HNO₃)
5 ml triethoxy(3-aminopropyl)silane (Sigma)
45 ml toluene

(I) | reflux 3 hr
    ↓ wash 5× with toluene and dry
+ 25 ml 1% glutaraldehyde (stock solution 25%, EM grade)
in 0.1 M potassium phosphate, pH 7.4

[44] H. H. Weetall, *Biochim. Biophys. Acta* **212**, 1 (1970).

(II) | 3 hr on ice
      ↓ wash 6× with 0.1 M potassium phosphate, pH 7.4
+ 6 ml of 100 μg/ml antibody in 0.1 M potassium phosphate, pH 7.4
(III) ↓ overnight at 4°
stop reaction by addition of W buffer,[44a] wash 5× with W buffer, store at 4° in W buffer

The antibody beads could be stored for at least 1 year without loss of activity in the EIA.

All aqueous solutions were made using quartz-distilled H_2O. Variation of the reaction times in the range of 0.5 to 5 hr for the first step, 1 to 15 hr for the second step, and 3 to 15 hr for the third step had no significant influence on the performance of the antibody beads in the EIA. Variation of the glutaraldehyde concentration between 0.03 and 2% changed neither the affinity nor the capacity of the antibody beads significantly, as determined by Scatchard analysis[45] of ^{125}I-labeled NGF binding (data not shown). Below 0.01% glutaraldehyde no specific binding was obtained.

Characterization of Covalently Coupled Antibodies

Affinity and capacity of the solid phase antibodies are determined by incubation with iodinated NGF (cf. Rohrer and Barde[39]) and Scatchard analysis[45] of the specific binding (Fig. 5). Unspecific binding was determined using nonspecific antibodies as solid phase. It did not exceed 5% of the total binding. The dissociation constant (K_D) of covalently coupled antibodies was 0.27 ± 0.04 nM (SEM) for the monoclonal 27/21 and 0.44 ± 0.16 nM for polyclonal sheep antibodies. This affinity was not significantly different from that of antibodies adsorbed to polystyrene (data not shown).

The Scatchard graph for the polyclonal antibodies is reasonably linear, indicating that the affinity distribution of the antibodies is relatively narrow. Stringent washing conditions during the affinity purification[19] of the antibodies probably lead to the loss of low-affinity antibodies. Several monoclonal antibodies obtained from two fusions did not exhibit usefully higher affinities than the mean affinity of the polyclonal antibodies. The capacity was 0.18 ± 0.04 ng NGF (SEM) for monoclonal 27/21 beads and 0.73 ± 0.08 ng for polyclonal sheep antibody beads. The latter value is near the value calculated (see above) for a monomolecular layer (1–3 ng NGF/bead assuming 2 NGF bound per antibody molecule), i.e., approaches the theoretical maximum.

The kinetic properties of the monoclonal antibody 27/21 were analyzed in detail. The time course of association was studied at room tem-

---

[44a] For composition, see standard EIA procedure.
[45] G. Scatchard, *Ann. N.Y. Acad. Sci.* **51**, 660 (1949)

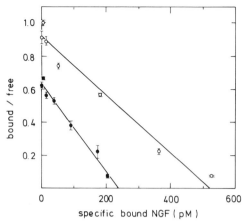

FIG. 5. Binding of ^{125}I-labeled NGF to specific antibodies. The antibodies were coupled to beads and incubated for 22 hr at room temperature with 2 pM to 8 nM ^{125}I-labeled NGF. Specifically bound to free NGF is shown as function of specifically bound NGF. (○) Sheep anti-NGF antibody (mean ± SEM); (●) monoclonal antibody 27/21 (mean ± SEM). Taken from Korsching.[17]

perature and 37° between 5 min and 32 hr using 2 nM ^{125}I-labeled NGF in W buffer. The association rate constant k_{+1} was determined assuming a second-order reaction (A + N ⇌ AN) from the equation

$$k_{+1}t = \frac{1}{K_D + N_0} \ln\left(\frac{AN_e}{AN_e - AN_t}\right)$$

where AN is the antibody–NGF complex concentration, N is the free NGF concentration (N_t approximated as N_0), index 0 is the concentration at time 0, index t is the concentration at time t, index e is the equilibrium concentration, and K_D is the dissociation constant.

The relatively low value for the association rate constant [k_{+1} (room temperature) = 2.2 × 10⁴ M^{-1} sec⁻¹, k_{+1} (37°) = 7.3 × 10⁴ M^{-1} sec⁻¹] indicates that long incubation times are required to reach the equilibrium, e.g., with 10 pg NGF/ml, 20 hr incubation at room temperature are needed to reach 35% of the equilibrium binding.

The time course of dissociation was studied at 37° between 20 and 300 min. After overnight preincubation at 37° with 1 nM ^{125}I-labeled NGF dissociation was initiated by 3 washes with W buffer and incubation at 37° in W buffer containing 380 nM unlabeled NGF. The dissociation rate constant k_{-1} was determined assuming a first-order decay (AN → A + N) and was found to be 3.2 × 10⁻⁵ sec⁻¹, yielding a half-life of 6.0 hr at 37°. It follows that washing periods of several hours at room temperature are acceptable.

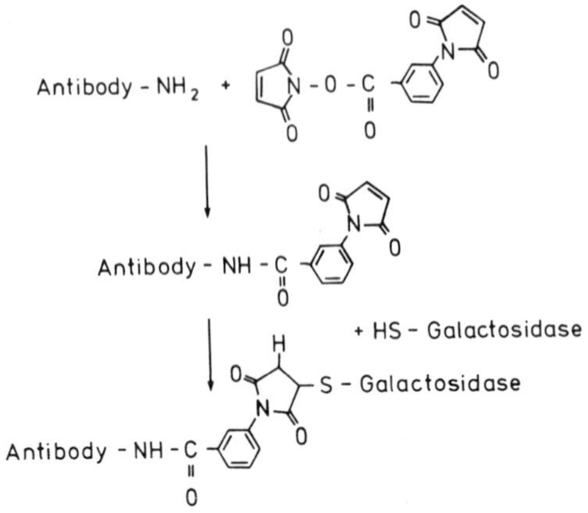

FIG. 6. Scheme of the coupling reaction for antibody and β-galactosidase.

Some batches of monoclonal antibody 27/21 beads showed poor performance in the EIA when low NGF concentrations (in the range of the standard curve) were analyzed, although both the capacity and affinity of these batches were found to be in the average range. This phenomenon probably results from the occupation by the solid phase antibody of both 27/21-binding sites present on the NGF dimer. A further indication that this is the case is that at high NGF concentrations (50–100 ng NGF/ml) the signal in the EIA is as high as usual. It would therefore be desirable to use two different monoclonal antibodies for the two sites of the EIA, whose binding sites on the NGF molecule do not overlap. However, although several monoclonal anti-NGF antibodies were tested for competition with 27/21 galactosidase conjugate in the EIA, no monoclonal with a binding site clearly different from that of 27/21 was found.

Coupling of the Second Antibody to β-Galactosidase

The heterobifunctional cross-linker m-maleimidobenzoyl-N-hydroxysuccinimide ester (MBS, Sigma) was used as coupling reagent according to O'Sullivan et al.[46] The reaction principle is shown in Fig. 6.

One milligram of antibody dissolved in 1 ml 0.1 M sodium phosphate, pH 7.0, 50 mM NaCl was rapidly mixed with 0.25 mg MBS in 10 μl tetrahydrofuran and incubated for 1 hr at 30° in a shuttle waterbath. Unreacted MBS was then removed by gel filtration over a BioGel P-30 column in 10

[46] M. J. O'Sullivan, E. Gnemmi, D. Morris, G. Chieregatti, A. D. Simmonds, M. Simmons, J. W. Bridges, and V. Marks, *Anal. Biochem.* **100,** 100 (1979).

mM sodium phosphate, pH 7.0, containing 50 mM NaCl. The antibody appears in the void volume. Under these conditions the absorption losses on the column amount to 10–30%, as compared to 50–70% on a Sephadex G-20 column. Immediately after elution of the activated antibody, 1 mg of β-galactosidase (*E. coli,* 600–800 U/mg, EIA quality, Boehringer-Mannheim) freshly dissolved in equilibration buffer was added. After 1 hr at 30° the reaction was terminated by addition of 1 M 2-mercaptoethanol to yield a final concentration of 2 mM. BSA and NaN$_3$ were added to final concentrations of 0.1 and 0.05% and the conjugate was stored at 4°. Performance in the EIA was unimpaired for at least 1 year. The conjugate should not be frozen, since enzyme activity is then reduced considerably.

Eighty to 100% enzyme activity was recovered after the coupling reaction, whereas the antibody activity decreased to 20–40%, as judged by performance in the protein A radioimmunoassay[47] using ^{125}I-labeled NGF, pansorbin (Calbiochem), and antibody concentrations of 0.01 to 1 μg/ml. Competition of enzyme-labeled antibodies with native antibodies showed that 30–50% of the remaining antibody activity was coupled to β-galactosidase. The overall coupling efficiency was around 50% for both antibody and enzyme, as judged by native, nonreducing agarose gel electrophoresis[48] at pH 8.6 [1% agarose, electroendoosmosis = -0.13, running buffer 92 mM Tris, 30 mM diethylbarbituric acid, 0.1% Triton X-100 (v/v), field strength 20 V/cm, run time 2 hr at 0°]. Under these conditions antibodies (pI around neutral), β-galactosidase (pI 4.6^{33}), and conjugate run in distinctly different positions. Proteins were fixed with 50 mM picric acid, 17% acetic acid (v/v), and stained with Coomassie Brilliant Blue for densitometric evaluation.

Optimization of the EIA

Ratio of Reactive Surface to Reaction Volume

When using antibodies coupled to wells as solid phase, the reactive surface/reaction volume ratio is not variable but determined by the size of the reaction volume and will increase with decreasing reaction volume. Since the background signal in the EIA is proportional to the reactive surface, the S/N ratio decreases by reduction of the assay volume, which is, however, desirable for technical reasons. In contrast, the surface/volume ratio can be manipulated at will, using beads as reactive surface.

[47] S. Jonsson and G. Kronvall, *Eur. J. Immunol.* **4**, 29 (1974).
[48] J. Miyake, H. Fehrnström, and B. Wallenborg, "Agarose Gel Electrophoresis with LKB 2117 Multiphor. Application Note 310" (LKB, ed.), pp. 1. LKB, Bromma, Sweden, 1977.

The optimal S/N ratio was obtained with a single bead/50 μl, when the specific signal was already more than half-maximal (Fig. 7). The same result was obtained for monoclonal antibodies testing 1 to 10 beads/50 μl (data not shown).

Concentration of the Second Antibody

The optimal dilution of the second antibody was determined for each batch of antibody–galactosidase conjugate. A typical example is shown in Fig. 8. Within a broad range the background signal depends linearly on the conjugate concentration, whereas a saturation curve is observed for the specific signal. The highest conjugate concentration still showing an optimal S/N ratio was used in the EIA.

Incubation Time with the Second Antibody

The reaction with the first antibody could in theory be performed until equilibrium is reached since then the maximal signal will be obtained. The situation in the second step of the EIA is more complicated since the second antibody is present in excess over the first antibody. Therefore most of the NGF bound to the solid phase after the first incubation step will end up being bound to two molecules of second antibody, and will thus no longer be detectable.

The EIA was therefore performed with different incubation times for the reaction with the second antibody. Saturation of the specific signal

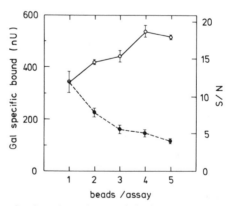

FIG. 7. Reactive surface/reaction volume ratio in the EIA. The EIA was performed under standard conditions with 1 to 5 sheep anti-NGF antibody beads per 50 μl NGF standard or EIA buffer, corresponding to a surface/volume ratio between 3.14 mm²/50 μl and 15.7 mm²/50 μl. (○) Specific signal ± SEM at 0.5 ng NGF/ml (left scale); (●) signal-to-noise ratio (right scale). Taken from Korsching.[17]

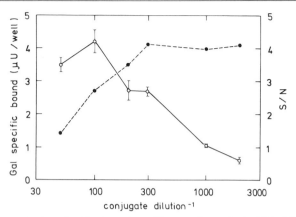

FIG. 8. Concentration of antibody–galactosidase conjugate in the EIA. Microtiter plates were coated with 1 μg/ml sheep anti-NGF antibodies. The EIA was performed essentially under standard conditions using 3–120 mU$_{MUG}$ galactosidase–antibody conjugate/assay (corresponding to dilution 2000^{-1}–50^{-1}). (○) Specific signal $\pm$ SEM at 0.5 ng NGF/ml (left scale); (●) signal-to-noise ratio (right scale). Taken from Korsching.[17]

was achieved after 1–3 hr, depending on the concentration of the second antibody (Fig. 9). Saturation was reached earlier with higher conjugate concentrations. Since the nonspecific signal increased with increasing incubation times (data not shown), a standard reaction time of 2 hr was

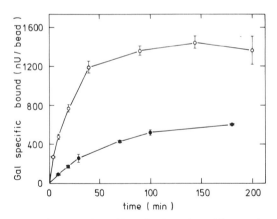

FIG. 9. Time course of the reaction with antibody–galactosidase conjugate. The EIA was performed under standard conditions using the monoclonal 27/21 with 0 and 0.5 ng NGF/ml. The specific signal is shown as a function of incubation time (5–200 min). (○) Using 120 mU$_{MUG}$ galactosidase/assay; (●) using 30 mU$_{MUG}$ galactosidase/assay. Taken from Korsching.[17]

chosen and the conjugate concentration was optimized for this incubation time (cf. Fig. 8).

Influence of pH

The EIA was performed under standard conditions, but with the pH of both EIA and washing buffer varying between 6.5 and 9.0 (steps of 0.5). The antibody 27/21 coupled to beads served as solid phase. Both the specific signal and the background signal did not change between pH 6.5 and 8.0 (data not shown). However, at more alkaline pH, the specific signal is reduced to two-thirds with a concurrent 4-fold increase of the background signal. Qualitatively, the same result was obtained with polyclonal sheep anti-NGF antibodies (data not shown).

Variation of Ionic Strength

An increase in ionic strength leads to a weakening of all polar interactions in the solution. Specific and nonspecific binding of antibody molecules may conceivably depend to a differing extent on polar interactions. The ionic strength in EIA and washing buffer was varied by addition of 0 to 5 M NaCl. Interestingly still one-quarter of the maximal specific binding is retained using 4 M NaCl. The results for 0 to 1 M NaCl are shown in Fig. 10. A medium salt concentration of 0.2 M resulted in an optimal S/N ratio.

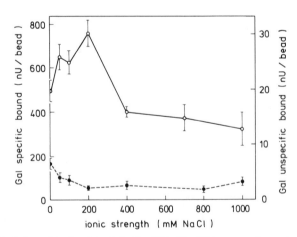

FIG. 10. Variation of ionic strength in the EIA. The EIA was performed using the 27/21 antibody under standard conditions, excluding rat serum in the second incubation step, but with 0 to 1 M NaCl in both EIA and washing buffer. (○) Specific signal ± SEM at 1.28 ng NGF/ml (left scale); (●) background signal ± SEM (right scale).

Influence of Nonionic Detergent

The EIA was performed under standard conditions, but with 0.1–0.25% Triton X-100 (Sigma), 0.1% Tween 20 (Serva, Heidelberg), or 0.2% octylglucoside (Sigma) each present in both EIA and washing buffer. The S/N ratio was comparable when using Triton X-100 or octylglucoside, but decreased slightly with Tween 20 (data not shown). Increasing the Triton X-100 concentration to 0.25% did not decrease the background signal further.

Addition of Rat Serum

Serum contains several carrier molecules which might be able to reduce nonspecific absorption. Since rat serum contains no NGF,[1] it could be added during either of the two incubation steps of the EIA. The specific signal was not influenced by addition of 5 or 10% rat serum, whereas the background signal decreased severalfold in all cases (data not shown). The optimal S/N ratio was obtained when 10% rat serum was added during the incubation with the antibody–galactosidase conjugate.

Standard EIA Procedure

The EIA procedure is shown in the following flow diagram:

1 bead coupled to anti-NGF or nonspecific antibodies
(0.5–2.5 ng antibody/bead)
+
50 μl NGF standard or tissue sample
in EIA buffer, pH 7.0, + protease inhibitors
(I) | 20 hr at room temperature
↓ wash 3× with 200 μl W buffer
+ 50 μl anti-NGF–antibody–β-galactosidase conjugate
in EIA buffer, pH 7.0, + 10% rat serum
(5–20 mU$_{MUG}$ galactosidase and 10–20 ng antibody per assay)
(II) | 2 hr at 37°
| wash 2× with 200 μl W buffer
| wash 1× with 150 μl 0.1 *M* sodium
↓ phosphate, pH 7.3, 1 m*M* MgCl₂
transfer of bead in new polystyrene tube (3.5 ml) containing
50 μl 200 μ*M* MUG in 0.1 *M* sodium phosphate, pH 7.3, 1 m*M* MgCl₂
(III) ↓ 20 hr at room temperature in the dark
stop by addition of 650 μl 0.15 *M* glycine/NaOH, pH 10.3
measure fluorescence at 364 nm excitation and 448 nm emission

Reagents

EIA buffer
50 m*M* Tris–HCl, pH 7.0

0.2 M NaCl
0.05% NaN$_3$
0.1% Triton X-100 (v/v)
1% bovine serum albumin (BSA, Sigma)
1% gelatin (calf skin, Sigma)

W buffer
    as EIA buffer, but without BSA

The EIA was performed either with monoclonal or with polyclonal antibodies. The same monoclonal could be used for both the first and second site of the EIA, since NGF is a homodimer (cf. Thoenen and Barde[8]). A standard curve in the range of 2.5 pg to 1.25 ng NGF/ml was measured for each assay (Fig. 11). Determinations were done in quadruplicate both for specific and nonspecific antibodies. The assay was performed in 96-well microtiter plates (Greiner), which were presaturated with W buffer. To avoid evaporation losses due to the small sample volumes used, the assay was performed in a water-saturated atmosphere. Since the diffusion of large molecules is fairly slow at room temperature, the incubations were done shaking.

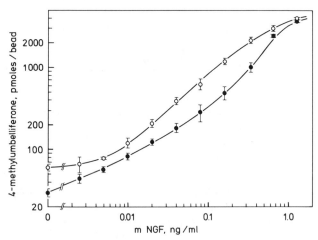

FIG. 11. Standard curves for the EIA. The EIA was performed under standard conditions excluding rat serum in the second incubation step. Bound galactosidase activity is shown as a function of NGF concentration (2.5 pg to 1.28 ng NGF/ml). (○) Polyclonal sheep anti-NGF antibodies, 24 hr incubation with MUG; (●) monoclonal antibody 27/21, 15 hr incubation with MUG. Taken from Korsching.[17]

Sensitivity of the EIA

The background signal amounted to 5×10^{-6} to 5×10^{-7} parts of the galactosidase activity added, depending on antibody and conjugate preparation. The detection limit was defined as the NGF concentration yielding an S/N ratio of 1. The detection limit is 5 pg NGF/ml (0.2 pg NGF/assay equivalent to 0.01 fmol/assay) for the monoclonal antibody 27/21 and double these values for the polyclonal sheep anti-NGF antibody. The maximal specific signal was 300 times higher than the background signal. The working range of the EIA therefore extends over more than 2 orders of magnitude.

Specificity of the EIA

The specificity of the antibodies used was determined by performing the EIA with 10 μg/ml of the following substances, which either resemble NGF with respect to isoelectric point, molecular weight (cytochrome c, aprotinin, β-bungarotoxin, Sigma), and amino acid sequence (bovine insulin,[49] Sigma), or were suspected as NGF contaminants (mouse immunoglobulin[23,24]). The resulting signals were not higher than background level, with the exception of mouse immunoglobulin using polyclonal sheep anti-NGF antibodies. The latter signal corresponds to 0.5% of anti-mouse IgG antibodies in the affinity-purified sheep anti-NGF antibodies. This small percentage will not become visible in an immunodiffusion assay. However, since the concentration of possible contaminating antigens in tissues or serum may by far exceed the extremely low NGF levels present in vivo,[1,13] false-positive results cannot be excluded when using only polyclonal antibodies for NGF determination. This is especially true since in the two-site EIA antibodies are present in great excess over the antigen. In comparison, monoclonal antibodies have the advantage of a unique specificity for a given epitope, however, this epitope may by chance also be present on a molecule otherwise unrelated to NGF. A combined approach using both polyclonal and monoclonal antibodies is likely to avoid either of these problems and was therefore generally used for NGF determination.

NGF Determination in Tissue and Blood

Tissues are minced, if necessary (e.g., nerve, skin), and glass/glass homogenized with 10 to 20 strokes at 0°. Homogenate concentrations

[49] M. N. Sabesan, J. Theor. Biol. 83, 469 (1980).

between 0.2 and 20% (w/v) could be used in the assay. The homogenization buffer contains 100 mM Tris–HCl, pH 7.0, 0.4 M NaCl, 0.1% NaN$_3$, 2% BSA, and 2% gelatin, but no detergent to avoid solubilization of NGF receptors and Fc receptors, which both could disturb the EIA. The high concentrations of BSA and collagen (main component of gelatin), both of which bind NGF,[50] should provide enough soluble binding sites for NGF to compete efficiently with adsorption of NGF to the insoluble tissue components. Protease inhibitors (final concentrations 4 mM EDTA, 2 mM benzamidine hydrochloride, 0.2 mM benzethonium chloride, 40 kallikrein U/ml aprotinin, 0.2 mM phenylmethanesulfonyl fluoride) were added to the homogenization buffer immediately before homogenization. The homogenate could be frozen and rethawed without effect on the NGF levels measured. The homogenate was centrifuged at 10^5 g for 10 min at 15° (Beckman ultracentrifuge, SW 27 rotor with adaptors, Ultra Clear tubes, Beckman) and the supernatant diluted 1 : 1 with 0.2% Triton X-100 to reach the composition of the EIA buffer used for the NGF standard solutions. To reduce adsorption losses both the homogenization vessels and the centrifuge tubes were saturated with protein by preincubation in W buffer.

The recovery of endogenous NGF was determined by adding 0.05–0.3 ng/ml mouse NGF to the homogenate and measuring the percentage of recovered NGF. For stringent control, it is important to add exogenous NGF in quantities corresponding to the range of endogenous NGF concentrations. Addition of the exogenous NGF to the supernatant of the centrifugation instead of adding NGF to the homogenate also yields erroneously high values for the recovery. The recovery is homogenate concentration and tissue dependent and varies between 40 and 100%.

Immediately after collecting blood samples, aprotinin and EDTA were added (final concentrations 100 U/ml and 5 mM, respectively). After 1 hr at room temperature, 10–15 hr at 4°, and centrifugation (10^4 g, 20 min, 4°) the serum was diluted 1 : 1 with 2× EIA buffer. For determination of recovery 0.2 ng/ml NGF was added to the blood.

Samples were analyzed in the EIA both with monoclonal and polyclonal anti-NGF antibodies. IgG from nonimmune sera of sheep and mouse, as well as 2 monoclonal antibodies directed against contaminants of the standard preparation of bovine NGF,[31] served as background controls. Beads with NGF specific and nonspecific antibodies exhibited approximately the same background signal when incubated in EIA buffer alone. In contrast, the background signal of beads coupled to nonspecific antibodies was 1.5 to 10 times higher (depending on tissue and homoge-

---

[50] R. R. Almon and S. Varon, *J. Neurochem.* **30**, 1559 (1978).

nate concentration) when they were incubated with tissue samples. This false-positive signal corresponds to up to 50 pg NGF/ml and had therefore to be subtracted from the total signal obtained with NGF-specific antibodies. A nonspecific, but kinetically stable, adsorption of sample components to antibody molecules is the most plausible explanation for this false-positive signal which becomes apparent owing to the sensitivity of the assay. Samples were only then considered NGF positive when the specific signal both with polyclonal and monoclonal antibodies amounted to at least three-quarters of the background signal and at the same time was within the range of the standard curve. This corresponds to a minimal amount of 3–6 pg NGF (assuming optimal recovery) needed to measure the NGF level in a tissue, since the detection limit in the standard curve is 5–10 pg NGF/ml. The detection limit for tissue NGF levels was 0.05 ng NGF/g wet weight. This means that NGF can still be determined when a 2×10^9-fold excess of other proteins is present.

Conclusion

The EIA presented here is sufficiently sensitive to analyze the physiologically relevant NGF levels. For example, target organs of rat sympathetic neurons contain between 0.2 and 2 ng NGF/g wet weight.[1] The use of both mono- and polyclonal anti-NGF antibodies together with the use of nonspecific antibody preparations as background control is essential to ensure the specificity of the NGF determination with the two-site enzyme immunoassay.

Acknowledgments

We are especially indebted to G. P. Harper for providing us with monoclonal antibodies against mouse NGF. We thank A. Brandhofer and F. Scholz-Marb for excellent technical assistance and D. Edgar for a critical review of the manuscript.

[16] Derivation of Monoclonal Antibody to Nerve Growth Factor

By KENNETH E. NEET, MICHAEL W. FANGER, and THOMAS J. BARIBAULT

Monoclonal antibodies to nerve growth factor (NGF) have been produced in three separate laboratories[1-3] and provide the potential for new insights into the functioning of this neuronotropic protein. NGF is essential for the differentiation of the neural crest during development and is required for the maintenance of sympathetic and sensory neurons in culture and *in vivo*.[4,5] Although NGF is not a mitogen, it shares many features with epidermal growth factor, insulin, and other protein hormones, such as interaction with a plasma membrane receptor, internalization, and pleiotropic signals to the target neuron.[4,5] The β-subunit of NGF is a noncovalent dimer of a 13,000-Da polypeptide chain and possesses the biological activity to stimulate neurons to generate neurites *in vitro*. The other subunits, α (25,600-Da monomer) and γ (26,000-Da monomer), of the oligomeric 7 S NGF from mouse submaxillary gland do not stimulate neurite outgrowth and have no known biological activity.[4,5] The 2.5 S form of NGF, which is studied in some laboratories, is a derivative of the β-subunit that is slightly shortened by proteolysis at both termini, but, for the purposes of this article, the 2.5 S NGF is functionally equivalent to βNGF. Immunological techniques, including radioimmunoassay and immunosympathectomy,[6] have been important in studying the distribution and action of NGF. Monoclonal antibodies are now adding new information.

Special problems arise in the production of monoclonal antibodies to NGF because the most readily available NGF preparation is of mouse origin, effectively limiting standard immunization of mice for hybridoma production.[7] Moreover, mouse NGF has relatively poor immunogenicity, even in rabbits. Other important considerations are the subunit nature of

[1] S. L. Warren, M. Fanger, and K. E. Neet, *Science* **210**, 910 (1980).
[2] A. Zimmerman, A. Sutter, and E. M. Shooter, *Proc. Natl. Acad. Sci. U.S.A.* **78**, 4611 (1981).
[3] R. Businaro, R. H. Butler, A. E. Rubenstein, and R. P. Revoltella, *J. Neurosci. Res.* **6**, 89 (1981).
[4] L. A. Greene and E. M. Shooter, *Annu. Rev. Neurosci.* **3**, 353 (1980).
[5] R. Levi-Montalcini, *Annu. Rev. Neurosci.* **5**, 341 (1982).
[6] R. Levi-Montalcini and P. U. Angeletti, *Pharmacol. Rev.* **18**, 619 (1966).
[7] G. Köhler and C. Milstein, *Nature (London)* **256**, 495 (1975).

the 7 S oligomer form of NGF and the apparent lack of cross-reactivity of human placental NGF to polyclonal antisera to mouse βNGF.[8a,b]

Monoclonal Antibodies to a Mouse Protein

The standard protocol for production of hybridomas involves the use of immunized mice.[7] With a mouse protein such as NGF, stimulation of the necessary immune response in a mouse is difficult, if not impossible, because of the lack of recognition of homologous proteins by the immune system. Therefore, Warren et al.,[1] Zimmerman et al.,[2] and Businaro et al.[3] obtained spleen cells from immunized rats and formed interspecies hybridomas with mouse myeloma cells (P3 × 63Ag8 or P3 × 63Ag8.6.5.3), thus obviating problems of lack of recognition of a "self" protein. Businaro et al.[3] also reported successful immunizations of male mice, although the sex of the animals was indicated in a later paper[9] to be female for the mice producing usable hybridomas. Immunity to NGF occurred in the latter case, presumably because the production of NGF in the submaxillary gland and in the circulation is sex limited with low NGF levels in adult, female mice. The hybridomas produced by the rat–mouse fusion are less stable than those by mouse–mouse fusions.

Immunization

The problem of the immunogenicity of βNGF has been addressed in three different ways, each based upon previous techniques of preparing antigen for producing polyclonal antisera to NGF or other weakly immunogenic proteins. Successful immunizations have been achieved by injection of βNGF polymerized with glutaraldehyde,[1] βNGF precipitated on alum, or 2.5 S NGF minced in polyacrylamide subsequent to sodium dodecyl sulfate (SDS)–gel electrophoresis.[3] In addition, immunization with 7 S NGF has also achieved production of monoclonal antibodies toward each of the three types of subunits.[10] These three published methods and our unpublished variations are described below.

Method A-1 (Warren et al.[1]). The β-subunit of NGF was polymerized with glutaraldehyde and 500 μg was injected ip in complete Freund's adjuvant into a 5-week-old, male Lewis rat. A booster injection of 300 μg of βNGF was given ip 3 weeks later in incomplete Freund's adjuvant. Spleen cells were collected 4 days later and used directly for fusion.

[8a] C. E. Beck and J. R. Perez-Polo, *J. Neurosci. Res.* **8**, 137 (1982).
[8b] P. Walker, M. E. Weichsel, Jr., and D. A. Fisher, *Life Sci.* **26**, 195 (1980).
[9] R. Businaro and R. Revoltella, *Clin. Immunol. Immunopathol.* **22**, 1 (1982).
[10] T. Stacey, M. W. Fanger, and K. E. Neet, unpublished observations.

Method A-2 (Stacey et al.[10]). The 7 S oligomer has been used directly as an antigen with an initial ip injection of 50 μg of 7 S NGF in complete Freund's adjuvant into a female Sprague–Dawley rat. Five weekly ip injections of 50 μg of 7 S NGF in incomplete Freund's adjuvant followed. Spleen cells were collected 4 days after the last injection and used for fusion with the NS-1 myeloma line. In this case hybrid cultures were screened for antibody toward 7 S NGF and its individual subunits.

Method A-3 (Stacey et al.[10]). A female Sprague–Dawley rat was injected ip with 50 μg of βNGF in complete Freund's adjuvant, followed by two more ip injections of 50 μg of βNGF in incomplete Freund's adjuvant at weekly intervals. Finally, three weekly injections were made with 50 μg of βNGF that had been previously complexed with an equimolar amount of an anti-βNGF monoclonal antibody (N60) and prepared in incomplete Freund's adjuvant. Spleen cells were collected 4 days after the last injection and used for fusion with the NS-1 myeloma line.

Method B (Zimmerman et al.[2]). After precipitation on alum, 100 μg of βNGF was administered ip with complete Freund's adjuvant to a 2-month-old, female Sprague–Dawley rat. Two weeks later 100 μg of βNGF was injected iv with no adjuvant. Spleen cells were collected 4 days later and used directly for fusion. The alum preparation of the antigen is done by mixing βNGF in phosphate-buffered saline with an equal volume of saturated $AlK(SO_4)_2$, titration to pH 7, and washing the protein–alum precipitate with saline.

Method C (Businaro et al.[3]). The 2.5 S NGF was further purified on 15% polyacrylamide gel electrophoresis in sodium dodecyl sulfate (0.1%) and 5–10 μg of protein, still contained in the acrylamide fragment, was injected ip weekly for 3 weeks into 3- to 4-month-old female BALB/c mice or into rats. Spleen cells for fusion were collected 3 days later.

Notes on Immunization Methods

We have attempted to reproduce the immunization of female mice with βNGF contained in acrylamide gel after electrophoresis by the procedure of Businaro et al.,[3] but without success.[10] Zimmerman et al.[2] also reported the inability to immunize female mice by their immunization schedule. These difficulties suggest that conditions have to be exactly correct to achieve success with this method. Attempts to obviate the problem of the stability of the interspecies hybrids by fusing immunized rat spleen cells with the Y-3 rat myeloma cell line have not been productive; viable clones were obtained but none that produced antibodies to NGF.[10]

Hybridoma Production

The fusions in each case used standard methods based on the protocol of Köhler and Milstein[7] for hybridoma cells. In all published procedures the P3 × 63Ag8 (IgG$_1$ producing) or P3 × 63Ag.6.5.3 (non-Ig producing) plasmacytoma cell lines have been used at a spleen cell to myeloma cell ratio of 5:1 to 11:1 with 30 to 35% polyethylene glycol 1000 or 6000. Unpublished experiments also successfully utilized the NS-1 (non-Ig producing) plasmacytoma line.[10] In one case, fusion of the mouse myeloma cells with circulating lymphocytes from immunized rabbits was reported.[3]

Screening for the desired antibody producing hybridomas involved standard, solid-phase immunological methods. Two groups[2,3] have used a radioimmunoassay (RIA) which employed plating the βNGF on microtiter plates, followed by blocking with serum albumin, washing, addition of the test cell supernatant, further washing, addition of ^{125}I-labeled protein A, and a final wash. The other report[1] utilized a solid phase enzyme-linked immunosorbent assay (ELISA) with alkaline phosphatase linked rabbit anti-rat IgG light chain. Approximately 0.05 to 0.1 μg of βNGF is initially plated on the polyvinyl microtiter plates (Cooke Laboratory). Microcomplement fixation has also been used to determine the titer of the supernatants.[3] Cloning and subcloning of producing cell lines were by standard methods.[7]

The production of the desired antibody has been accomplished in ascites fluid of pristane-treated mice[1,3] by standard inoculation procedures. Further purification of the antibody has been achieved by passage over protein A-affinity columns.[1,2] Fab fragments have been generated by papain digestion.[1,3] Zimmerman et al.[2] obtained a monoclonal antibody preparation substantially free of contaminating proteins for their binding studies by performing hybridoma fusions with a variant of the myeloma cell line (P3 × 63Ag8.6.5.3) that does not produce IgG heavy or light chains, maintenance of the growing hybridoma cells in serum-free medium, and protein A-affinity purification of the supernatant.

The efficiency of the production of monoclonal antibodies to βNGF by the different protocols is summarized in Table I. The majority of monoclonal antibodies obtained have been in the IgG class, although some IgM antibodies have also been obtained by Method C[3] and Method A-3.[10] Although, the rabbit–mouse fusions[3] appear to be the most effective, neither the stability of these clonal lines nor further studies on these monoclonal antibodies have been reported. Of the rat–mouse fusions, the procedure of Businaro et al. would appear to be the most effective (6% yield), but caution must be exercised based on the inability to reproduce this immunization in our laboratory (see above).
•

TABLE I
FUSION RESULTS IN MONOCLONAL ANTIBODY PRODUCTION TO NGF

Immunization (reference)	Total hybrid clones	Anti-NGF producing clones (% of total clones)	Stable lines
Method A-1 (Warren et al.[1])	720	20 (2.7%)	6[a]
Method A-2 (Stacey et al.[10])	300	8 (2.6%)	2[b]
Method A-3 (Stacey et al.[10])	325	2 (0.6%)	2[c]
Method B (Zimmerman et al.[2])	182	8 (4.4%)	2[d]
Method C (Businaro et al.[3])			
Rat fusion	133	40 (30%)	n.r.[e]
Mouse fusion	330	47 (14%)	n.r.
Rabbit fusion	420	175 (42%)	n.r.

[a] Stable for several years but lost subsequently.
[b] N60 is specific for βNGF; N175 is specific for γNGF; both are IgG.
[c] β171 and β7 are both IgM.
[d] MCβ-1 is an IgG.
[e] n.r., not reported. Four IgGs and two IgMs were studied in a subsequent paper.[9]

Properties

Various properties of the monoclonal antibodies produced by these procedures have been reported.

Bioassay. The monoclonal antibodies have been tested for the ability to inhibit neurite outgrowth from chick embryonic sensory or sympathetic neurons either in the ganglion explant assay[11] or the dissociated neuron assay.[12] The antibody tested by Warren et al.[1] and both antibodies tested by Zimmerman et al.[2] competitively inhibited NGF stimulation of neurite outgrowth in the dissociated cell assay with an estimated affinity constant for βNGF of 10^8 M^{-1} for clone-8[1] or 5×10^7 M^{-1} for MCβ-1.[2] Fab fragments from the clone-8 monoclonal antibody were also reported to inhibit in a similar manner.[1] Studies with the N60 monoclonal antibody[13] have also demonstrated competitive inhibition of βNGF stimulated neurite outgrowth with the primed PC12 cell bioassay.[14] Businaro et al.[3]

[11] R. Levi-Montalcini, H. Meyer, and V. Hamburger, *Cancer Res.* **14**, 49 (1954).
[12] L. A. Greene, *Neurobiology* **4**, 286 (1974).
[13] T. J. Baribault, K. E. Neet, and M. W. Fanger, unpublished observations.
[14] A. Rukinstein and L. A. Greene, *Brain Res.* **263**, 177 (1982).

reported data from 13 monoclonal antibodies which were tested at approximately equal antibody titer for the ability to inhibit neurite outgrowth at saturating 2.5 S NGF concentrations (10 ng/ml) in the qualitative ganglion explant assay. Both monoclonal antibodies from rat–mouse hybridomas, four of eight from mouse–mouse hybridomas, and one of three from rabbit–mouse hybridomas completely inhibited neurite extension. Of those that did not completely inhibit, four antibodies from mouse–mouse hybridomas and one from rabbit–mouse hybridomas produced partial inhibition (2+ on a 4+ scale). These results indicate that many monoclonal antibodies directed toward βNGF are capable of blocking its biological activity, either through steric hindrance or by reacting at or near the binding site of βNGF.

Competition with NGF Receptor Binding. Zimmerman et al.[2] reported a study of MCβ-1 monoclonal antibody inhibition of ^{125}I-labeled βNGF binding to dorsal root ganglia neurons. Competitive inhibition was observed that was consistent with both an affinity constant of $5 \times 10^7 M^{-1}$ and the bioassay results. Similar studies with the monoclonal antibody N60 (Method A-2) and βNGF interaction with pheochromocytoma (PC12) cells indicated competitive inhibition of βNGF binding, consistent with the inhibition of the bioresponse of the PC12 cells and an association constant of about 10^7 to $10^8 M^{-1}$.[13]

Affinity of Monoclonal Antibody for βNGF. The affinity of a highly purified monoclonal antibody for βNGF has been directly measured in solution by gel filtration and in the solid phase on microtiter plates.[2] For the solution studies 0.4 nM ^{125}I-labeled βNGF was incubated with concentrations of MCβ-1 antibody from 1.1 to 220 nM for 2 hr and then chromatographed on a 185 ml BioGel A-0.5m column; the proportion of complex formed was analyzed by Scatchard analysis. These gel filtration results indicated that two MCβ-1 molecules bound per βNGF dimer with an affinity constant of $5 \times 10^7 M^{-1}$, which was the basis for comparison to the bioassay and receptor competition studies described above. The solid phase analysis involved determination of the concentration dependency of ^{125}I-labeled βNGF to MCβ-1 monoclonal antibody bound to protein A-coated wells and subsequent Scatchard analysis. These studies yielded an association constant of $3 \times 10^9 M^{-1}$, nearly two orders of magnitude greater than the solution result, and was interpreted as the result of the multivalency of protein A and increased local concentrations of βNGF due to nonspecific binding.

Specificity of the Monoclonal Antibodies. Monoclonal antibodies toward βNGF do not cross-react with a series of related proteins. MCβ-1 antibody did not cross-react with insulin, proinsulin, or relaxin when used competitively in the protein A-solid phase assay.[2] Clone-8 monoclonal antibody and other βNGF specific antibodies subsequently produced in our laboratory did not cross-react with the α- or γ-subunits of NGF, with insulin, or with albumin as tested by the ELISA.[1,10] Twelve monoclonal antibodies tested by Businaro et al.[3] using the microcomplement fixation assay did not cross-react with insulin or serum albumin.

Warren et al.[1] reported that the clone-8 monoclonal antibody cross-reacted with partially purified human placental NGF and snake venom (Naja naja) NGF, two closely related proteins that have similar biological activities on neurons and also have homologous amino acid sequences.[15,16] Subsequent studies[17] on a series of monoclonal antibodies that were relatively instable revealed that one reacted with mouse and human NGF, but not snake NGF, and three reacted with only the mouse NGF initially used as the antigen. On the basis of this reactivity pattern and the inhibition of biological activity by the Fab fragments, Warren et al.[1] suggested that clone-8 monoclonal antibody interacted at or near the active site of βNGF. Furthermore, the cross-reactivity of two of these monoclonal antibodies with human placenta NGF is in agreement with the high homology (90% identical) observed between the amino acid sequences determined from the cDNA of these proteins[15] and suggest that they share some epitopes. These monoclonals detect similarities, perhaps fortuitously, that polyclonal antibodies do not,[8a,b] indicating that the monoclonal antibody approach can reveal relationships that are difficult to approach by classical means.

Immunosympathectomy. A passive immunization study of neonatal and prenatal mice has been performed with 10 monoclonal antibodies to NGF,[9] analogous to earlier immunosympathectomy studies with polyclonal antibodies.[6,18] In summary, 4 of 6 monoclonal antibodies caused a moderate to severe disease state after neonatal immunization as judged by clinical and histological analysis. By biochemical criteria, 3 of 4 monoclonal antibodies significantly decreased protein content, tyrosine monooxygenase activity, and cell number in superior cervical ganglia after neonatal (1 day) inoculation of mice. Prenatal treatment (day 17–18 of pregnancy) with the same antibodies demonstrated that 2 of 4 were capable of significant biochemical changes. The one monoclonal antibody that was effective in the neonatal and not in the prenatal situation was of the IgM class and presumably could not traverse the placental barrier. In every case, there was exact correlation between the ability of the anti-

[15] A. Ullrich, A. Gray, C. Berman, and T. J. Dull, *Nature (London)* **303**, 821 (1983).
[16] R. A. Hogue-Angeletti, W. A. Frazier, J. W. Jacobs, H. D. Niall, and R. A. Bradshaw, *Biochemistry* **15**, 26 (1976).
[17] S. L. Warren, M. W. Fanger, and K. E. Neet, unpublished observations.
[18] P. Gorin and E. M. Johnson, *Proc. Natl. Acad. Sci. U.S.A.* **76**, 5382 (1979).

body to inhibit neurite outgrowth in the *in vitro* ganglion explant assay and the ability to exert biological effects on the sympathetic nervous system. These studies quite clearly demonstrate the utility of examining subtle effects in a complex biological system with the use of monoclonal antibodies.

Related Monoclonal Antibodies

(1) A brief preliminary report has indicated the production of 12 monoclonal antibodies to human placental NGF that do not cross-react with mouse βNGF.[19] (2) A monoclonal antibody reactive to the γ-subunit has also been produced in our laboratory[10] utilizing similar methods and immunization with 7 S NGF; however, no further study of this reagent has been conducted. (3) A first report has described a monoclonal antibody to a PC12 cell surface molecule that enhances βNGF binding and might be directed toward the NGF receptor.[20] We have observed similar monoclonal antibodies in our laboratory.[10,13] (4) A precipitating monoclonal antibody to human melanoma βNGF receptor that inhibits βNGF binding has also been reported.[21]

Conclusions and Prospects

The procedures described above demonstrate that the problems inherent in production of monoclonal antibodies and immunological studies with mouse NGF can be essentially overcome in several effective ways. The results from species specificity analysis[1] and from the variety of effects of 12 antibodies on the bioassay[3] suggest that there are at least 3 epitopes on the 13,000-Da βNGF, one or more of which are shared with other species. Binding and bioassay analysis[1,2] present a consistent thermodynamic picture. A sensitive and highly specific immunoassay utilizing these monoclonal antibodies has not yet been reported,[21a] but could be readily developed from the RIA or ELISA techniques already described[1-3] or further refined by incorporation of the two-site RIA described for polyclonal antibodies.[22] These reagents should also be useful

---

[19] K. Werrbach-Perez, K. Hubner, and J. R. Perez-Polo, *Soc. Neurosci. Abstr.* **10,** 306.6 (1984).

[20] C. E. Chandler, L. M. Parsons, M. Hosang, and E. M. Shooter, *J. Biol. Chem.* **259,** 6882 (1984).

[21] A. H. Ross, P. Grob, M. Bothwell, D. E. Elder, C. S. Ernst, N. Marano, B. F. D. Ghrist, C. S. Slemp, M. Herlyn, B. Atkinson, and H. Koprowski, *Proc. Natl. Acad. Sci. U.S.A.* **81,** 6681 (1984).

[21a] But see M. Kasaian, Ph.D. dissertation, Case Western Reserve University (1987).

[22] K. Suda, Y. A. Barde, and H. Thoenen, *Proc. Natl. Acad. Sci. U.S.A.* **75,** 4042 (1978).

for affinity purification of NGF and for immunocytochemistry. More insight into the biochemistry of the NGF molecule, its interaction with responsive neurons, and its biological effects during development should be provided with these monoclonal antibodies.

Acknowledgment

This work was supported by research Grant USPHS, NIH, NS 17141.

[17] Isolation of Complementary DNA Encoding Mouse Nerve Growth Factor and Epidermal Growth Factor

By James Scott, Mark J. Selby, and Graeme I. Bell

Nerve growth factor (NGF) and epidermal growth factor (EGF) belong to a heterogeneous group of polypeptides that are responsible for the regulation of cell growth. NGF and EGF have no known structural homology or functional relationship, but are both synthesized at extraordinarily high levels in the submaxillary gland of the adult male mouse.[1,2] This tissue has therefore been an important source of mRNA for use in the preparation and characterization of NGF and EGF cDNAs.

In the peripheral nervous system NGF is necessary for this development and maintenance of sympathetic and sensory neurons.[1] It is also trophic for certain cholinergic neurons in the central nervous system.[3] NGF is produced by the targets of these nerve cells, taken up by the axons, and transported retrogradely to the cell body, where it acts to promote neurite outgrowth and neurotransmitter biosynthesis.[4] Mouse NGF is a polypeptide of 118 amino acid residues.[5] Its precursor is predicted from cDNA clones to be flanked at the amino-terminus by 187 amino acid residues and by two residues at the carboxy-terminus (Fig. 1).[6,7] A similar precursor has been predicted for human NGF.[7] EGF is a potent mitogen for cells of epithelial and epidermoid origin.[2] Administra-

[1] H. Thoenen and Y.-A. Barde, *Physiol. Rev.* **60**, 1284 (1980).
[2] G. Carpenter and S. Cohen, *Annu. Rev. Biochem.* **48**, 193 (1979).
[3] S. Korsching, G. Auburger, R. Heumann, J. Scott, and H. Thoenen, *EMBO J.* **4**, 1389 (1985).
[4] R. Heumann, S. Korsching, J. Scott, and H. Thoenen, *EMBO J.* **3**, 3183 (1984).
[5] R. H. Angeletti, M. A. Hermodson, and R. A. Bradshaw, *Biochemistry* **12**, 100 (1973).
[6] J. Scott, M. Selby, M. Urdea, M. Quiroga, G. I. Bell, and W. J. Rutter, *Nature (London)* **302**, 538 (1983).
[7] A. Ullrich, A. Gray, C. Bermann, and T. J. Dull, *Nature (London)* **303**, 821 (1983).

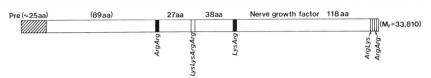

FIG. 1. Schematic representation of the mouse submaxillary preproNGF polyprotein. The basic di- and tetrapeptide possible cleavage sites are indicated: the sites at which cleavage is known to occur are indicated by solid boxes. The numbers in parentheses are tentative size designations: the others give accurate sizes.

tion of EGF accelerates eyelid opening, tooth eruption, and lung maturation. In addition EGF inhibits gastric acid secretion. EGF has 53 amino acid residues and has three disulfide bonds.[8] The precursor for EGF has been deduced from cDNA clones and has 1217 residues (Fig. 2).[9–11] EGF is flanked by polypeptides of 188 and 976 residues at its carboxy- and amino-termini, respectively. The amino-terminus of the precursor contains seven cysteine-rich peptides, that resemble EGF. Toward the carboxy-terminus is a 20 residue hydrophobic membrane spanning domain.[12] The midportion of the EGF precursor shares a 33% homology with the low-density lipoprotein receptor which extends over 400 amino acid residues.[13] These features suggest that EGF precursor could function as a membrane-bound receptor. The availability of cDNA probes for NGF and EGF has allowed other sites of synthesis to be determined.

Sources of NGF and EGF mRNA

The most abundant source of NGF and EGF mRNAs is the granular convoluted tubule cells in the adult male mouse submaxillary gland. The female mouse submaxillary gland contains one-tenth of the amount of mRNA found in the male gland.[6,14] mRNA from the male submaxillary gland was used to prepare cDNA probes for NGF and EGF, and these have been used to screen other tissues and cell lines for NGF and EGF mRNA (Tables I and II). NGF mRNA has been demonstrated in many of

[8] C. R. Savage, T. Inagami, and S. Cohen, J. Biol. Chem. 247, 7612 (1971).
[9] J. Scott, M. Urdea, M. Quiroga, R. Sanchez-Pescador, N. Fong, M. Selby, W. J. Rutter, and G. I. Bell, Science 221, 238 (1983).
[10] A. Gray, T. J. Dull, and A. Ullrich, Nature (London) 303, 722 (1983).
[11] S. Pfeffer and A. Ullrich, Nature (London) 313, 184 (1984).
[12] R. F. Doolittle, D. F. Fong, and M. S. Johnson, Nature (London) 307, 558 (1984).
[13] D. W. Russell, W. J. Schneider, T. Yamamoto, K. L. Luskey, M. S. Brown, and J. L. Goldstein, Cell 37, 577 (1984).
[14] L. B. Rall, J. Scott, G. I. Bell, R. J. Crawford, J. D. Penschow, H. D. Niall, and J. P. Coghlan, Nature (London) 313, 228 (1985).

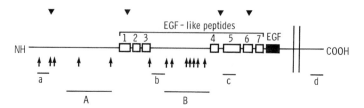

1217 amino acids

FIG. 2. Schematic representation of mouse submaxillary gland preproEGF. Triangles represent potential N-glycosylation sites. Arrows show dibasic residues. The vertical lines following the EGF moiety indicate the position of a possible membrane spanning domain in addition to EGF and the EGF-like peptides. Regions of homology in the precursor are shown (a, b, c, d: A and B).[12]

TABLE I

RELATIVE ABUNDANCE OF PREPRONGF mRNA IN
RAT TISSUES

Tissue	Relative prepronGF mRNA levels[a]
Submaxillary gland (male)	111
Submaxillary gland (female)	9.4
Vas deferens	6.8
Atrium	1.2
Ventricle	1.0
Superior cervical ganglion (male)	<0.6
Superior cervical ganglion (female)	<0.6
Skeletal muscle	<0.16

[a] The values were normalized to that of the ventricles.
From Heumann et al.[4]

the target tissues for sympathetic neurons at a concentration that reflects the density of innervation. NGF, but not NGF mRNA, has been found in nerve cells that innervate these tissues, thus confirming that NGF acts as a retrograde messenger between target tissues and innervating neurons.[3] Accordingly the vas deferens, atrium, ventricle, and spleen capsule have moderate amounts of NGF mRNA.[4,15] The hippocampus and cortex of the brain also contain NGF mRNA.[3] Mouse L cells have been demonstrated to produce NGF mRNA (D. Wion, P. Barrand, E. Dicou, J. Scott, and P. Brachet, unpublished).

Apart from the male mouse submaxillary the tissue with the greatest amount of EGF mRNA is in the kidney. This is unexpected as the kidney

[15] D. L. Shelton and L. F. Reichardt, Proc. Natl. Acad. Sci. U.S.A. **81,** 7951 (1984).

TABLE II
RELATIVE ABUNDANCES OF PREPROEGF mRNA IN
VARIOUS MOUSE TISSUES

Tissue	Relative preproEGF mRNA abundance[a]
Submaxillary gland (male)	2
Kidney (male, female)	1
Submaxillary gland (female)	0.2
Mammary gland (lactating)	0.01
Pancreas	0.008
Small intestine (duodenum)	0.004
Pituitary	0.003
Lung	0.002
Spleen	0.0002
Brain	0.002
Ovary	0.002
Liver	<0.002
Eye	<0.002
L cell	<0.002

[a] The values were normalized to that of the kidneys. From Rall *et al.*[14]

contains very little EGF protein. The explanation for this is that in the kidney EGF precursor is not processed, and antibody raised against native EGF does not readily recognize EGF within its precursor protein.[14] Low levels of EGF mRNA are found in many tissues including pancreas, duodenum, pituitary, lung, spleen, brain, and ovary.[14] None has been demonstrated in the fetus or extraembryonic membranes, and no cell line is known to produce EGF.

Preparation of mRNA

Intact poly(A) RNA was isolated from the submaxillary of adult male Swiss–Webster mice (Simonson). RNA was prepared from fresh tissue by homogenization with a tissuemizer (maximum speed, 30 sec) in freshly prepared 5 M guanidinium thiocyanate, 20 mM Tris–HCl (pH 7.5), 10 mM EDTA, and 2 M 2-mercaptoethanol (20 ml/g wet weight of tissue).[16] Cellular debris was removed by centrifugation for 10 min at 10,000 rpm in 30-ml Corex tubes (Du Pont Instruments). The dark brown solution was layered

[16] J. M. Chirgwin, A. E. Przybyla, R. J. MacDonald, and W. J. Rutter, *Biochemistry* **18,** 5294 (1979).

on 4 ml of 5.7 M CsCl, 5 mM EDTA (pH 8.0) in polyallomer tubes and centrifuged in an SW41 rotor at 36,000 rpm for 20 hr at 15°.[17] The contents of the tubes including the interface were aspirated by vigorous suction except for the bottom 3 ml of CsCl. The tubes were washed twice in guanidinium thiocyanate solution. After the final wash the last 3 ml of CsCl was aspirated and the tubes inverted. The bottom of the tubes containing the RNA pellet was cut off with a heated scapel blade. The pellet was transferred to a 50-ml Falcon tube and dissolved in 5 ml of 0.5% SDS, 1 M 2-mercaptoethanol. Extraction was performed twice with 5 ml phenol, buffered with 100 mM Tris–HCl (pH 7.5), 5 mM EDTA, and 5 ml chloroform : isoamyl alcohol (24 : 1), and once with 5 ml chloroform : isoamyl alcohol. One-tenth volume of 3 M sodium acetate and 2.5 volumes of ethanol were added to the aqueous phase; the RNA was precipitated at −20° overnight and stored as an ethanol precipitate. A yield of about 30 mg of RNA can be expected from 40 g of tissue. Poly(A) RNA was obtained by two cycles of chromatography on oligo(dT)-cellulose from which a yield of 1% was obtained.[18,19] Before use all glassware was treated for 6 hr at 260°. Reagents were dispensed from previously unused containers. Solutions were passed through Nalgene (Sybron) 0.2-μm filtration units. Plastic caps, polyallomer tubes, and all solutions were treated with 0.1% diethyl pyrocarbonate for 1 hr, and subsequently autoclaved for 30 min to remove the diethyl pyrocarbonate. Solutions containing Tris–HCl, which inactivates diethyl pyrocarbonate, were made up with diethyl pyrocarbonate-treated water and autoclaved. RNA was shown to be intact by visualization of the 28 S and 18 S ribosomal bands after glyoxylation and electrophoresis on 1.5% agarose gels.[20]

cDNA Synthesis

Two methods have been used for preparation of cDNA. Both methods have been described elsewhere in these volumes and only the general strategy will be outlined.[21,22] Single-stranded cDNA was prepared by oligo(dT) priming of reverse transcriptase and rendered double stranded with the Klenow fragment of DNA polymerase I.[21] The hairpin was

[17] V. Glison, R. Crkvenjakov, and C. Byus, *J. Biochem.* **13**, 2623 (1974).
[18] H. Aviv and P. Leder, *Proc. Natl. Acad. Sci. U.S.A.* **69**, 1408 (1972).
[19] J. A. Bantle, I. H. Maxwell, and W. E. Hahn, *Anal. Biochem.* **72**, 413 (1976).
[20] P. S. Thomas, this series, Vol. 100, p. 255.
[21] H. M. Goodman and R. J. MacDonald, this series, Vol. 68, p. 75.
[22] H. Land, M. Grez, H. Hansen, W. Lindenmaier, and G. Schutz, this series, Vol. 100, p. 285.

cleaved with S_1 nuclease and the cDNA inserted into the *Pst*I site of a plasmid vector using the dG : dC tailing method. To obtain clones encoding the amino-terminus of the EGF precursor an oligonucleotide was used to specifically prime the synthesis of the 5' EGF cDNAs. The single-stranded cDNA was tailed with dC and an oligo(dG) primer used to prepare double-stranded cDNA.[22] This process avoids the S_1 nuclease step which foreshortens 5' cDNA sequences. Subsequent cloning was by the dG : dC tailing method.

Gubler and Hoffman have recently described a procedure for synthesizing double-stranded cDNA which simplifies the construction of cDNA libraries and hence the cloning of low abundance mRNAs.[23] It combines the classical first-strand synthesis with RNase H–DNA polymerase I-mediated second-strand synthesis. The double-stranded cDNA is then tailed and cloned without further manipulation.

Screening of cDNA Libraries

Preparation and Design of Synthetic Oligodeoxynucleotides. The NGF and EGF cDNAs were identified using oligonucleotides synthesized by solid-phase phosphoramidite chemistry and purified on denaturing polyacrylamide gels as described by Urdea in this series.[24] Oligonucleotides can also be synthesized manually, purchased, or made with one of the commercial DNA synthesizers. The ideal oligonucleotide for direct probing of a cDNA library should be designed for maximum length, maximum dGdC content, and minimum degeneracy. For NGF a 32-fold degenerate 17-base oligonucleotide without ambiguity was used successfully.[6] This probe is of the minimum length that we have found useful for direct screening (Fig. 3). Shorter probes can be of value for specific priming of reverse transcription (Fig. 4) to make cDNA as described for NGF and EGF (see also below, primer extension).[7,11] For EGF 4 pools of 64-fold degenerate 20-base oligonucleotides were used as probes.[10] Subsequent to the identification of EGF cDNA clones using one of these pools, it was demonstrated that a single pool of 256-fold degenerate 20-base oligonucleotides could be used satisfactorily (Fig. 3). With probes of this degeneracy a length of 25 or more bases is, however, preferable. For longer probes unambiguous sequences can rarely be made. It is necessary to design long probes based on codon usage, use of nondestabilizing base choices, and avoidance of dCpdG dinucleotides, which are rare in verte-

[23] U. Gubler and B. J. Hoffman, *Gene* **25**, 263 (1983).
[24] M. Urdea, this series, Vol. 146 [3].

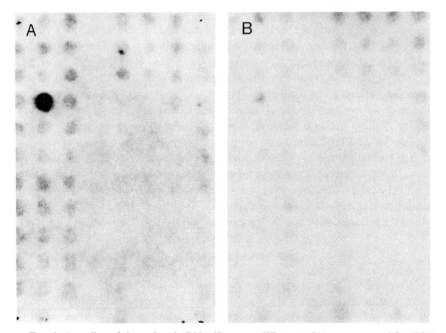

FIG. 3. A replica of the ordered cDNA library on Whatman 541 was screened for EGF with a 20-base 256-fold degenerate unambiguous oligonucleotide 3'-CCNCCNCANAC-^A_GTACGT^A_GTA-5'(A), and the same filter with a 15-base 48-fold degenerate unambiguous oligonucleotide 3'-TAT^A_GGT^A_GTACAC^A_GCAN-5'(B). No hybridization is seen with the 15-base oligonucleotide.

brate DNA.[25–27] With the availability of the mouse NGF and EGF mRNA sequences unambiguous probes can of course be synthesized.

Screening. When a cDNA library is to be screened for clones representing several different proteins of moderate abundance ($<0.01\%$ of the mRNA) preparation of a permanent ordered clone collection is useful for ease and sensitivity of screening.[28] Individual transformants are picked and grown in the separate wells of microtiter dishes. The storage media allows recovery of vector cells from all wells without supplementation even after several years of storage and repeated freezing and thawing.

[25] R. Grantham, C. Gautier, M. Gouy, M. Jacobzone, and R. Mercier, *Nucleic Acids Res.* **9**, r45 (1981).
[26] Y. Tawahaski, K. Kato, Y. Hayashizake, T. Wakabayashi, E. Ohtsuka, S. Matsuki, M. Ikehara, K. Matsubara, *Proc. Natl. Acad. Sci. U.S.A.* **82**, 1931 (1985).
[27] D. Barker, M. Schafer, and R. White, *Cell* **36**, 131 (1984).
[28] J. P. Gergen, R. H. Stern, and P. C. Wensink, *Nucleic Acids Res.* **7**, 2115 (1979).

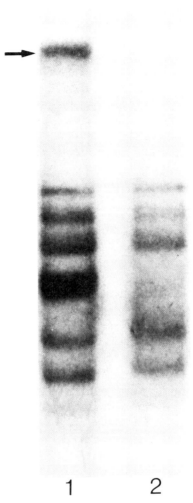

1   2

FIG. 4. A Northern blot of adult male (1) and female gland (2) submaxillary gland poly(A) RNA was screened with cDNA specifically primed with the 15-base 48-fold degenerate EGF oligonucleotide. The 15-base oligonucleotide primed low abundance EGF cDNA arrow synthesis together with cDNAs for several abundant submaxillary gland products. The cDNA preparation was prepared as described for direct sequence analysis from adult male mouse submaxillary gland poly(A) RNA, but labeled with [α-^{32}P]dCTP and used directly after removal of unincorporated label.

Replicas of this library on Whatman 541 filters can be used for numerous (5–10) separate screenings without significant loss of sensivity.

Storage and Replication. The cDNA library was initially plated at a density (<3000 colonies per 150 mm hard agar plate) so that individual colonies could be picked into 96-well (U-shaped wells) microtiter dishes using sterile toothpicks. The medium consisted of L-broth with 25 μg/ml tetracycline: K_2HPO_4 6.3 g, KH_2PO_4 1.8 g, sodium citrate 0.45 g, $MgSO_4 \cdot 7H_2O$ 0.09 g, $(NH_4)_2SO_4$ 0.9 g, and glycerol 44 g per liter,[29] and was added (150 ml/well) with a 12-channel multipipet. Cultures were grown to saturation in stacks of 10 dishes on top of damp paper towels and wrapped in Saran wrap. Plates are stored in plastic bags at −80°.

To prepare replica filters the microtiter plates were thawed and incubated for 2 hr at 37°. Condensation was removed with sterile paper tissues. An applicator with 96-flat prongs corresponding to the microtiter dish wells was used to lift the colonies from the dishes and apply these to L-broth (1.5% Bactoagar, 25 μg/ml tetracycline) plates. Gentle agitation of the applicator in the dishes was necessary to obtain a satisfactory replica. Between different microtiter plates the applicator was flamed in ethanol and cooled on the surface of an agar plate. Colonies were grown overnight and lifted intact onto sterile Whatman 541 filters marked with a Lab Marker permanent alcohol/waterproof pen (VWR Scientific Inc.) and placed on L-broth chloramphenicol (250 mg/ml) plates for amplification. A second filter was placed on top of the first and air bubbles removed by rubbing with a gloved finger. The plates were incubated at 37° overnight. This amplification increases the signal approximately 10-fold. The colonies were lysed and the DNA was denatured and immobilized by blotting successively on trays covered with 3MM paper which had been moistened, but not wetted, with 1 M NaOH for 5 min, 1 M Tris–HCl (pH 7.4), 1.5 M NaCl for 5 min and this step repeated (usually twice) until the filters were neutral. The filters were separated in 2 × SSC, 0.1% SDS (SSC is 0.15 M NaCl and 0.015 sodium citrate), and excess bacterial debris rubbed from the surface with a paper tissue. This step is essential to minimize nonspecific signal during screening. Finally the filters were baked at 80° for 3 hr.

Labeling of Oligonucliotides. Oligonucleotides were 5′-end labeled to high specific activity (usually in excess of 10^9 cpm/μg) with ^{32}P using [γ-^{32}P]ATP and polynucleotide kinase. The details of labeling and separation of unincorporated label are described elsewhere in this series.[30] The incorporated label is conveniently removed using Sep-Pak C_{18} cartridges

---

[29] D. S. Hogness and J. R. Simmons, *J. Mol. Biol.* **9**, 411 (1964).
[30] L. B. Rall, J. Scott, and G. I. Bell, this series, Vol. 146 [23].

(Waters Associates). The eluate from the cartridge can be used directly for hybridization, or dried in a Speed Vac concentrator.

Hybridization. With Whatman 541 filter paper no prehybridization was necessary. Hybridization was performed on single filters; positions were clear and duplicate screening unnecessary for the ordered library. The hybridization solution was 5 ml/filter of 5 × SSC, 40 mM sodium phosphate (pH 7.0), 2 × Denhardt's solution [100 × Denhardt's is 2% bovine serum albumin, 2% Ficoll 400, 2% poly(vinylpyrrolidone) 360], 0.1% SDS, and 300 μg/ml sonicated and heat-denatured salmon sperm DNA. Probe was added to a radioactive concentration of 5 × 10^6 cpm/ml and hybridization performed over 15 to 20 hr. Hybridization temperature and washing conditions were calculated according to the formula for DNA melting temperature (T_m) vs fraction GC (fgc) and total salt concentration (C_s): $T_m = 16.6 \log C_s + 41 fgc + 81.5$, and corrected for length by subtracting the molecular weight of a deoxynucleotide base pair of 650 Da divided by the length of the oligonucleotide. Washing conditions to remove unhybridized probe were also calculated by the above formula. As a rough guide the filters are washed in 2 × SSC at room temperature for 30 min and in the same solution at 50° for 30 min for an unambiguous oligonucleotide of 20 bases, or in 1 × SSC, 0.1% SDS for an oligonucleotide of 40 bases. The filters were exposed to Kodak X-Omat film for 3 to 15 hr with or without intensifying screens depending on the strength of the signal.

Positive hybridization denoted an NGF or EGF clone, which was picked directly from the ordered collection. The insert size was determined and positive Southern hybridization confirmed before the clone was grown in bulk for sequence analysis. To remove previous hybridizing probe, prior to use of a different probe, filters were immersed in boiling water, which was allowed to cool on the bench. Autoradiography is recommended to ensure that a previous probe has been removed.

To screen for cDNAs of low abundance (<0.01% mRNA) high-density screening of unordered colonies is necessary.[31,32] Results with nitrocellulose, nylon, or Whatman 541 are satisfactory for 10^4 colonies per 150 mm agar plate. At higher density, 5 × 10^4 colonies, nitrocellulose or nylon filters are preferable.

Primer Extension Analysis

The technique of primer extension can be used to obtain the 5' sequence of the mRNA directly as for NGF[6]; or for preparation of an

---

[31] D. Hanahan and M. Meselson, this series, Vol. 100, p. 333.
[32] D. Ish-Horowicz and J. F. Burke, *Nucleic Acids Res.* **9**, 2989 (1981).

enriched population of cDNA clones representing the 5′-end of the mRNA for screening and subsequent sequencing as for EGF.[10,11] *Direct sequence analysis:* For primer extension unambiguous synthetic oligonucleotides 3′-CCACCTCAAAACCGG-5′ were prepared corresponding to the 5′ sequence of available cDNA clones.[33] A convenient length for this oligonucleotide was 20 bases.

To sequence relatively low abundance mRNAs such as NGF (<0.1%), where 100–200 nucleotides were unavailable from the 5′ end of the mRNA, the chemical cleavage method of Maxam and Gilbert was most satisfactory (Figs. 5 and 6).[34] For more abundant mRNAs such as pancreatic amylase (>1% of mRNA) the dideoxynucleotide chain terminator method was also satisfactory.[35] The oligonucleotide was 5′-labeled with [γ-^{32}P]ATP and a full-length cDNA prepared using reverse transcriptase. As described above 0.5 μg of oligonucleotide was labeled with 1 to 2 mCi of [γ-^{32}P]ATP (ICN, crude. 7000 Ci/mmol) using polynucleotide kinase. The unincorporated label was removed on Sep-Pak columns and the labeled eluate dried in a Speedvac desiccator (Savant) in siliconized 1.5 ml Eppendorf tubes. The tubes were siliconized by dipping in dimethyldichlorosilane sol. 2% in 1,1,1-trichlorethane (BDH), washed in distilled water, and afterward baked at 100° for 3 hr.

To remove any contaminating RNase the primer was resuspended in 200 ml of diethyl pyrocarbonate-treated water and extracted twice with an equal volume of buffered phenol and chloroform : isoamyl alcohol (24 : 1), and once with an equal volume of chloroform : isoamyl alcohol. Ethanol precipitation was performed overnight at −20° by adding 20 μl 3 *M* sodium acetate, 5 μg of poly(dA), and 4 volumes of ethanol. The precipitate was washed with 200 μl of 70% ethanol made up with diethyl pyrocarbonate-treated water, and dried briefly in siliconized diethyl pyrocarbonate-treated tubes in a Speedvac desiccator. The primer and 20 μg of double passed poly(A) mRNA were heated to 90° for 5 min and annealed in 20 μl of 0.5 m*M* EDTA (pH 8.0) for 30 min at 60° in heat-treated 100 μl disposable micropipets (Drummond). The extension reaction was performed at 43° for 1 hr in 500 μl Eppendorf tubes and started by adding 30 μl of 50 m*M* Tris–HCl (pH 8.3), 50 m*M* KCl, 5 m*M* MgCl$_2$, 20 m*M* dithiothreitol, 500 m*M* dNTPs, 25 units reverse transcriptase (Life Sciences), and spun briefly.

To stop the reaction it was heated at 60° after adding 5 μl of 250 m*M*

[33] N. L. Agarwal, J. Brunstedt, and B. E. Noyes, *J. Biol. Chem.* **256**, 1023 (1981).
[34] A. M. Maxam and W. Gilbert, this series, Vol. 65, p. 499.
[35] F. Sanger, A. R. Coulson, B. F. Barrell, A. J. H. Smith, and J. Roe, *J. Mol. Biol.* **143**, 161 (1980).

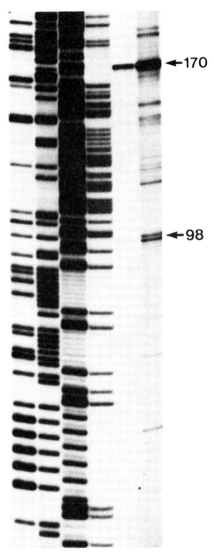

FIG. 5. Primer extension of NGF mRNA. The principal extension product was 170-bases using poly(A) RNA from the submaxillary glands of adult mice: male (right), female (left). A Maxam and Gilbert sequence ladder was used as a size marker.

EDTA (pH 8.0). The reaction was ethanol precipitated in 0.3 M sodium acetate and 2.5 volumes of ethanol, washed twice in 70% ethanol, dried briefly, and suspended in formamide dyes.[34] The extension product was electrophoresed in 8.3 M urea, 89 mM Tris-borate, 89 mM boric acid, 2

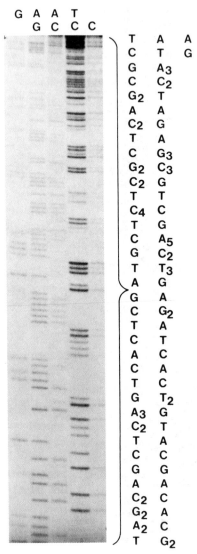

FIG. 6. Analysis of preproNGF 5' sequence (bases 1–107, Scott *et al.*[6]) direct from the 170-base primer extension product (Fig. 5).

mM EDTA (pH 8.0), on a 15% acrylamide gel of 1.5 mm thickness at 45 W power. After completion the gel is covered with Saran wrap and autoradiographed by covering with individually wrapped Kodak X-Omat film. The extension product (Fig. 5) was removed from the gel under brief long-wave UV irradiation, crushed by injecting through the nozzle of a 5-ml syringe, and eluted at 4° in 200 mM NaCl, 5 mM EDTA (pH 8.0) by shaking overnight. The eluate was filtered twice through 0.2-μm nitrocellulose filters (Schleicher and Schuel) and concentrated with n-butanol. The primed product was ethanol precipitated twice and washed twice in 70% ethanol and dried briefly before DNA sequencing by the chemical cleavage method (Fig. 6).

  cDNA Cloning of the Primer Extension Product. For cDNA cloning small reactions were performed. The oligonucleotide was used direct from synthesis[24] without labeling or phenol extraction. Of mRNA 10 μg was annealed in heated-treated 100-μm capillary tubes as described for the larger reaction. The reverse transcription reaction was performed in 500 μl Eppendorf tubes with 10 μCi of [α-^{32}P]dCTP. Chromatography[21] or precipitation in 2 M NH$_4$Cl with two volumes of ethanol was used to remove the first strand reaction mixture. Second strand DNA synthesis was performed by the Land procedure to preserve the 5′ end of the cDNA clones.[22] The Gubler and Hoffman method has also been used for second-strand synthesis.[23] To screen for clones a 5′ fragment including the site for the primer or an oligonucleotide made from DNA sequences 100–200 bases more 5′ to the original primer site was used.

Acknowledgments

  Mrs. T. Barrett and Mrs. A. Smith are thanked for typing the manuscript.

[18] PC12 Pheochromocytoma Cells: Culture, Nerve Growth Factor Treatment, and Experimental Exploitation

By LLOYD A. GREENE, JOHN M. ALETTA, ADRIANA RUKENSTEIN, and STEVEN H. GREEN

Introduction

  Over the past decade, the clonal PC12 cell line has become a widely used preparation for various model studies on chromaffin cells and neurons.[1] The PC12 line was originally cloned in 1975[1] from a transplantable rat adrenal medullary pheochromocytoma.[2] A major characteristic of the

[1] L. A. Greene and A. S. Tischler, *Proc. Natl. Acad. Sci. U.S.A.* **73**, 2424 (1976).
[2] S. Warren and R. Chute, *Cancer* **29**, 327 (1972).

cells is that they respond to the neuronotrophic agent nerve growth factor (NGF)[3] by slowly shifting from a proliferating chromaffin/pheochromocytoma cell-like phenotype to that of nonproliferating, neurite-bearing sympathetic-like neuron. The features of this NGF response as well as other properties and potential experimental advantages of the PC12 cell line have been reviewed elsewhere.[4]

Because the PC12 line presents certain potential difficulties in its successful use and since it is used by a growing number of laboratories, the aim of this chapter is to present various details regarding its culture and experimental exploitation. In particular, we shall consider methods for the maintenance and NGF treatment of PC12 cultures as well as for the generation and selection of PC12 cell mutants and for the preparation of a PC12 cell fraction which is highly enriched in neurites.

Culture of PC12 Cells

Culture Medium. Currently we grow and maintain PC12 cells in the originally described growth medium[1] consisting of 85% RPMI 1640 medium, 10% horse serum (heat-treated for 30 min at 56°), 5% fetal calf serum, 25 U/ml penicillin, and 25 μg/ml streptomycin. The nature of the horse serum used appears to be particularly important for the successful propagation of the cells; not all batches of horse serum from all suppliers yield adequate maintenance. We have had consistently satisfactory results with "donor-grade" horse serum purchased from KC Biologicals Inc. in Lenexa, KA. Serum from Flow Laboratories has also proved satisfactory. Pretesting of serum samples is suggested.

In the original description of the PC12 line,[1] it was noted that the cells could also be propagated, though with a lower growth rate, in growth medium in which RPMI 1640 medium was replaced by DMEM medium. Several laboratories have adapted PC12 cells to DMEM-based media and have made additional alterations in the growth conditions. If necessary to carry out, such adaptations should be made gradually, since in our experience large shifts in the growth medium lead to considerable cell death and to the selection of subpopulations. On the whole, however, it is recommended that the line be carried in the above described growth medium so as to avoid, as much as possible, variation of the cells from one laboratory to the next.

[3] R. Levi-Montalcini, *Harvey Lect.* **60,** 217 (1966).
[4] L. A. Greene and A. S. Tischler, *in* "Advances in Cellular Neurobiology" (S. Federoff and L. Hertz, eds), Vol. 3, p. 373. Academic Press, New York, 1982.

Substrate. Like cultured neurons and chromaffin cells, PC12 cells tend to strongly autoassociate and to attach poorly to tissue culture plastic or glass, but to attach well to substrates which are coated with collagen or with polylysine or polyornithine.[5,6] Because polylysine- and polyornithine-coated substrates appear to deteriorate progressively over time, collagen-coated substrates have proved to be more suitable for experiments in which the cells remain plated for more than 3 days. Collagen has been prepared in our laboratory from rat tails by a modification of the method of Bornstein.[7] Rat tails (freshly collected or stored frozen at $-20°$) are sterilized by immersion in 70% ethanol for 20 min and rinsed in sterile distilled water. Sections are then broken off and pulled from the tail, exposing sections of tendon. The latter are minced and washed in sterile distilled water and then transferred to centrifuge bottles in which they are extracted for 2–3 days at room temperature with 0.1% glacial acetic acid ($\sim$50–100 ml per tail). The material is then centrifuged at $\sim$12,000 g for 1 hr. The supernatant is collected and stored at $-20°$ in aliquots. Once thawed, the aliquots are stored at 4° for up to a month. All steps are carried out so as to assure sterility of the final preparation. Plastic tissue culture dishes are coated with the collagen by one of two procedures. In one method, drops (1 per 35 mm dish; 6 per 100 mm dish) of collagen-containing solution are placed on the dish, spread evenly with a bent-glass spreader, and the solution is allowed to air dry in a sterile hood. In the alternate procedure, the collagen is freshly diluted to a final volume of 30% ethanol, an appropriate volume (1 ml per 35 mm dish; 5 ml per 100 mm dish) is added to culture dishes, and the solution is air dried overnight in a sterile hood. The amount of collagen applied by these procedures is extremely important both for adequate cell attachment and neurite outgrowth. Stock collagen solutions should be empirically tested at various final concentrations (differing by serial 2-fold dilutions) to determine the optimal levels that should be employed. If too much collagen is applied, neurite outgrowth is retarded and collagen fibrils impede microscopic examination of the cells; if too little collagen is present, cell attachment is poor, the cells tend to form large clumps, and neurite outgrowth is sparse.

For certain types of short-term experiments, polylysine- or polyornithine-coated substrates may be especially useful. One case is growth of the cells on glass coverslips for fluorescence or other optical observations. Following previously described procedures,[5,6] cleaned ster-

[5] P. C. Letourneau, *Dev. Biol.* **44,** 77 (1975).
[6] E. Yavin and Z. Yavin, *J. Cell Biol.* **62,** 540 (1974).
[7] M. B. Bornstein, *Lab. Invest.* **7,** 134 (1958).

ile coverslips are exposed for 1 hr to a 50 μg/ml membrane-sterilized aqueous solution of poly(L-lysine hydrobromide) (Sigma Cat. P-1399). The coverslips are then washed 4 times with sterile water and allowed to air dry. PC12 cells remain well adhered to such a substrate for 1–3 days. If neurite-bearing cultures are desired, NGF-pretreated or "primed" cells can be used; these will regenerate long, well-attached processes within a day of plating on the polylysine-coated coverslips.[8] A second case is for short-term culture of the cells on a highly adhesive substrate. To prepare this, plastic tissue culture dishes are treated for 4 hr with a 1 mg/ml solution (1 ml/35 mm dish) of filter-sterilized poly(L-α-ornithine hydrobromide) (Type II-B, Sigma Cat. P-6012) in 150 mM sodium borate buffer, pH 8.4.[5] The dishes are then washed 4× with sterile water and air dried. PC12 cells plated for 1–2 days as monolayers on such a substrate can remain well attached throughout repeated washings. Also, if NGF-pretreated or "primed" cells are plated on dishes prepared in this manner, they do not regenerate neurites for at least 24 hr (A. Rukenstein, unpublished observations). This permits the study of attached NGF-treated cells without neurites.

For long-term culture of PC12 cells on glass coverslips, we have found collagen applied as above to be unsatisfactory due to peeling. Good results have been achieved by chemically bonding the collagen to the glass with 3-aminopropyltriethoxysilane as described by Gottlieb and Glaser.[9] Empirical testing is needed to establish the optimal amount of collagen preparation to be employed.

Growth and Maintenance. In our laboratory, cultures are fed three times weekly (Monday, Wednesday, and Friday) with complete growth medium. About 2/3 of the medium is exchanged; the presence of conditioned medium facilitates growth. Maintenance is in a 36° incubator in a water-saturated atmosphere containing 7.5% CO_2. When the cultures reach a near-confluent state, subculturing is carried out. The cells are detached from the dishes and dispersed into single cells and small clumps by repeated forceful aspirations of medium through a Pasteur pipet. The split ratio is usually 1 : 3 or 1 : 4, since proliferation is suboptimal at low cell densities. Because the doubling rate for PC12 cells is 2.5–4 days, this leads to about one subculturing per week.

NGF Treatment. There are several keys to obtaining optimal NGF-induced neurite outgrowth in PC12 cultures (Fig. 1). One is substrate. For long-term cultures, we have found ethanol-applied collagen to yield the most satisfactory results. As noted above, the amount of collagen applied

---

[8] L. A. Greene, *Brain Res.* **133,** 350 (1977).
[9] D. I. Gottlieb and L. Glaser, *Biochem. Biophys. Res. Commun.* **63,** 815 (1975).

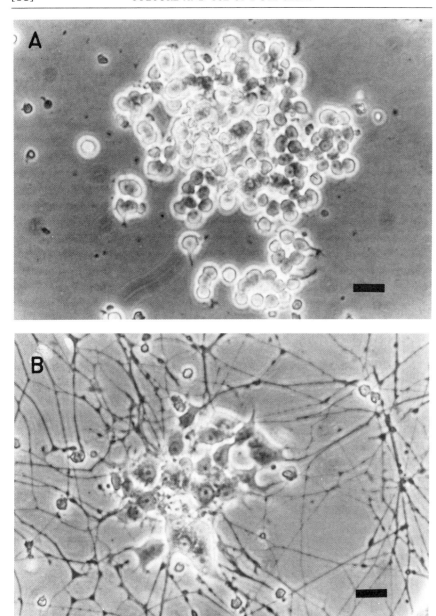

FIG. 1. Phase contrast photomicrographs of PC12 cells. (A) PC12 cells maintained without NGF; (B) cells exposed to 50 ng/ml NGF for 14 days. Bars represent 20 μm.

is critical and should be empirically determined by pretesting a range of dilutions. The second key is cell density. At very high cell densities neurite outgrowth is suppressed[10] while at too low of a cell density, cell survival is poor. Satisfactory results have been obtained at cell densities equivalent to 2–8 × 10^5 per 35 mm culture dish. For cell densities lower than this range, the use of medium conditioned by high density PC12 cell cultures has been found to be helpful. Exposure to NGF enhances PC12 cell autoadhesion. Minimization of the resultant cell clumping can be obtained if the cells are exposed to NGF in medium containing 1% horse serum. The NGF-treated cells are maintained quite adequately under these conditions and the decrease in serum does not appear to alter their properties. The NGF-exposed cultures are fed and maintained as described above with medium supplemented with 50 ng/ml ($\sim$2 nM) mouse salivary gland NGF.[11] During feeding or experimental manipulation, care should be taken to avoid tearing the collagen substrate on which the cells are growing lest the entire substrate peel from the dish.

It has been described in detail elsewhere that PC12 cells pretreated or "primed" with NGF can rapidly (i.e., within 18 hr) initiate or regenerate neurites.[8] This property has been exploited both experimentally and for the bioassay of NGF and other neuronotrophic factors.[8,10]

Cell Separation. PC12 cells are extremely adherent to each other and form clumps of cells which are difficult to dissociate. For certain types of experiments it is desirable to have suspensions of single cells or to obtain clones of cells each derived from a single cell. The following procedure[12] will produce suspensions in which over half of the cells are present as single cells and in which there are very few clumps of greater than four cells. Cell viability is not greatly reduced by this procedure.

1. Add EGTA to the medium in the dish to a final concentration of 1 mM. The EGTA can be made up as a 0.1 M stock. Normal medium (not Ca^{2+}-free) with serum should be used in order not to adversely affect cell viability. The cells are left in this medium for at least 6 hr and the treatment can be done overnight without a serious loss of viability.

2. Wash the cells three times with medium diluted 1 : 1 with water, but without serum.

3. Triturate the cells vigorously with a Pasteur pipet.

If PC12 cells are plated at very low density (<10^5 cells in a 100-mm dish), plating efficiency is very low although this can be mitigated some-

[10] L. A. Greene, D. E. Burstein, and M. M. Black, *Dev. Biol.* **91,** 305 (1982).
[11] W. C. Mobley, A. Schenker, and E. M. Shooter, *Biochemistry* **15,** 5543 (1976).
[12] S. H. Green, R. E. Rydel, J. L. Connolly, and L. A. Greene, *J. Cell Biol.* **102,** 830 (1986).

what by feeding the cells with medium conditioned by dense PC12 cell cultures. Therefore, if cells are to be plated at low density, treatments which reduce cell viability should be avoided. Dissociating the cells with trypsin is a problem in these circumstances. However, if suspensions of only single cells with no clumps are required, and a loss of plating efficiency is acceptable, then the cells may be trypsinized with 0.1–0.5% trypsin (Difco) for 10 min at 37°. Diluting the cells into medium containing serum prior to plating serves to inactivate the trypsin.

Suspension Cultures. PC12 cells may be grown in large numbers in spinner bottles using complete culture medium. Establishment of the cultures requires an initially high cell density >10^6 cells/ml. Expansion is best achieved by adding fresh medium to the cultures when the medium indicator shows a drop in pH. The cells may also be maintained in suspension in plastic nontissue culture petri dishes. Under these conditions, they form nonattached clumps and, if treated with NGF, as in spinner culture, will undergo "priming," but not neurite outgrowth.[10]

Spontaneous Variation. Any cell line undergoing numerous divisions may be expected to spontaneously generate phenotypic and genotypic variants. In some cases, culture conditions may select for such variants and they may ultimately predominate in the cell population. The PC12 line, being near-diploid, has remained relatively stable in culture, at least with respect to its major phenotypic properties. Nevertheless, spontaneously arising variants have been reported.[13] Several precautions may be followed to favor homogeneity of the line. First, after a given number of generation, cells should be replaced (using frozen stocks) with those from earlier passages. Our laboratory currently employs cells between passages 20 to 40. Second, cell growth conditions should be maintained as uniform as possible so as to avoid application of selective pressure. Third, it is recommended that all cultures be grown on a collagen or other suitably adhesive substrate. This avoids possible selection of variants with higher adhesivity than the parent cells.

Freezing and Long-Term Cell Storage. This can be accomplished by suspending the cells in a mixture containing 90% complete growth medium and 10% DMSO, and slow-freezing them in cryogenic vials at −70°. Long-term storage can be carried out in liquid nitrogen.

PC12 Cell Mutants

Mutagenesis. One potential major use of the PC12 line is the generation and selection of mutations that are altered in specific neuronal prop-

[13] D. E. Burstein and L. A. Greene, *Dev. Biol.* **94**, 477 (1982).

erties.[12-14] In order to obtain large numbers of single-gene mutations for genetic studies, mutagenesis must be employed. Chemical mutagenesis is often employed for this purpose. Optimal mutagenesis rates with chemical mutagens are achieved with lethality rates of 80–90%.[15] This rate can be reproducibly obtained by treatment of PC12 cultures with 15 mM ethylmethane sulfonate (EMS) (purchased from Sigma or Eastman) for 5–6 hr with 10^7 cells in a 100-mm dish. Other PC12 mutagenesis protocols have achieved satisfactory results with much lower EMS concentrations but using 24-hr treatments.[14] The EMS solutions and any equipment contaminated with EMS should be treated with 0.05 M mercaptoacetic acid in 1 M NaOH.

Following treatment with EMS, the cells are washed 5 or 6 times with medium and are replated the following day. After 10–14 days (3–5 cell divisions) the cells can be separated as described above and replated for screening.

We have quantified the efficiency of the mutagenesis for producing single-gene mutations by scoring for colonies resistant to 8-azaguanine (Sigma). These result from mutations in the HPRT gene. The frequency of resistant colonies increases from the spontaneous rate of 10^{-6} to 10^{-4} in mutagenized cells.[12]

Selection Techniques. Selection of drug-resistant cell clones, whether induced by mutation or by transfection with DNA carrying drug resistance markers, can be achieved by plating the cells at moderate density (~5 × 10^6 cells in a 100-mm dish) in the presence of the drug. The resistant cells form colonies over the time required for the nonresistant cells to die off so that the density should not be reduced to the point that the drug-resistant cells will die as well.

Certain selection techniques require the inspection of clonal cell colonies at low density. These can be obtained by separating cells as described above, plating at low density (5 × 10^4 to 10^5 cells in a 100-mm dish), and allowing the cells to grow into small discrete colonies (2–4 weeks). For instance, this approach has been used to screen for PC12 mutants that are defective in NGF-promoted neurite outgrowth.[12] When the cells are at low density, viability can be improved by reducing the number of times that the medium is changed from three times to once per week. Low density cultures can also be supplemented with medium conditioned by denser PC12 cell cultures to improve viability.

Colonies of mutant cells can be conveniently picked from plates with a

[14] M. A. Bothwell, A. L. Schechter, and K. M. Vaughn, *Cell* **21,** 857 (1980).
[15] F.-T. Kao and T. T. Puck, *Methods Cell Biol.* **8,** 23 (1974).

sterile, cotton-plugged Pasteur pipet, the tip of which has been drawn out in a flame to form a capillary 2–3 cm in length. Suction may be applied by mouth through a rubber tube or with a pipet bulb. Picked colonies should be plated into single wells of 96-well dishes to maintain high cell density. They can be transferred successively to 24-well dishes and then to 35-mm and 100-mm dishes as necessary.

Neurite-Enriched Cultures

One of the important advantages of the PC12 cell line is the large amount of homogeneous material it provides for experimental analysis. An example of the use of this asset is our attempt to understand the molecular basis of the regional specializations of neurite-bearing cells. To this end, we have devised techniques for obtaining preparations which are very highly enriched in neurites.[16] By culturing primed PC12 cells in small discrete clusters exposed to NGF for 2 to 6 weeks, large areas on the culture dish are produced which are covered only by growing neurites. The dense halo of neuritic processes that develops can be easily separated from the central cell body aggregate by means of a stainless steel microknife, thus allowing biochemical analysis of a neurite-enriched fraction. This approach is modified from that first described by Wood and Bunge[17] for separation of cultured ganglionic neuron cell bodies and neurites.

Preparation. PC12 cells are primed in medium containing 1% horse serum plus 50 ng/ml NGF on collagen-coated culture dishes. After 1 to 2 weeks, they are harvested from the dishes by vigorous trituration through a sterilized cotton-plugged glass pipet into complete growth medium containing 50 ng/ml NGF. The cells, sheared free of their neurites by this treatment, are pelleted by low-speed centrifugation and then resuspended in a small volume of NGF-containing complete growth medium. Approximately 20 μl of this cell suspension is loaded into small glass cylinders (8 mm long $\times$ 1.5 mm inside diameter) standing upright in collagen-coated culture dishes. The cylinders are individually cut from glass tubing and then ground flat on one end to provide a fluid-tight seal when placed on the culture dish. Two to three such cylinders can be placed in each 35-mm dish. The final concentration of cells should be 0.5–1.0 $\times$ 10^6 per ml so that a total of 10,000–20,000 cells will finally settle within the glass to form a circular plug at the bottom of the cylinder well. To avoid overestimating

[16] J. M. Aletta and L. A. Greene, *Soc. Neurosci. Abstr.* **11**, 759 (1985).
[17] P. Wood and R. P. Bunge, *Nature (London)* **256**, 662 (1975).

the number of cells loaded, the harvested primed cells must be resuspended immediately before loading each pair of cylinders in order to maintain a uniform dispersion of cells.

The cells are allowed to settle and attach overnight in a CO_2 incubator. The glass cylinders are then removed and NGF-containing complete medium is added to completely cover the cells. Care must be taken not to disrupt the clumps of cells, which at this point are rather weakly attached to the substrate. Rapid pipetting of the medium should be avoided and the cultures should be left in the incubator undisturbed for 24 hr before they are handled further. Finally, cultures are fed every other day in the same manner as typical monolayers of neurite-bearing cells, that is, with incomplete withdrawal of conditioned medium and gentle readdition of fresh prewarmed NGF-containing medium.

Separation of Neurites from Cell Bodies. Within a day after plating, neurites extend radially from the cell clusters, and approximately 2 weeks later they have grown roughly 1 mm in all directions. Branching and intermingling of neurites appear to continue as long as cultures are maintained, although extension, for the most part, seems to cease after 4 or 5 weeks, with the longest neurites reaching about 3.5 mm.

Separation of the central cell body cluster from the radiating neuritic network is accomplished with the aid of a transilluminating dissecting microscope by using a microknife to sever the connections at the base of the outgrowth. When a distinct cleft has been incised around the entire cell body cluster, the latter can be gently manipulated off the dish so that it floats freely. It can then be retrieved with a micropipet. The mass of neurites remaining on the dish is washed free of any cellular debris and may then be used for appropriate cellular or biochemical experiments. If maintained in NGF-containing medium, the isolated neurites appear to remain intact for at least 18 hr. With practice, the entire separation procedure can be performed in less than 15 min on two cell masses within the same 35-mm culture dish.

Acknowledgments

Supported by grants from the USPHS (NS16036), March-of-Dimes Birth Defects Foundation, Dysautonomia Foundation, and an American Cancer Society Institutional Grant (IN-14-Z). LAG was a Career Development Awardee of the Irma T. Hirschl Trust and JMA the recipient of a postdoctoral fellowship from the Muscular Dystrophy Association.

[19] Assay and Isolation of Glial Growth Factor from the Bovine Pituitary

By Jeremy P. Brockes

Glial growth factor (GGF) is a mitogenic growth factor identified by its action on cultured rat Schwann cells. Although the purified growth factor stimulates the division of other cell types, such as astrocytes[1,2] and fibroblasts,[1] the rat Schwann cell does not respond to a variety of other soluble growth factors[3–5] and is therefore central to the identification and purification of GGF. GGF is a basic protein of $M_r = 31K$ as determined both by the reactivity of four monoclonal antibodies raised against partially purified preparations,[6] and by the recovery of biological activity after SDS–polyacrylamide gel electrophoresis in the absence of reducing agents.[6] GGF activity on the rat Schwann cell has only been detected in extracts of neural tissue,[3] and it has been purified from the bovine pituitary.[1,6] One area of the bovine brain, the caudate nucleus, has an even higher level than the pituitary and the properties of the partially purified caudate activity are indistinguishable from those of pituitary GGF as evidenced by chromatographic and immunochemical criteria.[7,8] A similar molecule has been detected in extracts of the nervous system of all vertebrates investigated to date including human,[9] bovine,[3] rodent,[3] chick,[10] and amphibian[11] sources. There is some evidence that GGF may be involved in the nerve dependence of amphibian limb regeneration.[5,11]

In view of the importance of cultured rat Schwann cells for assaying GGF, this chapter initially describes the preparation of these cells and their purification by employing anti-Thy-1 antibodies to kill contaminating

[1] J. P. Brockes, G. E. Lemke, and D. R. Balzer, *J. Biol. Chem.* **255**, 8374 (1980).
[2] S. U. Kim, J. Stern, M. W. Kim, and D. E. Pleasure, *Brain Res.* **274**, 79 (1983).
[3] M. C. Raff, E. R. Abney, A. Hornby-Smith, and J. P. Brockes, *Cell* **15**, 813 (1978).
[4] J. L. Salzer, A. K. Williams, L. Glaser, and R. P. Bunge, *J. Cell Biol.* **84**, 739 (1980).
[5] C. R. Kintner, G. E. Lemke, and J. P. Brockes, *in* "Molecular Bases of Neural Development" (G. M. Edelman, W. E. Gall, and W. M. Cowan, eds.), pp. 119–138. Wiley, New York, 1985.
[6] G. E. Lemke and J. P. Brockes, *J. Neurosci.* **4**, 75 (1984).
[7] J. P. Brockes and G. E. Lemke, *in* "Development in the Nervous System" (D. R. Garrod and J. D. Feldman, eds.). Cambridge Univ. Press, London, 1981.
[8] J. P. Brockes, K. J. Fryxell, and G. E. Lemke, *J. Exp. Biol.* **95**, 215 (1981).
[9] J. P. Brockes, X. O. Breakefield, and R. Martusza, *Ann. Neurol.* **20**, 317 (1986).
[10] G. E. Lemke, Ph.D. thesis. California Institute of Technology, Pasadena, California, 1983.
[11] J. P. Brockes and C. R. Kintner, *Cell* **45**, 301 (1986).

fibroblasts; procedures for the Schwann cell DNA synthesis assay in microwells, and the purification of GGF from the bovine pituitary are also given.

Rat Schwann Cell Cultures: Preparation[12,13]

Materials

Dulbecco's modified Eagle's medium (DMEM, Gibco) + antibiotics + 10% fetal calf serum (FCS) (heat inactivated at 56° for 30 min)
DMEM without bicarbonate and buffered with HEPES (HMEM, Flow)
1% collagenase (Whatman CLS III) in HMEM
2.5% trypsin (Gibco), freshly thawed
Nylon mesh 125 (pore size 60 μm) sterilized with ethanol
5-ml plastic syringe, 21- and 23-gauge needles
2 sterile glass or plastic vials (approximately 20 ml)

Procedure

Sciatic nerves are dissected from 15–30 2-day-old rat pups under aseptic conditions and stored in HMEM in a sterile vial. Carefully decant the HMEM and resuspend the nerves in 2 ml HMEM + 0.08 ml collagenase + 0.25 ml trypsin. After incubation at 37° for 15 min, decant most of the medium and add 2 ml fresh HMEM with collagenase and trypsin. Repeat this procedure until the nerves have been digested for a total of three 15-min periods. After the final digestion add an equal volume of HMEM + 10% FCS, and take the suspension into a plastic syringe. Pass the nerves two or three times through an 18-gauge needle and twice through a 23-gauge needle. The cloudy suspension is passed through sterile nylon gauze to remove debris and centrifuged at 500 g for 10 min at room temperature. The cell pellet is suspended in DMEM + 10% FCS and the cells are grown in petri dishes or flasks in a humidified CO_2 incubator. For photographs of the appearance of cultures at various stages, see Brockes *et al.*[13]

Rat Schwann Cell Cultures: Maintenance and Purification[13]

Materials

Cytosine arabinoside (Sigma), 1 mM in HMEM
DMEM + antibiotics + 10% heat inactivated FCS

[12] J. P. Brockes, K. L. Fields, and M. C. Raff, *Nature (London)* **266,** 364 (1977).
[13] J. P. Brockes, K. L. Fields, and M. C. Raff, *Brain Res.* **165,** 105 (1979).

HMEM + antibiotics + 10% heat inactivated FCS
Bovine pituitary GGF (CM-cellulose fraction[1] approximately 2–3 mg/
ml), sterilized by filtration and stored as aliquots at −70°
Monoclonal IgM anti-mouse Thy-1.1 ascites fluid, diluted 1 : 100 in
HMEM + 10% FCS and stored as aliquots at −70°
Phosphate-buffered saline without Ca^{2+} or Mg^{2+} for tissue culture
(Gibco)
Trypsin/EDTA (Gibco) diluted in the above at 0.25% trypsin
Rabbit complement (Gibco) made up from lyophilized samples at
×2.5 and immediately stored as aliquots (0.25 ml) over liquid N_2

Procedure

After overnight growth of the cells dissociated from the nerve, add
cytosine arabinoside to a final concentration of 10^{-5} M. After 72 hr, wash
each flask or petri twice with 10 ml HMEM to remove the cytosine arabinoside, and add fresh DMEM + 10% FCS + 10 μg/ml GGF to stimulate
Schwann cell division. After 3–4 days remove the medium from the
flasks, wash the cells briefly with HMEM and remove them with 0.25%
trypsin/EDTA at 37°. Add an equal volume of HMEM + 10% FCS and
recover the cells by centrifugation at 500 g for 10 min. Remove the supernatant by aspiration, loosen the cell pellet by tapping the tube and add
0.25 ml of anti-Thy-1.1. After 10 min at 37° to allow reaction of antibody
with the cells, add 0.25 ml of freshly thawed rabbit complement and
continue the incubation for a further 30 min at 37°, while giving the tube
an occasional shake (do not vortex). Add 5 ml HMEM + 10% FCS,
recover the cells by centrifugation as before and plate out in DMEM +
10% FCS + GGF. If fibroblasts appear during the ensuing period in culture they can be removed by further treatment with anti-Thy-1.1 and
complement at the time of passaging the cells.[13]

Rat Schwann Cell Proliferation Assay[1,3]

Materials

Fibroblast-free Schwann cells cultured in the absence of GGF for 2–3
days
96-well flat-bottomed microwell plates for tissue culture (Flow or
Costar)
"Titertek" cell harvester (Flow) and accompanying glass fiber sheets
[^{125}I]Iododeoxyuridine ([^{125}I]IUdR, New England Nuclear), diluted to
20 μCi/ml in HMEM
0.05% trypsin/EDTA in phosphate-buffered saline without Ca^{2+} and
Mg^{2+} as above

Schwann cells are removed from flasks with trypsin, treated in suspension with anti-Thy-1.1 and complement as above, and plated into flat-bottomed 96-well plates at 7–10 thousand cells per microwell in 0.1 ml DMEM + 10% FCS. After 12–24 hr, test solutions are added in varying dilutions to the wells for 48 hr, and [^{125}I]IUdR is added for the final 15–24 hr of this period. The medium is removed by aspiration and replaced with 0.1 ml trypsin/EDTA for 10 min at 37° to loosen the cells. The labeled cells are harvested in rows of 12 wells onto glass fiber discs by use of the cell harvester. The discs are transferred to disposable plastic tubes and counted in a gamma counter. Assays are always conducted with a background (unstimulated) incorporation, generally in the range 100–300 cpm, and a maximal incorporation with optimal concentrations of CM-cellulose fraction GGF that stimulates 20- to 100-fold over the background depending on the batch of Schwann cells. Three microwells are usually averaged for each data point. It is worthwhile to select a batch of FCS that supports low background incorporation in the absence of GGF and a high stimulated incorporation since these two parameters vary independently with different batches of FCS.

When the stimulation of incorporation is plotted against the logarithm of protein concentration an irregular dose–response curve is generally obtained for crude extracts of tissues.[3,7] For partially purified fractions (CM-cellulose or later stages) the dose–response curve should be linear over most of the range (for example, see Fig. 1D). When analyzing peaks of activity from columns, it is clearly important to convert activity values into protein concentrations before selecting appropriate fractions to pool.

Isolation of GGF from Bovine Anterior Lobes[1,6]

For purification of GGF to homogeneity, is recommended that 10,000 lobes are used as starting material. If the growth factor is simply required in a partially purified form to stimulate Schwann cell or astrocyte division, then the CM-cellulose fraction may be conveniently purified on a much smaller scale, for example from 500 lobes. All steps are performed at 0–4° and all buffers contain 0.01% NaN$_3$. In general, GGF activity is stable to freezing and thawing, and to lyophilization in volatile buffers.

Preparation of (NH$_4$)$_2$SO$_4$ Fraction

Lyophilized anterior lobes (2000, from Pel-Freez Biologicals) are swollen for 1 hr in 15 liters of 0.15 M (NH$_4$)$_2$SO$_4$ before vigorous homogenization in a Waring Blender. After adjusting the pH to 4.5 by addition of 1 M HCl, the tissue is extracted with stirring for approximately 2 hr. After centrifugation for 15 min at 3000 rpm in a Beckman J6 centrifuge to remove large debris and pituitary fragments, the supernatant is filtered through several layers of cheesecloth and passed through a Sorvall RC-5B centrifuge fitted with a continuous flow attachment (TZ-28 rotor, flow rate of approximately 120 ml/min). The pH of the supernatant is adjusted to 6.5 by addition of 1 N NaOH, and $(NH_4)_2SO_4$ is added to a concentration of 200 g/liter. The precipitate is removed by continuous flow centrifugation as before, and the supernatant precipitated by addition of $(NH_4)_2SO_4$ to 250 g/liter (i.e., final concentration of 450 g/liter). The precipitate is collected by continuous flow centrifugation, dissolved in 300 ml 0.1 M sodium phosphate, pH 6.0 (P buffer) and extensively dialyzed against this buffer over 3–4 days.

The $(NH_4)_2SO_4$ fraction from 20,000 lobes (about 200 g protein) is applied to a 250 ml column of Whatman CM-52 carboxymethylcellulose equilibrated with P buffer. The column is washed with P buffer until the absorbance of the effluent at 280 nm is close to baseline, and then eluted with P buffer + 0.05 M NaCl. After washing, the bulk of the activity is eluted with P buffer + 0.2 M NaCl. Fractions are assayed in the microwell proliferation assay at approximately 10 μg/ml, and the fractions of highest specific activity are pooled and may be stored at $-20°$ without significant loss of activity. The CM-cellulose fraction should give maximal stimulation of Schwann cells at approximately 20 μg/ml, and is a convenient source of GGF activity for routine culture of Schwann cells.

Further Purification

The CM-cellulose fraction (1.09 g) is applied to a 200 ml CM-cellulose column equilibrated with P buffer + 0.025 M NaCl and eluted with a 2 liter linear gradient from 0.025 to 0.210 M NaCl in P buffer (Fig. 1A). The active fractions are pooled, concentrated by $(NH_4)_2SO_4$ precipitation at 560 g/liter, and the precipitate dissolved in 10 ml P buffer + 0.4 ml NaCl, prior to application to a column of AcA 44 Ultrogel (5 cm diameter × 90 cm). The AcA column is run at 80–90 ml/hr in P buffer + 0.4 M NaCl and fractions are assayed at 0.5 μg/ml. The active fractions (19 mg), eluting at a position corresponding to an apparent molecular weight of 3 × 10^4 (Fig. 1B), are pooled, applied to a 5.5 ml column of phosphocellulose P11 equilibrated with P buffer + 0.4 M NaCl, and then eluted with a 60 ml linear gradient from 0.4 to 0.85 M NaCl in P buffer (Fig. 1C). Fractions from the P11 column are assayed at 0.2 μg/ml, and may be concentrated in an Amicon pressure cell (UM10 membrane). An example of a proliferation assay on the P11 fraction is given in Fig. 1D.

GGF is purified from active fractions of the P11 column by electrophoresis in the absence of reducing agent on 12.5% SDS–polyacrylamide gels

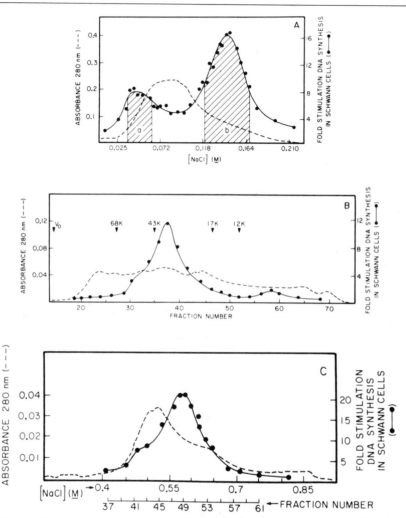

Fig. 1. Steps in the purification of GGF. (A) Gradient elution of the CM-cellulose fraction from CM-cellulose. Pool b is used for further purification. (B) Elution profile of the (gradient eluted) CM-cellulose fraction after gel filtration on AcA 44 Ultrogel. Fractions 36–40 are pooled for further purification. (C) Elution profile of the AcA 44 fraction from phosphocellulose. No activity is detectable in the flow-through. (D) Dose–response curve and further purification of the phosphocellulose fraction. The phosphocellulose fraction (fraction 49 of C) is diluted in medium and assayed in the Schwann cell proliferation assay. Inset: proteins of this fraction (1 μg) are analyzed on a 12.5% SDS gel followed by silver staining. The 31,000-Da GGΓ band accounts for approximately 10% of the protein and can be purified to apparent homogeneity by excision from the gel and elution[14] as shown in lane (ii).

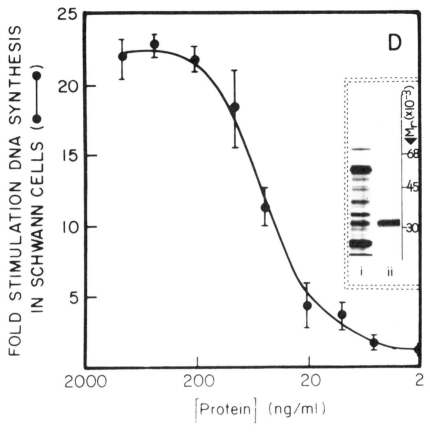

FIG. 1. (continued)

with 0.11% bisacrylamide. The proteins are visualized by staining with 2% Coomassie blue in 20 mM Tris–HCl, pH 6.8, and the GGF band at 31K is excised. The gel slice may be efficiently eluted by use of the electroelution procedure of Mendel-Hartvig[14] (see insert Fig. 1D). It can be estimated from dose–response curves that growth factor prepared in this way has lost approximately 95% of its biological activity as a result of the preparative electrophoresis steps. A summary of such a purification is given in Table I.

For purification of native growth factor that has not been exposed to SDS, reversed-phase HPLC columns have recently been used successfully, although it is important to avoid the use of trifluoroacetic acid as the ion pairing agent if biological activity is to be retained. The phosphocellu-

[14] I. Mendel-Hartvig, Anal. Biochem. 121, 215 (1982).

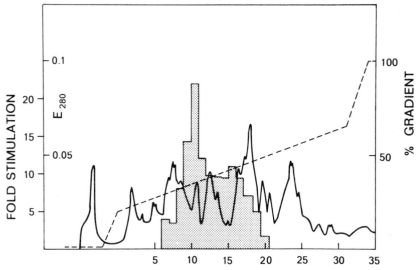

FIG. 2. Purification of GGF by reversed-phase HPLC. Elution profile of phosphocellulose fraction (0.9 mg) on an RPC 5/2 column, run as described in the text. Fractions 10 and 11 were pooled and rerun under the same conditions.

TABLE I
PURIFICATION PROCEDURE[a]

Step	Total protein (mg)	Recovery of activity (%)	Purification (fold)
Crude extract	4×10^5	—	—
(NH$_4$)SO$_4$ fraction	2.0×10^5	100	2
CM-Cellulose (batch)	1200	30	100
CM-Cellulose (gradient)	210	15	250
AcA 44	19	6.8	1250
Phosphocellulose	1.1	3.1	10,000
SDS–gel electrophoresis	0.08	0.45	100,000

[a] The activity was purified from 20,000 lyophilized bovine anterior lobes as described. The fold purification through the phosphocellulose step is determined from the displacement of dose–response curves. Details of the estimates made are given in Lemke and Brockes.[6]

lose fraction (1 mg) is dialyzed against 0.05 M ammonium formate (AF) pH 4.5 + 0.5 M NaCl, filtered through a 0.2-μm filter and injected onto a column of RPC 5/2 (Pharmacia, a C_8 column) connected to a Pharmacia FPLC apparatus. The column is washed with AF and then eluted with the programmed gradient shown in Fig. 2, which runs between AF and AF + 80% acetonitrile (adjusted to pH 4.5 with formic acid). After evaporation to dryness, the residues of fractions are dissolved in P buffer and an aliquot is assayed on rat Schwann cells. The peak fractions are pooled, and subjected to a second round of reversed-phase chromatography under identical conditions. This yields material that is homogeneous on SDS–gel electrophoresis.

[20] Purification of Glia Maturation Factor

By RAMON LIM and JOYCE F. MILLER

Although the activity of brain extract on fibroblast proliferation was detected in as early as 1939,[1] it was not until 1972 when the enhancing effect of the same on the development of glial cells was demonstrated.[2–4] We named this factor, an acidic protein endogenous to the adult brain, the glia maturation factor (GMF) for its profound effect on the phenotypic expression of cultured astroblasts. Since then a number of other functions have been ascribed to GMF. GMF is now known to be a mitogen as well as a maturation agent. GMF reverses some of the aberrant growth properties of glial tumor cells. It also stimulates astrocytes to produce second-order growth factors or hormones. When given to newborn rats, GMF promotes the recovery of brain tissue from injury. All in all, GMF appears to be an autoregulatory molecule produced by the brain and capable of controlling cell growth and development in the nervous system. For the most recent review up to December, 1983, see the article by Lim.[5]

[1] O. A. Trowell and E. N. Willmer, *J. Exp. Biol.* **16**, 60 (1939).
[2] R. Lim, W. K. P. Li, and K. Mitsunobu, *Soc. Neurosci. Abstr.* **2**, 181 (1972).
[3] R. Lim, K. Mitsunobu, and W. K. P. Li, *Exp. Cell Res.* **79**, 243 (1973).
[4] R. Lim and K. Mitsunobu, *Science* **185**, 63 (1974).
[5] R. Lim, *in* "Growth and Maturation Factors" (G. Gurroff, ed.), Vol. 3, Chap. 4, pp. 119–147. Wiley, New York, 1985.

Isolation of GMF from Bovine Brain[6,7]

Principle. GMF has no species specificity, but bovine brains are used as the source because of high specific activity and economy of cost. An aqueous extract at physiologic pH is first prepared, which is then fractionated with ammonium sulfate precipitation and column chromatography with DEAE-Sephacel, Sephadex G-75 and hydroxylapatite, yielding a 10,000-fold partially purified GMF. This preparation is useful for most biological studies. To obtain pure GMF, a final step with reversed-phase high-performance liquid chromatography (HPLC) is used. The results are summarized in Table I.

Reagents

Tris-buffered saline (TBS): 0.02 M Tris–HCl, 0.15 M NaCl, pH 7.45
Ammonium sulfate
Tris buffer: 0.02 M Tris–HCl, pH 7.45
Diethylaminoethyl(DEAE)-Sephacel: Pharmacia Inc.
Sephadex G-75: 40–120 μm bead size; Pharmacia Inc.
Hydroxylapatite: BioGel HT; Bio-Rad Laboratories
0.05 M potassium phosphate buffer: 0.05 M KH_2PO_4 adjusted to pH 7.45 with NaOH
0.2 M potassium phosphate buffer: 0.2 M KH_2PO_4 adjusted to pH 7.45 with NaOH
0.1 M ammonium formate
Acetonitrile: HPLC grade; Fisher Scientific. Dilute with water to desired concentration
Trifluoroacetic acid (TFA): HPLC grade; Pierce Chemical
Reversed-phase HPLC column: C_4 Vydac, 5 μm particle size, 300 Å pore size; The Separations Group, Hesperia, CA

Procedure

1. Beef brains are stored at $-20°$ for no more than a month.
2. Work out 4 brains at a time, each having a wet weight of about 350 g. Partially thaw the frozen brains by placing them in a 4° cold room overnight. Conduct all subsequent steps at 4°.
3. Peel off the meninges and surface blood vessels. Cut brain tissue into 1-cm cubes. Using a Waring blender, homogenize brain tissue in Tris-buffered saline (TBS) at a ratio of 1 g tissue/3 ml buffer, first at low speed setting for 30 sec, then at high speed for another 30 sec.

[6] R. Lim and J. F. Miller, *J. Cell. Physiol.* **119**, 255 (1984).
[7] R. Lim, J. F. Miller, D. J. Hicklin, and A. A. Andresen, *Biochemistry* **24**, 8070 (1985).

TABLE I
PURIFICATION OF GMF FROM BOVINE BRAIN[a]

Step	Protein recovered (mg)	Mitogenic activity		Morphological activity	
		Activity recovered (units)	Specific activity (units/mg)	Activity recovered (units)	Specific activity (units/mg)
Crude extract	69,430.00	69,430	1	69,430	1
(NH$_4$)$_2$SO$_4$ fraction	6,456.00	66,540	10	66,540	10
DEAE-Sephacel	1,788.00	65,370	37	65,370	37
Sephadex G-75	30.50	30,031	985	29,086	954
Hydroxylapatite	1.20	12,300	10,250	11,952	9,960
HPLC	0.03	3,100	103,333	2,990	99,667

[a] The data[6,7] are collected on the basis of 2.8 kg (wet weight) bovine brain as the starting material. One unit of activity is defined as that exhibited by 1 mg protein in the crude extract.

4. Centrifuge the homogenate for 1 hr at 20,000 g, using a Beckman JA-10 rotor at 10,000 rpm or type 19 rotor at 15,000 rpm. Clear the debris in the supernatant by pouring it through glass wool.

5. Add 25.8 g ammonium sulfate per 100 ml of the extract to achieve 45% saturation. Stir for 20 min and centrifuge to eliminate the precipitate. Add additional 15.6 g ammonium sulfate (per 100 ml of the original extract) to the supernatant to make 70% final saturation. Stir and centrifuge as before and take up the pellet in 100 ml of TBS.

6. Using Spectrapor-1 membrane (molecular weight cutoff 6000–8000), dialyze the solution against 10 liters of water overnight, then against 10 titers of Tris buffer for at least 6 hr. Discard the small amount of precipitate by centrifugation.

7. Apply the sample (approximately 120 ml) to a DEAE-Sephacel column, 2.5 × 37 cm, which has been preequilibrated with Tris buffer. Wash the charged column with 4 column volumes of Tris buffer and elute with 1300 ml of a linear gradient of 0 to 0.3 M NaCl in Tris buffer at 40 ml/hr (Fig. 1).

8. Pool the active fractions (about 380 ml) and combine the pool with that obtained from a parallel preparation to have a total volume of about 760 ml (equivalent to 8 brains). Concentrate the sample to 50 ml by Amicon PM-10 membrane filtration.

9. Apply the concentrate to a Sephadex G-75 column, 5 × 100 cm, which has been preequilibrated with TBS. Elute with TBS at 40 ml/hr (Fig. 2).

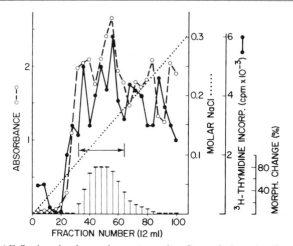

FIG. 1. DEAE-Sephacel column chromatography. Open circles, absorbance at 280 nm; solid circles, mitogenic activity expressed as increase in thymidine incorporation; vertical bars, morphologic activity expressed as percentage of cells responding; broken line, NaCl concentration. Only the profile of the gradient elution is shown. Double-headed arrow indicates fractions pooled for the subsequent step.

10. Pool the active fractions (about 575 ml) and apply directly to a hydroxylapatite column having a gel volume of 36 ml (5 cm diameter × 3 cm height), which has been preequilibrated with TBS. Wash with 2 column volumes of TBS. Remove the bulk of protein with 4.5 column

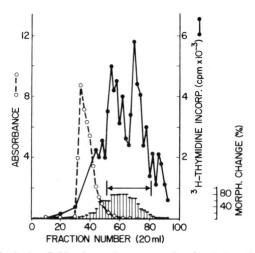

FIG. 2. Sephadex G-75 column chromatography. Symbols as in Fig. 1.

volumes of 0.05 M potassium phosphate buffer. Then elute GMF activity
with 280 ml of 0.2 M potassium phosphate buffer at a rate of 40 ml/hr
(Fig. 3).

11. Pool the active fractions (about 120 to 150 ml). Divide the sample
into 3 aliquots and treat each separately as follows. Dialyze for 5 hr
against 0.1 M ammonium formate and lyophilize. Take up the dried material in 2 ml of 20% acetonitrile containing 0.1% TFA (v/v). Filter the
sample with Millex GV membrane (0.22 μm; Millipore) to remove the
particulates. Apply to the HPLC column, 4.6 mm × 5 cm, containing
Vydac C_4 packing material, which has been preequilibrated with the same
acetonitrile/TFA solvent mixture. Wash the charged column with 20 ml of
the same solvent mixture. Elute with 40 ml of a linear gradient of acetonitrile (20 to 45%) containing 0.1% TFA, at a flow rate of 1.5 ml/min and
collecting 2 ml fractions (Fig. 4A). GMF activity should be detectable in
fractions No. 3, 4, and 5, with highest activity in No. 4, corresponding to
about 36% acetonitrile concentration. (Fraction No. 1 refers to that collected 20 ml following the start of gradient elution.)

12. Combine the corresponding fraction No. 4's from three parallel
HPLC runs and lyophilize. Redissolve the dried material in 2 ml of 20%
acetonitrile containing 0.1% TFA and rechromatograph in the same
HPLC column under identical conditions (Fig. 4B). (There is no need to
refilter the sample before application to column.) GMF should now
emerge as a single peak corresponding to fraction No. 4. This is the final
product. Overall purification is 100,000-fold (Table I).

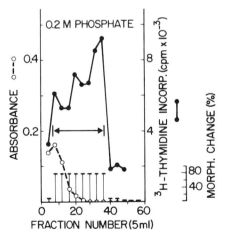

FIG. 3. Hydroxylapatite column chromatography. Symbols as in Fig. 1. Only the profile
of 0.2 M phosphate elution is shown.

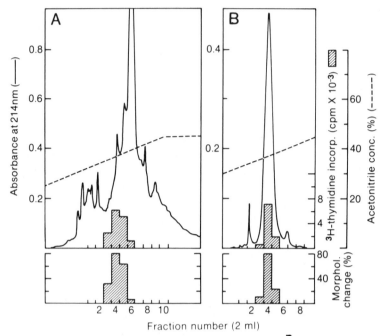

FIG. 4. Purification of GMF by repeated HPLC fractionation. (A) First HPLC run; (B) second HPLC run.

Comments

1. After applying sample to the hydroxylapatite column, avoid all subsequent contact with glass surfaces. Therefore, collect fractions in new, disposable polypropylene tubes, and use polypropylene disposable pipets and syringes for handling and for injection of samples into the HPLC column.

2. For bioassay of samples collected from DEAE-Sephacel, Sephadex G-75, and hydroxylapatite, first filter sterilize with Millex GV membrane (0.22 μm; Millipore), then apply samples directly to culture medium. Because the samples are mixed with the medium at 16-fold or higher dilution (see section on bioassay), the salt concentrations in the samples are compatible with cell survival. However, the salt concentration in the hydroxylapatite eluent is marginal for this purpose, and the samples may have to be dialyzed for some cells other than astroblasts.

3. For bioassay of samples collected from the HPLC column, it is not necessary to filter sterilize. To test these samples, first eliminate the ace-

tonitrile and TFA by lyophilization and then take up the dried material
with sterile TBS. It is advisable to include bovine serum albumin in the
TBS at 0.25 mg/ml to protect the GMF protein. If the HPLC fractions
have to be stored for a few days before processing, either for cell testing
or for further chemical work, do so at $-70°$ or lower temperature.

4. Perform all dialysis with Spectrapor-1 membrane having a cutoff
point of 6000 to 8000 molecular weight.

5. If the second run of HPLC fails to achieve homogeneity, a third run
is advisable and should complete the purification.

Properties of Purified GMF. The purified GMF shows a single protein
band in the SDS polyacrylamide gel under reducing conditions, corresponding to a molecular weight of 14,000 (Fig. 5). It has an isoelectric
point of pH 5.2. The preparation shows morphologic and mitogenic effects on astroblasts, with half-maximal activity at 8 ng/ml and full activity

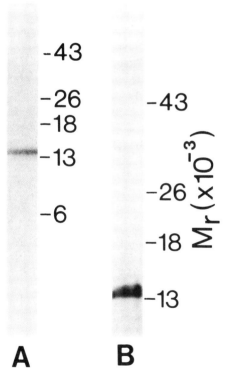

FIG. 5. SDS–polyacrylamide gel electrophoresis (PAGE) under reducing conditions.
Silver stained. (A) 10 ng GMF protein electrophoresed in 18% gel; (B) 100 ng GMF protein
electrophoresed in 12% gel. Molecular weight standards are indicated on the right.

at 40 ng/ml. GMF activity is destroyed by heating to 100° and by incubation with proteolytic enzymes.

Bioassay of GMF[6]

Bioassay is needed for monitoring GMF activity during purification. This is conducted on pure, secondary cultures of rat astroblasts. The assay detects morphologic and mitogenic effects of GMF.

Isolation of Rat Astroblasts[6]

Principle. Astroblasts are purified from a mixed culture of dissociated fetal brain cells. Contaminating cells are eliminated in the following manner: (1) neuroblasts and oligodendroblasts are outgrown by astroblasts; (2) many neurons do not survive under the culture conditions; (3) during subculture, remaining neuroblasts and oligodendroblasts are removed by differential attachment, i.e., they are washed out by medium change shortly after seeding. This is based on a time difference in cell attachment: it takes a shorter time for astroblasts than neurons and oligodendroglia to attach, since astroblasts adhere directly to the plastic surface whereas the latter only adhere to the spread-out astroblasts. This method does not eliminate fibroblasts, but they pose no problem because of their small number. This procedure yields astroblasts at over 90% purity as established by immunostaining for glial fibrillary acidic protein (astrocyte marker), neurofilament (neuronal marker), galactocerebroside (oligodendroglial marker), and Thy 1.1 (fibroblast marker).

Reagents

Tyrode salt solution: from Grand Island Biological Co. (GIBCO)
Calcium–magnesium-free Tyrode salt solution[8]: 8.00 g NaCl,
  0.30 g KCl, 0.05 g $NaH_2PO_4 \cdot H_2O$, 0.025 g KH_2PO_4,
  1.00 g $NaHCO_3$, and 2.00 g glucose per liter solution
Trypsin solution: 0.25% trypsin in Hanks' salt solution (U.S. Biochemical)
Trypsin/EDTA solution: 0.25% trypsin and 0.1% EDTA in Hank's salt solution (U.S. Biochemical)
Ham's F10 nutrient[9]: from K.C. Biological
Fetal calf serum: from K.C. Biological
Penicillin G
Streptomycin sulfate

[8] B. B. Garber and A. A. Moscona, *Dev. Biol.* **27**, 217 (1972).
[9] R. G. Ham, *Exp. Cell Res.* **29**, 515 (1963).

Procedure

1. Aseptically remove 18-day fetuses from one pregnant rat and place in sterile Tyrode salt solution at 4°. Take out the brains and tear away the meningeal layer and surface blood vessels.

2. Cut the cerebra into 1-mm cubes and wash 15 times with Tyrode solution.

3. Rinse the dissected tissue 2 times with 5 ml of calcium–magnesium-free Tyrode solution and then incubate with 5 ml of the same solution at 37° for 15 min.

4. Remove the liquid and incubate the tissue with 5 ml of trypsin solution at 37° for 15 min.

5. Rinse tissue with medium consisting of F10 nutrient containing 20% fetal calf serum. Add 2.5 ml of the medium and spin at 1000 g for 5 min, using a table top clinical centrifuge. Discard the supernatant. Take up the pellet in 2.5 ml of the medium and disperse the cells by trituration with an 18-gauge needle. (If the cells do not disperse easily, incubate with DNase for 10 min, using 0.05 mg/ml Tyrode solution.)

6. Distribute the cells into No. 3012 Falcon plastic culture flasks (25 cm^2) at a ratio of 1 brain per flask. Each flask should contain 5 ml of F10 supplemented with 20% fetal calf serum. Include penicillin and streptomycin at 50 units/ml and 100 μg/ml, respectively, in all media.

7. Incubate the cells at 37° for *2 days without disturbance*. This and all subsequent incubations should be in 5% CO_2 in air with saturated humidity. At the end of 2 days change medium to F10 containing 10% fetal calf serum (standard medium). Renew medium every 2 or 3 days.

8. The cells should be confluent in 1 week, containing a mixture of neuroblasts, oligodendroblasts, and astroblasts, and are ready to be subcultured. Cover the cells with cold trypsin/EDTA solution and remove the excess liquid after 1 to 2 min. With the remaining trypsin/EDTA, all the cells will round up and detach from the flask in about 5 to 10 min. Take up the cells in the "standard medium" and seed at 1 : 4 split, i.e., seed the content of one flask into two 8-well trays (Lux No. 5218), each well having dimensions of 3 × 3.5 cm and taking 2 to 2.5 ml medium. *Change medium between 10 to 20 hr* after seeding to eliminate contaminating neuroblasts and oligodendroblasts. Renew the medium in 2 or 3 days. Four to five days after seeding, the astroblasts will be confluent and ready for testing.

Comments

1. Disturbing the dissociated cells in primary culture during the first 2 days will result in excessive cell aggregation and poor adhesion to the flask surface.

2. To achieve high yield of astroblasts with low contaminating cells, during subculture the interval between seeding and initial medium change should be optimized with each batch of serum and probably with the strain of rats used.

Cell Testing[6]

Principle. This procedure concurrently examines process outgrowth and proliferation (as increase in DNA synthesis) of rat astroblasts after stimulation by GMF.

Reagents

Ham's F10 nutrient[9]: from K.C. Biological
Fetal calf serum: from K.C. Biological
[*methyl*-³H]Thymidine: 50–80 Ci/mmol; New England Nuclear
Tris-buffered saline (TBS): 0.02 M Tris–HCl, 0.15 M NaCl, pH 7.45
0.9% (w/v) NaCl
10% (w/v) trichloroacetic acid (TCA)
5% (w/v) trichloroacetic acid (TCA)
95% (v/v) ethanol
Scinti-Verse-I scintillation counting fluid: from Fisher Scientific

Procedure

1. Use confluent rat astroblasts grown in 8-well trays (see above). The culture should appear as a carpet of homogeneous polygonal cells under phase-contrast microscope. Replace the medium in each well (3 × 3.5 cm) with 2 ml of F10 containing 5% fetal calf serum. Add 0.125 ml of test sample containing GMF. Incubate at 37° in 5% CO_2 in air and saturated humidity.

2. After 24 hr, renew the medium and restimulate the cells with GMF sample. Add 0.2 μCi of tritiated thymidine.

3. Twenty four hours later, examine the cells for morphological changes. Any cell having at least one process longer than the diameter of the cell body is scored as a positive response. Estimate the percentage of cells responding (morphologic effect of GMF).

4. Harvest the cells for isotope counting in the following manner. Wash the monolayer twice with TBS. Add 20 μl of 0.9% NaCl and wipe the cells on a cotton swab, using one swab for each well. Secure with a modeling compound the opposite end of the cotton swab in a microtiter plate. By holding and inverting the microtiter plate, soak the cotton swabs collectively in a beaker (stirred with magnetic bar) containing cold 10% TCA for 20 min. Follow this with a 10-min soak in cold 5% TCA and another 10-min soak in 95% ethanol.

5. Air dry the swabs and immerse the tip in 5 ml of Scinti-Verse-I for counting. Express the results as increase in cpm over wells where GMF is replaced with the buffer vehicle. This represents the mitogenic effect of GMF.

Comments

1. This procedure enables us to perform morphologic assessment and isotope counting all in one setting. Since morphologic assay is semiquantitative, it is important to detect the mitogenic effect in the same test sample. On the other hand, mitogenicity does not necessarily imply GMF activity because of the presence of other growth factors in the brain which also act on astroblasts. The real GMF activity should be both morphologic and mitogenic.

2. Although morphologic changes can be seen 24 hr after the first GMF stimulation, the changes are more stable and the scores more reproducible 24 hr after the second stimulation.

3. In each assay always include a negative control (blank) where the GMF sample is replaced with the buffer vehicle such as Tris-buffered saline, and a positive control such as a sample of crude beef brain extract which is known to exert full GMF activity.

4. For unknown reasons, not every batch of astroblasts respond well to GMF, and interassay variability can go as high as 20%. However, intraassay variability is usually within 5%. For best comparison, samples should be tested on the same batch of cells in the same assay.

5. Astroblasts aged in culture may lose their responsiveness to GMF.

Section IV

Transferrin, Erythropoietin, and Related Factors

[21] Isolation of Transferrin Receptor from Human Placenta

By PAUL A. SELIGMAN and ROBERT H. ALLEN

Introduction

Iron is transported in plasma by the 80,000 M_r transport protein, transferrin.[1] Cellular iron uptake is mediated by specific binding sites for transferrin found on the cell surface.[1] This cell surface protein, called the transferrin receptor, was first described on hemoglobin-producing cells[2] and placental cells,[3] both of which have been known to have high iron requirements. The availability of large amounts of placental tissue not only has proved useful in measurements of transferrin binding to isolated cell membrane preparations,[3,4] but also permits the isolation and purification of the human transferrin receptor. This chapter details the methodology employed for the assessment of transferrin binding to placental membrane preparations. The chapter mainly describes methodology for purifying the receptor by sequentially utilizing two different affinity chromatography steps as the major purification steps. Some of the properties characterizing the isolated receptor protein will then be discussed briefly. Recently, the finding of high densities of transferrin receptors on the surface of proliferating cells has generated considerable interest.[5] Therefore, characterization of the receptor may better elucidate the functional significance of this protein as it relates to cellular proliferation.

Assay

Measurements of ^{125}I-Labeled Transferrin Binding to Solubilized Receptor Preparations

The purification of transferrin receptor from human placenta requires a method for assaying transferrin receptor in various detergent solubilized

[1] P. Aisen and E. B. Brown, in "Progress in Hematology" (E. B. Braun, ed.), p. 25. Grune and Stratton, New York, 1975.
[2] J. H. Jandle and J. H. Katz, J. Clin. Invest. **42,** 314 (1963).
[3] P. A. Seligman, R. B. Schleicher, and R. H. Allen, J. Biol. Chem. **254,** 9943 (1979).
[4] H. D. Wada, P. E. Hass, and H. H. Sussman, J. Biol. Chem. **254,** 12629 (1979).
[5] P. A. Seligman, in "Progress in Hematology" (E. B. Brown, ed.), Vol. XIII. p. 131. Grune and Stratton, New York, 1983.

preparations. The solubilized receptor assay utilizes ammonium sulfate precipitation to separate free ^{125}I-labeled transferrin from ^{125}I-labeled transferrin bound to receptor. The ^{125}I-labeled transferrin is prepared by iodination of iron saturated transferrin using the chloramine-T method.[6] The preparation remains stable for 2 to 4 weeks at 4° and the stability is longer if aliquots are frozen at 0° and each thawed once and then kept at 4°.

Assay Method

Triton X-100 solubilized receptor (see receptor preparation method for solubilization below) in 580 μl of 10 mM Tris–Hcl, pH 8.0, in 150 mM NaCl is incubated at 37° for 30 min with ^{125}I-labeled transferrin (2–20 ng) in 20 μl of the same buffer. After incubation at 37° for 30 min the mixture is cooled to 4° for 5 min in ice water and 400 μl of cold saturated ammonium sulfate is added to a final concentration of 40%. This mixture is then centrifuged at 25,000 g for 30 min at 4°. After centrifugation, depending on the amount of detergent present in the preparation, at least a portion of the ammonium sulfate protein precipitate is found floating on top of the supernatant. The best way, therefore, to uniformly remove supernatant is to quickly aspirate each tube in exactly the same manner with a Pasteur's pipet attached to a vacuum suction device. Tubes containing the precipitated pellet are then counted for ^{125}I radioactivity in a gamma counter. Specifically bound ^{125}I-labeled transferrin is determined by subtracting precipitated ^{125}I-labeled transferrin counted in an identical assay mixture which contains a 1000-fold excess of nonradioactive transferrin.

Assay Results

[125]I-Labeled transferrin binding to a fixed amount of solubilized receptor preparation is a saturable process using increasing amounts of ^{125}I-labeled transferrin as shown in Fig. 1. Figure 1 shows that saturation is reached utilizing purified receptor but saturation is also reached using crude receptor preparations. Although the ^{125}I-labeled transferrin binding is specific and saturable using this assay, the amount of receptor present in any assay mixture cannot be absolutely quantitated for the following reasons: (1) preparations with differing amounts of Triton X-100 could not be compared since an increased amount of Triton X-100 in the assay mixture causes more floatation of precipitate and therefore a greater chance of loss of precipitated radioactivity using the vacuum suction device. (2) The 40% concentration of ammonium sulfate may dissociate,

---

[6] F. C. Greenword, *in* "Modern Trends in Endocrinology" (H. Gardiner-Hill, ed.), Vol. 3, p. 288. Butterworths, London, 1967.

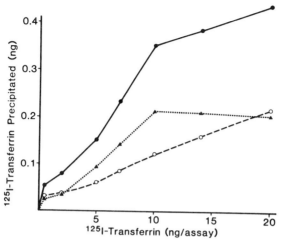

FIG. 1. Nonspecific precipitated ^{125}I-labeled transferrin (○) is subtracted from total ^{125}I-labeled transferrin precipitated (●) to determine specifically bound ^{125}I-labeled transferrin to receptor (▲). In this assay, increasing amounts of ^{125}I-labeled transferrin and the same amount of purified transferrin receptor (40 ng) are added. Adapted from Seligman et al.[3]

or does not precipitate all of the transferrin–receptor complex, and therefore the amount of complex precipitated may also vary from preparation to preparation.

However, the assay does allow for determining the presence of receptor, and during the purification process (see below) similar preparations of receptor can be compared to each other qualitatively. Better quantitation of solubilized transferrin receptor can be determined by using another assay for solubilized receptor developed by Sussman and colleagues[4] (see [23], this volume) which utilizes polyethylene glycol precipitation. Results of all solubilized binding assays, however, are complicated by the fact that varying amounts of endogenous transferrin are found in different placental preparations. This is not surprising since it appears that the receptor is highly saturated with transferrin in vivo. This latter assessment is based on the fact that transferrin has a relatively high concentration in serum (25 μM) which is higher than the K_a of transferrin binding to the receptor (1 nM^{-1}).[1,4,5] Since binding studies, particularly calculation of the association constant, are affected by the amount of endogenous transferrin present, it is possible to remove the transferrin by incubating cell suspensions in dilute buffer containing no transferrin and allowing dissociation of the transferrin to occur at 37°.[7] This step can be repeated a

---

[7] D. P. Witt and R. C. Woodsworth, Biochemistry 17, 3913 (1978).

number of times until essentially all of the transferrin has been dissociated.

Radioimmunoassay for Transferrin

It is possible to account for endogenous transferrin by simply measuring the amount of transferrin present in the preparation using a radioimmunoassay. To 410 μl of 10 mM KPO$_4$, pH 7.5, 150 mM NaCl, containing 10 to 500 ng of nonlabeled human transferrin the following are added: (1) 20 μl of 10 mM KPO$_4$, pH 7.5, 150 mM NaCl containing human ^{125}I-labeled transferrin (2 ng), and (2) 70 μl of control rabbit serum containing 0.2 μl of rabbit anti-human transferrin antiserum. After 45 min of incubation at 37°, 500 μl of cold saturated ammonium sulfate is added and the mixture centrifuged at 30,000 g for 20 min at 4°, and 500 μl of supernatant is counted for ^{125}I in a gamma counter. Under these conditions, approximately 85% of the ^{125}I-labeled transferrin is precipitated in the absence of unlabeled human transferrin. This value falls significantly and progressively to 15% as the amount of unlabeled transferrin is increased. The assay can be modified by utilizing polyethylene glycol 6000 (final concentration 4.5%) instead of ammonium sulfate to precipitate the antigen antibody complex.

Figure 2 shows the results of a radioimmunoassay of a crude Triton X-

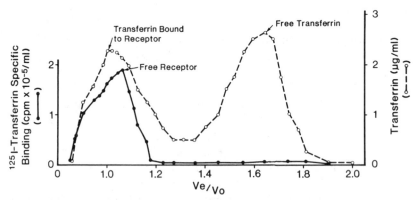

FIG. 2. Six milliliters of crude solubilized placental homogenate subjected to gel filtration on Sephadex G-200 in 0.1% Triton X-100. Fractions are assayed separately for both immunoreactive transferrin using a immunoassay and ^{125}I-labeled transferrin specific binding utilizing the solubilized receptor binding assay. The higher molecular weight immunoreactive transferrin peak represents transferrin bound to receptor and elutes as a slightly higher molecular weight than the peak of ^{125}I-labeled transferrin binding which presumably identifies receptor molecules less fully saturated with endogenous transferrin. Adapted from Seligman et al.[3]

100 solubilized placental preparation applied to Sephadex G-200. The first high M_r peak of immunoreactive transferrin elutes at the void volume and makes up 40% of the total immunoreactive transferrin. This second lower M_r peak contains the rest of the immunoreactive transferrin and elutes at the apparent M_r of free transferrin. Assays for solubilized transferrin receptor using ^{125}I-labeled transferrin binding ability showed a single peak of binding which eluted just after the void volume and indicates that the first immunoreactive transferrin peak represents solubilized receptor saturated with endogenous transferrin. The rationale of the purification scheme, illustrated below, therefore, is to isolate initially the saturated form of the transferrin receptor.

Purification of the Human Placental Transferrin Receptor

Preparation of Placental Homogenate and Solubilization with Triton X-100

Two placenta with a combined weight of 1500 g should be cut into approximately 20 g pieces and aliquots homogenized in a Waring Blendor for 1 min in 1.5 volumes of 10 mM KPO$_4$, pH 7.5, containing 150 mM NaCl. The mixture is then centrifuged at 25,000 g for 90 min and the supernatant is discarded and the pellets are resuspended in 1500 ml of the same buffer. This homogenate can then be solubilized immediately or can be stored frozen up to 7 days with no loss of receptor activity. For solubilization the resuspended pellet (containing 1500 ml) is reconstituted with the addition of 500 ml of 10 mM KPO$_4$, pH 7.5, containing 150 mM NaCl and 4% Triton X-100 (final Triton concentration 1%). Solubilization is continued first with homogenization using a Waring Blendor for two 30-sec bursts. Then the mixture is put in a stirrer in a cold room and sonication is performed using a large sonicator probe for two 5 min bursts. The mixture is then allowed to sit in the cold room stirring for another 2 hr and is then centrifuged at 25,000 g for 30 min. The pellet is discarded and the supernatant containing the solubilized membrane preparation is then utilized for the rest of the purification steps.

Ammonium Sulfate Precipitation

To the supernatant, solid ammonium sulfate (31.5 g/100 ml) is added and stirred for 10 min at 4°. The sample is then centrifuged at 20,000 g for 20 min. Since the Triton X-100 present causes floatation of the precipitate, clear plastic centrifuge bottles are used and the supernatant is carefully aspirated using a vacuum suction device. The precipitate is resuspended in 600 ml of 10 mM KPO$_4$, pH 7.5, containing 150 mM NaCl and 0.1% Triton X-100 and dialyzed against 4 liters of the same solution for 16 hr with the dialyzate changed at 4 and 12 hr. The ammonium sulfate step is performed to concentrate the sample and more importantly, separate free transferrin which remains in the supernatant, from transferrin bound to receptor which is precipitated. This transferrin bound receptor preparation is then subjected to the first affinity chromatography step which utilizes anti-human transferrin-Sepharose as described below.

Immunochromatography on Antihuman Transferrin-Sepharose

In order to prepare the antihuman transferrin-Sepharose column, antibody to human transferrin was prepared in rabbits[8] and partially purified with ammonium sulfate precipitation.[9] The partially purified antibody containing 5 mg of protein per ml and 60 ml of 50 mM KPO$_4$, pH 8.15, was coupled to 60 ml of Sepharose 2B with a coupling efficiency of 95%.[10]

The dialyzed ammonium sulfate preparation is centrifuged in small portions (about 2 to 300 ml) at 10,000 g for 20 min to remove any remaining precipitate which might hinder the flow of the column. The supernatant is then applied to a 2 × 30 cm column containing 60 ml of anti-human transferrin-Sepharose which is in 10 mM KPO$_4$, pH 7.5, 150 mM NaCl (equilibration buffer) and 0.1 Triton. The entire immunochromatography step is performed at 4°. After approximately 20 to 300 ml of the solution has been applied, the column is washed with a variety of buffers in the following order. Wash (1) 200 ml of equilibration buffer, wash (2) 100 ml of 10 mM KPO$_4$, pH 7.5, 500 mM NaCl (high salt wash), wash (3) 100 ml of equilibration buffer, wash (4) 200 ml of 20 mM glycine/NaOH, pH 10.0, 500 mM NaCl and 0.5% Triton, and wash (5) 100 ml of 10 mM glycine/HCl, pH 2, 150 mM NaCl. Wash 4 elutes the receptor from transferrin which is bound to the antitransferrin antibody on the column. This wash contains most of the receptor activity measured by ^{125}I-labeled transferrin binding. Wash 5 is performed to remove all of the bound transferrin so the affinity column can be reused for subsequent batches of the ammonium sulfate preparation as described above. Wash 4 containing the eluted receptor activity is then subjected to ammonium sulfate precipitation with solid ammonium sulfate (31.5 g/100 ml) added. This second ammonium sulfate precipitation is mainly performed to concentrate the partially purified receptor solution. The ammonium sulfate pellet is resuspended in 50

[8] R. L. Burger and R. H. Allen, J. Biol. Chem. 249, 7220 (1974).
[9] N. Harbee and A. Ingild, Scand. J. Immunol. 2 (Suppl. 1), 162 (1973).
[10] R. L. Burger, S. Waxman, H. S. Gilbert, C. S. Mehlman, and R. H. Allen, J. Clin. Invest. 56, 1262 (1975).

ml of equilibration buffer plus 0.1% Triton and dialyzed as described above for the first ammonium sulfate precipitation.

Affinity Chromatography on Human Transferrin-Sepharose

The dialyzed ammonium sulfate pellet, containing partially purified receptor without significant amounts of transferrin bound, is then applied to a column containing 15 ml of human transferrin-Sepharose in equilibration buffer. The column is then washed with the following buffers. Wash (1) 100 ml of equilibration buffer, wash (2) 50 ml of 10 mM KPO$_4$, pH 7.5, 1 M NaCl (high salt), wash (3) 20 ml of equilibration buffer to remove the high salt buffer, and wash (4) for 40 ml of 50 mM glycine-NaOH, pH 10.0, 1 M NaCl, and 1.0% Triton. Wash 4 which elutes most of the receptor is then concentrated using ammonium sulfate precipitation (31.5 g/100 ml) and resuspended in 3 ml of equilibration buffer plus 0.1% Triton and dialyzed. The final solution after dialysis contains about 1 mg of protein as measured by a minor modification[11] of the method of Lowry[12] for solutions containing Triton X-100.

Properties of the Purified Protein

Purity of the final transferrin receptor preparation can be assessed using polyacrylamide gel electrophoresis utilizing both a 0.1% Triton X-100 system[13] and sodium dodecyl sulfate system (SDS).[14] The SDS system requires removal of Triton X-100 from the preparation using an ethanol extraction method.[15] The purified receptor preparation run on polyacrylamide disc gels in 0.1% Triton X-100 migrates as a single broad protein band which is easily distinguishable from another gel run with purified transferrin. Since the preparation run on this form of gel electrophoresis maintains its ability to bind ^{125}I-labeled transferrin, an identical gel containing receptor protein can be cut into small slices which are put in separate tubes containing equilibration buffer and 0.1% Triton. The eluate from each gel slice can be then assayed for ^{125}I-labeled transferrin binding ability and a single peak of binding ability is found which coincides with the position of the transferrin receptor protein band found on the gel stained for protein.

Receptor protein detritonized receptor protein run on SDS–poly-

[11] C. Wang and R. L. Smith, *Anal. Biochem.* **63**, 414 (1975).
[12] O. H. Lowry, N. J. Rosebrough, A. L. Farr, and R. J. Randall, *J. Biol. Chem.* **193**, 265 (1951).
[13] T. Kawasaki and G. Ashwell, *J. Biol. Chem.* **251**, 1296 (1976).
[14] U. K. Laemmli, *Nature (London)* **227**, 680 (1970).
[15] C. E. Frasch and E. C. Gotschlich, *J. Exp. Med.* **140**, 87 (1974).

acrylamide gel electrophoresis shows an approximately 180,000 M_r band in the absence of, and a 90,000 M_r band in the presence of 2-mercaptoethanol. On SDS gels, a faint band at 80,000 M_r may also be seen which presumably represents transferrin which contaminates the purified receptor preparation. The transferrin contamination, which arises from residual endogenous transferrin, or transferrin eluted from the final affinity column, represents only 10 to 20 ng/μg of purified receptor protein. Identical SDS gels stained for carbohydrate show that both the unreduced 180,000 M_r species and the 90,000 M_r subunit both stain positively for carbohydrate with Schiff's reagent.

Preparation of Polyclonal Antisera to Purified Human Placental Transferrin Receptor

Antibody to the purified human transferrin receptor has been raised in both rabbit and chicken. The rabbit antibody is prepared as previously described[8] and the chicken antibody is raised by giving three weekly injections of 100 μg of protein intradermally with all injections mixed 1 : 1 with complete Freund's adjuvant. At the fourth week blood is removed from a wing vein every 48 hr. The chicken antisera is partially purified by

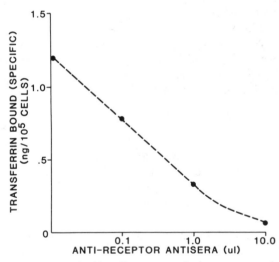

FIG. 3. Specificity of polyclonal antihuman transferrin receptor antisera prepared in chicken is shown by inhibition of specific [125I]-labeled transferrin binding to human WI-38 fibroblasts. The antisera was partially purified as described in the text and is specific for inhibition of transferrin binding since 20 μl of control partially purified chicken sera had no effect on binding. The data are expressed measuring nanograms of transferrin bound to 10^6 fibroblasts as described previously.[17]

means of ammonium sulfate precipitation[9] and is applied to an affinity column containing human transferrin-Sepharose. This latter step is used to remove the small amount of anti-human transferrin antibody which is found in the antisera. Using the radioimmunoassay for human transferrin,[3] it can be determined that greater than 99% of this antibody is removed. This partially purified chicken antibody inhibits the functional transferrin receptor since human fibroblasts incubated with antibody and ^{125}I-labeled transferrin show a greater than 90% inhibition of specific ^{125}I-labeled transferrin binding as compared with cells incubated with control chicken serum (Fig. 3). Additional experiments using human reticulocytes have shown that the antibody specifically inhibits transferrin-mediated iron uptake utilizing transferrin-^{59}Fe by reticulocytes. The polyclonal chicken anti-human transferrin antibody can be utilized in a radioimmunoassay for human transferrin receptor on various human cells preparations as described previously.[17]

[16] J. P. Secrest and R. L. Jackson, this series, Vol. 28, p. 54.
[17] J. L. Frazier, J. H. Caskey, M. Yoffe, and P. A. Seligman, *J. Clin. Invest.* **69**, 853 (1982).

[22] Receptor Assay with Radiolabeled Transferrin

By JOHN H. WARD and JERRY KAPLAN

Transferrin is an iron-binding protein (M_r 80,000) which binds two molecules of ferric iron in the presence of an anion. Diferric transferrin binds to specific receptors present on the surface of maturing erythroid cells and a variety of other cultured cell lines. The diferric transferrin–transferrin receptor complex is internalized into an acidic, membrane-bound, nonlysosomal compartment (the endosome) where iron is released.[1,2] At the acidic pH of the endosome, apotransferrin remains bound to the receptor and the apotransferrin–transferrin receptor complex is recycled to the cell surface. At the neutral pH of the cell surface, apo-

[1] J. van Renswoude, K. R. Bridges, J. B. Harford, and R. D. Klausner, *Proc. Natl. Acad. Sci. U.S.A.* **79**, 6186 (1982).
[2] J. Lamb, F. Ray, J. H. Ward, J. P. Kushner, and J. Kaplan, *J. Biol. Chem.* **258**, 8751 (1983).

transferrin dissociates from the receptor, where it is free to bind iron and initiate another cycle of iron uptake.[3,4] Transferrin receptors can be measured on cultured cells with either radiolabeled transferrin or with specific antireceptor antibodies. This chapter reviews the method of assaying cellular transferrin receptors by measuring the binding of ^{125}I-labeled diferric transferrin to cells.

Saturation of Transferrin with Iron

Transferrin can be purchased from commercial sources or purified from serum.[5] Since transferrin receptors have a much higher affinity for diferric transferrin than for apotransferrin at neutral pH (pH 7.2–7.7)[3,6–8] it is essential to fully saturate transferrin with iron. This is easily accomplished by the method of Larrick and Cresswell.[9] Transferrin, 10 mg/ml, is dissolved in 10 mM sodium bicarbonate containing 0.1 mg/ml ferric ammonium citrate. This solution provides the anion required for iron binding as well as an excess of iron. The solution is incubated at 20° with gentle stirring for 4 hr by which time it should be a deep orange color. The solution is then dialyzed extensively against phosphate-buffered saline (PBS) at 4° to remove excess iron. The dialyzate may be sterilized by Millipore filtration and saved for months without loss of activity. The degree of iron saturation can be assessed by the A_{465}/A_{280} ratio, which should be 0.046 for fully saturated diferric transferrin.[1] Polyacrylamide gel electrophoresis in the presence of 8 M urea may also be used to determine if full saturation has been achieved.[10]

Iodination of Diferric Transferrin

In our laboratory, we have used two methods for iodinating diferric transferrin. The first uses chloramine-T. This method reliably yields ^{125}I-labeled diferric transferrin of high specific activity.[11,12] We also employ

[3] A. Dautry-Varsat, A. Ceichanover, and H. F. Lodish, *Proc. Natl. Acad. Sci. U.S.A.* **80,** 2258 (1983).

[4] R. D. Klausner, G. Ashwell, J. van Renswoude, J. B. Harford, and K. R. Bridges, *Proc. Natl. Acad. Sci. U.S.A.* **80,** 2263 (1983).

[5] G. Sawatzki, V. Anselstetter, and B. Kubanck, *Biochem. Biophys. Acta* **667,** 132 (1981).

[6] J. H. Jandl and J. H. Katz, *J. Clin. Invest.* **42,** 314 (1963).

[7] S. P. Young and P. Aisen, *Hepatology* **1,** 114 (1981).

[8] J. H. Ward, J. P. Kushner, F. A. Ray, and J. Kaplan, *J. Lab. Clin. Med.* **103,** 246 (1984).

[9] J. W. Larrick and P. Cresswell, *Biochim. Biophys. Acta* **583,** 483 (1979).

[10] D. G. Makey and U. S. Seal, *Biochim. Biophys. Acta* **453,** 250 (1976).

[11] J. H. Ward, J. P. Kushner, and J. Kaplan, *J. Biol. Chem.* **257,** 10317 (1982).

[12] J. H. Ward, J. P. Kushner, and J. Kaplan, *Biochem. J.* **208,** 19 (1982).

1,3,4,6-tetrachloro-3α,6α-diphenylglycoluril (Iodogen), a mild solid-phase oxidizing agent which appears to cause less oxidation-related damage than does chloramine-T.[13] The ^{125}I-labeled diferric transferrin derived by the chloramine-T method yields a functional ligand which has a nonspecific binding of only 5–15% of total counts bound when binding to cultured HeLa cells or human fibroblasts is assayed. Nonetheless, ^{125}I-labeled diferric transferrin generated with Iodogen yields values for nonspecific binding of less than 5%. For this reason, and because of simplicity, we prefer to use Iodogen to label transferrin.

Preparation of Iodogen Tubes. This procedure is a slightly modified version of the method of Wiley and Cunningham.[14] Five milligrams of Iodogen (Pierce Chemical Co.) is dissolved in 12 ml of chloroform. Aliquots of 110 μl are placed in 12 × 75 mm borosilicate glass tubes. The tubes are then placed in a vacuum for 10 min. The vacuum is released and reapplied for an additional 10 min. This procedure is repeated until a thin white film of Iodogen coats the bottom of the tube. These tubes are then stored in a vacuum at 20° and may be used up to 6 months after their preparation.

Iodination Procedure. All procedures are done at 4°. An Iodogen tube prepared as above is rinsed once with 500 μl of cold phosphate-buffered saline. The following solutions are then added in order: diferric transferrin (10 mg/ml in PBS) (60 μl), phosphate-buffered saline (10 μl), and Na[125]I in 0.1 M NaOH (1 mCi/10 μl) (30 μl).

After addition of Na[125]I, the contents are gently mixed and incubated on ice for 12 min. The mixture is then applied to a 200 × 7 mm disposable plastic column of Sephadex G-10 equilibrated with PBS. One milliliter fractions are collected, and 5-μl aliquots are monitored for radioactivity. The ^{125}I-labeled diferric transferrin should begin to elute by fraction 3 and the entire sample should elute by fraction 6. Free iodine usually appears by fraction 9 or 10. The fractions containing ^{125}I-labeled diferric transferrin are pooled and the specific activity of the labeled protein determined. The resulting protein concentration is usually 1–3 μM with a specific activity of 0.06–0.37 μCi/pmol (100–600 cpm/fmol when counted on a Packard Auto-Gamma 5780 with an ^{125}I counting efficiency of 74%). About 90% of the starting material (600 μg of diferric transferrin) is recovered as ^{125}I-labeled diferric transferrin. Smaller amounts may be iodinated keeping in mind two points. First, the total reaction volume should be less than 100 μl since this is the volume exposed to Iodogen in the tubes

[13] P. R. P. Salacinski, C. McLean, J. E. C. Sykes, V. V. Clement-Jones, and P. J. Lowry, *Anal. Biochem.* **117**, 136 (1981).
[14] H. S. Wiley and D. D. Cunningham, *J. Biol. Chem.* **257**, 4222 (1982).

prepared as above. Second, a ratio of 200 μg of diferric transferrin per mCi of ^{125}I seems to be optimal. If the specific activity of the ^{125}I-labeled diferric transferrin is too high, the time of incubation may be decreased or the ratio of ^{125}I to transferrin may be reduced. ^{125}I-Labeled diferric transferrin prepared by this method migrates as a single band on 9% SDS polyacrylamide gels and is completely precipitable by trichloroacetic acid. When stored at 4° it retains its use as a functional ligand for at least 2 months. If nonspecific binding progressively increases, we discard the batch, as this is evidence of contamination or degradation.

Transferrin Receptor Assay Using ^{125}I-Labeled Diferric Transferrin

Transferrin receptors are present on maturing erythroid cells, a variety of neoplastic cells, mitogen-stimulated lymphocytes, and many other cultured cell lines.[15-17] ^{125}I-Labeled diferric transferrin may be used to assay transferrin receptors on cells grown either in monolayer culture or in suspension. The temperature of the assay is critical, since a large component of a cell's transferrin receptors is present inside the cell, even if the cell has had no prior exposure to diferric transferrin.[2,18] At 0°, only surface receptors are assayed. At 37°, both surface and internal transferrin receptors are detected. The number of transferrin receptors which are expressed on a given cell line may vary depending on the proliferative state.[19] For this reason, the culture conditions and their relation to ^{125}I-labeled diferric transferrin binding should be carefully defined for each cell type to be assayed.

Incubation Procedure. If the cells to be studied are growing in monolayer culture, the media is removed, and the cells are incubated in serum-free media at 37° for 30 min to allow any endogenous transferrin to dissociate from the transferrin receptor. On HeLa cells, the time required for 50% of the bound ligand to dissociate from transferrin receptor at 37° is 11.6 min.[11] After 30 min the majority of endogenous ligand will no longer be cell associated. Similar rapid rates of dissociation are seen at 37° in other cell types such as human skin fibroblasts. If the cells to be studied

[15] R. M. Galbraith, P. Werner, P. Arnaud, and G. M. P. Galbraith, *Cell. Immunol.* **49,** 215 (1980).
[16] T. A. Hamilton, H. G. Wada, and H. H. Sussman, *Proc. Natl. Acad. Sci. U.S.A.* **76,** 6406 (1979).
[17] J. H. Ward, J. P. Kushner, and J. Kaplan, in "Current Hematology/Oncology" (V. Fairbanks, ed.), Vol. 3, pp. 1–50. Wiley, New York, 1984.
[18] M. Karin and B. Mintz, *J. Biol. Chem.* **256,** 3245 (1981).
[19] J. L. Frazier, J. H. Caskey, M. Yoffe, and P. A. Seligman, *J. Clin. Invest.* **69,** 853 (1982).

are of human origin and are grown in calf serum, such a step is probably unnecessary since the human transferrin receptor does not measurably interact with bovine transferrin.[12,20]

Assay of Receptors at 0°. To measure surface transferrin receptors, cells on 35 mm plates are then incubated with phosphate-buffered saline containing 4 mg/ml of bovine serum albumin and varying concentrations of ^{125}I-labeled diferric transferrin. The K_d of the transferrin receptor varies from 5 to 20 nM, depending on the cell type. Generally, concentrations varying from 2 to 100 nM are sufficient to generate reproducible Scatchard plots. A second set of duplicate plates is incubated with the same constituents with the addition of 1.25×10^{-5} M nonradioactive diferric transferrin to the incubation media. The final volume in both sets of plates is 1.0 ml. The plates are then incubated on ice (0°) for 90 min. This time period should allow for equilibrium to be reached at the concentrations of ^{125}I-labeled diferric transferrin suggested above. The plates are rapidly washed four times with 2 ml of 4° PBS per wash. One milliliter of 1% sodium dodecyl sulfate is added to the plates, and the solubilized cells are transferred to counting vials.

At each specific concentration of ^{125}I-labeled diferric transferrin, the counts bound in the presence of high concentrations of unlabeled diferric transferrin (nonspecific binding) are subtracted from the counts bound in the absence of excess diferric transferrin. This difference represents the specific binding of ^{125}I-labeled diferric transferrin at a given concentration of radioligand. The nonspecific binding using ^{125}I-labeled diferric transferrin made by the Iodogen method is usually less than 5%. On occasion, if cells have been in culture for prolonged periods of time, the background may increase. The protein concentration of the individual samples is determined by the method of Lowry.[21] The specific activity of each sample may then be expressed as fmol bound per μg cellular protein. The maximum binding capacity (i.e., "receptor number") can be determined by the method of Scatchard.[22] Each transferrin receptor consists of a dimer with a 90,000 Da subunit and each subunit binds one molecule of diferric transferrin.[23,24] Therefore, the number of diferric transferrin molecules bound does not strictly represent transferrin receptor number. Also, it has been

[20] R. G. Sephton and A. W. Harris, *Int. J. Nucl. Med. Biol.* **8**, 333 (1981).
[21] O. H. Lowry, N. J. Rosebrough, A. L. Farr, and R. J. Randall, *J. Biol. Chem.* **193**, 265 (1951).
[22] G. Scatchard, *Ann. N.Y. Acad. Sci.* **51**, 660 (1949).
[23] C. A. Enns and H. H. Sussman, *J. Biol. Chem.* **256**, 9820 (1981).
[24] C. Schreider, R. Sutherland, R. A. Newman, and M. F. Greaves, *J. Biol. Chem.* **257**, 8516 (1982).

reported that the number of immunologically recognizable transferrin receptors may be different than the number which is functionally able to bind diferric transferrin.[19]

The above procedure is also applicable to cells maintained in suspension. For most lines, 1×10^6 cells are used per sample and are incubated in 1500 μl polypropylene microcentrifuge tubes in a final incubation volume of 1000 μl. At completion of the incubation period, the cells are washed twice by centrifugation in an Eppendorf centrifuge, and the cell pellet solubilized in 1 ml of 1% sodium dodecyl sulfate. It is particularly important that individual protein concentrations are determined because differing amounts of protein may be lost during the washing procedure. Standardizing the bound radioactivity to the amount of protein greatly decreases the variability of the assay.

Assay of Receptors at 37°. If cells are incubated in the presence of ^{125}I-labeled diferric transferrin at 37°, cell-associated ligand can be found on the cell surface as well as inside the cell in a membrane-bound, nonlysosomal compartment (the endosome). The internalized transferrin is eventually returned to the media undegraded.

When ^{125}I-labeled diferric transferrin binding is measured at 37° in a 5% CO_2 atmosphere, Eagle's minimum essential media with Earle's salts is used in place of phosphate-buffered saline. If cells are incubated with ^{125}I-labeled diferric transferrin at 37° for 60 min, a steady state is achieved. Cells can rapidly be cooled, washed, and the specific binding of ^{125}I-labeled diferric transferrin determined. This represents both surface and intracellular ligand.

If it is desired to determine the percentage of transferrin in each compartment, the acid stripping procedure of Karin and Mintz may be used to remove surface bound counts.[18] Cells which have been incubated at 37° are rapidly cooled to 0° and washed to remove unbound ligand as previously described. Cells are then incubated with 1.5 ml of 0.5 M NaCl, 0.2 M acetic acid for 8 min at 0°. This is removed and saved. Each plate is then washed with 1 ml of this "acid-stripping" solution, and this wash is pooled with the initial 1.5 ml. The radioactivity in the pooled sample is determined and represents surface bound ligand. The cell-associated or "acid-resistant" radioactivity is then determined. This represents ligand which has been internalized.[2,18] Using the acid-stripping technique in HeLa cells, the surface spillover is 4.9% as determined by the method of Wiley and Cunningham. The rate at which diferric transferrin is internalized can then be determined.[14]

[23] Radioimmunoassay of Transferrin Receptor

By Susan E. Tonik and Howard H. Sussman

Introduction

Human cells in tissue culture have been shown to require iron for normal growth.[1] This requirement is met by the internalization of iron-loaded transferrin via its specific receptor.[2,3] The presence of the transferrin receptor (TR) has been demonstrated in cells in culture,[4-9] and also in reticulocytes,[10-13] the syncytial trophoblastic membranes of the placenta,[14,15] leukemic cells,[16] and microsomal preparations of neoplastic tumors.[17,18] All of these studies have employed either a radioligand binding technique or immunofluorescence. These techniques are not appropriate for adequate quantitation in many instances. A constraint of the binding assay is that it can accurately measure the receptor only under the condition of significant unsaturation with ligand. This is rarely the case with the transferrin receptor because of the normally high levels of endogenous ligand present in most samples and the high affinity of transferrin for its receptor.[19] The other problem with the binding assay, as well as indirect immunofluorescence, is the requirement that the binding activity be preserved through any processing conditions for samples in an experi-

[1] D. Barnes and G. Sato, *Cell* **22,** 649 (1980).
[2] M. Karin and B. Mintz, *J. Biol. Chem.* **256,** 3245 (1981).
[3] J. H. Jandl and J. H. Katz, *J. Clin. Invest.* **42,** 314 (1963).
[4] T. A. Hamilton, H. G. Wada, and H. H. Sussman, *Proc. Natl. Acad. Sci. U.S.A.* **76,** 6406 (1979).
[5] J. A. Fernandez-Pol and D. J. Klos, *Biochemistry* **19,** 3904 (1980).
[6] J. W. Larrick and P. Cresswell, *J. Supramol. Struct.* **11,** 579 (1979).
[7] G. M. P. Galbraith, R. M. Galbraith, and W. P. Faulk, *Cell. Immunol.* **49,** 215 (1980).
[8] N. J. Verhoef and P. J. Noordeloos, *Clin. Sci. Mol. Med.* **52,** 87 (1977).
[9] B. S. Stein and H. H. Sussman, *J. Biol. Chem.* **258,** 2668 (1983).
[10] M. Steiner, *Biochem. Biophys. Res. Commun.* **94,** 861 (1980).
[11] F. M. van Bockxmeer and E. H. Morgan, *Biochim. Biophys. Acta* **468,** 437 (1977).
[12] B. Ecarot-Charrier, B. L. Grey, A. Wilczynska, and H. M. Schulman, *Can. J. Biochem.* **58,** 418 (1980).
[13] E. Baker and E. H. Morgan, *Biochemistry* **8,** 1133 (1969).
[14] H. G. Wada, P. E. Hass, and H. H. Sussman, *J. Biol. Chem.* **254,** 12629 (1979).
[15] T. T. Loh, D. A. Higuchi, F. M. van Bockxmeer, C. H. Smith, and E. B. Brown, *J. Clin. Invest.* **65,** 1182 (1980).
[16] J. W. Larrick and G. Logue, *Lancet* **2,** 862 (1980).
[17] J. E. Shindelman, A. E. Ortmeyer, and H. H. Sussman, *Int. J. Cancer* **27,** 329 (1981).
[18] W. P. Faulk, B.-L. Hsi, and P. J. Stevens, *Lancet* **2,** 390 (1980).
[19] H. Tsunoo and H. H. Sussman, *J. Biol. Chem.* **258,** 4115 (1983).

mental protocol. Direct immunofluorescence does not require binding activity but cannot give quantitative results for comparison of samples. Radioimmunoassays give quantitative measurements of receptor and are generally not affected by the presence of ligand.

A radioimmunoassay to measure the human transferrin receptor has been established in our laboratory which can employ both monoclonal and polyclonal, monospecific antibodies.[20] This method allows for highly sensitive and specific measurements to be made of the levels of TR present in human cells from a variety of sources. One other laboratory to date has reported an RIA method for measuring the human transferrin receptor[21] (see also Seligman and Allen [21], this volume). Their assay also employs a polyclonal, monospecific antibody, and they find good correlation between the results of radioimmunoassay and those of a standard ligand-binding assay for human reticulocytes and cells in culture. Presented in this chapter are the methods used for the preparation and establishment of the RIA as performed in our laboratory.

Purification of the Transferrin Receptor

Sequential Affinity Chromatography

The transferrin receptor can be purified by the method of Seligman et al.[22] (see Seligman and Allen [21], this volume). For the first affinity step, we employed our own goat-produced antihuman transferrin serum linked to CNBr-Sepharose 4B (Pharmacia) although anti-human transferrin is commercially available. The second affinity step utilizes human transferrin (Calbiochem) linked to Sepharose 4B.

Purity of the receptor produced by this procedure can be assessed either by Coomassie blue staining or by silver staining on a sodium dodecyl sulfate (SDS)/8% (w/v) polyacrylamide gel (PAGE) run by the method of Laemmli[23] with modifications by Wada et al.[24] Under reducing conditions, a single, M_r 94,000 band is observed.

One-Step Affinity Chromatography

We have also used the following procedure for purifying the receptor. This single-step procedure involving binding to transferrin-Sepharose can

[20] C. A. Enns, J. E. Shindelman, S. E. Tonik, and H. H. Sussman, Proc. Natl. Acad. Sci. U.S.A. 78, 4222 (1981).
[21] J. L. Frazier, J. H. Caskey, M. Yoffe, and P. A. Seligman, J. Clin. Invest. 69, 853 (1982).
[22] P. A. Seligman, R. B. Schleicher, and R. H. Allen, J. Biol. Chem. 254, 9943 (1979).
[23] U. K. Laemmli, Nature 227, 680 (1970).
[24] H. G. Wada, P. E. Haas, and H. H. Sussman, J. Biol. Chem. 254, 12629 (1979).

be performed if the raw material contains little unbound transferrin to compete for binding. This is accomplished using a preparation of placental brush border membranes[25] which are solubilized in 10 mM Tris, 150 mM NaCl, 2% Triton X-100, pH 7.5, at 4° for 1 hr, then centrifuged at 79,000 g for 30 min. The supernatant is incubated with the transferrin-Sepharose overnight at 4°.

The resin is washed sequentially with 25 volumes twice of wash (1) 10 mM Tris, 150 mM NaCl, 0.1% Triton X-100, pH 7.5, wash (2) wash 1 plus 100 mM ethylenediaminetetraacetic acid (EDTA), pH 7.0, wash (3) 50 mM sodium acetate, 0.1% Triton X-100, 100 mM EDTA, pH 5.0, wash (4) same as wash 2, and wash (5) same as wash 1 plus 1 mM EDTA and 10% dioxane, pH 7.5. The majority of receptor recovered will be in wash 5 and the second half of wash 4. These washes are pooled and dialyzed against 4 liters of wash 1 overnight, and concentrated as in the previous purification. This material will have some transferrin contamination but is adequate for use in the RIA both as a label and standard provided that the antibody is monospecific for the receptor.

This purification most closely mimics the cell's own process for removing iron from transferrin and dissociating the ligand–receptor complex.[26] The extreme conditions are necessary due to the strong affinity of the receptor for diferric transferrin.[19]

Preparation and Evaluation of Antiserum

Goat Anti-Human Transferrin Receptor Antiserum

A goat is immunized with at least 50 μg of purified receptor mixed with 4 ml of Freund's complete adjuvant, and is injected subcutaneously at multiple sites. This injection process is repeated at 4 weeks. Two weeks after the booster injection, the animal is exsanguinated, the serum separated and stored frozen with better stability achieved at $-70°$.

The IgG fraction of the antiserum is purified and concentrated from an aliquot of the antiserum by a three-step sequential sodium sulfate precipitation using 18, 12, and 12% (w/v) Na_2SO_4, respectively, performed at 37° with stirring for 2 hr at each step.[27] Each precipitation is separated by centrifugation at 1500 g for 20 min at room temperature. The supernatants are discarded and the pellets redissolved in distilled deionized water at

[25] C. H. Smith, D. M. Nelson, B. F. King, T. M. Donohue, S. M. Ruzycki, and L. K. Kelley, *Am. J. Obstet. Gynecol.* **128**, 190 (1977).
[26] A. Dautry-Varsat, A. Ciechanover, and H. F. Lodish, *Proc. Natl. Acad. Sci. U.S.A.* **80**, 2258 (1983).
[27] R. A. Kekwick, *Biochem. J.* **34**, 1248 (1940).

37°; the solution volumes achieved at each step are approximately 40, 20, and 10% of the original volume, respectively. The final solution is dialyzed against 3 changes of phosphate-buffered saline (PBS: 10 mM sodium phosphate, 150 mM NaCl, pH 7.5) at 4° for 36 hr.

This IgG fraction of the antiserum is then adsorbed with an aliquot of human transferrin-linked Sepharose (see purification section) to remove any antitransferrin activity resulting from minor contamination of the immunizing antigen with transferrin. The resin and IgG preparation are rotated overnight at 4°. The nonadsorbed IgG supernatant antiserum is stored in aliquots at −20°.

A similar procedure was employed by Frazier et al.[21] to raise antiserum in a chicken host (see Seligman and Allen [21], this volume).

Monoclonal Antibodies[28]

Transferrin receptor for immunization is prepared as in the one-step ligand-binding procedure previously described but using an acetone powder extract of human placenta[28] as the source material instead of a membrane preparation and using the eluate from wash 3 which contains some ligand–receptor complex. This material is used to immunize (50 μg per injection) three BALB/c mice at 2 week intervals: twice intraperitoneally in Freund's complete adjuvant and once intravenously in PBS. Four days after the intravenous booster injection, the spleens are removed and the spleen cells divided into equal parts and fused in the presence of 50% polyethylene glycol 1500 with NS-1 myeloma cells. Hybrid cells are selected in 0.1 mM hypoxanthine, 0.4 mM aminopterin, and 13 μM (HAT) medium at 37° in a 10% CO_2 90% air humidified atmosphere. Dulbecco's modified Eagle's medium (Gibco), with 20% horse serum (Gibco), 1 mM sodium pyruvate, penicillin (100 μg ml), and streptomycin (100 IU/ml), is used as culture medium. After 2 weeks in culture, supernatants from wells with cell growth are tested for antibody production by the binding of ^{125}I-labeled transferrin receptor (TR). The assay is carried out in 1 ml final volume of assay buffer (10 mM Tris, 150 mM NaCl, 10 mM EDTA, 0.5% bovine serum albumin, 0.2% Triton X-100, 0.01% NaN_3, pH 7.5) containing 20 μl of a suspension of conditioned medium from wells to be tested (20 μl of new medium is used for the nonspecific control) and 5000 cpm of ^{125}I-labeled TR. This is incubated overnight at room temperature and precipitated the next day for 1 hr at 4° with 100 μl of *Staphylococcus aureus* cells (Pansorbin, Calbiochem) previously washed with assay

[28] B. Nikinmaa, C. A. Enns, S. E. Tonik, H. H. Sussman, and J. Schroder, *Scand. J. Immunol.* **20,** 441 (1984).

buffer. These tubes are centrifuged at 4000 g for 10 min; the supernatants aspirated and the radioactivity in the pellets is counted in a gamma counter (Gamma 6000, Beckman Instruments). Clones whose media precipitate more than twice nonspecific levels of ^{125}I-labeled TR, using this method of detection, are considered positive. Positive primary clones can be subcloned and cultured for further investigations.

Evaluation of Specificity of Antiserum

The specificity of the antiserum for the transferrin receptor can be tested in the following manner.[28] K562 cells (5 × 10[6]) are washed three times in PBS and resuspended in 2 ml of the same buffer. The cells are surface-labeled with Na[125]I (New England Nuclear) by a lactoperoxidase-catalyzed iodination reaction, using the method of Wada et al.[24] After surface iodination, the cells are solubilized with Triton X-100 and centrifuged at 79,000 g for 30 min. Aliquots of radioactive lysates (1 × 10[6] cpm) are incubated with 50 μl of supernatant for 1 hr at 0°. At the same time 100 μl of S. aureus and 5 μl of rabbit anti-mouse immunoglobulin (Dakopatts) are incubated, after which these reagents are combined, incubated overnight, and washed extensively with buffer. The pellet is eluted with 100 μl of 6% SDS/10% glycerol/0.1 M Tris–HCl, pH 6.8. The eluate is divided in equal parts, after which one part is reduced with 5 μl 2-mercaptoethanol and boiled for 1 min before electrophoresis.

The samples are then subjected to SDS–PAGE (8% acrylamide) as described in the purification section. A sample treated identically but with the monospecific polyclonal IgG fraction of antiserum is used as a control. If the antibody in each sample evaluated reacts with the TR, autoradiograms reveal identical single protein bands of M_r 94,000. These protein bands are then excised with a razor blade from the gels, and each band is cut into 4 equal pieces and subjected to partial proteolytic digestion with S. aureus protease by the method of Cleveland et al.[29] Autoradiograms of these peptide maps should reveal identical patterns of digestion to the control, confirming the identity of the antigen precipitated by each antiserum as being the human transferrin receptor.

Evaluation of Affinity

The binding affinity of each monoclonal antibody or of the polyclonal goat antiserum can be evaluated by the method of Muller[30] from RIA

---

[29] D. W. Cleveland, S. G. Fisher, M. W. Kirschner, and U. K. Laemmli, J. Biol. Chem. **252**, 1102 (1977).
[30] R. Muller, this series, Vol. 92, p. 589.

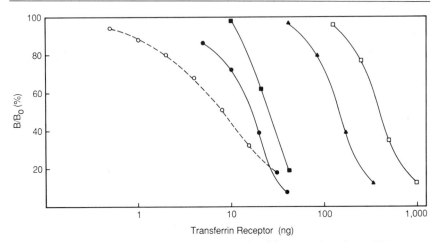

FIG. 1. Examples of inhibition curves generated by different antisera in the RIA system. (○) Monospecific polyclonal goat antihuman transferrin receptor IgG. (●) Monoclonal antibody ID9. (■) Monoclonal antibody IIG8D7. (▲) Monoclonal antibody IIB2A9. (□) Monoclonal antibody IIF9C3.

inhibition curves such as those in Fig. 1. This calculation is in the form

$$K_a = 8/3(I_{50} - T) \qquad (1)$$

where I_{50} is the molar concentration of antigen needed to produce a 50% inhibition of binding of labeled antigen in the presence of the quantity of antibody necessary for 1/2 maximal binding of label in the absence of competition. T is defined as the molar concentration of the radiolabeled antigen. Affinity constants for monoclonal antibodies in the range K_a 10^9 to 10^{10} have been reported,[29] and the polyclonal antiserum has a K_a of 1.4×10^{11}.

Iodination

The purified transferrin receptor is labeled by a lactoperoxidase cata- lyzed reaction based on the method of Thorell and Johansson.[31] This is a very gentle procedure and is the least sensitive to interference by the presence of detergent (Triton X-100). A purified receptor preparation con- taining at least 40 μg/ml and not more than 1% Triton X-100 should be used.

[31] J. I. Thorell and B. G. Johansson, *Biochim. Biophys. Acta* **251,** 363 (1971).

Example: 75 μl of TR (40 μg/ml in column buffer; see below)
            20 μl of 500 mM Na$_2$HPO$_4$, pH 7.5
            25 μg of lactoperoxidase (5 μl of 5 mg/ml stock)
            1.5 mCi of Na[125]I (New England Nuclear)

The reaction volume should not exceed 120 μl. If the purified material is an order of magnitude more concentrated than in this example, deionized water may be used to bring the final volume up to 100 μl. The reaction is initiated with 5 μl of freshly diluted 9 mM hydrogen peroxide. Mix gently and incubate for 2.5 min. Add another 5 μl of the diluted peroxide and continue incubation for an additional 2.5 min. The reaction is stopped by the addition of 500 μl of 500 mM sodium phosphate, pH 7.5.

This mixture is immediately chromatographed on a 1 × 50 cm column of Sephacryl S-300 (Pharmacia) which had been equilibrated with column buffer (10 mM Tris, 150 mM NaCl, 0.1% Triton X-100, pH 7.5) and which had been washed previously with 250 μg of bovine serum albumin (BSA). Fractions of 250 μl are collected into tubes containing 50 μl of 5% BSA in column buffer.

An example of the elution profile of the radiolabeled TR from this column is shown in Fig. 2. An aggregated form of the receptor elutes off first followed by the receptor dimer which appears as a shoulder on a larger peak of Na[125]I associated with Triton micelles. The free Na[125]I elutes off the column last.

Radiolabeled fractions are tested for the presence of the transferrin receptor by the ability to bind to the specific antibody by the technique described for the testing of monoclonal antibodies. One microliter of each fraction is incubated with 0.1 μl of the monospecific IgG preparation [normal goat serum (NGS) is used as the nonspecific control] in 1 ml of RIA buffer (column buffer plus 0.5% BSA and 10 mM EDTA, pH 7.5) overnight at room temperature. The precipitation step is identical to that previously described, and the radioactivity in the pellets is counted. For each fraction, the nonspecific (NGS) radioactivity is subtracted from the total (anti-TR) radioactivity and divided by the total amount of radioactivity added per tube to give the percentage specific ^{125}I-labeled TR bound. Both the aggregated receptor in the early fractions and the receptor dimer are immunoreactive (Fig. 3), but the majority of the radioactivity is in the dimer fractions with a specific activity of 1–3 μCi/μg. This is the material to be used in the radioimmunoassay.

If an *S. aureus* precipitation is used in the RIA, it is necessary to preclear the ^{125}I-labeled TR by adsorption to 100 μl of Pansorbin prior to use. This lowers the nonspecific binding from approximately 10 to 2% or less.

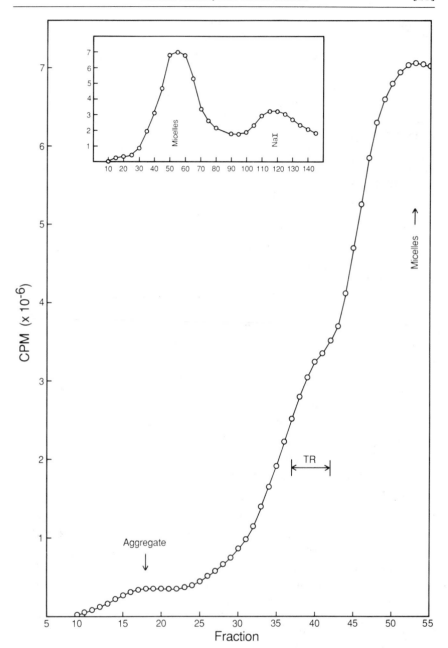

FIG. 2. Sephacryl S-300 column purification of ^{125}I-labeled transferrin receptor. Total radioactivity is measured in each 250 μl fraction.

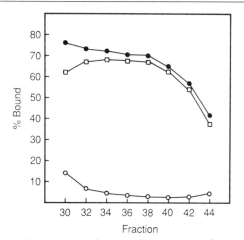

Fig. 3. Evaluation of radiolabeled fractions for presence of the transferrin receptor. (●) Total radioactivity precipitated by specific antiserum to the transferrin receptor. (○) Nonspecific radioactivity precipitated by *S. aureus*. (□) Specific ^{125}I-labeled TR (total minus nonspecific).

Iodination of the transferrin receptor by the chloramine-T method of Mehdi and Nussey[32] has not been successful when a preparation of receptor whose concentration of Triton X-100 is in excess of 1% is employed because there is such strong labeling of the Triton X-100 as to prevent detection of the labeled receptor. This method might be feasible at Triton concentrations of 0.1% or less but has not been examined in our laboratory. Frazier *et al.*[21] have successfully used the method of Bolton and Hunter[33] for iodination of this purified TR preparation that contains 0.1% Triton X-100.

Radioimmunoassay

The antibody concentration is chosen to precipitate 50% of the maximum bindable ^{125}I-labeled TR. This allows for an acceptable signal/noise ratio as well as for the required condition of antigen excess.

Material for standards need not be as pure as that for iodination if the curve generated by the more crude preparation is identical to that of purified TR. We have been able to use a solubilized placental extract as described in the previous purification of the transferrin receptor. In our experience, this material is stable at 4 or −70° but is less stable at −20°.

[32] S. Q. Mehdi and S. S. Nussey, *Biochem. J.* **145**, 105 (1975).
[33] A. E. Bolton and W. M. Hunter, *Biochem. J.* **133**, 529 (1973).

The range of sensitivity of the RIA depends on the affinity of the antibody, but should be between 1 and 100 ng for a good antibody ($K_a \geq 10^{11}$). Standards are run at 1, 2, 4, 8, 16, 32, and 64 ng/tube. Ten times concentrated standards are made by serial dilution from a stock solution, and 100 μl is added to each assay tube. Standards are assayed in duplicate as is a reagent blank (no standard, no primary antibody) for nonspecific binding (NSB). Four tubes not containing standard or sample (zero tubes) are run in every assay. The buffer for this RIA is 10 mM Tris, 150 mM NaCl, 10 mM EDTA, 0.5% bovine serum albumin, and 0.01% NaN$_3$, pH 7.5. A final concentration of 0.1% Triton X-100 is required in all tubes to maintain the solubility of the TR. For example, if the $10\times$ standards as well as the samples are in buffer containing 1% Triton X-100, then adding 100 μl of this would produce the 0.1% Triton X-100 final concentration when diluted to the final 1 ml reaction volume. However, in the zero tubes and NSB, 100 μl of 1% Triton X-100 must be added separately because no other detergent is present.

As with other RIAs, any sample can only be compared to a standard curve that is assayed containing the same buffering components as the sample. For example, conditioned tissue culture media can be assayed if the standard curve contains fresh medium that is TR free. Samples requiring dilution must also follow this procedure. Problems arising from this requirement are discussed in the section dedicated to the limitations of the assay.

In addition to antibody and standard or sample, 5000 cpm of ^{125}I-labeled TR is incubated in each tube. The level of radioactivity is chosen based on a compromise between the necessary signal strength and the low specific activity of the radiolabeled antigen. The volume is brought up to 1 ml with RIA buffer, and the tubes are mixed. The incubation is carried out overnight at room temperature (or at 37° for 1 hr[21]).

The separation of bound from free antigen on the second day can be achieved in a variety of ways depending on time, expense, and the type of samples assayed. The quickest procedure is the addition of *S. aureus* as previously described. This requires only a 1-hr incubation at 4° and is adequate for samples that have little intrinsic IgG such as solubilized cells from tissue culture. The quantity of *S. aureus* employed is also enough to overcome the IgG present in solubilized tumor samples, particularly because these often need extensive dilution.

For samples such as conditioned tissue culture media (containing 10–20% mammalian serum) or whole serum, this technique can only be applied if the samples had been previously adsorbed on enough *S. aureus* to remove their protein A-reactive IgG. This must be performed prior to RIA and can be very expensive and tedious if there are many samples to be assayed.

For large assays of samples containing IgG, a second antibody technique is used. Carrier IgG of the same species as the primary antibody is added to an amount equivalent to 1–2 μl of whole serum. This is followed by the second antibody (i.e., an IgG preparation of rabbit anti-goat IgG) which has been titrated to be in excess. The tubes are mixed and again allowed to incubate overnight at room temperature.

Frazier et al.[21] also employed a second antibody, but incubated the assay for only 1 hr at 37°. This was followed by the addition of ammonium sulfate to a final concentration of 10% to precipitate the antigen–antibody complex.

The centrifugation is identical for either precipitation method: 4000 g for 10 min. A fixed-angle rotor is preferable because it allows for easy visualization of the pellet during aspiration of the supernatant. The supernatants are decanted if ammonium sulfate is used for the precipitation of the complex. The radioactivity of the pellets is measured by gamma counting and the calculations are as follows: (1) subtract the NSB from each tube, (2) average the values for the zero tubes (0_{ave}), (3) divide all other results by 0_{ave} to get percentage bound/bound$_0$ (B/B_0), (4a) plot standards on semilog paper and approximate a smooth sigmoidal curve (Fig. 4) or (4b) plot standards on log–logit paper and calculate a least-squares fit for a straight line, and (5) read samples off standard curve and correct for any dilution factors.

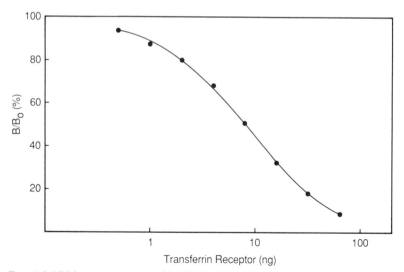

FIG. 4. Inhibition curve generated by TR standards in the RIA procedure. All points are performed in duplicate per assay. This curve is the average of seven sequential assays obtained from one TR-iodination. The standard error for each concentration is less than 3 percentage points.

Limitations and Interpretation

As with many other techniques, the major limitation in the RIA of the transferrin receptor lies in the trade-off that must be made between sensitivity and specificity. The most sensitive assay is totally dependent on an antibody of high affinity. This almost always means a polyclonal antiserum. Our goat anti-human TR has an affinity of 1.4×10^{11} but also has the general problems of polyclonal antisera. This means that although there are different species of antibody IgG present in this serum which react with the TR, these have different binding affinities and react to different parts of the molecule, some to the unbound form, some to the TR–transferrin complex, and undoubtedly, some to inactive forms (breakdown products or precursor molecules). In practice, the presence of different species of the TR in different proportions in various samples has a tendency to alter the shape of the inhibition pattern produced by multiple dilution (i.e., less sensitive to changes in concentration than the standard). This inability to obtain the dilution curves as the standard makes it impossible to determine an absolute value for the concentration of TR for such samples, however, the results are quantitatively valid if all are read from the same percentage B/B_0 (different dilution factors) for the same type of sample. For example, the TR levels of 10 flasks of the same cell line in tissue culture grown under different conditions can be compared quantitatively.

A common method in other RIAs to avoid this problem (and the similar problem of other interfering proteins) is to use material similar to that of the sample in the standard curve but which is free of the test antigen. This is, unfortunately, not possible in the assay for the TR because it is ubiquitously present on proliferating cells which require iron for growth. We have found no human cell line in continuous culture that can be used in the standard curve to correct for the effects of other cellular proteins because all tested cell lines express the transferrin receptor. The transferrin receptors can be adsorbed out with the polyclonal antiserum, but this will also adsorb out all the other cross-reacting proteins as well, thus defeating the purpose.

Using a monoclonal antibody to the transferrin receptor for the RIA will avoid the problems associated with cross-reactive materials and different antigen species because a monoclonal antibody reacts with only a single antigenic determinant. The limit to the use of monoclonal antibodies in general is that many of these have lower binding affinities and therefore are less sensitive in an RIA by a factor of 10–100. Use of these lower affinity antibodies in the RIA for the TR might be satisfactory to measure the high levels of TR in tumors, but not the lower levels present in most normal tissues. Much larger volumes of cells from tissue culture would also need to be processed for assay.

The optimal RIA for the transferrin receptor would employ a high-affinity monoclonal antibody, preferably not to the site of ligand binding so that both bound and free receptor can be measured. This should enable one to detect the presence of even low levels of TR under varying conditions while not having interference from other cross-reactive substances. High-affinity monoclonal antibodies are the exception rather than the rule at present and thus will be very costly either in terms of production or commercial purchase. For most of the purposes so far encountered in our work, the simpler and less expensive polyclonal, high-affinity, monospecific antiserum can be used provided that its limitations are understood and accepted.

Acknowledgment

This work was supported by NIH Grant CA 13533 from the National Cancer Institute.

[24] Monoclonal Antibodies to Transferrin Receptor and Assay of Their Biological Effects

By Ian S. Trowbridge, Jayne F. Lesley, Derrick Domingo, Roberta Schulte, Carol Sauvage, and Hans-Georg Rammensee

Transferrin, the major serum iron-binding protein, is an essential growth factor for most mammalian cells in tissue culture,[1] and the expression of transferrin receptors on the cell surface is coordinately regulated with cell growth.[2-5] For most cell types, transferrin is the major source of iron, and transferrin-mediated iron uptake takes place via receptor-medi-

[1] D. Barnes and G. Sato, Cell 22, 649 (1980).

[2] J. W. Larrick and P. Cresswell, J. Supramol. Struct. 11, 421 (1979).

[3] G. M. P. Galbraith, R. M. Galbraith, and W. P. Faulk, Cell. Immunol. 49, 215 (1980).

[4] T. A. Hamilton, H. G. Wada, and H. H. Sussman, Proc. Natl. Acad. Sci. U.S.A. 75, 6406 (1979).

[5] I. S. Trowbridge and R. A. Newman, in "Antibodies to Receptors: Probes for Receptor Structure and Function" (M. F. Greaves, ed.), pp. 235–261. Academic Press, London, 1984.

ated endocytosis.[6–9] After release of iron under acid conditions within the endosomal membrane compartment, apotransferrin remains bound to the transferrin receptor and is recycled back to the cell surface. The expression of transferrin receptors on growing cells and their obligatory requirement for transferrin suggests that this iron transport system and iron itself are important for cell growth. This chapter describes the derivation of monoclonal antibodies against the transferrin receptor and assay of their biological effects. Antibodies that block receptor function have proved to be useful in dissecting the role of transferrin receptors and iron uptake in cell growth.

Derivation and Characterization of Monoclonal Antibodies against the Transferrin Receptor

Immunization. Monoclonal antibodies against the transferrin receptor of human,[10–14] murine,[15,16] and rat[17] cells have been obtained. With one exception, all these antibodies were derived from fusions of spleen cells from mice or rats immunized with viable whole xenogeneic cells, usually cultured hematopoietic cell lines or activated lymphocytes. The human transferrin receptor is an immunodominant cell surface antigen in the mouse, and hybridomas producing monoclonal antibodies against the receptor can be expected to be found by screening 100 to 200 culture supernatants from fusions of spleen cells of mice immunized weekly with $1–2 \times 10^7$ cells for several weeks. However, it is now possible, and preferable, to immunize animals with purified transferrin receptor as the receptor can be isolated by affinity chromatography on monoclonal antibody-

[6] A. Dautry-Varsat, A. Ciechanover, and H. F. Lodish, *Proc. Natl. Acad. Sci. U.S.A.* **80**, 2258 (1983).

[7] J. van Renswoude, K. R. Bridges, J. B. Harford, and R. D. Klausner, *Proc. Natl. Acad. Sci. U.S.A.* **79**, 6186 (1982).

[8] R. D. Klausner, G. Ashwell, J. van Renswoude, J. B. Harford, and K. R. Bridges, *Proc. Natl. Acad. Sci. U.S.A.* **80**, 2263 (1983).

[9] C. R. Hopkins and I. S. Trowbridge, *J. Cell Biol.* **97**, 508 (1983).

[10] I. S. Trowbridge and M. B. Omary, *Proc. Natl. Acad. Sci. U.S.A.* **78**, 3039 (1981).

[11] R. Sutherland, D. Delia, C. Schneider, R. Newman, J. Kemshead, and M. Greaves, *Proc. Natl. Acad. Sci. U.S.A.* **78**, 4515 (1981).

[12] J. W. Goding and G. F. Burns, *J. Immunol.* **127**, 1256 (1981).

[13] B. F. Haynes, M. Hemler, T. Cotner, D. Mann, G. S. Eisenbarth, J. L. Strominger, and A. S. Fauci, *J. Immunol.* **127**, 347 (1981).

[14] I. S. Trowbridge and F. Lopez, *Proc. Natl. Acad. Sci. U.S.A.* **79**, 1175 (1982).

[15] I. S. Trowbridge, J. Lesley, and R. Schulte, *J. Cell. Physiol.* **112**, 403 (1982).

[16] J. F. Lesley and R. J. Schulte, *Mol. Cell. Biol.* **4**, 1675 (1984).

[17] W. A. Jefferies, M. R. Brandon, A. F. Williams, and S. V. Hunt, *Immunology* **54**, 333 (1985).

Sepharose[10] or human transferrin-Sepharose[18] columns with a yield of 300–600 μg/10^{10} cells. A single subcutaneous injection of 40 μg of the purified human receptor in complete Freund's adjuvant is sufficient to immunize mice. Four to six weeks after the initial injection, mice are bled and sera tested for antireceptor antibodies by immunprecipitation. Mice with positive sera are boosted with an injection of 40 μg of purified receptor in saline and fusions are carried out 3–5 days later.

Initial Screening of Hybridomas. If whole cells are used for immunization, it is more convenient to screen hybridoma supernatants for antibodies that bind to the cell surface using a trace-labeled indirect antibody assay[19] then assay supernatants positive in this test by immunoprecipitation. If purified receptor is used, the frequency of hybridomas producing monoclonal antibodies against the transferrin receptor is sufficiently high to screen hybridoma supernatants directly by immunoprecipitation.

Identification of Antitransferrin Receptor Antibodies

Two criteria are used to identify monoclonal antibodies against the transferrin receptor unequivocably.

SDS–PAGE Analysis of Immunoprecipitates. The transferrin receptor is a disulfide-bonded homodimer (M_r of subunit = 95,000). Thus, antibodies are first tested for their ability to immunoprecipitate a labeled species of the appropriate M_r from detergent lysates of surface-iodinated cells. Analysis of immunoprecipitates under reducing and nonreducing conditions by SDS–PAGE on 7.5% gels gives the result shown in Fig. 1a which is characteristic of the transferrin receptor of all species tested.

Coprecipitation of Transferrin–Transferrin Receptor Complex. The second and most important criterion used to define antitransferrin receptor antibodies is the coprecipitation of labeled transferrin as a complex with the receptor.[10] Transferrin binds to its receptor with a sufficiently high affinity ($K_d = 2$–$7 \times 10^{-9}\ M$)[5] that the receptor–ligand complex can be solubilized in nonionic detergents. Thus, after ^{125}I-labeled transferrin is allowed to bind to cells, antibodies against the transferrin receptor have been shown to specifically immunoprecipitate the radiolabeled receptor–transferrin complex from detergent lysates of cells.[10,11,15,17]

1. Human transferrin (Miles) is iodinated using the chloramine-T method.[20] Transferrin [20 μg in 20 μl 0.15 M NaCl–0.01 M sodium phos-

[18] I. R. van Driel, P. A. Stearne, B. Grego, R. J. Simpson, and J. W. Goding, *J. Immunol.* **133**, 3220 (1984).

[19] R. J. Morris and A. F. Williams, *Eur. J. Immunol.* **5**, 274 (1975).

[20] P. G. McConahey and F. J. Dixon, *Int. Arch. Allergy Appl. Immunol.* **29**, 185 (1975).

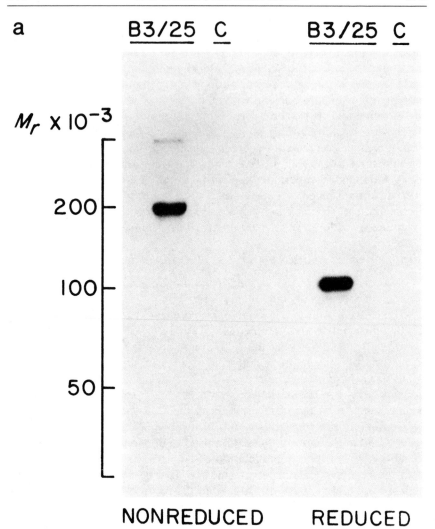

NONREDUCED   REDUCED

FIG. 1. Identification of monoclonal antibodies against the transferrin receptor. (a) SDS–PAGE analysis on a 7.5% acrylamide gel of the molecular species precipitated by monoclonal antibody B3/25 from a detergent lysate of CCRF-CEM cells labeled by lactoperoxidase catalyzed iodination. The human transferrin receptor precipitated by B3/25 antibody has an M_r of 190,000 under nonreducing conditions and 95,000 after reduction. (b) The specific immunoprecipitation by B3/25 monoclonal antibody of ^{125}I-labeled transferrin (M_r = 77,000) complexed to the human transferrin receptor. The coprecipitation assay was performed as described in the text using CCRF-CEM cells. The control precipitation shown was carried out with T29/33 monoclonal antibody which reacts with human T200 (leukocyte-common) glycoprotein, a major cell surface antigen of the leukemic cells.

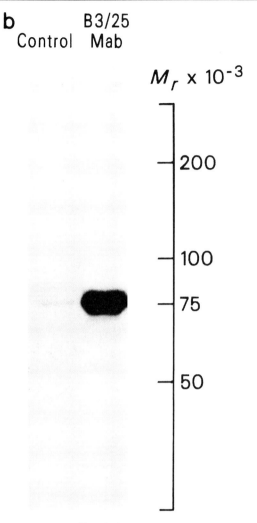

FIG. 1. (*continued*)

phate buffer, pH 7.2 (PBS)] is mixed with 20 μl 0.5 M sodium phosphate buffer (pH 7.2), 1 mCi carrier-free Na^{125}I (New England Nuclear), and 20 μl of freshly made chloramine-T (1 mg/ml in H$_2$O). After incubation at 23° for 3 min, 20 μl of L-tyrosine (0.4 mg/ml in H$_2$O) is added. After 2 min, 50 μl of fetal bovine serum is added as carrier and the ^{125}I-labeled transferrin separated from free Na^{125}I by chromatography on a BioGel P-10 column (10 × 0.4 cm) equilibrated in PBS.

2. For the coprecipitation assay, cells are washed three times in serum-free RPMI 1640 medium prewarmed to 37°. Cells (1×10^7 in 1.0 ml of RPMI 1640 medium) are then incubated with 50 μl of ^{125}I-labeled transferrin (5×10^7 cpm; specific activity 5–10 μCi/μg) for 60 min at 37° in a CO_2 incubator. Afterward, cells are washed three times with 1.0 ml of PBS by centrifugation for 5 sec in an Eppendorf microfuge and solubilized in 1.0 ml of 1% NP-40 in PBS for 10 min at 4°. After removal of nuclei by centrifugation for 2 min in a microfuge, 100 μl samples of the cell lysate are incubated with 20 μl of antibody-Sepharose beads (5 mg of antibody per ml of packed beads prepared by cyanogen bromide coupling[21]) for 15 min. The beads are then washed three times with 0.4 ml of 0.5% deoxycholate–0.5% NP-40–0.05% SDS in PBS. Under these conditions, antibodies against the transferrin receptor will bind 20- to 100-fold more radioactivity than a control antibody against another major cell surface antigen such as leukocyte T200.[22] In addition, it is necessary to rule out the possibility that the antibody reacts directly with transferrin itself. Immunoprecipitation procedures using fixed *S. aureus*[23] can be substituted for antibody-Sepharose beads, and as commercial transferrin preparations contain trace amounts of other proteins, it is advisable to analyze the radiolabeled species bound to the antibody by SDS–PAGE (Fig. 1b). It should be noted that even an antibody such as 42/6 which competes with transferrin for binding to the human transferrin receptor[14] coprecipitates sufficient ^{125}I-labeled transferrin to be identified by this procedure.

Screening for Antibodies That Block Transferrin Receptor Function

The monoclonal antibodies against the human transferrin receptor that were first described did not inhibit receptor function.[10–13] To identify such antibodies, hybridoma supernatants from a fusion of spleen cells from mice immunized with purified receptor were screened by immunoprecipitation and then assayed for their capacity to inhibit transferrin binding to human cells.[14]

Inhibition of Transferrin Binding. Human transferrin (Miles) is ^{125}I-labeled as described earlier. CCRF-CEM cells[24] [50 μl containing 1–2 $\times$ 10^7 cells in 0.1% bovine serum albumin in PBS containing 15 mM NaN$_3$ (BSA–PBS)] are incubated for 45 min at 4° with 50 μl of hybridoma supernatant in microtiter wells. Cells are then pelleted by centrifugation

[21] P. Cuatrecasas, *J. Biol. Chem.* **245**, 3059 (1970).
[22] M. B. Omary, I. S. Trowbridge, and H. A. Battifora, *J. Exp. Med.* **152**, 842 (1980).
[23] M. B. Omary and I. S. Trowbridge, *J. Biol. Chem.* **256**, 12888 (1981).
[24] G. E. Foley, H. Lazarus, S. Farber, B. G. Uzman, B. A. Boone, and R. E. McCarthy, *Cancer* **18**, 522 (1965).

at 1500 g and washed three times with 150 μl of BSA–PBS. Approximately 2 × 10⁶ cpm of ¹²⁵I-labeled transferrin (specific activity 5–10 μCi/μg) in 50 μl of BSA–PBS is added to each well and cells are incubated for 45 min at 4°. After washing three times, cells are resuspended in 150 μl of BSA–PBS, 100 μl of the cell suspension removed, and radioactivity determined in a gamma counter. As a positive inhibition control, cells are preincubated with 50 μl of unlabeled human transferrin (15 μg/ml). Incubations are carried out in the cold in the presence of NaN₃ to prevent antibody-mediated antigenic modulation.

Two points should be emphasized. First, antibodies such as 42/6[14] and 43/31[25] which inhibit transferrin binding were detected using this assay even though the tissue culture supernatants tested contain 10% horse serum. Thus, 300–500 μg of horse transferrin is present during the first incubation. Second, it is important to realize that as the transferrin receptor mediates iron uptake via receptor-mediated endocytosis, antibodies may interfere with other steps in the process than the initial binding of transferrin to its receptor. Antibodies against the murine transferrin receptor that inhibit iron uptake but do not block transferrin binding have already been described.[15,16] Such antibodies appear to block receptor function at least in part by interfering with endocytosis of the receptor–ligand complex.[26]

Inhibition of Cell Growth. In order to identify antibodies that do not inhibit transferrin binding but nevertheless block receptor function, hybridoma supernatants are routinely tested to see whether they inhibit the growth of cell lines known to be sensitive to antibodies that block transferrin receptor function. Two murine leukemic cell lines, AKR.1[16] and SL-2,[27] and two human leukemic cell lines, CCRF-CEM and RPMI 8402,[28] are good indicator cells in the growth inhibition assay. Cells are set up at an initial cell density such that they will still be growing exponentially at the end of the assay (5 × 10⁴/ml). Cells are incubated in duplicate wells of Linbro 24-well plates in tissue culture medium containing either 0, 25, or 50% (v/v) supernatant from heavy cultures of hybridoma cells. After 3 or 5 days growth for mouse cell lines and human cell lines, respectively, the cells are harvested and counted using a Coulter counter. Supernatant from cultures of 42/6[14] and RI7 208[15] hybridomas are used as controls. As 42/6 monoclonal antibody does not cross-react with the murine transferrin receptor and, likewise, RI7 208 does not cross-react with the human

²⁵ R. Taetle, J. M. Honeysett, and I. Trowbridge, *Int. J. Cancer* **32**, 343 (1983).
²⁶ J. F. Lesley and R. J. Schulte, *Mol. Cell. Biol.* **5**, 1814 (1985).
²⁷ I. D. Bernstein, M. R. Tam, and R. C. Nowinski, *Science* **207**, 68 (1980).
²⁸ J. Minowada, G. Janossy, M. F. Greaves, T. Tsubota, B. I. S. Srivastava, S. Morikawa, and E. Tatsumi, *J. Natl. Cancer Inst. (U.S.)* **60**, 1269 (1978).

receptor, it is convenient to use the two antibodies as reciprocal positive and negative controls. Thus, 42/6 monoclonal antibody blocks the growth of human cells but not mouse cells, whereas RI7 208 monoclonal antibody blocks the growth of mouse cells but not human cells.

Biological Effects of Monoclonal Antibodies against
the Transferrin Receptor

Several monoclonal antibodies against human and murine transferrin receptors that have been selected using either transferrin binding or growth inhibition assays when purified can be shown to specifically inhibit transferrin-mediated iron uptake and cell growth in a dose-dependent manner.[14-16,25,26] The concentration of antibody required for maximum effect will vary depending on the antibody and the target cells used, but is usually within the range of 2–10 μg/ml. The most important observations that have been made are as follows:

1. Of the antibodies tested so far, only antireceptor antibodies of the IgM or IgA class markedly inhibit transferrin-mediated iron uptake and cell growth. Thus, of the rat anti-murine transferrin receptor antibodies, RI7 208 and REM 17 antibodies are IgMs and block the growth of mouse cells, whereas RI7 217, RR24, and R234-14 antibodies are IgGs and have little or no effect on cell growth.[15,16,26] Similarly, 42/6, an IgA, is the only murine antibody against the human transferrin receptor that has been shown to efficiently inhibit cell growth.[14] Antibodies such as B3/25,[10] T56/14,[10] OKT9,[11,12] and 5E9,[13] all of which are IgGs, do not block the growth of human cells. Monoclonal antibody, 43/31, a murine IgG, is unusual because it was scored as positive when screened for its ability to inhibit transferrin binding. However, although this antibody slows cell growth, it is less effective than 42/6 monoclonal antibody.

2. Cells vary widely in their sensitivity to growth inhibition by anti-transferrin receptor antibodies. For example, antireceptor antibodies can completely inhibit the in vitro growth of normal lymphocytes,[29] hematopoietic progenitor cells,[25] and many leukemic and lymphoma cell lines[14-16] but either have no effect or only weakly inhibit the growth of mouse L cells,[15] the human diploid embryonic lung fibroblast cell line, Flow 2000,[30] the human melanoma cell lines, M21[31] and SK-MEL 28,[32] KB[33] cells, and

[29] J. Mendelsohn, I. Trowbridge, and J. Castagnola, *Blood* **62**, 821 (1983).
[30] B. J. Walthall and R. G. Ham, *Exp. Cell Res.* **134**, 303 (1981).
[31] T. F. Bumol, Q. C. Wang, R. A. Reisfeld, and N. O. Kaplan, *Proc. Natl. Acad. Sci. U.S.A.* **80**, 529 (1983).
[32] J. Fogh, J. M. Fogh, and T. Orfeo, *J. Natl. Cancer Inst.* (*U.S.*) **59**, 221 (1977).
[33] J. A. Hanover, M. C. Willingham, and I. Pastan, *Cell* **39**, 283 (1984).

the human carcinoma cell lines, Caco-2,[32] HT-29,[32] and SK-HEP-1.[32] It is not possible to predict from iron uptake studies which cells will be sensitive to the growth inhibitory effects of the antibodies as inhibition of iron uptake by antireceptor antibodies may be as great in resistant cells as in cells sensitive to the growth inhibitory effects of the antibodies.[15] Further, cells that can grow in the presence of the antibodies display similar numbers of receptors as cells that are killed and binding studies do not reveal significant differences in the affinity of antibody binding.

3. Inhibition of the growth of some cells by antireceptor antibodies can be overcome by addition of soluble iron in the form of ferric-nitrilotriacetic acid or ferric-fructose. However, the capacity of iron chelates to restore growth in the presence of antireceptor antibodies varies widely. Inhibition of the growth of a human myeloid leukemic cell line, KG-1, by the antitransferrin receptor antibody, 42/6, is completely prevented by 20 μM ferric-nitrilotriacetic acid whereas inhibition of the growth of human peripheral blood lymphocytes is only partially reversed under the same conditions.[25,29] At the other end of the spectrum, the growth of CCRF-CEM cells is completely inhibited by 42/6 monoclonal antibody even in the presence of soluble iron.[14] The ability of ferric complexes to restore the growth of cells grown in the presence of antireceptor antibodies provides direct evidence that the antibodies inhibit growth by depriving cells of iron. This is consistent with the observation that the human myeloid cell lines, HL-60 and KG-1, can be adapted to growth in transferrin-free medium and, under these conditions, their growth becomes refractory to inhibition by 42/6 monoclonal antibody.[34] Cells may differ in their capacity to utilize soluble iron and this may account for the failure of ferric chelates to consistently overcome the effects of antireceptor antibodies. However, it is not possible to rule out that, in the instances when cell growth cannot be restored by iron chelates, antireceptor antibodies may not only be blocking iron uptake but also interfering with other processes required for cell growth. It can be envisaged, for example, that antitransferrin receptor antibodies could interfere indirectly with the receptor-mediated uptake of other nutrients or growth factors.

Procedures

Purification of Monoclonal Antibodies. Purified antibodies are used in all biological assays. IgM antibodies are purified from serum and ascitic fluid by dialysis against 5 mM sodium phosphate buffer (pH 5.0) overnight. The precipitated IgM is collected by centrifugation, washed three

---

[34] R. Taetle, K. Rhyner, J. Castagnola, D. To, and J. Mendelsohn, *J. Clin. Invest.* **75,** 1061 (1985).

times with the same buffer, and then dissolved in PBS. Monoclonal antibodies of other classes are purified by ammonium sulfate precipitation and fractionation on DEAE cellulose (see P. Parham, this series, Vol. 92, p. 110).

Preparation of ^{59}Fe*–transferrin.* Transferrin-mediated iron uptake is measured using ^{59}Fe–human transferrin. Transferrin is labeled with ^{59}Fe by the method of Bates and Schlabach.[35] Apotransferrin is incubated with a 10% molar excess of ^{59}Fe–nitrilotriacetic acid complex (molar ratio 1 : 2) in the presence of bicarbonate ions. It is convenient to use 1 mCi of $^{59}FeCl_3$ and, depending on its specific activity, vary the amounts of nitrilotriacetic acid and apotransferrin accordingly.

As a typical example, 1 mCi of $^{59}FeCl_3$ (16.2 mCi/mg, 0.37 μmol) in 0.1 ml of 0.5 M HCl was obtained from New England Nuclear. A 2-fold molar excess of nitrilotriacetic acid (20 μl of 7.2 mg/ml in H_2O, 0.75 μmol) was added. Human apotransferrin (Behringwerke 13.5 mg, 0.17 μmol) was dissolved in 200 μl of 0.1 M NaClO$_4$–0.5 M Tris–HCl (pH 7.5) and immediately before the addition of the ^{59}Fe–nitrilotriacetic acid complex, 10 μl of 0.1 M NaHCO$_3$ was added to give a final concentration of 5 mM bicarbonate. After addition of the ^{59}Fe–nitrilotriacetic acid complex, the solution turns brown and is left at room temperature for 30 min. The incubation mixture is then applied to a fresh Sephadex G-25 column poured in a Pasteur pipet and equilibrated with 0.1 M NaClO$_4$. Fractions (0.2 ml) are collected and the protein peak that can be detected visually is recovered and passed down a second Sephadex G-25 column equilibrated with PBS. Protein concentration is determined using the values of $E_{280}^{1\%} = 11.0$ for apotransferrin, $E_{280}^{1\%} = 14.0$ and $E_{465}^{1\%} = 0.58$ for fully saturated transferrin.

Assay of Transferrin-Mediated Iron Uptake. Suspension cells are incubated at a cell density of 5×10^6 per ml in 2 ml of serum-free tissue culture medium (Dulbecco's modified Eagle's or RPMI 1640) in 35-mm petri dishes. Cells treated with antibody are preincubated with 50 μg/ml antitransferrin receptor monoclonal antibody for 30 min at 37° prior to the addition of labeled transferrin. ^{59}Fe–transferrin (25 μg in 100 μl of PBS, specific activity approximately 8×10^3 cpm/μg) is then added, cells incubated at 37° in a CO_2 incubator, and duplicate 100 μl samples of cell suspension removed at various times (0, 2, 4, 6, and 24 hr) into 1.0 ml cold BSA–PBS. Cells are then pelleted by centrifugation for 5 sec in an Eppendorf microfuge, washed twice with 1.0 ml of BSA–PBS, and the radioactivity retained in the cell pellet counted in a gamma counter.

For monolayer cultures, 5×10^5 cells are plated in 30-mm tissue cul-

---

[35] G. W. Bates and M. R. Schlabach, *J. Biol. Chem.* **218**, 3228 (1973).

ture dishes and cultured overnight. Duplicate dishes are set up for each time point. The next day, the tissue culture medium is removed and monolayers washed three times with 1 ml of serum-free medium. Iron uptake is then measured exactly as described for suspension cultures except that 1 ml of medium is used per dish. At the end of incubation, cell monolayers are washed three times with 1 ml of BSA–PBS, cells removed from the dishes with versene, and cell-associated radioactivity determined.

Assays for Cell Growth and Cloning Efficiency. Antireceptor antibodies are tested for their effects on growth rate and plating efficiency. Mouse cells are grown in Dulbecco's modified Eagle's medium containing 10% horse serum. Human cells are grown in RPMI 1640 or Dulbecco's modified Eagle's medium containing 10% fetal calf serum. For cell growth assays, cells are plated in duplicate in 35-mm dishes and incubated with purified monoclonal antibody (usually concentrations ranging from 0 to 20 μg/ml). Cell number is determined each day by counting in a Coulter counter. For clonal assays, 24 2-fold serial dilutions of cells (beginning with 2×10^5 cells per well, 8 wells per dilution) are made in 100 μl of tissue culture medium. An equal volume of medium is then added containing 20 μg/ml of monoclonal antibody. Wells positive for growth are counted 2, 3, and 4 weeks after plating. When a sensitive cell line, such as the mouse lymphoma AKR.1[16] was used as the indicator cell, it was found that IgM antibodies that greatly inhibit growth in bulk culture reduced the cloning frequency of cells to 10^{-4}–5×10^{-5} of that of untreated cells. In contrast, IgG antibodies against the murine transferrin receptor that have little effect on growth in bulk culture delayed the appearance of colonies by about 1 week but only reduced cloning efficiency by 50%. To determine the cloning efficiency of untreated cells and cells treated with the IgG antibodies accurately, 24 wells of each of 8 cell concentrations (between 6 and 0.5 cells per well) are plated in 96-well plates and the proportion of negative wells at each cell concentration is plotted. The slope of the regression line is fitted by the method of least squares and used to determine the cloning frequency.

Competition Binding Assays. The effects of purified antireceptor monoclonal antibodies on transferrin binding are assayed by competition binding studies. Transferrin is ^{125}I labeled for these studies and is prepared as follows: to human transferrin (1 mg in 0.25 ml PBS) is added to 25 μl of 0.5 M sodium phosphate buffer (pH 7.0) containing 80 μg/ml NaI, 2 mCi Na[125]I, 12.5 μl dimethyl sulfoxide, and 50 μl of chloramine-T (2 mg/ml). After 3 min at room temperature, the ^{125}I-labeled transferrin is purified on a BioGel P10 column. The ^{125}I-labeled transferrin prepared by this procedure has a specific activity of approximately 5×10^5 cpm/μg.

Assays for ^{125}I-labeled human transferrin binding to lymphoma cells in the presence of saturating amounts of monoclonal antibody (100 μg/ml) are performed at 4° in V-bottomed microwells containing 10^6 cells per well. Binding at each concentration of transferrin (0.1–10 μg/ml) is determined in triplicate and nonspecific binding at each concentration of ^{125}I-labeled transferrin is determined by measuring binding in the presence of 100 μg/ml of unlabeled transferrin.[16] Cells are incubated at 4° for 45 min and then centrifuged at 1500 rpm in a Beckman Model TJ-6 centrifuge. The cells are then washed three times with 150 μl of BSA–PBS and transferred to tubes for radioactivity counting in a gamma counter.

For blocking experiments, dilutions of unlabeled antibody (concentration range 0–50 μg/ml) are added to cells in triplicate followed by the addition of a near-saturating concentration of ^{125}I-labeled human transferrin (4 μg/ml). Determinations are made in triplicate and inhibition of binding by unlabeled human transferrin is used as a positive control.

These assays clearly distinguish 42/6 monoclonal antibody which inhibits the binding of transferrin to human cells and RI7 208 which blocks growth but does not inhibit the binding of transferrin to mouse cells.[16,26]

Applications of Monoclonal Antibodies That Block Transferrin Receptor Function

Monoclonal antibodies against the transferrin receptors of human and murine cells that block function have been used in a variety of different studies. It will be evident from the examples described that the capacity of such antibodies to directly interfere with cell growth can be exploited to analyze a diverse range of problems.

Selection of Mutant Cell Lines Resistant to Antitransferrin Receptor Antibody

The derivation of cell surface antigen mutants of cultured hematopoietic cell lines and their genetic and biochemical analysis provides a general approach with which to study the biosynthesis and regulation of specific cell surface molecules (reviewed in Ref. 36). Selection of mutants has most frequently been carried out using antibody and complement to kill wild-type cells bearing the specific antigen. The capacity of monoclonal antibodies against the transferrin receptor to inhibit cell growth provides an alternative method of negative selection that has been used successfully to derive a series of resistant mutant cell lines.[16]

[36] R. Hyman, *Biochem. J.* **225**, 27 (1985).

Four murine thymic lymphoma cell lines have been used for the selection of transferrin receptor mutants: AKR1, BW5147, S49, and EL4.[16] All cells are grown in Dulbecco's modified Eagle's medium supplemented with 10% horse serum. For antibody selection, parental cells are mutagenized by culturing overnight in 1, 0.7, or 0.5 mg of ethylmethane sulfonate per ml, removed from the mutagen the next day, and observed for cell death. Cultures in which about 75% of the cells died are chosen for selection. One or two weeks after mutagenization, replicate cultures containing 1–2 × 10⁶ cells are grown in the presence of 10 μg of purified RI7 208 monoclonal antibody per ml. Cells stop growing within 1 day but remain metabolically active longer and it is sometimes necessary, if the cultures turn acid, to suck off and replenish the cultures with fresh medium. After several days, most cells begin to die but 1–3 weeks later, vigorous growth in the cultures resumes. Cells from these cultures are then cloned by limiting dilution in the presence of 10 μg/ml of monoclonal antibody.

The properties of the mutants obtained by selection with RI7 208 monoclonal antibody are described in detail by Lesley and Schulte.[16] In short-term growth assays, the cells grow equally well in the presence or absence of the antibody, whereas the growth of the parental cell line is completely inhibited. The cloning efficiency of the mutant cells is not reduced by the presence of 10 μg of RI7 208 monoclonal antibody per ml. Transferrin-mediated iron uptake into wild-type lymphoma cells is inhibited by greater than 90% by saturating amounts of antibody. In contrast, iron uptake by mutant cells is only partially blocked and substantial iron uptake still occurs in the presence of antibodies.

Some insight into the phenotype of the mutant cells has been gained from genetic and biochemical analysis. Hybrids were made between independent mutants and between mutant and wild-type cells. There was no indication of complementation between different mutants, and hybrids between a sensitive parental line and an antibody-resistant line were intermediate in sensitivity suggesting resistance is codominant. This genetic analysis is consistent with the possibility of an alteration in the transferrin receptor structural gene. Biochemical studies provided further evidence that this is the case and showed that probably the defect in the mutant cells affects only one copy of the transferrin receptor gene, consistent with the frequency with which mutants were obtained from nonmutagenized AKR1 cells (~10⁻⁵). In particular, it was shown by sequential immunoprecipitation studies that RI7 208 monoclonal antibody only reacts with about 75% of the surface transferrin receptors of the mutant cells. As the transferrin receptor is a disulfide-bonded dimer, this result is interpreted as showing that the mutant cells may make both altered receptor polypeptides lacking the RI7 208 antigenic site and wild-type molecules that then randomly associate to form dimers on the cell surface.

Negative Selection of Antigen-Specific Cells

An important practical use of antireceptor antibodies that block function is to selectively kill proliferating antigen-specific cells. The induction of cytotoxic T lymphocytes (CTL) specific for allogeneic cells in a primary mixed lymphocyte culture (MLC) has been used as a model system to study how effective antireceptor antibodies are in eliminating antigen-activated lymphocytes. As described in detail by Rammensee et al.,[37] two antibodies that block iron uptake by the murine transferrin receptor, RI7 208 and REM 17.2, have been tested for their ability to kill proliferating antigen-specific cells. It was found that both antibodies specifically inhibited the generation of effector CTL in a dose-dependent manner if they were present for the entire period of culture. At a concentration of 15 μg/ml of RI7 208 antibody, the generation of CTL was completely inhibited.

Restimulation of cells from primary MLC demonstrated that antitransferrin receptor antibodies directly and selectively kill the CTL precursors specific for the alloantigen used to stimulate the cultures. Cells from cultures treated with antireceptor antibody only respond weakly to the same alloantigen but respond normally to an unrelated alloantigen. This is in marked contrast to the normal situation in which a much stronger secondary CTL response is obtained after restimulation with the same alloantigen compared to a primary response to an unrelated one.

Studies to optimize the depletion of antigen-specific cells showed that REM 17.2 antibody is more effective than RI7 208 and that the most complete depletion of CTL precursors was obtained by prolonged culture with the antibody (7 days) in the presence of rat concanavalin A (Con A) supernatant (culture supernatant from rat spleen cells stimulated with concanavalin A) as a source of growth factors including interleukin-1 and -2 (IL-1 and IL-2).

It appears that treatment with antitransferrin receptor antibody offers a method of selectively killing antigen-activated T lymphocytes that is as good or better than existing procedures, such as treatment with bromodeoxyuridine and light.[38]

---

[37] H.-G. Rammensee, J. Lesley, I. S. Trowbridge, and M. J. Bevan, Eur. J. Immunol. **15**, 687 (1985).

[38] D. C. Zoschke and F. H. Bach, Science **170**, 1404 (1970).

Immunotherapy with Monoclonal Antibodies That Block Transferrin Receptor Function

The effects of the *in vivo* administration of monoclonal antibodies against the transferrin receptor have been investigated with regard to their potential use as antitumor agents. Transferrin receptors are found more abundantly expressed on tumors than on most normal tissues.[39] However, it is clear that transferrin receptors are expressed on normal tissues at limited sites in sufficient numbers to be detected by immunohistological techniques.[17,39,40] Thus, two questions are of paramount importance: one is whether the growth of tumor cells can be inhibited *in vivo* and the second is what effects antitransferrin receptor antibodies have on normal tissues.

The availability of rat monoclonal antibodies against the murine transferrin receptor provides the opportunity to study these questions in a syngeneic mouse tumor model system. It has been shown that RI7 208 monoclonal antibody can inhibit the *in vivo* growth of a transplantable AKR/J leukemia and prolong the survival of tumor-bearing mice.[41,42] Under the conditions of these experiments, the *in vivo* administration of RI7 208 monoclonal antibody was not associated with detectable toxicity as judged by histological examination of tissues and assay of hematopoietic progenitor cells in the bone marrow. Further work is now required to assess the effects of chronic administration of antitransferrin receptor antibodies on normal tissues and to enhance their inhibitory effects on tumor cell growth *in vivo*.

Acknowledgments

We thank Ami Koide for her help in preparation of the manuscript. The work described was supported by NIH Grants CA-34787 and CA 25893.

[39] K. C. Gatter, G. Brown, I. S. Trowbridge, R.-E. Woolston, and D. Y. Mason, *J. Clin. Pathol.* **36**, 539 (1983).

[40] W. A. Jefferies, M. R. Brandon, S. V. Hunt, A. F. Williams, K. C. Gatter, and D. Y. Mason, *Nature (London)* **312**, 162 (1984).

[41] I. S. Trowbridge, in "Monoclonal Antibodies and Cancer" (B. D. Boss, R. Langman, I. Trowbridge, and R. Dulbecco, eds.), pp. 53–61. Academic Press, New York, 1983.

[42] I. S. Trowbridge, R. A. Newman, D. L. Domingo, and C. Sauvage, *Biochem. Pharmacol.* **33**, 925 (1984).

[25] Molecular Cloning of Receptor Genes by Transfection

By ALAN MCCLELLAND, MICHAEL E. KAMARCK, and
FRANK H. RUDDLE

The isolation of eukaryotic genes based on expression and selection offers an appealing alternative to cloning strategies which rely on mRNA or protein purification. The method which we have applied to the isolation of the transferrin receptor gene represents an extension of techniques which have been used to isolate genes encoding biochemically selectable markers such as thymidine kinase[1,2] and cellular transforming genes[3] which confer altered cell morphology on recipients. Since the expression of a surface antigen does not generally provide a direct selection, we adopted a two-step selection which exploits the phenomenon of cotransfer.[4] Transfection of mouse Ltk⁻ cells with the herpes simplex virus thymidine kinase (HSV-tk) gene and HAT medium selection provides an efficient way of isolating large numbers of cells which have stably incorporated and express exogenous DNA applied as a calcium phosphate precipitate.[5] Cotransfection of the HSV-tk gene with an excess of genomic DNA results in a population of tk⁺ cells which contain between 10^3 and 10^4 kilobases (kb) of unselected DNA. Thus one human genome equivalent of 3×10^6 kb will be represented in 300 to 3000 individual transformants. A useful feature of the L cell recipient line is that it does not necessarily pose a barrier to the expression of genes which are not normally expressed in these cells.[6]

The pool of L-cell transformants is then screened for expression of the surface antigen of interest and positive cells are selected. In the case of the transferrin receptor this was achieved by indirect immunofluorescence using a human receptor specific monoclonal antibody and a fluorescent second antibody.[7] Positive cells were selected using the fluorescence

---

[1] P.-F. Lin, S.-Y. Zhao, and F. H. Ruddle, *Proc. Natl. Acad. Sci. U.S.A.* **80**, 6528 (1983).

[2] H. D. Bradshaw, *Proc. Natl. Acad. Sci. U.S.A.* **80**, 5588 (1983).

[3] C. Shih and R. A. Weinberg, *Cell* **29**, 161 (1982).

[4] M. Wigler, R. Sweet, G.-K. Sim, B. Wold, A. Pellicer, E. Lacy, T. Maniatis, S. Silverstein, and R. Axel, *Cell* **22**, 777 (1979).

[5] M. Wigler, S. Silverstein, L.-S. Lee, A. Pellicer, T. Cheng, and R. Axel, *Cell* **11**, 223 (1977).

[6] C. Hsu, P. Kavathas, and L. Herzenberg, *Nature (London)* **312**, 68 (1984).

[7] L. C. Kuhn, A. McClelland, and F. H. Ruddle, *Cell* **37**, 95 (1984).

activated cell sorter (FACS) as described in detail below. If the antigen to be selected is not normally expressed on L-cells, it may not be necessary to use a species-specific monoclonal antibody.

Since the original pool of primary transfectants contain several thousand kilobases of foreign DNA, a second round of gene transfer and selection is required to eliminate irrelevant sequences prior to cloning in a bacterial vector. The presence of donor sequences is monitored by Southern blotting using a species-specific repetitive DNA probe. A genomic library is constructed from a suitable secondary transformant and is screened using the same probe. The final step in this procedure is the reintroduction of the cloned sequences into L-cells to test for their ability to express the surface antigen.

Methods and Results

Preparation of High-Molecular-Weight DNA

The following protocol results in high-molecular-weight genomic DNA suitable for the transfection of mammalian cells. The volumes given are for 10^8 cultured cells and will result in yields of up to 1 mg of DNA. All solutions, glassware, and dialysis tubing should be autoclaved to ensure sterility of the final DNA solution.

The cells are initially washed twice in phosphate-buffered saline (PBS). Overlay the cells with 20 ml of 10 mM Tris, pH 7.6, 10 mM NaCl, 1 mM EDTA, 0.5% SDS, and add proteinase K (Type XIV Sigma) to 100 μg/ml. The cell lysate is incubated at 37° for 12–16 hr with gentle mixing (e.g., on a roller bottle rack). Transfer the lysate to a 50-ml disposable tube and add an equal volume of phenol (BRL Ultrapure) equilibrated with 10 mM Tris, pH 7.6, 10 mM NaCl, 1 mM EDTA. Gently mix by inversion for 10 min and then centrifuge at 10,000 g for 10 min in a swinging bucket rotor to separate the phases.

The aqueous phase is transferred to a dialysis bag and dialyzed overnight at 4° against 50 mM Tris, pH 7.6, 100 mM NaCl, 10 mM EDTA. Add pancreatic RNase (heated to 100° for 10 min to inactivate DNase) to a concentration of 100 μg/ml and incubate at 37° for 3 hr. SDS is then added to 0.5% and proteinase K to 100 μg/ml and the incubation continued for a further 3 hr. Extract twice with phenol as above and then twice with chloroform. The DNA at this point will probably be too dilute for transfection or gel analysis. It can be concentrated by transferring the solution to a dialysis bag which is placed in a tray of PEG 8000 (Sigma). When the volume has reduced by one-half, knot the bag to reduce its size. Repeat this process until an estimated DNA concentration of 300 μg/ml or greater is achieved. Dialyze the solution against 10 mM Tris, pH 7.6, 10 mM NaCl, 1 mM EDTA at 4° for 12–16 hr.

Cells and Selection Procedures

Mouse L-cells are maintained in alpha minimal essential medium (αMEM, Gibco, Grand Island, NY) supplemented with 10% fetal calf serum (FCS) and the antibiotics penicillin (100 U/ml) and streptomycin (100 μg/ml). We have used the mouse thymidine kinase-deficient cell line Ltk$^-$ for the generation of cell pools containing human genomic DNA. Ltk$^-$ cells offer the advantages of efficient transformation frequencies, and a convenient selection system with undetectable levels of spontaneous reversion. Although transfection frequencies of 1 in 300 cells per 50 pg of selectable marker can be achieved with mouse L-cells,[7] subclones may vary dramatically both in transfection efficiency and growth rate. For this reason sublines should be established and transfection efficiencies assessed with selectable plasmid DNA prior to cotransfection experiments.

The selectable marker for the transfection of Ltk$^-$ cells is the plasmid pTKX-1 which contains the 3.5 kb HSV-tk BamHI fragment cloned in pBR322.[8] Cells stably transfected with this plasmid will survive treatment with HAT (1 × 10^{-4} M hypoxanthine, 4 × 10^{-7} M aminopterin, 1.6 × 10^{-5} M thymidine). L-cell transformants can also be obtained using the plasmid pSV2-neo which carries the bacterial neomycin resistance gene. Selection is exerted by the addition of 400 μg/ml of the antibiotic G418[9] to the culture medium.

Cotransfection with Plasmid and Genomic DNA

Twenty-four hours prior to gene transfer the cells should be plated at a density of 2.6 × 10^6 cells per 150-cm^2 tissue culture flask. A total of 20 flasks are used in each transformation experiment to ensure the generation of greater than 20,000 independent clones. For each flask 200 ng of plasmid DNA is mixed with 80 μg of human genomic DNA and the volume is adjusted with 10 mM Tris pH 7.6, 1 mM EDTA, and 2.5 M calcium chloride to 1.2 ml and 250 mM calcium chloride. The DNA is carefully mixed and is added dropwise to an equal volume of 280 mM NaCl, 25 mM HEPES, 1.5 mM Na$_2$HPO$_4$, pH 7.12. Constant mixing is achieved by bubbling a gentle stream of air through the solution using a 1-ml sterile plastic pipet attached to an air source. A DNA calcium phosphate precipi-

[8] L. W. Enquist, G. F. Van de Woude, M. Wagner, J. R. Smiley, and W. C. Summers, *Gene* **7**, 335 (1979).

[9] F. Colbere-Garapin, F. Horodniceanu, P. Kourilsky, and A. Garapin, *J. Mol. Biol.* **150**, 1 (1981).

tate forms immediately, but the tube should be left undisturbed for 30 min at 20° to allow additional precipitate to form. The precipitate is dispersed by shaking the tube until large clumps are no longer visible and is added directly to the culture medium in the recipient cell flask which is incubated immediately at 37°.

The medium is replaced with fresh nonselective medium at 20 hr. At 60 hr after transfection, selective medium is added and is changed every 2–3 days. Small colonies appear 8–10 days after transfer and after 2 weeks should be detached and replated in new flasks. This will considerably reduce the problems caused by cellular debris during the cell sorter selection. Frozen cell stocks should be established as soon as possible in order to minimize the impact of faster growing colonies on the population. Independent pools of HAT-resistant mass populations should be initially established if the recovery of independent transfectants is desirable. Since the genomic DNA used for transfection consists of fragments greater than 100 kb in size, the ratio of 200 ng of plasmid to 80 μg of genomic DNA represents a molar ratio of approximately 1 to 40 or more. This ensures not only that the majority of tk$^+$ colonies contain unselected genomic DNA, but that the transformants contain sufficient DNA that the entire genome can be represented in a few thousand transformants.[10]

Fluorescence Labeling

The transfectants are divided into pools of approximately 5000 colonies which are analyzed separately by the FACS. This ensures the isolation of independent transfectants. Cell pools are expanded to greater than 10^7 cells prior to sorter analysis. The cells are detached from the flask by incubation in PBS pH 7.4 containing 0.03% EDTA for 5 min at room temperature. The cells are washed twice in ice cold αMEM containing 2% heat-inactivated serum. Three to five $\times$ 10^6 cells are incubated with saturating amounts of antibody in 50 μl αMEM with 2% serum for 1 hr at 4°. Control staining of 1 $\times$ 10^6 cells is performed with irrelevant myeloma or hybridoma antibodies. The cells are washed twice and then incubated for 1 hr at 4° with a fluorescein-conjugated second antibody directed against the first antibody. Propidium iodide is added at 0.5 μg/ml during this incubation. Finally, the cells are centrifuged through 3 ml of horse serum and resuspended at 2 $\times$ 10^5 cells/ml of medium.

FACS Parameters for Analysis and Sorting

Fluorescence analysis and sterile cell sorting is performed on a FACS IV (Becton-Dickinson FACS Systems, Mountain View, CA). Fluorescein

[10] L. C. Kuhn, J. A. Barbosa, M. E. Kamarck, and F. H. Ruddle, *Mol. Biol. Med.* **1**, 335 (1983).

and propidium iodide dyes are excited by an Argon-ion laser producing 400 mW at 488 nm. A Zeiss 580 nm dichroic mirror (46 63 05) is used to separate fluorescence signals to two photomultiplier tubes (PMT). Signals to the fluorescein PMT (QL30, EMI Gencon Inc., Plainview, NY) were further blocked with a 530/30 dichroic filter (Becton-Dickinson), and signals to the rhodamine/PI PMT (QL20, EMI) by dichroic filter 625/35 (Becton-Dickinson). Logarithmic conversion of PMT voltage for histogram display was performed with FACS electronics. Cell sorting was at rates of 2000 cells/sec with an 80 μm nozzle at transducer frequencies of 20 kHz.

Live/Dead Cell Discrimination

For analysis and sorting it is useful to "gate out" dead cells in order to lower the background due to nonspecific staining.[11] While cell viabilities of greater than 98% can be achieved, dead cells which have taken up fluorescence nonspecifically will obscure specific immunofluorescence in the first analysis. These cells can be excluded from the fluorescence histogram by "gating" only those cells which display a distinctive scatter profile and exclude 0.5 μg/ml propidium iodide which is taken up by dead cells.[12]

Figure 1 presents an example of a two-dimensional contour analysis of forward scatter and propidium iodide staining of a transfected cell population. Dead cells possess distinguishable scatter and fluorescence from the majority of live cells in the population. "Gating" is accomplished by analyzing only those cells within the region between the horizontal lines and left of the arrows labeled "A." Because live/dead discrimination of L-cells can be accomplished by scatter alone (area below lines), we regularly gate based only on this parameter. By eliminating propidium iodide staining an additional fluorescence photomultiplier tube is available for antigen analysis using additional antibodies. A comparison of gated and ungated transfected cells analyzed by indirect immunofluorescence is shown in Fig. 2. It demonstrates that the effect of removing dead cells from the analysis is to decrease the background in the region of the curve where signals from specifically stained cells are present.

Sterile Cell Sorting

Even with efficient live cell gating, antigen-positive cells which occur at frequencies of 0.05–0.1% in the population do not appear as a distinct

[11] M. R. Loken and L. A. Herzenberg, *Ann. N.Y. Acad. Sci.* **254**, 163 (1975).
[12] J. L. Daugl, D. R. Parks, V. T. Oi, and L. A. Herzenberg, *Cytometry* **2**, 395 (1982).

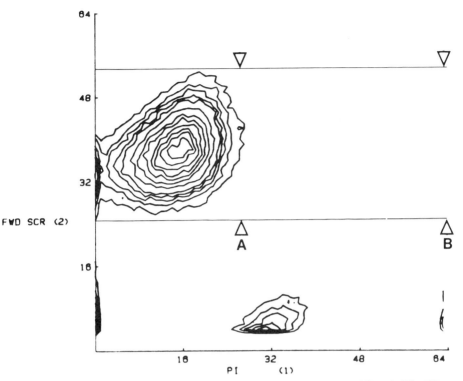

FIG. 1. FACS contour plot distributions of forward scatter and propidium iodide (PI) signals produced by a transfected cell population. Two-dimensional contour plots were generated by the Consort 40 computer using Becton-Dickinson software. FACS optical configuration is described in the text. Dead cells appear in the lower center of the plot with significantly lower scatter signals and higher PI staining. They constitute approximately 5% of the cell population. Living cells are bounded by the horizontal lines and can be further bracketed by vertical PI gates set at arrow "A" or "B" (see text).

peak on the first analysis (Fig. 2). It is therefore necessary to sort the brightest 0.5% of the cells to assure recovery of antigen-positive cells and to obtain sufficient cell numbers so that reanalysis can be done in 7 to 10 days. Cells to the right of the arrow in the gated population in Fig. 2 represent this fraction and would be sorted in a typical analysis. Sterility is assured by washing all FACS tubing with detergent and ethanol prior to sorting. Cells are sorted initially into αMEM, with 10% FCS and antibiotics in 35 × 10-mm petri dishes. Figure 3a shows reanalysis of this population of transfectants following the initial enrichment. In this example the original population was sorted simultaneously for the expression of human HLA and 4F2 antigens. It can be seen that two groups of antigen-positive cells are visible after the first sort, and that each is present in a

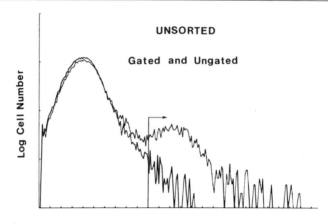

Fig. 2. Indirect immunofluorescence and FACS analysis of unsorted L-cell transfectants. Cells were stained with a mixture of anti-HLA-A,B,C and anti-4F2 antibodies and fluoresceinated second antibody. Detection of fluorescence was made either on live cell gated or ungated populations. FACS parameters are described in the text and cell numbers are presented on a log scale for easier visualization. The sorting window defined by the arrow includes 0.5% of the gated cells and approximately 5% of the ungated cells.

different fraction of the population. These cells were then resorted, again using a mixture of both antibodies to produce the population of cells shown in Fig. 3b, c, and d. After these two selections, transfectants stained with a single antibody can be subcloned using the FACS single cell deposition system (Becton-Dickinson). Subcloning by limiting dilution into 96-well microtiter dishes can also be done once a population is sufficiently homogeneous by sorting. The data presented in Fig. 3 demonstrate that simultaneous selection of different primary antigen transfectants can be achieved by using mixtures of antibodies. This considerably improves the overall efficiency of the procedure since the expansion of sorted cells is time consuming.

Following cell cloning, primary transformant DNA is analyzed by filter hybridization for the presence of donor repetitive sequences. The large amount of DNA incorporated in the majority of primary transformants necessitates the use of this material as donor in a second round of gene transfer and cell sorting, which is performed identically to the first round.

Gene Isolation from Secondary Transformants

Reiterated interspersed DNA elements such as the human Alu family provide a means of detecting the foreign DNA in primary and secondary

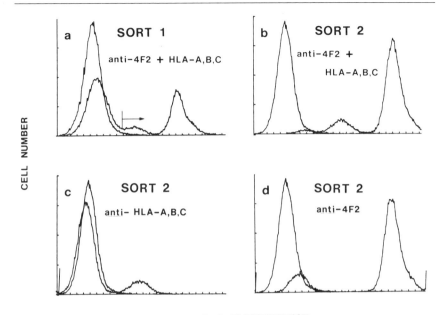

RELATIVE FLUORESCENCE

FIG. 3. Indirect immunofluorescence and FACS analysis of L-cell transfectants sorted for HLA-A,B,C and 4F2 expression. (a) SORT 1: cells were expanded into T-75 cm² flasks and reanalyzed with gating using a mixture of anti-4F2 and anti-HLA-A,B,C antibodies. The control curve was labeled without the first antibody. This population was resorted using the window designated by the arrow. (b) SORT 2: indirect immunofluorescence with anti-HLA and anti-4F2 compared to control (c) Indirect immunofluorescence of cell population shown in b with anti-HLA alone compared to control. (d) Indirect immunofluorescence of cell population shown in b with anti-4F2 alone compared to control.

transformants.[13] Southern blot analysis of the DNA of several independent secondary transformants should reveal the presence of common fragments corresponding to the selected sequences. Repetitive DNA probes can be made using either cloned sequences or by direct labeling of genomic DNA. If a cloned probe is used, the vector should have no homology to pBR322 (e.g., M13 mp8), since the transfectants contain integrated plasmid sequences. The use of genomic DNA as a probe offers the advantage that all families of highly repeated sequences are represented, thereby increasing the likelihood of detecting the foreign DNA. Genomic DNA is labeled to specific activities of 2×10^8 dpm/μg or greater by nick translation[14] using [^{32}P]dCTP (>3000 Ci/mmol). Due to the

[13] J. F. Gusella, C. Keys, A. Varsanyi-Breiner, F.-T. Kao, C. P. Jones, T. T. Puck, and D. Houseman, *Proc. Natl. Acad. Sci. U.S.A.* **77**, 2829 (1980).
[14] P. W. J. Rigby, M. Dieckman, C. Rhodes, and P. Berg, *J. Mol. Biol.* **113**, 237 (1977).

initial viscosity of high-molecular-weight DNA it is useful to mix the reaction by gentle agitation every 30 min. The rate of nucleotide incorporation is monitored by TCA precipitation to determine the optimum reaction time which may be longer than required for plasmid DNA. An example of a Southern blot analysis of several primary and secondary L-cell transfectants which express the human transferrin receptor gene is shown in Fig. 4. The probe used in this experiment was nick translated total human DNA. Hybridization was for 40 hr at 42° in 50% formamide, 5× SSC, 10× Denhardt's solution, 0.1% SDS, 10% dextran sulfate, 100 μg/ml sonicated salmon sperm DNA, and 10 ng/ml of probe DNA. The filter was washed three times at room temperature for 5 min in 2× SSC, and four times at 65° for 15 min in 0.1× SSC, 0.1% SDS. Autoradiography was for 7 days at −70° using an intensifying screen (Dupont Cronex Lightning-Plus). Shown in lanes A and B are DNA samples from two independent primary transfectants selected for expression of the human transferrin receptor using the species specific monoclonal antibody OKT9.[15] We estimate from dot blot titrations that these lines each contain approximately 3000 kb of human DNA.[10] The enrichment for specific human sequences obtained by generating secondary transformants is illustrated by the samples in lanes C to I. They represent independent transferrin receptor expressing lines derived by transfection with the DNAs shown in lanes A and B. The bulk of the integrated DNA in the secondaries is of mouse origin, but a discrete set of human specific fragments can be seen. Furthermore a subset of these fragments is shared by all of the lines, indicating that they represent the coding sequences and closely linked flanking sequences of the selected gene.

Since repetitive elements are widely distributed throughout the human genome, the majority of human genes will be detected by this approach. However, in some cases a gene may be selected which has no closely associated repeats such as the T8 gene.[16,17] In this situation the secondary transformant can still represent a useful starting point for cDNA cloning by subtractive hybridization. A more direct approach which may be used in most cases is to construct a phage or cosmid library from the DNA of a secondary transformant and to isolate the human specific clones by repetitive sequence hybridization. It is advisable to screen a library represent-

[15] R. Sutherland, D. Delia, C. Schneider, R. Newman, J. Kemshead, and M. Greaves, *Proc. Natl. Acad. Sci. U.S.A.* **78**, 4515 (1981).

[16] D. R. Littman, Y. Thomas, P. J. Maddon, L. Chess, and R. Axel, *Cell* **40**, 237 (1985).

[17] P. Kavathas, V. P. Sukhatme, L. A. Herzenberg, and J. R. Parnes, *Proc. Natl. Acad. Sci. U.S.A.* **81**, 7688 (1984).

A B C D E F G H I C

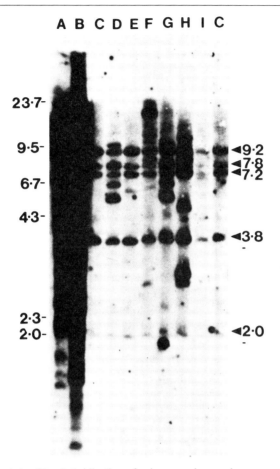

FIG. 4. Analysis by filter hybridization of primary and secondary mouse transfectants expressing the human transferrin receptor. Ten micrograms of DNA from two primary (lanes A and B) and seven secondary L-cell transformants (lanes C to I) was restricted with EcoRI, transferred to nitrocellulose and hybridized to nick-translated human DNA. The positions of λHindIII marker fragments are indicated in kb. The arrows indicate shared fragments of human origin in secondary transformants. [Reprinted from Kuhn et al.,[7] Cell 37, 95–103 (1984); copyright 1984 by MIT Press.]

ing at least 10 genome equivalents since there is only one copy of the selected gene per cell. Cross-reaction between the human repetitive probe and related mouse sequences is minimized by washing the filters at high stringency after hybridization (e.g., in 0.1× SSC at 65°). Under these conditions strongly hybridizing plaques which contained the human transferrin receptor gene were readily distinguished from weakly cross-hybridizing mouse clones. Of course optimum hybridization and washing conditions will vary for different genes, and can be determined empirically by Southern blotting of transfectant DNA prior to library screening. Potential positive clones should be checked using both human and mouse repetitive DNA probes to ensure their origin. The restriction map of the cloned region is compared to the Southern blot pattern of the secondary transfectant DNA in order to ascertain that all of the gene has been recovered.

To further confirm the identity of the cloned DNA it is reintroduced into Ltk⁻ cells by cotransfection and the transfectants are analyzed for expression of the selected antigen using the cell sorter. Depending on the size of the gene and the distribution of the individual clones, it may be necessary to transfect with more than one recombinant in order to obtain expression. For example, the transferrin receptor gene is 33 kb long and two overlapping phages were required in order to produce an active gene. An expressing gene can be produced either by *in vitro* reconstruction using infrequent restriction sites, or by cotransfer of overlapping clones.[7] In the latter case expression is presumably the result of recombination between the input DNA during transfection.

Summary

We have described a transfection method for the isolation of surface antigen genes which requires no mRNA or protein purification. Application of this technique results in the recovery of the entire gene in a single step since selection for expression of genomic DNA forms the basis of the procedure. Based on our results with the transferrin receptor gene and other systems[7,10] it is evident that large transcription units can be transferred and expressed in mouse L-cells. This size consideration represents a major advantage over the use of cosmid shuttle vectors for genomic DNA expression. In the case of genes which code for very long mRNAs this method may also have advantages over cDNA expression systems. Although we have described methods for FACS isolation of transfectants based on the binding of species specific antibodies to surface antigens, other methods of identifying transfected cells could be employed. For example, in combination with an appropriate assay, sib selection of recipient cells could be used to identify genes encoding secreted products. Ligand binding assays could be used for receptors which are not expressed on the host cell. Finally, the development of cDNA expression vectors which produce membrane-associated products would extend this methodology to genes not normally expressed at the cell surface.

Acknowledgments

The authors are grateful to John Hart for his expert technical assistance and to Suzy Pafka for help with the figures. This research was supported by NIH Grant GM09966. Special thanks go to a private benefactor who wishes to be unnamed.

[26] Transferrin: Assay of Myotrophic Effects and Method for Immunocytochemical Localization

By George J. Markelonis and Tae H. Oh

Transferrin and Muscle

Our interest in transferrin (Tf) arises from the fact that in differentiating embryonic chicken skeletal muscle cultures, the protein influences both the initial proliferative stage of myogenesis which occurs just prior to myoblast fusion,[1] and the later, maturational stage of myogenesis which occurs after fusion has been completed.[1-3] This latter maturational stage of development is characterized by the appearance of specialized gene products specific to the maturing myotubes.[4,5] Transferrin must be present in the growth medium throughout all stages of development in order for muscle cells to differentiate *in vitro*,[1] and promotes the maintenance of mature myotubes in culture for prolonged periods.[6] In this chapter, we will review several useful assays for assessing the myotrophic effects of Tf or other growth factors upon skeletal muscle cells and will review the method for localizing Tf by immunocytochemistry.

Purification of Transferrin

We have purified Tf from chicken serum[3] and this protocol is detailed in Fig. 1. This two-column purification procedure[7] takes advantage of two physicochemical characteristics of Tf: (1) the polypeptide is a glycoprotein and binds to concanavalin A-agarose in the presence of Mn^{2+} (Fig. 1,

[1] T. H. Oh and G. J. Markelonis, *Proc. Natl. Acad. Sci. U.S.A.* **77**, 6922 (1980).
[2] G. J. Markelonis, T. H. Oh, M. E. Eldefrawi, and L. Guth, *Dev. Biol.* **89**, 353 (1982).
[3] T. H. Oh and G. J. Markelonis, *J. Neurosci. Res.* **8**, 535 (1982).
[4] A. Shainberg, G. Yagil, and D. Yaffe, *Dev. Biol.* **25**, 1 (1971).
[5] A. I. Caplan, M. Y. Fiszman, and H. M. Eppenberger, *Science* **221**, 921 (1983).
[6] G. J. Markelonis and T. H. Oh, *Proc. Natl. Acad. Sci. U.S.A.* **76**, 2470 (1979).
[7] T. H. Oh and G. J. Markelonis, in "Growth and Maturation Factors" (G. Guroff, ed.), Vol. 2, p. 56. Wiley, New York, 1984.

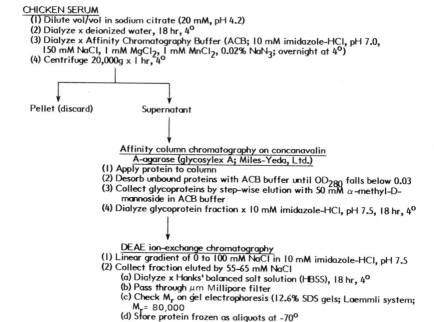

FIG. 1. Purification protocol for serum transferrin. Modified by permission from Oh and Markelonis.[7]

Affinity column)[3,8,9]; and (2) Tf has a pI of approximately 5.75[6,10] and will bind to an anion-exchange matrix at neutral pH (Fig. 1, DEAE column).[8] The protein eluted by 55–65 mM NaCl is checked for purity ($M_r = 80,000$) on 12.6% SDS vertical gel electrophoresis using the discontinuous system of Laemmli.[11] Since the protein purified by this procedure is virtually iron free, the apoprotein is dialyzed against 1 mM ferric citrate in the presence of 10 mM NaHCO$_3$ in order to obtain diferric-Tf.[9] The resulting diferric-Tf can then be dialyzed at neutral pH against a physiological salt solution overnight at 4°, sterilized by filtration through a 0.45-μm filter, and stored frozen at −70°. To check iron saturation, iron is estimated using the ferrozine assay.[12]

[8] G. J. Markelonis and T. H. Oh, *J. Neurochem.* **37**, 95 (1981).
[9] G. J. Markelonis, R. A. Bradshaw, T. H. Oh, J. L. Johnson, and O. J. Bates, *J. Neurochem.* **39**, 315 (1982).
[10] F. W. Putnam, in "The Plasma Proteins" (F. W. Putnam, ed.), Vol. 1, p. 60. Academic Press, New York, 1975.
[11] U. K. Laemmli, *Nature (London)* **227**, 680 (1970).
[12] D. D. Harris, *J. Chem. Educ.* **55**, 539 (1978).

The procedure outlined above would be useful for purifying Tf from the sera of rats, mice, guinea pigs, etc. However, both human and chicken Tf can be obtained from commercial sources. We routinely purchase chicken diferric-Tf from the Sigma Chemical Co. of St. Louis, MO (conalbumin, type II; iron-complex; #C 0880). This product is usually contaminated with albumin, but this contaminant can be removed by chromatographing the Tf on an affinity column of rabbit anti-chicken-Tf IgG conjugated to Sepharose 4B.[13] However, for tissue culture purposes, the commercial product is suitable even without further purification. We should point out that the iron-free preparation from Sigma has little biological effect on cultured muscle cells.[3] This Tf preparation migrates at a faster M_r than the diferric form and may be denatured during commercial purification.[3]

Skeletal Muscle Cell Culture

Myogenic cells are obtained from trypsin-dissociated breast muscles of 12-day-old chicken embryos (Spafas, Inc.).[1,14] The myogenic cells (5 × 10^5) are plated on Linbro plastic dishes (35 × 15 mm; FB-6-TC) precoated with acid-soluble collagen (50 μg/dish; Calbiochem). Cultures are then maintained in 1.0 ml of culture medium at 37° in a humidified atmosphere of 95% air/5% CO_2. Cells are plated for 24 hr in a standard culture medium consisting of 87.5% (v/v) Dulbecco's modified Eagle's medium (DMEM), 10% (v/v) horse serum (heat inactivated and Millipore filtered), and 2.5% chicken embryo extract.[1] After 24 hr, the control cultures are grown in standard culture medium supplemented with 35 μg/ml BSA (3.75 × 10^{-7} M) while experimental cultures are grown in standard medium containing 30 μg/ml chicken Tf.

Myotrophic Assay: Myogenesis

The bioassay of Tf provides a unique challenge because we have shown that myogenesis will not ensue in the absence of this protein.[1] Thus, even control cultures must contain some Tf in order for normal development to proceed. For convenience then, we grow all cells in a standard culture medium which contains residual Tf as a component of embryo extract[1] and simply test the effect of an increased level of exogenous Tf added to the basal level already present. However, the use of a

[13] G. J. Markelonis, T. H. Oh, L. P. Park, C. Y. Cha, C. A. Sofia, J. W. Kim, and P. Azari, *J. Cell Biol.* **100**, 8 (1985).

[14] C. Richler and D. Yaffe, *Dev. Biol.* **23**, 1 (1970).

Tf-free culture medium can rapidly and unmistakably demonstrate the requirement of this protein for myogenesis.[1] For example, if one plates cells for 24 hr in standard culture medium and then replaces the medium with 90% DMEM/10% horse serum, only cultures containing added Tf will continue to proliferate and fuse to form myotubes; Tf-free cultures show an arrest of myogenesis as evidenced by a failure to proliferate or to fuse (Fig. 2). This simple, direct morphologic assay is rapid and quite specific. This assay relies on the fact that the Tf in horse serum does not

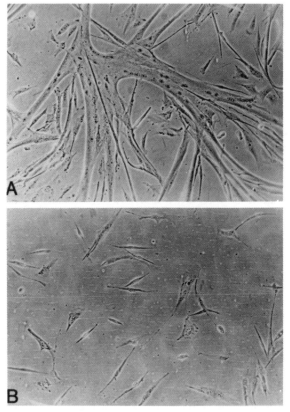

Fig. 2. Effect of transferrin on myogenesis. Cultures were plated for 24 hr in standard culture medium at which time the medium was changed to 90% DMEM/10% horse serum. Tf (30 μg/dish) was added to some dishes while control cultures were free of Tf (i.e., neither embryo extract nor Tf was added). (A) Culture grown for 6 days in the presence of Tf (30 μg/dish). (B) Six-day-old culture which was free of Tf after the initial, 24-hr plating period. Note that myogenesis has been completely arrested in the culture grown without Tf (B).

cross-react with the transferrin receptor on chicken myogenic cells.[15] In a similar manner, it is unlikely that chicken Tf will cross-react with most nonavian transferrin receptors. Therefore, it is more appropriate to use human diferric-Tf in cell cultures of rodent or other mammalian tissues.

Myotrophic Assay: DNA Synthesis

The effect of Tf upon the incorporation of labeled thymidine into the DNA of myoblasts is a useful, but nonspecific assay. Depending on the plating density of myogenic cells, myoblasts will incorporate radioactive thymidine into DNA in the presence of Tf for at least several days. This uptake and incorporation of [³H]thymidine thus serves as an indicator for the stimulation of *myoblast proliferation* by Tf. However, one must bear in mind that some fibroblasts are also present and may contribute to the total incorporation observed. Thus, it is useful to measure [³H]thymidine incorporation into myoblasts approximately 30 hr after plating at a time when myoblast proliferation is at a peak and fibroblast numbers are low.

The method for assaying the uptake and incorporation of [³H]thymidine into myoblast DNA is as follows[16]:

1. Twenty-four hours after plating cells, Tf (30 μg/dish) or BSA (35 μg/dish) is added to experimental or control cultures, respectively. Cytosine arabinoside (10^{-5} M) should be added to some cultures as a further control.

2. At 30–36 hr after plating, add 1.0–5.0 μCi [*methyl*-³H]thymidine (e.g., 7 Ci/mmol) to each experimental and control culture and incubate at 37°.

3. At timed intervals, wash the cultures thrice with 1.0-ml portions of Hanks' balanced salt solution (HBSS) containing 5 mM thymidine. Remove the wash solutions by suction–aspiration.

4. Place the cultures on ice and extract twice with 1.0 ml ice-cold 10% trichloroacetic acid for 5 min. Aspirate the TCA using suction.

5. Wash cultures three times with absolute ethanol–diethyl ether (3 : 1, v/v) in a well-ventilated room and then air dry the cultures. (Note: Be certain to pretest the culture dishes for resistance to ethanol–ether. This mixture dissolves certain plastics!)

[15] E. Ozawa, I. Kimura, T. Hasegawa, I. Ii, K. Saito, Y. Hagiwara, and T. Shimo-oka, *in* "Muscular Dystrophy: Biochemical Aspects" (S. Ebashi and E. Ozawa, eds.), p. 53. Japanese Scientific Press, Tokyo, 1983.

[16] T. H. Oh, *Exp. Neurol.* **50,** 376 (1976).

6. Add 1.0 ml 2 N NaOH and swirl gently until the alkali covers the bottom of the culture dish. Allow the precipitate to dissolve at room temperature.

7. Add an aliquot of the dissolved precipitate (10–100 μl) to 10 ml of a toluene-based scintillation cocktail such as Aquasol (New England Nuclear); add a few microliters of glacial acetic acid to clarify the mixture and count the sample.

It should be noted that Tf has a rapid and pronounced effect on myoblast proliferation in culture as measured by the incorporation of [³H]thymidine into DNA. Using a paradigm similar to that shown above, it is also possible to determine the effect of Tf on RNA and protein synthesis in prefusion myoblasts by simply using [G-³H]uridine or L-[U-¹⁴C]leucine as tracers.[17,18] The incorporation of [³H]uridine into RNA and L-[¹⁴C]leucine into protein are best carried out 48 hr after plating has occurred at a time just prior to or concomitant with myoblast fusion.

Myotrophic Assay: Acetylcholinesterase Activity

Tf increases the specific activity of a number of proteins which are highly enriched in myotubes but deficient in fibroblasts. Acetylcholinesterase (AChE) and acetylcholine receptors (AChRs) are two examples of such proteins.[2] These assays are useful because they specifically measure the influence of Tf on proteins specific to muscle cells, and because both AChE activity and AChRs are significantly elevated just after myoblast fusion into myotubes has occurred. Thus, these proteins are amplified by growth factors even further and are indicative of the stimulatory effect of Tf on the *maturational phase* of myogenesis.

The AChE assay used in our laboratory is a radiometric assay[2] which is a modification of the procedure of Johnson and Russell[19,20]:

1. Cultures are washed with HBSS and AChE is quantitatively extracted by solubilizing the cells in 1.0 ml 1% Triton X-100 in 10 mM Tris–HCl, pH 7.2.[20] The cell extracts are centrifuged at 10,000 g for 30 min at 4° and AChE activity is determined on the supernatant.

2. The following solutions are prepared:
   a. Labeled acetylcholine (ACh) stock solution is prepared by adding 1.5 ml of dilute acetic acid (pH 3.5) to a vial containing 250

---

[17] G. J. Markelonis and T. H. Oh, *Exp. Neurol.* **58**, 285 (1978).
[18] G. J. Markelonis, T. H. Oh, and D. Derr, *Exp. Neurol.* **70**, 598 (1980).
[19] C. D. Johnson and R. L. Russell, *Anal. Biochem.* **64**, 229 (1975).
[20] R. L. Rotundo and D. M. Fambrough, *J. Biol. Chem.* **254**, 4790 (1979).

μCi [^{14}C]acetylcholine iodide ([*acetyl*-1-^{14}C], specific activity, 1.0–5.0 mCi/mmol, New England Nuclear).

b. Unlabeled ACh stock solution is prepared by adding 69.2 mg acetylcholine iodide (Sigma) and 29.1 mg iso-OMPA (tetraisopropylpyrophosphoramide, Koch-Light Ltd.), a pseudo-cholinesterase inhibitor, to 25 ml of sodium phosphate buffer (100 mM, pH 7.0).

c. Working substrate solution is prepared just before use by mixing 0.1 ml of [^{14}C]ACh stock solution (a) with 7.4 ml of ACh-iso-OMPA stock solution (b); 30 μl of the working substrate solution (c) diluted to a final assay volume of 100 μl gives an ACh concentration of approximately 3 mM, an iso-OMPA concentration of 10^{-4} M, and contains 66.7 nCi[^{14}C]ACh (148,000 dpm). All of the above stock and working solutions are stored at $-20°$ and defrosted prior to use.

3. Assays are carried out in 6.0-ml glass, miniscintillation vials (Wheaton Glass). To duplicate assay tubes, 40 μl of cell homogenate and 30 μl deionized water are added. Enzyme blanks contain 40 μl of homogenizing buffer solution in place of enzyme solution. The reaction is initiated by the addition of 30 μl of working substrate solution (c) (see above) to assay tubes. Each assay tube is capped, vortexed on a rotary mixer, and placed in a shaking water bath at 26°. After 30–120 min, tubes are removed from the water bath, 100 μl of stopping buffer[19] is added, and the tubes are recapped and vortexed.

4. Four milliliters of toluene–isoamyl scintillation cocktail[19] is added to each tube and the tubes are vortexed vigorously. Tubes are centrifuged at 3000 rpm (1000 g) for 10 min to separate the aqueous and organic phases and the level of [^{14}C]acetate in the solvent phase is then determined in a scintillation counter equipped with automatic dpm conversion. ^{14}C counting efficiency is greater than 94% in our assays; extraction efficiency for [^{14}C]acetate averages 71–73%. Counts from duplicate enzyme assay tubes should agree to within ±2%.

5. To convert disintegrations per minute (dpm) of [^{14}C]acetate liberated into the mass of ACh hydrolyzed, the following formula can be used: nmol ACh hydrolyzed/assay time/40 μl enzyme = total dpm sample/ 147,900 × extraction efficiency^{-1} × Y nmol ACh/assay tube where Y equals 300 nmol unlabeled ACh + Z nmol [^{14}C]ACh and Z equals 66.7 nCi/assay tube × nmol/nCi (specific activity of [^{14}C]ACh).

The AChE assay described is linear with respect to time and with respect to enzyme concentration. It is well suited for multiple determinations and we have found it to be most helpful in assaying the myotrophic activity of Tf in myotube cultures.[2] AChE activity can be expressed per culture dish or per mg noncollagen protein.[21,22]

Myotrophic Assay: Acetylcholine Receptors

Transferrin causes an increase in the number of acetylcholine receptors (AChRs) in treated cultures and also promotes the aggregation of these receptors into "hot spots."[2] Like AChE, AChRs are absent from fibroblasts and thus may serve as a specific measure of growth effects on the maturational phase of myogenesis. The method used for titrating and assaying AChRs is based on a protocol adapted from several sources[23–25]:

1. A stock vial (250 μCi) of ^{125}I-labeled-α-bungarotoxin (^{125}I-α-BTx; 16.0–16.5 μCi/μg; New England Nuclear) is diluted with 3% BSA in DMEM so that a 20-μl aliquot (sufficient for one culture dish) contains 120 ng (15 pmol) ^{125}I-α-BTx. The diluted solution is apportioned into 1.0-ml conical, plastic Eppendorf tubes and stored at $-20°$. In our hands, the shelf-life of the protein is 4 weeks before significant breakdown occurs. Tubes are defrosted only once and the unused portion of iodinated protein is discarded.

2. The standard culture medium is aspirated from cultures and replaced with 1.0 ml 90% DMEM/10% heat-inactivated horse serum.

3. Several control dishes (nonspecific binding controls) are preincubated in 200 μM d-tubocurarine (Boehringer-Mannheim)[25] at 37° for 60 min. This toxin blocks all available AChRs on the cell surface.

4. All cultures (experimentals and controls) are saturated with 15 nM ^{125}I-α-BTx (20 μl diluted stock/culture dish) at 37° for 30 min. This concentration of ^{125}I-α-BTx and this incubation time are optimal for saturating AChRs when 35-mm culture dishes are used.[24]

5. After labeling, the cell cultures are washed extensively with 1.0-ml portions of 90% DMEM/10% horse serum and twice with HBSS. Cells are then dissolved in 1.0 ml Triton X-100 in 10 mM Tris–HCl, pH 7.2[23] and the extract is counted in a gamma counter. Nonspecific binding should be carefully monitored; values of 11–14% are not unusual, but lower values (5%) are more acceptable. Also, one should remember that ^{125}I-α-BTx has

[21] D. H. Rifenberick, C. L. Koski, and S. R. Max, *Exp. Neurol.* **45**, 527 (1974).

[22] O. H. Lowry, N. J. Rosebrough, A. L. Farr, and R. J. Randall, *J. Biol. Chem.* **193**, 265 (1951).

[23] P. N. Devreotes and D. M. Fambrough, *J. Cell Biol.* **65**, 335 (1975).

[24] J. Patrick, S. F. Heinemann, J. Lindstrom, D. Schubert, and J. H. Steinbach, *Proc. Natl. Acad. Sci. U.S.A.* **69**, 2762 (1972).

[25] T. M. Jessell, R. E. Siegel, and G. D. Fischbach, *Proc. Natl. Acad. Sci. U.S.A.* **76**, 5397 (1979).

a propensity to adhere to glassware and plastics. Therefore, manipulation of the protein via pipetting, etc. should be reduced to a minimum.

Immunocytochemical Localization of Transferrin

Recent interest has focused on the immunocytochemical localization of Tf in muscle and nonmuscular tissues.[13,26-29] This technique is so popular and so informative because it enables one to visualize a peptide or polypeptide at the cellular or subcellular level and to make estimations of the distribution of the protein within a particular organ or tissue. The many advantages of this technique are well discussed by Sternberger.[30]

Successful peroxidase–antiperoxidase (PAP) immunocytochemistry depends on a number of factors.[30] In our laboratory, we have found that fresh reagents of high quality must be used, that the source and titer of the antibody used for visualization must be carefully considered, that the nature and method of fixation are often critical, and that sources of endogenous peroxidatic activity must be completely quenched. Our procedure[26] is adapted from that originally described by Sternberger[30] and is summarized in Fig. 3. In reference to this protocol, the following points should be made:

1. Reagents are obtained from the following sources:
   a. Sternberger-Meyer Immunocytochemicals, Inc.: 2° antibodies (goat anti-rabbit IgG serum, rabbit anti-goat IgG serum, rabbit anti-mouse IgG serum), peroxidase–antiperoxidase reagent (PAP), control sera (preimmune rabbit, preimmune goat, and preimmune mouse)
   b. Cappel Laboratories: 2° antibodies
   c. Aldrich Chemicals: 3,3'-diaminobenzidine (DAB)
   d. Fisher Chemicals: 30% H_2O_2 (ACS grade)
   e. Baker Chemicals: absolute methanol (ACS grade)
2. Rabbit antitransferrin serum (1 : 1000 in PBS) is used as the 1° antibody. This polyclonal serum shows excellent reactivity to Tf. Goat antitransferrin serum is also available in our laboratory, but this antiserum has a slightly lower titer than the rabbit antiserum. By contrast, a mouse

[26] T. H. Oh, C. A. Sofia, Y. C. Kim, C. Carroll, H. H. Kim, G. J. Markelonis, and P. J. Reier, *J. Histochem. Cytochem.* **29**, 1205 (1981).

[27] W. A. Jeffries, M. R. Brandon, S. V. Hunt, A. F. Williams, K. C. Gatter, and D. Y. Mason, *Nature (London)* **312**, 162 (1984).

[28] K. Mollgard and M. Jacobsen, *Dev. Brain Res.* **13**, 49 (1984).

[29] R. Matsuda, D. Spector, J. Micou-Eastwood, and R. C. Strohman, *Dev. Biol.* **103**, 276 (1984).

[30] L. Sternberger, "Immunocytochemistry." Wiley, New York, 1979.

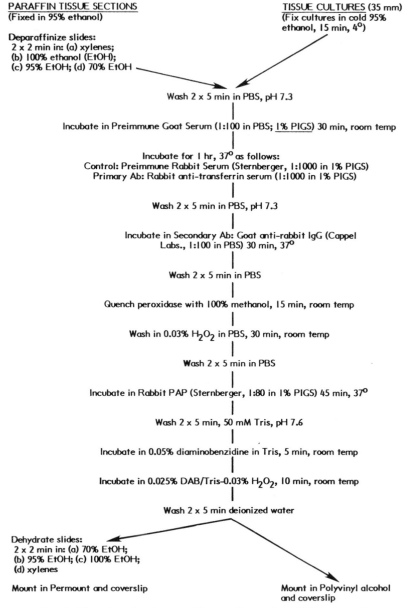

PARAFFIN TISSUE SECTIONS
(Fixed in 95% ethanol)

Deparaffinize slides:
2 x 2 min in: (a) xylenes;
(b) 100% ethanol (EtOH);
(c) 95% EtOH; (d) 70% EtOH

TISSUE CULTURES (35 mm)
(Fix cultures in cold 95%
ethanol, 15 min, 4°)

Wash 2 x 5 min in PBS, pH 7.3

Incubate in Preimmune Goat Serum (1:100 in PBS; 1% PIGS) 30 min, room temp

Incubate for 1 hr, 37° as follows:
Control: Preimmune Rabbit Serum (Sternberger, 1:1000 in 1% PIGS)
Primary Ab: Rabbit anti-transferrin serum (1:1000 in 1% PIGS)

Wash 2 x 5 min in PBS, pH 7.3

Incubate in Secondary Ab: Goat anti-rabbit IgG (Cappel
Labs., 1:100 in PBS) 30 min, 37°

Wash 2 x 5 min in PBS

Quench peroxidase with 100% methanol, 15 min, room temp

Wash in 0.03% H_2O_2 in PBS, 30 min, room temp

Wash 2 x 5 min in PBS

Incubate in Rabbit PAP (Sternberger, 1:80 in 1% PIGS) 45 min, 37°

Wash 2 x 5 min, 50 mM Tris, pH 7.6

Incubate in 0.05% diaminobenzidine in Tris, 5 min, room temp

Incubate in 0.025% DAB/Tris-0.03% H_2O_2, 10 min, room temp

Wash 2 x 5 min deionized water

Dehydrate slides:
2 x 2 min in: (a) 70% EtOH;
(b) 95% EtOH; (c) 100% EtOH;
(d) xylenes

Mount in Permount and coverslip

Mount in Polyvinyl alcohol
and coverslip

FIG. 3. Flow chart for the peroxidase–antiperoxidase (PAP) protocol.

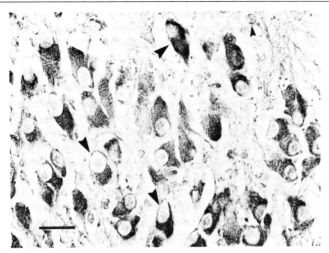

FIG. 4. Immunocytochemical localization of transferrin in adult chicken spinal cord. Adult chicken neural tissue was fixed in 95% ethanol, embedded into paraffin, cut into 8-μm sections, and stained by PAP as indicated in Fig. 3. Large arrowheads, ventral horn neurons in the ventral gray matter of the spinal cord; small arrowheads, staining in glial (oligodendrocyte?) cell. Bar = 20 μm.

monoclonal antitransferrin antibody (2EI2/47; IgG) with a titer of 1:250,000 shows poor reactivity against fixative-denatured antigen presumably because the epitope is labile. Therefore, this antibody is not particularly useful for PAP immunocytochemistry.

3. We have found that ethanol and methanol serve as excellent fixatives for the PAP of transferrin in chicken tissues.[26] For example, cold ethanol (95%) gives excellent preservation of Tf in tissue from developing chickens. The fixed tissue is processed by the method of Sainte-Marie[31] and 8-μm paraffin sections are used.

4. We have found that incubation of tissue with absolute methanol followed by 0.03% H_2O_2 in PBS just prior to the addition of PAP reagent (see Fig. 3) is required to reduce the background to negligible levels and to remove nonspecific staining.

Figure 4 demonstrates the immunocytochemical localization of Tf in a section of embryonic chicken spinal cord. The presence of Tf is indicated by the dark reaction product found within neurons and the weak staining found in glial (oligodendrocytes?) cells (see arrowheads). Control slides incubated with an appropriate dilution of preimmune rabbit serum show no staining and are not shown as a result.

[31] G. Sainte-Marie, J. Histochem. Cytochem. 10, 250 (1962).

Summary

1. Primary cultures of dissociated embryonic chicken skeletal muscle cells provide an ideal model for investigating the effects of growth factors such as Tf because these cells undergo a highly integrated pattern of differentiation and maturation.

2. The trophic effects of a growth factor such as Tf can be assessed on muscle cultures by the determination of such parameters as acetylcholinesterase and acetylcholine receptors. These proteins are specific to the cultured myotubes, appear in high levels following fusion of myoblasts into myotubes, and are relatively easy to assay.

3. Tf and other growth factors are internalized by a receptor-mediated mechanism (see Trowbridge et al. [24] and Seligman and Allen [21], this volume). These growth factors can be localized to specific tissues by immunocytochemistry at the light or electron microscopic level. This information on cellular distribution could be very useful in assessing the pattern of growth and differentiation with regard to the particular growth factor under study.

Acknowledgments

The authors would like to thank Dr. V. Kemerer and Dr. L. Guth for help with the AChE assay, Ms. T. L. Dion and Ms. S. L. Hobbs for technical assistance, and Mrs. E. DeLong and Mrs. E. Ayers for editorial assistance. This work was supported in part by NIH Grants NS 15013 and NS 20490.

[27] Cell Culture Assay of Biological Activity of Lactoferrin and Transferrin

By Shuichi Hashizume, Kazuhiko Kuroda, and Hiroki Murakami

Introduction

A red protein, later named lactoferrin, was first found in bovine milk by Sørensen and Sørensen[1] in 1939. It has been isolated from bovine and human milk by several workers.[2-6] Lactoferrin had been considered to be

[1] M. Sørensen and S. P. L. Sørensen, C. R. Trav. Lab. Carlsberg 23, 55 (1939).
[2] M. L. Groves, J. Am. Chem. Soc. 82, 3345 (1960).
[3] B. Johansson, Acta Chem. Scand. 14, 510 (1960).
[4] R. Grüttner, K. H. Schäfer, and W. Schröter, Klin. Wochenschr. 38, 1162 (1960).
[5] J. Montreuil, J. Tonnelat, and S. Mullet, Biochim. Biophys. Acta 45, 413 (1960).
[6] B. Blanc and H. Isliker, Bull. Soc. Chim. Biol. 43, 929 (1961).

a specific milk protein until its occurrence in the sputum of bronchitis patients was reported by Biserte *et al.*[7] in 1963. The presence of lactoferrin was also demonstrated in the intracellular compartment and in various other biological fluids, such as tears, saliva, urine, and seminal fluid.[8-10]

Lactoferrin shares with transferrin the property of reversibly binding two atoms of iron, and presents structural and functional homology with transferrin. Both are monomeric glycoproteins, having two similar oligosaccharide chains,[11,12] with molecular weights of about 8×10^4. Both molecules possess two independent metal binding sites, each of which can bind a ferric ion (Fe^{3+}) together with a bicarbonate anion.[13-15] The complete amino acid sequence of human serum transferrin[16] and partial amino acid sequence of lactoferrin[17,18] have been reported. In the regions where comparisons are possible between human lactoferrin and transferrin amino acid sequences (376 positions), 187 (49.7%) identical residues exist.[16] In spite of these similarities, however, no immunological cross-reactivity has been demonstrated between lactoferrin and transferrin of the same species.[3,5,6] In addition, the complex of ferric ion and lactoferrin is more stable at low pH than that of ferric ion and transferrin,[2,3,5] and the binding constant for ferric ion is 300 times larger for lactoferrin than for transferrin.[19] Altogether, lactoferrin is considered to be fairly similar to transferrin.

Transferrin was first demonstrated by Hayashi and Sato[20] to be an essential growth factor for rat pituitary cell line, GH_3, in serum-free medium and is now used for the culture of various cell lines in synthetic

[7] G. Biserte, R. Havez, and R. Cuvelier, *Expo. Annu. Biochim. Med.* **25**, 85 (1963).
[8] P. L. Masson, J. F. Heremans, and C. Dive, *Clin. Chim. Acta* **14**, 735 (1966).
[9] A. L. Schade, C. Pallavicini, and U. Wiesman, *Protides Biol. Fluids* **16**, 633 (1968).
[10] P. L. Masson, J. F. Heremans, and E. Schonne, *J. Exp. Med.* **130**, 643 (1969).
[11] L. Dorland, J. Haverkamp, B. L. Schut, J. F. G. Vliegenthart, G. Spik, G. Strecker, B. Fournet, and J. Montreuil, *FEBS Lett.* **77**, 15 (1977).
[12] J.-P. Prieels, S. V. Pizzo, L. R. Glasgow, J. C. Paulson, and R. L. Hill, *Proc. Natl. Acad. Sci. U.S.A.* **75**, 2215 (1978).
[13] P. Querinjean, P. L. Masson, and J. F. Heremans, *Eur. J. Biochem.* **20**, 420 (1971).
[14] J. Jollès, J. Mazurier, M.-H. Boutigue, G. Spik, J. Montreuil, and P. Jollès, *FEBS Lett.* **69**, 27 (1976).
[15] P. Aisen and I. Listowsky, *Annu. Rev. Biochem.* **49**, 357 (1980).
[16] R. T. A. MacGillivray, E. Mendez, J. G. Shewale, S. K. Sinha, J. Lineback-Zins, and K. Brew, *J. Biol. Chem.* **258**, 3543 (1983).
[17] M.-H. Metz-Boutigue, J. Mazurier, J. Jollès, G. Spik, J. Montreuil, and P. Jollès, *Biochim. Biophys. Acta* **670**, 243 (1981).
[18] J. Mazurier, M.-H. Metz-Boutigue, J. Jollès, G. Spik, J. Montreuil, and P. Jollès, *Experientia* **39**, 135 (1983).
[19] P. Aisen and A. Leibman, *Biochim. Biophys. Acta* **257**, 314 (1972).
[20] I. Hayashi and G. H. Sato, *Nature (London)* **259**, 132 (1976).

medium.[21-25] Iron plays an important role in cell growth and transferrin transports iron to the growing cells.[15] Models for this transport and for the cycling of transferrin have been proposed.[26,27]

It has been reported that lactoferrin may play a role in iron transport in the intestine, and transferrin may not,[28] though lactoferrin and transferrin are similar as described above. It has also been reported that lactoferrin binds to a specific macrophage receptor without competition with transferrin.[29,30] However, there had been no reports about lactoferrin as a growth factor until Hashizume et al.[31] reported that lactoferrin was an essential growth factor for lymphoma cell lines in serum-free medium. Subsequently it was reported that lactoferrin stimulated the growth of human colon adenocarcinoma cell line, HT29, in the presence of a low concentration iron solution.[32]

In this chapter, we describe the methods for cell culture in serum-free medium supplemented with lactoferrin, and compare the biological activities of lactoferrin with those of transferrin.

Cell Culture Assay

Purification of Lactoferrin from Human Milk

All subsequent procedures are carried out at 0–5°. Fresh human milk (30 ml) is defatted by centrifugation at 20,000 g for 20 min. The defatted milk (26 ml) is dialyzed overnight against 1 liter of 50 mM Tris–HCl, pH 8.0/50 mM KCl and applied to a column (0.7 × 25 cm) of Blue-Sepharose CL-6B (Pharmacia Fine Chemicals) previously equilibrated with the above buffer. The column was washed with 30 ml of the same buffer, and elution was performed with a gradient formed between the equilibration

[21] S. E. Hutching and G. H. Sato, *Proc. Natl. Acad. Sci. U.S.A.* **75,** 901 (1978).
[22] H. Murakami and H. Masui, *Proc. Natl. Acad. Sci. U.S.A.* **77,** 3464 (1980).
[23] D. Barnes and G. Sato, *Cell* **22,** 649 (1980).
[24] D. Barnes and G. Sato, *Anal. Biochem.* **102,** 255 (1980).
[25] M. Darmon, G. Serrero, A. Rizzino, and G. Sato, *Exp. Cell Res.* **132,** 313 (1981).
[26] A. Dautry-Varsat, A. Ciechanover, and H. F. Lodish, *Proc. Natl. Acad. Sci. U.S.A.* **80,** 2258 (1983).
[27] R. D. Klausner, G. Ashwell, J. van Renswoude, J. B. Harford, and K. R. Bridges, *Proc. Natl. Acad. Sci. U.S.A.* **80,** 2263 (1983).
[28] T. M. Cox, J. Mazurier, G. Spik, J. Montreuil, and T. J. Peters, *Biochim. Biophys. Acta* **588,** 120 (1979).
[29] B. Markowetz, J. L. van Snick, and P. L. Masson, *Thorax* **34,** 209 (1979).
[30] E. J. Campbell, *Proc. Natl. Acad. Sci. U.S.A.* **79,** 6941 (1982).
[31] S. Hashizume, K. Kuroda, and H. Murakami, *Biochim. Biophys. Acta* **763,** 377 (1983).
[32] M. Amouric, J. Marvaldi, J. Pichon, F. Bellot, and C. Figarella, *In Vitro* **20,** 543 (1984).

buffer and 3 M KCl in equilibration buffer (30 ml each), and fractions of 2.5 ml were collected. Two peaks were obtained during elution. The minor peak, eluted just after starting the gradient, had low growth-stimulatory activity and was discarded. The major peak (25 mg protein) of lactoferrin with high activity eluted around 1 M KCl and was pooled. Up to this step, electrophoretically pure lactoferrin can be obtained. However, in order to eliminate minor contaminants such as human serum albumin, the pooled major peak (5 ml) of the Blue-Sepharose step was dialyzed overnight against 1 liter of 10 mM Tris–HCl, pH 7.2/100 mM KCl. and applied to a column (0.7 × 6 cm) of DEAE-Sephadex A-50 (Pharmacia Fine Chemicals) previously equilibrated with the dialysis buffer. Lactoferrin, which passed through the column, was collected. Lactoferrin was not adsorbed to the column and contaminants were adsorbed under this condition. During this step, there was no loss of lactoferrin. The lactoferrin obtained (about 2.5 mg/ml) was dialyzed overnight against 2 liters of phosphate-buffered saline and was used for further experiments.

Purification of Lactoferrin from Bovine Milk

The purification is essentially carried out as described for human milk except that a step of acid precipitation at pH 4.6[33] is inserted prior to the Blue-Sepharose column chromatography step. Since the bovine milk contains 1/100 to 1/3 times as much lactoferrin as human milk,[34,35] the starting volume of the milk was increased to 300 ml.

Chemicals

Human transferrin, bovine insulin, and human serum albumin were purchased from Sigma Chemical Co. Ethanolamine, selenium, and 2-mercaptoethanol were from Wako Pure Chemical Industries, Ltd., Tokyo. Fetal calf serum (FCS) was from Armour Pharmaceutical Co.

Cells and Cell Culture

Human lymphoma cell lines [Bri 7, K 562, AL-60, RPMI 8226, NL, HYON, Jijoye (P-2003), CCRF-SB, WIL2-NS, CCRF-HSB-2 and CCRF-CEM], mouse lymphoma cell lines (NS-1, MPC 11, P3U1 and X63.6.5.3), and adhesive cell lines (human stomach cancer cell line, MKN-45, human colon cancer cell line, C-1, and human melanoma cell line, Bowes) were purchased from Dainippon Pharmaceutical Corp. Ltd. Stock lymphoma

[33] J. M. Ley and R. Jenness, *Arch. Biochem. Biophys.* **138**, 464 (1970).
[34] B. Reiter, *J. Dairy Res.* **45**, 131 (1978).
[35] R. Jenness, *J. Dairy Sci.* **63**, 1605 (1980).

cells were ordinarily maintained in a 1:1 mixture of Ham's F12 and Dulbecco's modified Eagle's media with 15 mM HEPES and 1.2 g/liter NaHCO$_3$ (SFFD), supplemented with 10% FCS. Adhesive cells were in 10% FCS-supplemented RPMI 1640. All stocks were grown in 60-mm dishes at 37° in a humidified atmosphere of 5% CO$_2$/95% air and subcultured every 3–5 days.

Assay of Growth-Stimulatory Activity with Lymphoma Cell Lines

Exponentially growing cells were harvested from serum-containing medium by low-speed centrifugation and washed with SFFD. The cells were inoculated into 2 ml of SFFD supplemented with lactoferrin or other factors at 1 × 10^5 cells per 35-mm dish. Cells were counted with a cell counter (Toa Medical Electronics Co. Ltd., Tokyo) at 3 days of growth, unless otherwise stated. Determinations were made from triplicate cultures. Growth ratio is expressed as percentage growth of cells in lactoferrin medium over that in serum medium.

Growth of Lymphoma Cell Lines with Human Lactoferrin Only

Nine human B cell, two human T cell, and four mouse B cell lines were examined for growth in SFFD supplemented with lactoferrin. Cell numbers were measured in log phase of growth. As shown in Table I, human B and T cell lines except for K 562 proliferated in the lactoferrin-supplemented SFFD while mouse cell lines did not. When heat stability of lactoferrin was examined, the growth-stimulatory activity of lactoferrin was fully maintained at temperatures up to 60° and completely lost above 80°.[31] The stimulatory activities of lactoferrin and serum with each cell line varied somewhat upon repeated experiments, but the percentage growth of each cell line in lactoferrin medium over that in serum medium remained almost constant. The variability in the stimulatory activity is considered to be due to the variation of physiological conditions of cells in each experiment, even if the same culture conditions were used.

Growth of Lymphoma Cell Lines with Various Growth-Stimulatory Factors in Addition to Human Lactoferrin or Transferrin

Mouse lymphoma cell lines, MPC 11 and NS-1 cells, proliferate in SFFD supplemented with a combination (ITES)[36] of insulin (I), transferrin (T), ethanolamine (E), and selenium (S) or another combination

---

[36] H. Murakami, H. Masui, G. H. Sato, N. Sueoka, T. P. Chow, and T. Kano-Sueoka, *Proc. Natl. Acad. Sci. U.S.A.* **79**, 1158 (1982).

TABLE I

GROWTH OF HUMAN AND MOUSE LYMPHOMA CELL LINES WITH
HUMAN LACTOFERRIN[a]

	Cell number ($\times 10^5$/dish)		Growth ratio (%)
Cell lines	In lactoferrin	In serum	
Human			
B cell lines			
Bri 7	5.5 ± 1.8	10.2 ± 3.1	54
K 562	0.3 ± 0.2	3.5 ± 2.4	9
AL-60	2.1 ± 0.4	2.8 ± 0.6	75
RPMI 8226	5.2 ± 1.3	10.0 ± 2.6	52
NL	2.3 ± 0.5	7.8 ± 1.3	29
HYON	0.9 ± 0.3	3.1 ± 0.5	29
Jijoye (P-2003)	2.6 ± 0.6	11.0 ± 1.7	24
CCRF-SB	2.5 ± 0.5	11.2 ± 3.1	22
WIL2-NS	7.4 ± 1.8	10.3 ± 2.1	72
T cell lines			
CCRF-HSB-2	1.2 ± 0.3	4.3 ± 1.1	28
CCRF-CEM	1.6 ± 0.4	6.7 ± 0.9	24
Mouse			
B cell lines			
NS-1	0.4 ± 0.1	15.8 ± 2.0	3
MPC 11	0.2 ± 0.2	17.2 ± 2.4	1
P3U1	0.0 ± 0.1	17.5 ± 2.7	0
X63.6.5.3	0.0 ± 0.0	18.0 ± 3.5	0

[a] AL-60 cells were counted at days 5 in a log phase. Lactoferrin was added to SFFD at a final concentration of 35 μg/ml. Each value presents three replicate determinations with standard deviation. Cell number was obtained by subtracting the control cell number in SFFD. Growth ratio indicates percentage growth of cells in lactoferrin-containing medium over that in serum-containing medium. (Data taken from Ref. 31.)

(MITES)[37] of 2-mercaptoethanol (M) in addition to ITES. Accordingly, growth of human and mouse lymphoma cell lines was next examined in the substituted media (ILES and MILES) with human lactoferrin (L) instead of transferrin (T) in the above defined media, as well as the original media. For most human cell lines, ILES and MILES were more effective than ITES and MITES (Table II). In the case of mouse cell lines, NS-1 cells multiplied in ILES and MILES, although lactoferrin alone did

[37] T. Kawamoto, J. D. Sato, A. Le, D. B. McClure, and G. H. Sato, Anal. Biochem. **130**, 445 (1983).

TABLE II
GROWTH OF HUMAN AND MOUSE LYMPHOMA CELL LINES WITH SEVERAL
FACTORS IN ADDITION TO HUMAN LACTOFERRIN OR TRANSFERRIN[a]

Cell lines	Cell number ($\times$ 10⁵/dish)			
	In ILES	In MILES	In ITES	In MITES
Human				
B cell lines				
Bri 7	4.8	5.3 (52)	−0.2	0.5
K 562	0.2	0.1	0.8 (23)	0.7
AL-60	2.6 (93)	2.3	1.0	0.8
RPMI 8226	5.6 (56)	5.3	1.7	1.5
NL	2.4	2.6 (33)	0.5	0.4
HYON	2.5	2.7 (87)	2.3	2.3
Jijoye (P-2003)	4.0	4.6 (42)	2.5	2.3
CCRF-SB	2.8	3.6 (32)	1.3	1.2
WIL2-NS	6.1 (59)	5.5	1.6	1.3
T cell lines				
CCRF-HSB-2	1.3 (30)	1.3 (30)	0.6	0.8
CCRF-CEM	1.2 (18)	1.1	0.7	0.8
Mouse				
B cell lines				
NS-1	7.5	7.9 (50)	5.8	5.7
MPC 11	1.8	2.3	13.7	14.3 (83)
P3U1	1.0	1.1 (6)	0.5	0.4
X63.6.5.3	0.6	0.7 (4)	0.2	0.2

[a] Each growth factor was added to SFFD in the indicated combinations at final concentrations of 5 μg/ml insulin (I), 35 μg/ml lactoferrin (L), 20 μM ethanolamine (E), 2.5 $\times$ 10⁻⁹ M selenium (S), 10 μM 2-mercaptoethanol (M), and 35 μg/ml transferrin (T). Values in parentheses indicate percentage growth of cells in indicated media over that in serum-containing medium. (Data taken from Ref. 31.)

not stimulate growth, indicating that insulin, ethanolamine, and selenium may be required for the proliferation of NS-1 cells. MPC 11 cells grew well in ITES and MITES, but not in ILES and MILES. None of the above media supported the growth of P3U1 or X63.6.5.3 cells.

Growth of Lymphoma Cell Lines with Bovine Lactoferrin

In large-scale cultures of human lymphoma cell lines human lactoferrin is required in considerable quantity. Substitution of bovine lactoferrin for human lactoferrin would make large-scale culture of cells feasible if it possessed growth-stimulatory activity similar to that of human lactofer-

TABLE III
GROWTH-STIMULATORY ACTIVITY OF
BOVINE LACTOFERRIN[a]

| | Cell number ($\times$ 10^5/dish) in bovine lactoferrin concentration | |
Cell lines	6 μg/ml	35 μg/ml
Human		
B cell lines		
Bri 7	3.1 (129)	3.2 (91)
K 562	0.0	0.2 (85)
AL-60	1.2 (89)	1.3 (87)
RPMI 8226	4.7 (104)	5.0 (106)
NL	1.7 (100)	2.0 (116)
HYON	0.6 (100)	0.9 (113)
Jijoye (P-2003)	2.1 (117)	2.5 (147)
CCRF-SB	0.3 (19)	2.0 (80)
WIL2-NS	5.4 (91)	7.1 (92)
T cell lines		
CCRF-HSB-2	0.9 (100)	1.2 (100)
CCRF-CEM	1.2 (86)	1.6 (100)
Mouse		
B cell lines		
NS-1	0.2 (67)	0.5 (125)
MPC 11	0.4 (300)	2.4 (967)
P3U1	0.0	0.0
X63.6.5.3	0.0	0.0

[a] Bovine lactoferrin was added to SFFD at the indicated concentration. Values in parentheses present percentage growth of cells in bovine lactoferrin-containing medium over that in human lactoferrin-containing medium. (Data taken from Ref. 31.)

rin. Bioassay indicated that bovine lactoferrin possesses almost the same level of activity as human lactoferrin for all human and mouse lymphoma cell lines except MPC 11 cells (Table III). In the case of MPC 11 cells, the activity with bovine lactoferrin was several times higher than that with human lactoferrin.

Assay of Growth-Inhibitory Activity with Adhesive Cell Lines

Exponentially growing stock cells are trypsinized with 0.25% trypsin solution and washed by low-speed centrifugation. The cells are inoculated at 1×10^5 cells per 35-mm dish and cultivated for 4 days. After 4 days culture, the cells were detached with 0.25% trypsin solution. The cell numbers were determined as described above.

The effect of lactoferrin or transferrin, which was added before or after the adhesion of the cells, on the growth was examined in RPMI 1640 supplemented with 1.0 mg/ml human serum albumin (HSA medium) or 5% FCS (FCS medium), unless otherwise stated. Effects on the growth after the adhesion of the cells were examined after overnight preculture in the FCS-containing medium and washing with RPMI 1640 prior to the addition of lactoferrin or transferrin.

Growth of Stomach Cancer Cell Line, MKN-45, Colon Cancer Cell Line, C-1, and Melanoma Cell Line, Bowes, with Lactoferrin or Transferrin

Lactoferrin had a growth-inhibitory activity on MKN-45 and C-1 cells at a concentration of more than 10 μg/ml, but transferrin had almost no effect, when added to the HSA medium before the adhesion of the cells. In FCS medium, lactoferrin also had a small inhibitory activity at 50 μg/ml or higher, but transferrin had no effect. Effects of lactoferrin or transferrin on the growth of Bowes cells were examined in SFFD supplemented with 9 μg/ml of insulin, 100 μ*M* ethanolamine, and 10^{-8} *M* selenium (IES). Bowes cells did not adhere to the dish in IES supplemented with 1 or 35 μg/ml of lactoferrin, but adhered and grew well in IES supplemented with the same concentrations of transferrin.

Cell Culture Assay of Lymphoma Cell Line, WIL2-NS, in Iron-free
    SFFD Supplemented with Lactoferrin or Transferrin

Iron-containing standard SFFD supported growth of WIL2-NS cells better than iron-free SFFD at all concentrations of lactoferrin and transferrin (Fig. 1). Optimum levels of lactoferrin and transferrin were about 30 and 100 μg/ml, respectively, in standard and iron-free SFFD. Iron (Fe^{3+})-saturated lactoferrin or transferrin, and apolactoferrin or apotransferrin were prepared as described below, and their effects in iron-free SFFD were next examined. Iron-saturated lactoferrin or transferrin had more growth-stimulatory effect than apolactoferrin or apotransferrin in iron-free SFFD, but all had same effect in standard SFFD. These data indicate that the growth-stimulatory activity of lactoferrin or transferrin is enhanced by the addition of iron and suggest that the transport of iron is at least a part of their functions.

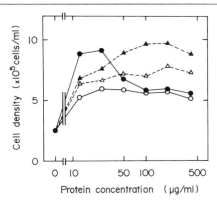

FIG. 1. Effect of iron in SFFD on the growth of WIL2-NS cells. Effects of varied concentrations of lactoferrin (○,●) or transferrin (△,▲) were examined in iron-free SFFD (open symbols) and iron-containing standard SFFD (closed symbols).

Assay of Binding of $^{59}Fe^{3+}$-Saturated Lactoferrin or Transferrin to Lymphoma Cell Lines

Five cell lines with different characteristics for growth in lactoferrin- or transferrin-supplemented medium were selected for binding assay. Bri 7 grows well in lactoferrin-supplemented medium but poorly in transferrin-supplemented medium. WIL2-NS grows in both media supplemented with lactoferrin and transferrin. K 562 and MPC 11 grow well in transferrin-supplemented medium but poorly in lactoferrin-supplemented medium. P3U1 does not grow in medium supplemented with lactoferrin or transferrin.

Preparation of $^{59}Fe^{3+}$-Saturated Lactoferrin or Transferrin

Apolactoferrin and apotransferrin were prepared as described.[38,39] The preparation of $^{59}Fe^{3+}$-saturated lactoferrin or transferrin was essentially carried out according to Klausner et al.[40] as follows. $^{59}FeCl_3$ (0.1 mCi), 100 μl, was added to 10 μl of 1 M disodium nitrilotriacetate. This solution was combined with 250 μl of 10 mg/ml of apolactoferrin or transferrin in 0.25 M Tris–HCl, pH 8.0, containing 10 μM $NaHCO_3$, and incubated for

[38] P. L. Masson and J. F. Heremans, *Protides Biol. Fluids* **14**, 115 (1966).
[39] H. Huebers, E. Csiba, B. Josephson, E. Huebers, and C. Finch, *Proc. Natl. Acad. Sci. U.S.A.* **78**, 621 (1981).
[40] R. D. Klausner, J. van Renswoude, G. Ashwell, C. Kempf, A. N. Schechter, A. Dean, and K. R. Bridges, *J. Biol. Chem.* **258**, 4715 (1983).

1 hr at room temperature. This ^{59}Fe-saturated lactoferrin or transferrin was then passed through a Sephadex G-25 column. The ratios of $A_{465\ nm}/A_{280\ nm}$ for the lactoferrin and transferrin were 0.045 and 0.046, respectively. The specific activities for ^{59}Fe-saturated lactoferrin and transferrin were 362 and 341 cpm/pmol protein, respectively. The molar ratios of Fe/lactoferrin and Fe/transferrin were calculated to be 1.7 and 1.6, respectively, when based upon the specific activity of ^{59}Fe.

Binding Assay

The assay was essentially carried out according to Klausner et al.[40] Cells at the log phase were harvested, and cell viability, as judged by the exclusion of 0.2% nigrosine, was always more than 90%. Cells were washed twice with RPMI 1640 containing 10 mM HEPES and 10 mg/ml bovine serum albumin by low-speed centrifugation, and then suspended at a concentration of about 10^7 cells/ml in above medium. The suspended cells, 0.5 ml, were incubated with ^{59}Fe^{3+}-saturated lactoferrin or transferrin. After the incubation, 1 ml of the same medium was added and the cells were pelleted by centrifugation at 1000 rpm for 1 min in a refrigerated centrifuge (HITACHI, 05PR-22). The pellet was washed twice with the medium and counted by Auto-gamma Counter (Packard, PGD).

Association of ^{59}Fe-Saturated Lactoferrin or Transferrin with Lymphoma Cell Lines, Bri 7

The assay was carried out at 0 and 37° (Fig. 2). At 0°, binding of ^{59}Fe-saturated lactoferrin to Bri 7 cells proceeded gradually up to 60 min and then reached a plateau, while the binding of ^{59}Fe-saturated transferrin was more rapid and reached a plateau within 20 min as reported in experiments using ^{125}I-labeled transferrin.[40]

When the incubation was performed at 37°, the ^{59}Fe-saturated lactoferrin associated with Bri 7 increased gradually up to 40 min and then reached a plateau at 64% of that attained at 0°. In contrast, cell-associated counts from ^{59}Fe-saturated transferrin continued to rise almost linearly over a period of 120 min at 37° and this level was seven times higher than that attained at 0°. Therefore, the uptake of Fe^{3+} ion from Fe^{3+}-lactoferrin is considered to be different from that of Fe^{3+}-transferrin. Similar phenomena were also observed with the other cell lines.

Dissociation Constant (K_d) and Binding Sites per Cell of Lymphoma Cell Lines

The incubation was carried out at 0° for 90 min to complete the equilibrium of the binding, and equilibrium binding data were subjected to Scat-

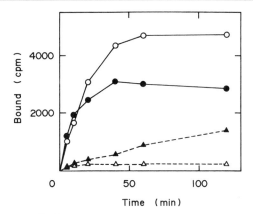

FIG. 2. Association of $^{59}Fe^{3+}$-saturated lactoferrin or transferrin to Bri 7 cells. Cells (8 × 10^6 cells) were incubated with lactoferrin (○,●) or transferrin (△,▲) at a final concentration of 35 μg/ml at 0° (open symbols) and 37° (closed symbols). At the indicated times, samples were assayed as described in the text.

chard analysis.[41] ^{59}Fe-saturated lactoferrin displayed K_d values of about 10^{-6} M with more than 10^6 binding sites per cell. On the other hand, K_d values for ^{59}Fe-saturated transferrin were about one order lower, at 10^{-7}–10^{-8} M, and the receptor numbers were several times lower, at less than 10^6 sites per cell. Although there were distinct differences in K_d values and binding sites/cell between lactoferrin and transferrin for each cell line, these differences did not show a pattern among the five cell lines described above, exhibiting different characteristics for transferrin- and lactoferrin-stimulated growth.

Binding Competition between Lactoferrin and Transferrin

Bri 7, WIL2-NS, and K 562 lymphoma cells (3 × 10^6 cells) were incubated with $^{59}Fe^{3+}$-saturated lactoferrin or transferrin at a final concentration of 35 μg/ml at 0° for 90 min. In addition to these ^{59}Fe-saturated materials, nonradioactive Fe-saturated lactoferrin, apolactoferrin, Fe-saturated transferrin, or apotransferrin was added at a final concentration of 315 μg/ml, and the amount of cell-associated radioactivity was determined.

^{59}Fe-saturated lactoferrin bound to the cell lines without competing with Fe-saturated transferrin and apotransferrin, and ^{59}Fe-saturated transferrin also did not compete with Fe-saturated lactoferrin and apolactoferrin. These data suggest that lactoferrin receptors are specific for lactoferrin and different from transferrin receptors.

[41] G. Scatchard, *Ann. N.Y. Acad. Sci.* **51,** 660 (1949).

Conclusion

1. Lactoferrin has a growth-stimulatory activity for lymphoma cell lines in serum-free medium, and this effect is somewhat different from that of transferrin.

2. The growth-stimulatory activity of lactoferrin and transferrin for lymphoma cell lines is enhanced by the addition of iron to iron-free medium.

3. Lactoferrin has a marked growth-inhibitory activity for all adhesive cell lines tested, while transferrin has a growth-stimulatory or small growth-inhibitory activity.

4. $^{59}Fe^{3+}$-saturated lactoferrin binds gradually to lymphoma cell lines at 0°, while the binding of $^{59}Fe^{3+}$-saturated transferrin is more rapid.

5. $^{59}Fe^{3+}$-saturated lactoferrin associated with lymphoma cell lines at 37° reaches a plateau within 40 min, while the associated counts from $^{59}Fe^{3+}$-saturated transferrin continued to rise almost linearly over a period of 120 min.

6. $^{59}Fe^{3+}$-saturated lactoferrin displays K_d values of about 10^{-6} M with more than 10^6 binding sites per lymphoma cell. On the other hand, K_d values for $^{59}Fe^{3+}$-saturated transferrin are about one order lower at 10^{-7}–10^{-8} M and the receptor numbers are less than 10^6 sites per cell.

7. Lactoferrin binds to lymphoma cell lines without competition from transferrin.

[28] Biological Activity of Human Plasma Copper-Binding Growth Factor Glycyl-L-histidyl-L-lysine

By LOREN PICKART and STEVE LOVEJOY

Background and Preparation of GHL and GHL-Cu

Biological Actions of GHL and GHL-Cu

Glycyl-L-histidyl-L-lysine : copper(II) (GHL-Cu) is a growth factor isolated from human plasma. The peptide portion of the complex, GHL (glycyl-L-histidyl-L-lysine), has an affinity for copper(II) equivalent to that of the copper transport site on albumin and addition GHL to culture medium presumably results in a significant conversion into GHL-Cu by chelation of ionic copper from the culture medium. GHL is used for *in vitro* culture of a diverse variety of cells and organs[1-3] where it appears to function as a transporter of ionic copper(II) and possibly other transition metal ions in serum-free or low-serum media.[4,5] In addition, GHL-Cu possesses significant superoxide dismutase-like activity (rate constant about 5×10^8 or about 20 to 25% of activity of enzymatic Cu,Zn-superoxide dismutase on a molar basis), an activity associated with tissue-protective and antitrauma effects and which may serve as the basis of some of GHL's diverse biological actions. In general, GHL works best with cell cultured on biological substrata (e.g., collagen, fibronectin) and often in cases where GHL gives a marginal effect, the use of GHL-Cu produces a marked response. Rapidly growing cells produce GHL or a similar peptide and recent work suggests a physiological role for GHL-Cu related to the processes of wound healing and tissue repair (e.g., angiogenesis, neurogenesis, chemoattraction of capillary endothelial and mast cells, acceleration of wound healing).

Structure–Activity Relationships in GHL and GHL-Cu

Physicochemical studies using a variety of techniques have indicated that GHL binds copper(II) to form a tridentate, triaza complex with copper being chelated to the α-amino nitrogen of glycine, the nitrogen in the amide bond between the glycyl and histidyl residues, and an unprotonated nitrogen in the imidazole ring of the histidyl residue (see Fig. 1).[5-11] The lysyl side chain is essential for activity. Analogs of that lack this residue such as glycyl-L-histidine and glycyl-L-histidyl-L-glycine form a similar

[1] L. Pickart, *In Vitro* **17**, 459 (1981).
[2] L. Pickart, in "Chemistry and Biochemistry of Amino Acids, Peptides, and Proteins" (B. Weinstein, ed.), p. 99. Dekker, New York, 1981.
[3] L. Pickart, *Lymphokines* **8**, 425 (1983).
[4] L. Pickart, J. Freedman, J. Loker, W. J. Peisach, C. M. Perkins, R. E. Stenkamp, and B. Weinstein, *Nature (London)* **288**, 715 (1980).
[5] J. Fernandez-Pol, in "Microbiology 1983" (D. Schlessinger, ed.), p. 313. American Society of Microbiology, Washington, D.C., 1983.
[6] S. Lau and B. Sarkar, *Biochem. J.* **199**, 649 (1981).
[7] J. Freedman, L. Pickart, B. Weinstein, W. B. Mims, and J. Peisach, *Biochemistry* **21**, 4540 (1982).
[8] E. Y. Kwa, B. S. Lin, N. J. Rose, B. Weinstein, and L. Pickart, *Pept. Struct. Funct.* **8**, 805 (1983).
[9] J. P. Laussac, R. Haran, and B. Sarkar, *Biochem. J.* **209**, 533 (1983).
[10] C. Perkins, R. Stenkamp, N. Rose, B. Weinstein, and L. Pickart, *Inorg. Chim. Acta* **67**, 93 (1984).
[11] M. J. A. Rainier and B. M. Rode, *Inorg. Chim. Acta* **92**, 1 (1984).

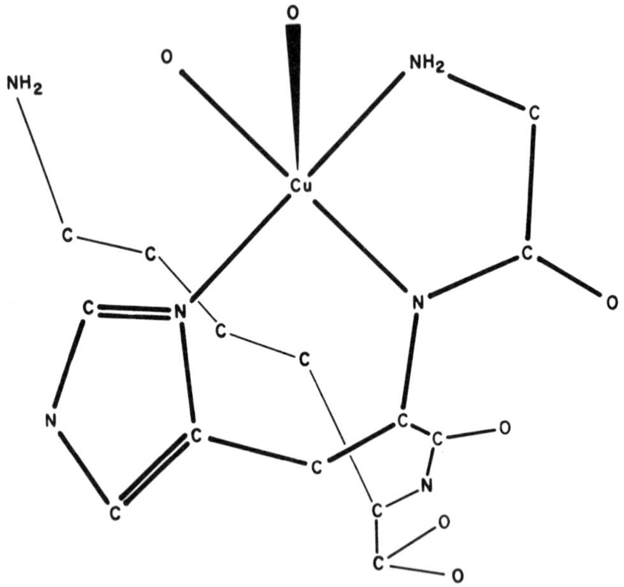

FIG. 1. Proposed structure of GHL-Cu in solution. This figure is based on X-ray analysis of GHL-Cu crystals and finds the copper ion bound to three nitrogen of GHL plus two oxygens in water. Other procedures (proton magnetic resonance, electron spin resonance, electron spin echo, minimum energy computer modeling) suggest essentially the same structure persists in solution. However, it is probable in physiological milieu that an amino acid such as histidine or cysteine displaces an oxygen and binds to the fourth equatorial bonding orbital of copper.

complex with copper(II)[8,11] but lack bioactivity.[12-14] Biological activity is retained after attachment of hydrophobic groups (e.g., octyl, benzyl) to the carboxyl-terminus of GHL.[15]

Affinity measurements indicate that GHL-Cu should be a reasonably stable complex *in vivo* under physiologic conditions. GHL has an amino acid structure similar to the copper ion transport site on human albumin[4] and has an affinity for Cu(II) equivalent to that of the copper ion transport site on plasma albumin (pK 16.2).[6] Measurements of the pK between GHL and copper(II) have given pK values of 14.9,[16] 16.2,[11] and 16.4[6] by potenti-

[12] L. Pickart and M. M. Thaler, *FEBS Lett.* **104**, 119 (1979).
[13] M. V. Williams and Y. Cheng, *Cytobios* **27**, 19 (1980).
[14] T. Poole and B. Zetter, *Cancer Res.* **43**, 5857 (1983).
[15] S. Lovejoy, Ph.D. thesis. Department of Chemistry, University of Washington, Seattle, Washington, 1985.
[16] P. M. May, J. Whittaker, and D. R. Williams, *Inorg. Chim. Acta* **80**, L5 (1983).

ometric titration, 16.3 by electron spin resonance spectroscopy,[11] and 16.5 with a copper-ion selective electrode.[17] It is likely GHL-Cu is even more stable than these pK values suggest since the fourth (unbound) equatorial orbital on copper in GHL-Cu can bind to other ligands and further stabilize the complex. For example, the pK of the GHL-Cu–histidine complex is 29.0.[6]

Preparation and Handling of GHL and GHL-Cu

GHL is widely available from commercial sources. Detailed information on the synthesis of GHL and analogs plus copper(II) complexes of these compounds is available in published theses by Lovejoy,[15] Loker,[18] Perkins,[19] and Kwa.[20] Commercial GHL is about 95% pure and useful for most cell culture purposes, but often includes small amounts of mildly neurotoxic materials (as measured by behavior after intracranial injection, tail flick assays, and gripping ability of mice on spinning disks). Most of this material can be removed by dissolving GHL in glass-distilled water (50 mg/ml), centrifuging at 20,000 g for 1 hr at 3°, then lyophilizing the supernatant. This removes poorly water-soluble material (probably GHL that was not completely deblocked of protecting groups during the final synthetic step). The supernatant is lyophilized, then passed through a Sephadex G-10 column at 3° in a solvent of 0.5% acetic acid in glass distilled water. The main peak that elutes behind the solvent front (monitored by absorption at 254 nm) is lyophilized to dryness.

GHL-Cu is prepared by combination of purified GHL with equimolar cupric acetate, followed by neutralization with 0.1 N sodium hydroxide, and centrifugation at 5000 g for 30 min at 3° to remove insoluble material [usually excess copper(II) as its hydroxide]. The supernatant is passed through a G-10 column in a solvent of glass-distilled water and the elution peak absorbing at 600 nm collected and lyophilized to obtain GHL-Cu. Crystalline GHL-Cu is prepared by dissolving purified GHL (30 mg, 88 μmol) in an aqueous solution copper(II) acetate (0.3 ml, 0.3 M). Ethanol (1.26 ml) is added and vessel walls scratched to initiate crystallization of dark blue-purple crystals. The mother liquor is decanted and the crystals

[17] A. Avdeef, L. Pickart, and B. Weinstein, unpublished observations.
[18] W. J. Loker, Ph.D. thesis. Department of Chemistry, University of Washington, Seattle, Washington, 1981.
[19] C. Perkins, Ph.D. thesis. Department of Biological Structure, University of Washington, Seattle, Washington, 1982.
[20] E. Kwa, Ph.D. thesis. Department of Chemistry, University of Washington, Seattle, Washington, 1983.

dissolved by addition of distilled water (0.2 ml). Ethanol (0.4 ml) is slowly introduced to reach a cloud point. After standing, dark purple-blue octahedral crystals formed that were isolated by decanting the mother liquor.[10] Microanalysis to determine amino acid and copper content confirms the composition of the product.

Both GHL and GHL-Cu are stable for more than 4 years when stored at −20° in a desiccator over $CaCl_2$. GHL is hygroscopic, hence must be weighed rapidly. Aqueous solutions (50 mg/ml) in buffer (pH 7.4) are stable at 3° for at least 6 months. The use of reasonably concentrated solutions (>1 mg/ml) during sterile filtration and for dilution is recommended because of absorption to glass and plastic surfaces at low (<20 μg/ml) concentrations. Often GHL is more effective on sparse cell cultures; this may reflect endogenous synthesis of adequate quantities of GHL by rapidly growing cells.[14]

The Uses of GHL in Culture Systems

Neuronal Cells

GHL facilitates the growth of cultured central and peripheral neurons plus promoting axonal and dendrite outgrowth. The use of GHL in low-serum medium also promotes the growth and survival of neurons while decreasing the glial cell content, thus markedly increasing the neuron to glial cell ratio in culture.

For cultivation of chick cerebral neurons,[21,22] a medium composed of Eagle's minimum essential medium (MEM) supplemented with 500 mg% glucose, 1% fetal calf serum, and GHL (200 ng/ml) is used. Embryonic cerebral cells are obtained from 7-day-old cerebral hemispheres and plated onto Petri dishes covered with denatured collagen and covered with medium. After 4 days in this medium, there is strong nerve fiber differentiation and axonal outgrowth while glial cell growth is greatly suppressed resulting in an essentially pure neuronal culture.

GHL has also been used to stimulate neuronal outgrowth of chick embryo ganglion trigeminale cells cultured in a similar medium but with 5% placental serum and GHL (10 ng/ml).[23]

---

[21] M. Sensenbrenner, G. G. Jaros, G. Moonen, and P. Mandel, *Neurobiology* **5**, 207 (1975).
[22] M. Sensenbrenner, G. G. Jaros, G. Moonen, and B. J. Meyer, *Experientia* **36**, 660 (1980).
[23] G. Lindner, G. Gross, W. Halle, and P. Mandel, *Z. Mikrosk.–Anat. Forsch.* **93**, 820 (1979).

Glomerular Kidney Cells

Primary glomerular kidney cells may be cultured in media containing high concentrations of calf serum or alternately on fibronectin in a defined medium containing GHL.[24] Use of GHL as the kidney cell growth factor eliminates some of the dedifferentiation of glomerular cells observed with serum-containing medium.

Young guinea pigs (Hartley, 300 g) are used as the source of glomeruli, of which 2000 are added to 9.6 cm² wells in 2 ml medium. Culture medium consists of Waymouth's MB 752/1 medium supplemented with sodium pyruvate (1%), nonessential amino acids (1%, Gibco), penicillin–streptomycin (100 units/ml), GHL (50 μg/ml), and fibronectin (10 μg/ml). During the several days it takes glomeruli to attach, 1 ml of medium in the well is replaced daily with fresh medium. After attachment, cells are fed three times weekly. Four distinct types of well-differentiated glomerular cells [I, II(a), II(b), II(c)] are observed under these culture conditions with type II(a) becoming predominant.

Organ Culture on Collagen Pads

GHL is used for the culture of a variety of organs. Organ pieces (human liver,[25,26] human kidney,[27] human placental tissue,[28] normal and neoplastic endometrium,[29] plus human kidney cells[27]) are cultured on gelatin (collagen) foam slices (Spongostan, Ferrosan, Sweden) in plastic Petri dishes using Parker's 199 medium supplemented with GHL (20 ng/ml). Tissue explants of 1 to 1.5 mg are used. In this medium, liver explants demonstrate linear production of export proteins (albumin, orosomucoid, α_1-antitrypsin, protein HC) for 10 days while kidney explants remain viable for 20 days and kidney cells for 35 days.

In Vitro Model of Fibroplasia

This *in vitro* model of fibroplasia permits simultaneous quantification of fibroblast proliferation, migration, and collagen synthesis in a GHL-

[24] T. D. Oberley, P. J. Murphy, B. W. Steinert, and R. M. Albrecht, *Virchows Arch. B* **41**, 145 (1982).

[25] S. Eriksson, R. Alm, and B. Astedt, *Biochim. Biophys. Acta* **542**, 496 (1978).

[26] L. Tejler, S. Eriksson, A. Grubb, and B. Astedt, *Biochim. Biophys. Acta* **542**, 506 (1978).

[27] B. Astedt, G. Barlow, and L. Holmberg, *Thromb. Res.* **11**, 149 (1977).

[28] L. Holmberg, I. Lecander, B. Persson, and B. Astedt, *Biochim. Biophys. Acta* **544**, 128 (1978).

[29] L. Svanberg and B. Astedt, *Experientia* **35**, 818 (1979).

based culture medium and facilitates the study of the influence of various healing and inflammatory factors on the events in fibroplasia.[30]

Fibroblasts are prepared from tendon plugs (2 mm diameter) prepared from flexor digitorium profundus tendons removed from the long digit of female Leghorn chickens. Plugs are placed in culture wells and first covered with 10 μl thrombin solution (2 mg/ml in DMEM), then 50 μl fibrinogen solution (3 mg/ml), allowed to clot for 30 min at 37° in a humidified atmosphere, then covered with 1 ml of medium used to maintain fibroblasts in a quiescent state. This medium consists of Dulbecco's modified Eagle's medium to which are added GHL (200 ng/ml), aprotinin (100 kIU/ml, Trasylol), and ascorbate (0.1 mM). After 48 hr, exogenous factors to be assayed (e.g., calf serum, platelet lysate) are added. Fresh ascorbate is added every 24 hr.

Fibroblast migration is quantified by measurement of cellular outgrowth from the tendon plug by planimetry of photomicrographs. Fibroblast proliferation was measured as pulsed incorporation of [^{125}I]iododeoxyuridine (2000 Ci/mmol) into DNA in the presence of 10^{-5} M 5-fluorodeoxyuridine. Collagen synthesis is measured by the incorporation of [^{3}H]proline into cellular protein for 6 hr followed by treatment with cold 5% trichloroacetic acid to precipitate protein and radioactivity of the precipitate determined. Treatment with collagenase to lyse collagen followed by another trichloroacetic acid precipitation permits a correction for proline incorporation into noncollagen protein.

Immunologically Related Cells

The inclusion of GHL in culture media used for assay of cellular immune function often permits the elimination or a marked reduction in the serum requirement. For example, good lymphocyte transformation responses can be obtained in medium with 20 ng/ml GHL, eliminating the normally required 5 to 10% serum in medium. This serum reduction can minimize many of the experimental complications associated with serum use. For eosinophils, macrophages, and mast cells, the serum requirement is reduced from 10 to 20% down to 1%.

Lymphocyte Transformation.[31] Lymphocytes obtained from spleen tissue of antigen-sensitized rats are preincubated with RMPI 1640 medium supplemented with 5% fetal calf serum for 45 min, washed 3 times with medium to remove serum, then cultured in medium supplemented with GHL (20 ng/ml), L-glutamine (290 μg/ml), and antibiotics. Lymphocyte

[30] M. F. Graham, R. F. Diegelman, and I. K. Cohen, *Proc. Soc. Exp. Biol. Med.* **176,** 302 (1984).

[31] A. Haque and A. Capron, *Nature (London)* **299,** 361 (1982).

transformation is performed in microtiter plates using 0.1 ml medium containing 0.5×10^6 cells. Antigens are added to the wells and after 4 hr pulsed with 1 μCi [^{3}H]thymidine. Cells are collected on a Titertek cell collector (Skatron) and incorporated radioactivity determined in a scintillation counter.

Eosinophil Preparations.[32,33] Eosinophils are collected from normal rats by rinsing peritoneal cavities in culture medium [Eagle's MEM, 1% heat-inactivated fetal calf serum, GHL (20 ng/ml)] plus 25 IU/ml calcium heparinate. Eosinophil-rich fractions are prepared by centrifugation at 1800 g at 4° and washed twice in the medium, then added to plastic Petri dishes for 2 hr at 37° in a 5% CO_2 atmosphere. The nonadherent cells are collected and pooled. This population consists of 47% eosinophils and 9% mast cells and is maintained in the GHL-containing medium during further purification steps and during assay procedures.

Macrophage Cytotoxicity.[34,35] Macrophage cytotoxicity toward ^{51}Cr-labeled schistosomula was determined by collecting normal rat macrophages by peritoneal washings. The cells were cultured in Eagle's MEM with GHL (20 ng/ml) and 1% heat-inactivated fetal calf serum at 37° for 2 hr. Nonadherent cells are removed by three washings with medium. The adhering macrophage cells are cultured overnight in the medium. Macrophages are activated by incubation with MEM medium containing 20% serum from infected rats for 3 to 6 hr, after which the labeled schistosomula are added. Macrophage cytotoxicity is determined by the percentage of chromium released after an overnight incubation.[34]

A GHL-containing medium (Hanks–Wallace medium, 2 mM glutamine, 20 ng/ml GHL) is also effective for studies of IgE stimulation of human alveolar macrophages. The endpoint in IgE stimulation is the release of lysosomal β-glucuronidase and neutral proteases.[35]

Mast Cell Function.[36] Mast cell degranulation induced by nonspecific degranulators or anaphylactic antibodies reacting with antigen is quantified by incubation of rat or mouse mast cells in a medium consisting of Eagle's MEM, GHL (20 μg/ml), calcium heparinate (50 U/ml), and glutamine (2 mM). The mast cells are incubated for 30 min at 37° in the medium plus 2 μCi [^{3}H]serotonin for 10^6 cells, then washed 3 times to remove unincorporated medium. The percentage of radioactivity released

[32] M. Capron, J. Rousseaux, C. Mazingue, H. Bazin, and A. Capron, *J. Immunol.* **121**, 2518 (1978).
[33] M. Capron, A. Capron, G. Torpier, H. Bazin, D. Bout, and M. Joseph, *Eur. J. Immunol.* **8**, 127 (1978).
[34] M. Joseph, J. P. Dessaint, and A. Capron, *Cell. Immunol.* **34**, 247 (1977).
[35] M. Joseph, A. B. Tonnel, A. Capron, and C. Voisin, *Clin. Exp. Immunol.* **40**, 416 (1980).
[36] C. Mazingue, J. P. Dessaint, and A. Capron, *J. Immunol. Methods* **21**, 65 (1978).

into the medium by after addition of degranulating substances gives the measure of degranulation.

Hepatocytes and Hepatoma Cells

The viability of hepatocytes, when cultured in low-serum medium (0.5 to 1.0% fetal calf serum in 90% Eagle's MEM and 10% Swim's S-77 medium) is enhanced, and the growth of hepatoma cells (HTC$_4$ or AH-130) stimulated, by addition of GHL (20 ng/ml).[37–39] This action on hepatoma cells is more pronounced when the GHL is prechelated to copper(II) and iron(II) suggesting the primary role of GHL in this system is transition metal ion transport.[40] The addition of GHL (2 ng/ml) to isolated rat hepatocytes cultured in L-15 medium results in a 27% increase in α_1-macroglobulin synthesis while reducing transferrin synthesis 29%.[41]

Mast Cell Migration Assay

GHL is a potent chemoattractant for mast cells.[14] Plastic tissue culture dishes are pretreated with a 1% aqueous solution of bovine serum albumin, then 2.5 ml of 2.4% agarose is mixed with a 2× concentrated tissue culture medium 199 supplemented with 0.2% albumin at 50°, is poured into the dishes. After cooling to room temperature, a gel forms. With an Ouchterlony punch and template, 3-mm wells are punched 3 mm apart in a concentric circle around a center well. Rat mast cells (3 × 10^4) are added to the center well in 10 μl medium and the outer wells receive solutions of GHL or other chemoattractant substances. After 4 hr, the dishes are fixed in 10% formalin, dried, and stained with Bismark brown. Migration patterns are quantified using a stage micrometer at 20–30× magnification to measure linear distance.

Ascaris suum and Litomosoides carinii Larvae

The culture of these parasitic organisms through their third to fourth stage of morphogenesis can be achieved using serum supplemented medium. However, their cultivation in serum-free GHL-containing medium greatly facilitates the collection of stage-specific somatic and excretory/

[37] L. Pickart and M. M. Thaler, Nature (London) New Biol. **243,** 85 (1973).
[38] L. Pickart, L. Thayer, and M. M. Thaler, Biochem. Biophys. Res. Commun. **54,** 562 (1973).
[39] T. Aoyagi and H. Umezawa, Proc. FEBS Meet. **61,** 89 (1980).
[40] L. Pickart and M. M. Thaler, J. Cell. Physiol. **102,** 129 (1980).
[41] F. M. Fouad, M. A. E. Fattah, R. Scherer, and G. Ruhenstroth-Bauer, Z. Naturforsch. **36c,** 350 (1980).

secretory antigens that are used for the development of antiparasite vaccines.

Ascaris suum larvae are collected from the lungs of rabbits 10 days after infection with 100,000 embryonated eggs. Third-stage larvae are incubated in Medium 199 supplemented with GHL (14 ng/ml), penicillin (300 units/ml), and streptomycin (0.3 mg/ml) under an atmosphere of $N_2-CO_2-O_2$ (90 : 5 : 5) at 37°. After 12 days the larvae grow in length and moult to the fourth developmental stage. Larval proteins isolated from used growth medium are collected, and a larval antigen isolated that produces immunological protection from *Ascaris* infection.[42,43]

Inocula of *L. carinii* (stage 3) are recovered from rats 7 days after exposure of rats to mites infected with the nematode. Larvae are collected aseptically by lavaging the pleural cavities of the rats with warm Lebovitz's L-15 medium. The larvae progress from stage 3 to stage 4 after 4 to 10 days culture in L-15 medium containing 2 μg/ml GHL.[44]

T-Strain Mycoplasma (Ureaplasma urealyticum)

The incorporation of GHL into the basic indicator broth used for clinical detection of the T-strain mycoplasma *Ureaplasma urealyticum* associated with urinary tract infections results in an increased initial growth rate and higher final titer of the organism. This improved detection broth consists of PPLO broth (2.1 g, Difco), yeast extract (0.1 g, Difco), 0.4% bromothymol blue solution (1 ml), horse serum (10 ml), urea (0.025% w/v), and GHL (20 ng/ml).[45]

Coelomomyces punctatus

GHL added to basic media stimulates the initial growth and development of the fungi, *Coelomomyces punctatus*, known to parasitize and cause high levels of mortality in natural populations of mosquito larvae. The fungi are isolated from the gut of mosquito larvae of *Anopheles quadrimaculatus* and *Cyclops vernalis*, then cultured in Mitsuhashi–Maramorosch insect tissue culture medium which has been conditioned by growing Varma's *Anopheles stephensi* cells in it for 3 weeks, and supplemented with 20% fetal calf serum and GHL (20 ng/ml).[46]

[42] B. E. Stromberg, P. B. Khoury, and E. J. L. Soulsby, *Int. J. Parasitol.* **7,** 149 (1977).
[43] B. E. Stromberg, *Int. J. Parasitol.* **9,** 307 (1979).
[44] P. D. Nelson, D. J. Weiner, B. E. Stromberg, and D. Abraham, *J. Parasitol.* **68,** 971 (1982).
[45] J. A. Robertson, *J. Clin. Microbiol.* **7,** 127 (1978).
[46] J. M. Castillo and D. W. Roberts, *J. Invertebr. Pathol.* **35,** 144 (1980).

GHL as a Component of Growth-Promoting Mixtures

Primary Human Tumor Lines

The establishment of cells from primary human tumors (mammary, ovarian, cervical) for over than 3 months with a greater than 80% success rate was achieved using a medium composed of Earle's salts, MEM vitamins (1×), MEM nonessential amino acids (2×), MEM essential amino acids (4×), L-glutamine (4 mM), sodium pyruvate (1 mM), and 10–20% fetal calf serum plus the growth factors GHL (10 ng/ml), insulin (80 miU/ml), transferrin (10 ng/ml), fetuin (5 μg/ml) and L-thyroxine (10^{-7} M).[47]

Rat Thyroid Follicular Cells

Rat thyroid cells are cultured in Ham's F12M medium and 0.5% calf serum plus six hormones; GHL (10 ng/ml), insulin (10 μg/ml), hydrocortisone (10 nM), transferrin (5 μg/ml), somatostatin (10 ng/ml), and thyrotropin (10 mU/ml). This combination produced diploid growth of the thyroid cell line for greater than 175 generations with maintenance of normal morphology, secretion of thyroglobulin, and ability to concentrate iodine from the culture medium.[48]

Human KB and HeLa Cells

The growth rate of the human tumor cell lines, KB and HeLa, cultured on growth-limiting levels (0.5%) of dialyzed fetal calf serum in Eagle's minimal essential medium plus nonessential amino acids (1×) was stimulated by the addition of GHL (250–500 ng/ml) and bovine serum albumin (fatty acid free, 6 mg/ml). These two additions produced a cellular growth rate in 0.5% serum equivalent to that normally obtained with 5% serum.[13]

Erythropoietin Production in Rats

In normal and hepatectomized rats, the intravenous injection of GHL (100 μg) produces about a 40% rise in erythropoietin, the hormone that increases red blood cell production.[49] This increase is similar to that pro-

[47] W. E. Simon and F. Holzel, *J. Cancer Res. Clin. Oncol.* **94**, 307 (1979).
[48] F. S. Ambesi-Impiombato, L. A. M. Parks, and H. G. Coon, *Proc. Natl. Acad. Sci. U.S.A.* **77**, 3455 (1980).
[49] B. A. Naughton, G. K. Naughton, P. Lui, G. B. Zuckerman, and A. S. Gordon, *J. Surg. Oncol.* **21**, 97 (1982).

duced by about 500 μg glucagon. In this system, insulin administration was without effect.

These effects may reflect a linkage between GHL's growth and viability enhancing actions on liver and kidney cells and the production of erythropoietin by the fetal liver and adult kidney.

Uses of GHL-Cu

Induction of Angiogenesis Effectors and New Capillary Formation

For the induction of new capillary formation in the rabbit eye model, GHL-Cu is incorporated into a slow-release plastic polymer (Elvax), then inserted into a 2 × 3 mm surgical pocket in the lower cornea of New Zealand rabbits (2 to 3 kg).[50] The amount of GHL-Cu in the polymer is calculated to introduce 20 μg into the piece inserted into the corneal pocket. After 80 hr, budding of capillaries begins and by 7 to 8 days dense capillaries surround the implant.

GHL-Cu induces synthesis of proteins that act as potent chemoattractants toward capillary endothelial cells and a family of 5 to 6 angiogenic-related polypeptides with molecular weights ranging from 35 to 66 × 10³ by 76 to 80 hr after application and immediately prior to endothelial cell migration influx.[51] These proteins are obtained from corneal tissue at this time by anesthetizing the rabbit, freezing the corneas with a dichloromethane spray, excising the tissues immediately underlying the GHL-Cu impregnated pellet, and storing them at −80° until use. An extract of this tissue is prepared by chopping the thawing samples into fragments about 1 mm³ and incubating them overnight at 37° in Dulbecco's MEM or in phosphate-buffered saline. The volume of liquid is adjusted to obtain, after 15 hr of extraction followed by centrifugation (2000 g for 15 min), a supernatant containing 1.5 mg/ml protein.

The chemoattractant activity in this supernatant is assayed by determining the passage of cells across pores against a gradient of migration effector. The Boyden chamber has an upper well of 200 μl in size and a lower well of 40 μl which are separated by a poly(vinylpyrrolidone) nucleopore filter (Nucleopore Corp.) 13 mm in diameter precoated with type I collagen. A 5-μm micropore size filter is used for capillary endothelium. Both chambers are filled with DMEM with 1% fetal calf serum. About 10⁵ cells in 100 μl medium are placed in the upper chamber and the migration

[50] K. Raju, G. Alessandri, and P. Gullino, J. Natl. Cancer Inst. (U.S.) 69, 1183 (1983).
[51] K. Raju, G. Alessandri, and P. Gullino, Cancer Res. 44, 1579 (1984).

effector (i.e., tissue proteins) placed in the lower chamber and incubated for 3 hr at 37° in 5% carbon dioxide. At the end of the migration, the cells are fixed in 10% formalin in PBS and stained with Wright's stain. Cells on the underside of the filter are counted in ×200 microscope fields.

The angiogenic related proteins in this supernatant are purified by passage through a gelatin : Sepharose 4B (Pharmacia) column (0.7 × 20 cm, 2 to 3 ml bed volume) in 0.02 M sodium phosphate, 0.15 M sodium chloride, pH 7.4. After application the column is washed with the buffer to remove nonadhering proteins, then eluted with a solution containing 0.05 M sodium acetate and 1.0 M sodium bromide, pH 5.0. The eluate is dialyzed against 50 mM ammonium acetate to remove salts, then lyophilized. Proteins are reduced with 10% 2-mercaptoethanol then separated by SDS–polyacrylamide gel electrophoresis using a 3.3% spacer gel and a 5 or 7.5% separating gel.

Acceleration of Wound Healing

Administration of GHL-Cu accelerates the healing and closure of superficial wounds in rats, mice, and pigs.[52,53] For a wound in rats caused by the removal of a circular patch of skin (2 cm diameter), an effective procedure is the infiltration of 50 μg of GHL-Cu (in 50 μl phosphate buffered saline, pH 7.4) into the skin surrounding the wound margin after wounding, and 24 and 48 hr later. The enhancement of reepithelialization and wound closure becomes statistically significant by 15 days. At 25 days after surgery, 60% of the GHL-Cu treated wounds are fully healed while no controls are fully healed. In pigs, one treatment with GHL-Cu accelerated the healing of square (2.5 cm) wounds on the upper back in comparison with contralateral control wounds on the same animal. After 21 days, one treatment with GHL-Cu reduced the remaining wound size 64.9% (±22.1) in 14 wounds on eight pigs observed for this period ($p = 0.0023$ for pooled data). Pig wounds covered with autologous skin grafts or dehydrated pigskin responded best to GHL-Cu, but the healing of collagen pad covered wounds and uncovered wounds was also enhanced. Multiple treatments of pig wounds with GHL-Cu produced a more pronounced healing effect. In mice, daily swabbing of linear superficial incision wounds with GHL-Cu (100 μg/ml) produces a significant acceleration of wound closure by the fifth day.

This healing action presumably arises both from GHL-Cu's tissue-protective superoxide dismutase activity that minimizes tissue damage

---

[52] L. Pickart, *U.S. Patent Appl.* Serial Number 694,430, Jan. 24, 1985.
[53] L. Pickart, D. Downey, S. Lovejoy, and B. Weinstein, *Proc. Int. Conf. Superoxide Dismutase*, **4th,** Sept. 1985.

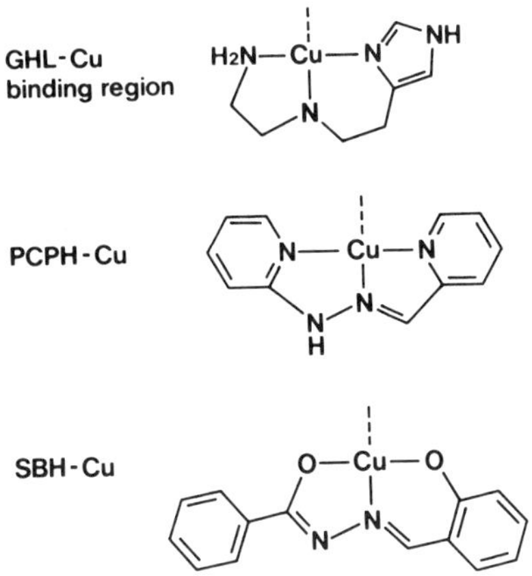

FIG. 2. Comparison of proposed structure of the GHL-Cu copper-binding region and the structures of PCPH-Cu and SBH-Cu. The PCPH-CU and SBH-Cu structures are based on X-ray data plus spectroscopic and nuclear magnetic resonance measurements. In all compounds a near-planar ring structure surrounds the copper. Other small ligands (amino acids, salts) are likely to bind to unsaturated bonding orbitals of copper.

after wounding and its chemoattractant actions on mast and capillary endothelial cells that produce an enhanced accumulation of these cells in the wound area; thusly stimulating neovascularization and nutrient flow into the wound.

Growth-Inhibitory Aroylhydrazone Analogs of GHL-Cu

Aroylhydrazone analogs, in which the copper-binding region is structurally similar to GHL-Cu, are potent mitotic inhibitors with antitumor and immunosuppessive actions (see Fig. 2).[54–58]

The most potent of these analogs, PCPH-Cu [pyridine-2-carboxylaldehyde-2'-pyridylhydrazonatocopper(II) dichloride] is prepared by reacting PCPH (0.40 g in 40 ml ethanol, Aldridge Chemical, Milwaukee, WI) with CuCl$_2$ · 2H$_2$O (0.34 g in 40 ml ethanol). On standing at room tempera-

[54] L. Pickart, W. H. Goodwin, W. Burgua, T. B. Murphy, and D. K. Johnson, *Biochem. Pharmacol.* **32**, 3868 (1983).

[55] L. Pickart, W. H. Goodwin, and W. Burgua, *J. Cell Biol.* **7A**, 175 (1983).

[56] D. K. Johnson, T. B. Murphy, N. J. Rose, W. H. Goodwin, and L. Pickart, *Biochem. Pharmacol.* **32**, 3868 (1983).

[57] A. A. Aruffo, T. B. Murphy, D. K. Johnson, N. J. Rose, and V. Shoemaker, *Inorg. Chim. Acta* **67**, L25 (1982).

[58] E. J. Blantz, F. A. French, J. R. Doamaral, and D. A. French, *J. Med. Chem.* **13**, 1124 (1970).

ture, green needlelike crystals are deposited, filtered off, washed with ethanol, and dried under vacuum.

For use, PCPH-Cu is dissolved in water (2 mg/ml) at 37°, then mixed with growth medium to obtain desired concentrations for testing. Glass pipets and glassware must be used because of the propensity of PCPH-Cu to absorb to plastics, a problem especially acute at concentrations below 1 ng/ml. For most cell lines, a concentration of 1 ng/ml produces a halving of DNA synthesis.

When used as an antitumor agent against fibrosarcoma MCA-1511 in mice, PCPH-Cu (100 μg in 0.05 ml aqueous buffer) is injected directly into the tumor twice weekly for 1 month. This results in the killing of about one-third of implanted tumors. Intraperitoneal administration is partially effective, but intravenous injection by tail vein is difficult due to the rapid development of necrosis at the infusion site. After 1 month of injection of PCPH-Cu into the tumor area, the mice in which the tumors have disappeared resume a normal pattern of weight gain and growth.

Salicylate benzoyl hydrazonate copper(II) (SBH-Cu) is a similar mitotic inhibitor, being about one-tenth as potent as PCPH-Cu. SBH-Cu is synthesized by dissolving benzoylhydrazone (20 mmol) in an ethanol : water mixture (1 : 3, v/v, 40 ml). A solution of salicylaldehyde (2.44 g, 20 mmol) in ethanol (20 ml) is added to the hydrazide solution with stirring, then placed on a steam bath for 20 min. When cooled to room temperature, crystals of salicylate benzoylhydrazone form are filtered off and dried under vacuum. Recrystallization from ethanol yields the pure product. Copper is complexed to the hydrazone by dissolving the hydrazone crystals (0.50 g, 2.1 mmol) in boiling 95% ethanol (20 ml). This solution was then added to a solution of copper(II) chloride dihydrate in boiling ethanol (20 ml) forming a deep green-brown solution. Cooling overnight yielded black, rodlike crystals that were filtered off, washed with ethanol (2 × 10 ml) and diethyl ether (2 × 10 ml), then dried under vacuum.

[29] Isolation of Human Erythropoietin with Monoclonal Antibodies

By RYUZO SASAKI, SHIN-ICHI YANAGAWA, and HIDEO CHIBA

Mature erythrocytes have a limited life span and cannot proliferate. The recruitment of these cells is achieved by differentiation and maturation of the erythroid precursor cells. All types of blood cells, including lymphocytes, originate from common pluripotent stem cells. After stem cells differentiate into a specific lineage (or commitment), further differentiation and maturation are mainly regulated by lineage-specific humoral factors. In the erythroid pathway, two major classes of nonhemoglobinized progenitor cells are present before proerythroblasts commence globin synthesis. Burst-forming unit erythroid (BFU-E) is an early progenitor and is closely related to the stem cell. Colony-forming unit erythroid (CFU-E) is a progeny of BFU-E and is probably the direct precursor of the proerythroblast. Some protein factors have been implicated in the erythroid pathway. Factors which act on BFU-E or possibly more primitive cells have been found in various sources,[1-18] but their physiological significance and molecular properties are poorly understood. A factor referred to as erythroid-potentiating activity was isolated from a leukemic cell line.[19] This factor enhanced the growth of human BFU-E and CFU-E. Erythropoietin (EPO) promotes the growth of CFU-E and the maturation process leading to the hemoglobinization of the erythroblasts. The presence of a serum protein that enhances hemoglobin accumulation in late

[1] M. T. Aye, *J. Cell. Physiol.* **91**, 69 (1977).

[2] G. R. Johnson and D. Metcalf, *Proc. Natl. Acad. Sci. U.S.A.* **74**, 3879 (1977).

[3] A. A. Axelrad, D. L. McLeod, S. Suzuki, and M. M. Shreeve, *in* "Differentiation of Normal and Neoplastic Hematopoietic Cells" (B. Clarkson, P. A. Marks, and J. E. Till, eds.), p. 155. Cold Spring Harbor Lab., Cold Spring Harbor, New York, 1978.

[4] M. J. Murphy, Jr. and A. Urabe, *in* "In Vitro Aspects of Erythropoiesis" (M. J. Murphy, Jr., ed.), p. 189. Springer-Verlag, New York, 1978.

[5] G. Wagemaker, *in* "In Vitro Aspects of Erythropoiesis" (M. J. Murphy, Jr., ed.), p. 44. Springer-Verlag, New York, 1978.

[6] D. G. Nathan, L. Chess, D. G. Hillman, B. Clarke, J. Breard, E. Merler, and D. E. Housman, *J. Exp. Med.* **147**, 324 (1978).

[7] N. N. Iscove and M. Schreier, *Exp. Hematol.* **7** (Suppl. 6), 4 (1979).

[8] K. S. Zuckerman, *Exp. Hematol.* **8**, 924 (1980).

[9] J. I. Kurland, P. A. Meyers, and M. A. S. Moore, *J. Exp. Med.* **151**, 839 (1980).

[10] L. I. Gordon, W. J. Miller, R. F. Branda, E. D. Zanjani, and H. S. Jacob, *Blood* **55**, 1047 (1980).

[11] E. D. Werts, R. L. DeGowin, S. K. Knapp, and D. P. Gibson, *Exp. Hematol.* **8**, 423 (1980).

[12] P. P. Dukes, A. Ma, and D. Meytes, *Exp. Hematol.* **8** (Suppl. 8), 128 (1980).

[13] J. L. Ascensao, N. E. Kay, T. Earenfight-Engler, H. S. Koren, and E. D. Zanjani, *Blood* **57**, 170 (1981).

[14] P. N. Porter and M. Ogawa, *Blood* **59**, 1207 (1982).

[15] T. Okamoto, A. Kanamaru, H. Hara, and K. Nagai, *Exp. Hematol.* **10**, 844 (1982).

[16] N. Dainiak and C. M. Cohen, *Blood* **60**, 583 (1982).

[17] K. S. Zuckerman, V. R. Patel, and D. D. Goodrum, *Exp. Hematol.* **11**, 475 (1983).

[18] K. S. Zuckerman and M. Haak, *Br. J. Haematol.* **55**, 145 (1983).

[19] C. A. Westbrook, J. C. Gasson, S. E. Gerber, M. E. Selsted, and D. W. Golde, *J. Biol. Chem.* **259**, 9992 (1984).

erythroblasts has been indicated.[20] Of the factors that may be involved in the regulation of erythropoiesis, erythropoietin is the best characterized. Furthermore it is widely accepted that this hormone is the primary regulator of erythropoiesis. The kidney is the primary site of EPO production in adults and the liver in fetal and neonatal mammals (reviewed in Ref. 21). Normal human serum contains small amounts of EPO (30 ± 10 mU/ml).[22] The mechanism of EPO production is not well understood, but oxygen concentration plays a role in regulating EPO production. Hypoxia and anemia usually cause elevation of the blood EPO level and EPO appears in the urine of some anemic humans. Extracts from plasma and urine of anemic animals and humans have been used as a starting material to purify EPO. Miyake et al.[23] first reported the isolation of EPO from urine of patients with aplastic anemia using conventional purification procedures. However, laborious and time-consuming purification procedures, low yields, and limited supplies of starting materials make it worthwhile to develop a rapid isolation method. High yields of pure EPO are necessary to study its structure, mechanism of action, and metabolism.

A stable hybridoma clone that secretes monoclonal antibody against human EPO has been established.[24] We describe here a very simple isolation method of EPO from urine of aplastic anemic patients using this monoclonal antibody.[25] The method described here yields pure EPO with a high-recovery rate within approximately 14 days.

Assay for Erythropoietin

Several methods have been developed for measuring EPO activity. In vivo assays with starved rats or artificially polycythemic mice use the EPO-dependent enhancement of ^{59}Fe incorporation into the hemoglobin-synthesizing cells[26]; starvation or polycythemia makes the animals more sensitive to the exogenous hormone by decreasing endogenous hormone levels. In in vitro assays, the cells from the hemopoietic tissues (marrow, fetal liver, or spleen) are cultured and the EPO-induced biochemical events of the responsive cells are measured. (See also Sawyer et al. [30],

[20] G. Krystal, Exp. Hematol. 11, 18 (1983).
[21] J. W. Fisher, Proc. Soc. Exp. Biol. Med. 173, 289 (1983).
[22] R. D. Lange, J. P. Chen, and C. D. R. Dunn, Exp. Hematol. 8, 197 (1980).
[23] T. Miyake, C. K.-H. Kung, and E. Goldwasser, J. Biol. Chem. 252, 5558 (1977).
[24] S. Yanagawa, S. Yokoyama, K. Hirade, R. Sasaki, H. Chiba, M. Ueda, and M. Goto, Blood 64, 357 (1984).
[25] S. Yanagawa, K. Hirade, H. Ohnota, R. Sasaki, H. Chiba, M. Ueda, and M. Goto, J. Biol. Chem. 259, 2707 (1984).
[26] E. Goldwasser and M. Gross, this series, Vol. 37, p. 109.

this volume.) One of the *in vitro* assays is based on the fact that EPO stimulates the rate of ^{59}Fe incorporation into heme.[27] The hormone-induced increase in the rate of [^{3}H]thymidine[28,29] or ^{125}I-labeled deoxyuridine[30] incorporation into the DNA is an index of another *in vitro* assay. Scoring the number of erythroid colonies formed dependent on EPO, when the hemopoietic tissue cells are cultured in semisolid medium, is an *in vitro* assay method which does not use radioactive compounds.[31–33] *In vivo* methods are the most specific for EPO, but they are laborious and time consuming. Therefore we usually measure EPO activity by *in vitro* methods using fetal mouse liver cells[27,28,34] but for some samples the *in vivo* method with starved rats was used.[26] Sheep plasma EPO step-III (Connaught Medical Research Laboratories, Canada, 4 units/mg of protein) or aplastic anemia urine (Toyobo Co., Ltd., Japan, 67 units/mg of protein) was used as a standard.

Urine Concentrate

For selection of urine from anemic patients with high EPO titers, the activity was assayed by counting erythroid colonies that were formed dependent on EPO in *in vitro* suspension culture of 13-day-old mouse fetal liver cells.[34] Urine (>0.5 unit/ml of urine) from anemic patients was collected in a 10-liter bottle containing about 10 g of sodium azide, frozen at −20° immediately after discharge, and stored until use. The urine was thawed, filtered under suction, and concentrated by ultrafiltration on a hollow-fiber device (Amicon) with a nominal molecular weight cutoff of 10,000 at 4°. The urine concentrate was lyophilized. One liter of urine gave approximately 0.5 g of dry materials.

Purification of Erythropoietin as Antigen

To prepare the material for immunization of mice, EPO was purified from the lyophilized material from 600 liters of urine, on columns of

[27] J. R. Stephenson and A. A. Axelrad, *Endocrinology* **88**, 1519 (1971).
[28] N. C. Brandon, P. M. Cotes, and J. Espada, *Br. J. Haematol.* **47**, 461 (1981).
[29] G. Krystal, *Exp. Hematol.* **11**, 649 (1983).
[30] C. D. R. Dunn and L. Gibson, *Exp. Hematol.* **11**, 590 (1983).
[31] J. R. Stephenson, A. A. Axelrad, D. L. McLeod, and M. M. Shreeve, *Proc. Natl. Acad. Sci. U.S.A.* **68**, 1542 (1971).
[32] N. N. Iscove, F. Sieber, and K. H. Winterhalter, *J. Cell. Physiol.* **83**, 309 (1974).
[33] T. Okamoto, A. Kanamaru, H. Hara, and K. Nagai, *Am. J. Hematol.* **12**, 179 (1982).
[34] S. Yanagawa, H. Narita, R. Sasaki, H. Chiba, N. Itada, and H. Okada, *Agric. Biol. Chem.* **47**, 1311 (1983).

DEAE-cellulose, phenyl-Sepharose, hydroxylapatite, Sephadex G-100, and sulfopropyl-Sephadex in that order.[24] Eighty-eight-fold purification with 4.5% recovery of activity was achieved. The specific activity based on absorbance at 280 nm (assuming $E_{280\,nm}^{1\%} = 10$) was about 5100 units/mg protein with the *in vitro* assay methods using radioiron[27] and [^{3}H]thymidine.[34] Analysis of the purified preparation by SDS–polyacrylamide gel electrophoresis showed a main band with MW 32,000 and some minor bands. Measurement of EPO activity in the extract from sliced gels revealed that the activity does not reside in the main band but in the faint band with MW 35,000.

One of the biggest advantages of the hybridoma technique is that one can select hybridomas producing a specific antibody to the desired antigen even when immunizations are performed with impure antigen preparations. We prepared hybridomas by using the partially purified EPO preparation as an antigen with the hope that hybridomas producing antibodies to EPO would occur. However, this trial was unsuccessful, but it yielded a number of hybridomas that produced antibodies against contaminating proteins. These antibodies were used to remove the contaminants. Some stable hybridomas with high production of the antibodies were selected and the antibodies were prepared from ascitic fluid of mice. Two columns were prepared containing Affi-Gel 10 (Bio-Rad) to which the antibodies from hybridomas directed against contaminants were fixed. One column contained antibodies against three identified contaminants (MW 37,000, 40,000, and 62,000) and those produced by seven hybridomas directed against unidentified antigens. The other column contained the antibody against the main contaminant having MW 32,000. The EPO preparation (5100 units/mg of protein) was applied on the first column. EPO was fully recovered in the flow-through fractions. The fractions were applied on the second column and EPO appeared again in the flow-through fractions without loss of the activity. The specific activity based on absorbance at 280 nm was 23,000 units/mg of protein with the *in vitro* assay methods using ^{59}Fe and [^{3}H]thymidine. Analysis of the preparation with SDS–polyacrylamide gel electrophoresis indicated quite effective removal of the main contaminant (MW 32,000 protein) as well as other contaminants, and EPO protein was recognized as a main band with MW 35,000. Because some other contaminants were still detected, however, the EPO preparation was further purified with preparative SDS–polyacrylamide gel electrophoresis and EPO was extracted from the ground gel. Approximately 650-fold purification was achieved with these purification methods and 1 mg of protein was obtained from 600 liters of urine. The specific activity of the final preparation was about 38,000 units/mg of protein by the *in vitro* ^{59}Fe and [^{3}H]thymidine methods and also by the *in vivo* assay with starved rats. This preparation was used as the antigen for immunization of mice. In some cases, the erythropoietin in the ground polyacrylamide gel was directly injected into mice without extraction.

Hybridomas Producing Monoclonal Antibodies to Erythropoietin

The immunization protocol and the methods of hybridoma production including fusion procedure, cloning, and antibody production have been described.[24] BALB/c mice were used for immunization and NS-1 myeloma cells for fusion. Hybridomas directed against erythropoietin are prepared from spleen cells of mice that were immunized with the EPO in the ground polyacrylamide gel. Two mice receiving the EPO in the gel suspension developed sera that bound EPO. Four mice receiving the EPO extracted from the gel were all negative in terms of EPO-binding sera and therefore were not used for fusion. Finally three hybridoma clones that produced immunoglobulin that bound EPO were established. One (S1) of these clones has been quite stable in high antibody production. The subclass of the mouse immunoglobulin produced by S1 clone is IgG_1.

Isolation of Erythropoietin and Its Characterization

Preparation of Immunoadsorbent. BALB/c mice received intraperitoneal injections of 0.5 ml of sterile pristane 10 and 3 days before 0.5 ml (10^7 cells) of the hybridoma cells (S1 clone) was given intraperitoneally. Ten to fifteen days after cell injection, 5–18 ml of ascitic fluid was obtained from each mouse. The cells and other insoluble materials in the fluid were removed by centrifugation and the supernatant was diluted with 10 mM phosphate-buffered saline (PBS), pH 7.4 to a protein concentration of 12 mg/ml. The antibody was pelleted by addition of 0.9 volume of saturated ammonium sulfate solution and by subsequent centrifugation. The pellet was dissolved in and dialyzed thoroughly against 0.2 M NaHCO$_3$, pH 8, containing 0.3 M NaCl. The protein in the dialyzate was fixed on Affi-Gel 10 to construct an immunoadsorbent column.[35]

Gel Electrophoresis and EPO Activity in Extracts of Sliced Gels. SDS–polyacrylamide gel electrophoresis was performed by the method of Laemmli[36] and stained with silver according to the manufacturer's directions (Bio-Rad). In some cases, EPO activity of extracts from sliced gels was measured. The unstained gel was sliced into 1-mm lengths. The gel pieces were put into test tubes containing PBS and 0.1% bovine serum

---

[35] R. Sasaki, H. Ohnota, S. Yanagawa, and H. Chiba, *Agric. Biol. Chem.* **49,** 2671 (1985).

[36] U. K. Laemmli, *Nature (London)* **227,** 680 (1970).

albumin (0.5 ml/sliced gel) and ground with a glass rod. After incubating the suspensions overnight to extract protein, they were centrifuged to obtain clear supernatants. The activity in the supernatants was measured by the [^{3}H]thymidine method.

Purification with a Monoclonal Antibody Column. The lyophilized urine concentrate (350 g) was dissolved in 700 ml of PBS and dialyzed extensively. Insoluble materials were removed by centrifugation and the supernatant (690 ml) was heated at 100° for 3 min after the addition of solid SDS (final concentration 2%). The SDS-treated solution was cooled and left overnight at 0°. The precipitated SDS was removed by centrifugation and the supernatant (680 ml) was loaded at room temperature on an immunoadsorbent column (3.2 × 2.5 cm) containing 716 mg of the EPO-directed monoclonal antibody fixed on Affi-Gel 10. The column was extensively washed and then developed with the pH 2.5 buffer. As shown in Fig. 1, purification with the immunoadsorbent column was tremendously effective; most of the protein in the urine concentrate emerged in the flow-through fractions without being adsorbed, and EPO was eluted sharply by the pH 2.5 buffer. About 2600-fold purification was achieved by this single step and 75% of the activity was recovered (see Table I). The presence of

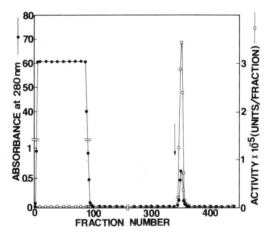

FIG. 1. Purification of EPO from human urine concentrate with an immunoadsorbent column. After the sample was applied on an immunoadsorbent column, the column was extensively washed with 1 liter of PBS, 500 ml of 10 m*M* NaP$_i$, pH 7.4, containing 0.5 *M* NaCl, and 500 ml of 0.15 *M* NaCl, in this order, and then eluted by reverse flow of 0.2 *M* acetic acid, pH 2.5, containing 0.15 *M* NaCl (the vertical line with arrowhead). The eluted fractions (112 ml) were immediately neutralized by adding 5.3 ml of 3.4 *M* Tris. The flow rate was 24 ml/hr during application of the sample and elution with the pH 2.5 buffer, while it was 200 ml/hr during washing of the column. The volume of one fraction was 8 ml. From Yanagawa *et al.*[25]

TABLE I
PURIFICATION OF EPO FROM URINE CONCENTRATE

	Protein[a] (mg)	Activity in vitro (units × 10⁻³)	Yield (%)	Specific activity in vitro[b] (units/mg)	Purifi- cation	Specific activity in vivo[b] (units/mg)
Urine concentrate	34 × 10	792	100	23	1	14
Immunoadsorbent column	10	594	75	59,400	2,580	59,000
Sephadex G-100	5.7	500	63	88,000	3,830	81,600

[a] Urine concentrate from about 700 liters of human urine was used as the starting material. Protein was measured with Coomassie brilliant blue binding assay using ovalbumin as a standard.[38]

[b] EPO activity was measured with the in vitro [³H]thymidine incorporation using fetal liver cells[28,34] and with the in vivo ⁵⁹Fe incorporation method using starved rats.[26]

SDS in heat treatment of the urine concentrate is needed for EPO to be retained on the column. EPO mostly appeared in the flow-through fractions with the heat treatment in the absence of SDS. EPO eluted from the immunoadsorbent column was not adsorbed on a newly prepared column but when subjected to the SDS treatment it was adsorbed again.

EPO purified using the immunoadsorbent column was recognized as a main band with MW 35,000 on SDS–polyacrylamide gel electrophoresis, but some other protein bands were seen (data not shown) including a MW 20,000 protein which was a main contaminant. Furthermore, this preparation was faintly tinged with brown.

Purification Using a Sephadex G-100 Column. The EPO fraction eluted from the immunoadsorbent column was extensively dialyzed against distilled water and lyophilized. The dried material was dissolved in 2 ml of PBS, pH 6.9, and the clear solution was loaded on a Sephadex G-100 column (1.2 × 130 cm) equilibrated with PBS and the column was developed at 4° with a speed of 6 ml/hr. The purification with the Sephadex G-100 column revealed two peaks of protein (Fig. 2A). By measuring the activity in the pooled preparation of each peak it was found that most of the EPO activity was recovered in the second peak (see Table I) and that about 5% of the total recovered activity appeared in the void fractions (the first peak). The faint brown appeared in the void fractions, while the pooled EPO preparation was quite transparent. Figure 2B shows the SDS–polyacrylamide gel electrophoresis of the Sephadex G-100 fractions. Contaminants including the main contaminant, the MW 20,000 protein, were not detected in any fractions of the second peak (Fig. 2B, lanes

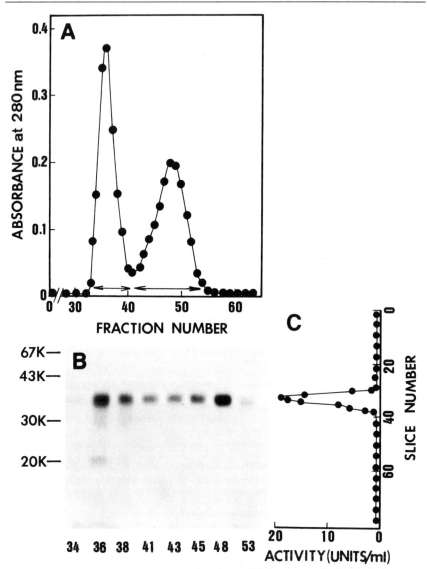

FIG. 2. Sephadex G-100 chromatography of the EPO fraction purified with the immuno-adsorbent column. The volume of one fraction was 2 ml. (A) shows the elution pattern of protein based on absorbance at 280 nm. The horizontal lines with arrowheads indicate fractions pooled after elution. The pooled fractions were designated the first peak (or the void fraction) and the second fraction (or the EPO fraction). (B) shows SDS–polyacrylamide gel electrophoresis of fractions (20 μl of each). Lane numbers correspond to the fractions of Sephadex G-100 chromatography. (C) shows EPO activity of the extracts from sliced gels. The other gel to which fraction 48 was applied was sliced without staining and EPO was extracted for assay. From Yanagawa et al.[25]

43–53). As is most clearly seen in fraction 48 (Fig. 2B, lane 48), however, they contained faint but detectable protein components on the leading side of the EPO main band with MW 35,000. Activity measurements in the extracts from sliced gels indicated that there is EPO activity in the leading side as well as in the main band (Fig. 2C), suggesting heterogeneity of the EPO protein. The Western blotting technique[37] showed that the components in the leading side are indeed EPO protein; the monoclonal antibody against EPO reacted with these components (Fig. 3, lane 2). The EPO protein with MW 35,000 binds properly with the antibody. Protein species which migrated somewhat faster than the EPO main band and bound with the antibody were also present in the EPO preparation before Sephadex G-100 column chromatography (Fig. 3, lane 1). Thus, the final EPO preparation contains small amounts of EPO species with mobilities larger than the MW 35,000 EPO. Treatment with sialidase converted the main MW 35,000 band to one with larger mobility (Fig. 3, lane 4), which suggests that electrophoretic heterogeneity of EPO is due to a variable amount of sialic acid attached to the EPO protein.

The main MW 20,000 contaminant and others in the EPO preparation eluted from the immunoadsorbent column appeared in the void fractions on Sephadex G-100 chromatography (Fig. 2B, lanes 36 and 38). It was rather surprising that the main component in the void fractions was a MW 35,000 protein. Furthermore, it was shown that this component was indeed EPO protein; EPO activity was found in the extracts from sliced gels containing the MW 35,000 component after SDS–polyacrylamide gel electrophoresis of fraction 36 (not illustrated) and Western blotting revealed the binding of this component with the monoclonal antibody against EPO (Fig. 3, lane 3). Interestingly, the MW 20,000 protein did not bind with the antibody. Activity measurement of fraction 36, in which there should be no contamination of EPO from the second peak, showed about 1/16 of the specific activity of the second peak. More than 50% of the protein of fraction 36, however, appears to be the MW 35,000 protein (Fig. 2B, lane 36) from the intensity of protein staining. Therefore, it seems that EPO molecules in the void fractions have much lower activity than those in the second peak. In spite of the occurrence of the undefined EPO species on Sephadex G-100 chromatography, purification of human urinary EPO with an immunoadsorbent and a Sephadex G-100 column provides pure EPO in a high yield (Table I).

Characterization. The Coomassie brilliant blue binding method[38] for protein determination gave a value of $E_{280 \text{ nm}}^{1\%} = 13.1$ for pure EPO using

---

[37] W. N. Burnette, Anal. Biochem. 112, 195 (1981).
[38] M. Bradford, Anal. Biochem. 72, 248 (1976).

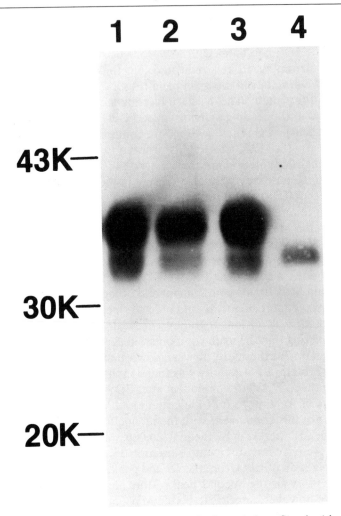

FIG. 3. Identification of EPO with the Western blotting technique. Samples (about 2 μg of protein of each) were subjected to the electrophoresis and the fractionated proteins were transferred to nitrocellulose paper. The blot was reacted with monoclonal antibody against EPO (20 μg/ml), then the peroxidase-conjugated anti-mouse IgG. The EPO protein was visualized by incubating the blot with the substrates (H_2O_2 and 4-chloro-1-naphthol) of peroxidase. Lane 1, fraction eluted from the immunoadsorbent column with the pH 2.5 buffer (see Fig. 1); lane 2, the second peak from the Sephadex G-100 column chromatography (see Fig. 2A); lane 3, the first peak; lane 4, the sialidase-treated second peak. The second peak (0.4 ml) from the Sephadex G-100 column was brought to about pH 5.5 by adding 0.4 M acetate, pH 5.2, and incubated with 0.1 unit of sialidase for 5 hr at 37°. The sialidase was inactivated by heating at 100° for 5 min. From Yanagawa et al.[25]

ovalbumin as a standard and the method by Lowry *et al.*[39] gave a value of $E_{280 \text{ nm}}^{1\%}$ = 12.6 using bovine serum albumin as a standard. The specific activity of purified EPO was 88,000 units/mg of protein by *in vitro* methods and 81,600 by the *in vivo* method (see Table I). A specific activity of 70,400 units/mg of protein has been reported for EPO isolated from human urine with conventional purification procedures by Miyake *et al.*[23] using $E_{280 \text{ nm}}^{1\%}$ = 8.5.

Sialidase treatment (see legend of Fig. 3) of EPO completely abolished *in vivo* activity, while *in vitro* activity increased 1.3-fold. Similar results have been reported with EPO partially purified from the plasma of anemic sheep.[40,41] Loss of *in vivo* activity could be caused by the hepatic removal of asialoglycoprotein from the circulation.[42] Analysis of the carbohydrates of the EPO yielded fucose, galactose, mannose, *N*-acetylglucosamine, and sialic acid.[43]

Comments

The immunoadsorbent column does not bind EPO in the untreated urine concentrate but binds EPO in the SDS-treated one. Treatment with SDS was needed for the purified EPO to be retained again on the immunoadsorbent column. It appears, therefore, that the lack of binding of EPO in the untreated urine concentrate with the antibody is not attributable to the interaction of the EPO with other components in the concentrate but to a property of the antibody. The hybridoma secreting this antibody was constructed by using spleen cells from a mouse receiving its primary immunization with an antigen-containing ground SDS–polyacrylamide gel. Therefore, this antibody may be directed against an epitope which is buried in the interior of the native EPO molecule, inaccessible to the antibody, but exposed in the presence of SDS. It is also possible that the antibody recognizes a structure formed by the binding of SDS to EPO. The binding of EPO to the immunoadsorbent column can occur only in the presence of SDS but there appears to be no problems concerning the eluted EPO. The purified EPO no longer binds with the antibody, indicating that the native conformation of EPO has been regained. The immunoadsorbent column is so extensively washed before the elution of the

[39] O. H. Lowry, N. J. Rosebrough, A. L. Farr, and R. J. Randall, *J. Biol. Chem.* **193**, 265 (1951).
[40] W. A. Lukowsky and R. H. Painter, *Can. J. Biochem.* **50**, 909 (1972).
[41] E. Goldwasser, C. K.-H. Kung, and J. Eliason, *J. Biol. Chem.* **249**, 4202 (1974).
[42] A. G. Morell, R. A. Irvine, I. Sternlieb, I. H. Scheinberg, and G. Ashwell, *J. Biol. Chem.* **243**, 155 (1968).
[43] Unpublished results.

EPO that, if SDS remains in the eluted EPO, the amount would be very small.

The presence of the EPO molecule with a low specific activity in the void fractions of Sephadex G-100 chromatography is puzzling. The void fractions may contain associated forms of EPO. Association of EPO may result in reducing the activity. The MW 20,000 protein also may appear in void fractions as associated forms without interaction with EPO molecules. It seems, however, that the binding of the MW 20,000 protein to the antibody column is mediated by EPO, since the MW 20,000 protein fractionated by gel electrophoresis was unable to bind with the monoclonal antibody (see Fig. 3, lane 3). An intriguing hypothesis is that the occurrence of high-molecular-weight species of EPO with reduced activity is caused by association of EPO with the MW 20,000 protein, which is a physiological regulator of erythropoiesis.

The pure EPO is now manufactured using the procedures described here by Snow Brand Milk Products Co., Ltd., Japan and is available from Toyobo Co., Ltd., Japan.

The isolation and sequence analysis of genomic and cDNA clones of human EPO have been reported.[44] The mature EPO was composed of 166 amino acids and three sites of N-linked glycosylation were assigned. A secretory leader peptide of 27 hydrophobic amino acids was predicted.

[44] K. Jacobs, C. Shoemaker, R. Rudersdorf, S. D. Neill, R. J. Kaufman, A. Mufson, J. Seehra, S. S. Jones, R. Hewick, E. F. Fritsch, M. Kawakita, T. Shimizu, and T. Miyake, *Nature (London)* **313**, 806 (1985).

[30] Large-Scale Procurement of Erythropoietin-Responsive Erythroid Cells: Assay for Biological Activity of Erythropoietin[1]

By Stephen T. Sawyer, Mark J. Koury, and Maurice C. Bondurant

Erythropoietin (EP), the hormone which is the normal regulator of erythropoiesis, has been purified and characterized[2] (see also Sasaki *et al.*

[1] Supported by National Institutes of Health grants AM-31513 and AM-15555 by the Veterans Administration. M. J. K. is a scholar of the Leukemia Society of America.
[2] T. Miyake, C. Kung, and E. Goldwasser, *J. Biol. Chem.* **252**, 5558 (1977).

[29], this volume). The gene for EP has been cloned,[3] and pure EP from recombinant material is commercially available. Until recently, the knowledge of the molecular events associated with the differentiation of erythroid progenitor cells or the mechanisms by which EP controls these events has been retarded by the lack of a uniform cell population which is responsive to the hormone. This report describes a method of procuring a large number of erythroid progenitor cells which have the characteristics critical for the study of mechanism by which EP acts: (1) high purity, (2) adequate cell number, (3) synchrony of differentiation, and (4) responsiveness to the hormone. The central feature of this method is that a population of erythroid progenitor cells is generated in the spleens of mice as a result of infection with the anemia-inducing strain of Friend virus.

The anemia-inducing strain of Friend virus (FVA) causes mouse erythroid progenitor cells to proliferate and to differentiate partially. Infection of bone marrow or spleen cells *in vitro* and subsequent culture in semisolid medium leads to the growth of large erythroid colonies without the addition of EP to the culture medium.[4,5] This same type of cellular response to infection appears to be responsible for the massive erythroblastosis which occurs in adult mice at 2 to 3 weeks following infection[6] (reviewed in Ref. 7). This acute erythroblast proliferation increases by 10-fold or greater the splenic nucleated cell number, and it results in virtually complete replacement of the other nucleated cell types in the spleen with the erythroblasts. The large majority of erythroblasts derived from FVA-infected progenitors fail to progress as far as hemoglobin synthesis and erythrocyte maturation, and yet they are sensitive to physiological levels of EP and will complete maturation into erythrocytes *in vitro* in response to the hormone.

In earlier studies, we employed manual isolation of FVA-induced erythroid colonies from cultures of infected bone marrow cells to obtain these EP-sensitive erythroblasts.[5,8,9] This isolation procedure is very time consuming for studies which require large numbers of cells. More recently we have used spleens from FVA-infected mice to provide a source

[3] K. Jacobs, C. Shoemaker, R. Rudersdorf, S. D. Neill, R. J. Kaufman, A. Mufson, J. Seehra, S. S. Jones, R. Hewick, E. F. Fritsch, M. Kawakita, T. Shimizu, and T. Miyake, *Nature (London)* **313**, 806 (1985).

[4] W. D. Hankins and D. Troxler, *Cell* **22**, 693 (1980).

[5] M. J. Koury, M. C. Bondurant, D. T. Duncan, S. B. Krantz, and W. D. Hankins, *Proc. Natl. Acad. Sci. U.S.A.* **79**, 635 (1982).

[6] M. J. Koury, S. T. Sawyer, and M. C. Bondurant, *J. Cell. Physiol.* **121**, 526 (1984).

[7] S. Ruscetti and L. Wolff, *Curr. Top. Microbiol. Immunol.* **112**, 21 (1984).

[8] M. Bondurant, M. Koury, S. B. Krantz, T. Blevins, and D. Duncan, *Blood* **61**, 751 (1983).

[9] S. T. Sawyer and S. B. Krantz, *J. Biol. Chem.* **259**, 2769 (1984).

of large numbers of EP-sensitive cells like those isolated from *in vitro* cultures. These splenic erythroid cells have been used for sequential studies of biochemical and molecular changes which occur during the terminal erythroid differentiation induced by EP.

Isolation of Splenic Erythroblasts

The Virus. The Friend virus is a retroviral complex consisting of a replication-defective spleen focus-forming virus (SFFV) and a replication-competent murine leukemia virus (MuLV). The SFFV component of the virus is responsible for the erythroblastosis. There are two strains of SFFV which when complexed with MuLV are termed Friend virus anemia-inducing (FVA) and Friend virus polycythemia-inducing (FVP) (reviewed in Ref. 7). The anemia-inducing strain of Friend virus (FVA) which we use is designated $SFFV_A/FRE$ cl-3/MuLV(201). This virus complex was prepared by Dr. David Troxler by first isolating a $SFFV_A$-infected nonproducer cell clone of the Fisher rat embryo (FRE) cell line and initiating virus production by infecting the cells with biologically cloned Friend helper murine leukemia virus, F-MuLV(201).[4,10] For routine use, the virus stock is maintained at high titer by serial passage of infectious plasma in BALB/c mice. Mice are injected via tail vein with approximately 10^4 spleen focus-forming units[10] of FVA. Blood is collected from infected animals at 3 weeks postinfection into tubes containing heparin and the blood cells are removed by centrifugation at 600 g for 15 min. The plasma is distributed into small tubes and stored at $-70°$ without filter sterilization.

FVA-Induced Erythroblastosis (Fig. 1). Mice are injected with FVA via the tail vein as described for virus propagation. At 13 to 18 days after infection the mice are killed and their spleens removed. The spleens should weigh 1000–1800 mg as compared to 100–200 mg in uninfected mice. A single cell suspension of spleen cells in Iscove's modification of Dulbecco's medium (IMDM) is made by passing the spleen through a nylon mesh bag taken from human blood administrations sets (Fenwal Labs., Deerfield, IL). Then cells are vigorously pipetted to further disrupt clumps of cells. The volume of the suspension is adjusted to 50 ml and the nucleated cells are counted in a hemacytometer after diluting the cell suspension 1 : 10 in a 0.1% aqueous solution of methylene blue.

The nucleated cell counts range from 1.0 to 2.0 × 10^9 cells per infected spleen as compared to only 1.5 to 2.0 × 10^8 cells per uninfected spleen.

---

[10] D. H. Troxler, S. K. Ruscetti, D. L. Linemeyer, and E. M. Scolnick, *Virology* **102**, 28 (1980).

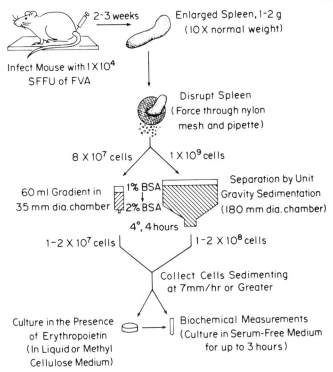

Fig. 1. Procurement of erythropoietin-responsive erythroid cells from mice infected with the anemia strain of Friend virus.

Although the spleens will enlarge further during the third and fourth week following infection, spleens with greater than 2×10^9 nucleated cells often have necrotic, blood filled areas and are not used. By the fourth week of infection, the animals become anemic and more of the splenic erythroblasts begin to differentiate spontaneously, presumably in response to higher EP levels in the animals. At this point the cells are unsuitable as a system for analyzing EP effects.

Purification of the Erythroid Cells by Unit Gravity Sedimentation. Although the spleens of FVA-infected animals contain about 90% immature erythroblasts as the composition of nucleated cells, there are also a significant number of mature erythrocytes present which are derived from the blood. The erythroid population also contains a few percent of more mature erythroblasts which give a slight positive staining reaction with benzidine suggesting some degree of heme synthesis. There are a few percent remaining lymphocytes and other cell types present. To reduce further these unwanted cell types in the spleen cell population, we routinely subject the cells to velocity sedimentation at unit gravity as described in detail by Miller and Phillips.[11]

For unit gravity sedimentation of high cell numbers, we use a STA-PUT cell separator (Johns Scientific, Toronto, Ontario) with a 180-mm-diameter sedimentation chamber, and we use a slight modification of the method of Miller and Phillips. While those authors used gradients of serum in culture medium, we use a linear gradient of 1 to 2% deionized bovine serum albumin (BSA) in IMDM. Briefly, 0.8 to 1.0 × 10⁹ nucleated spleen cells (counted in methylene blue) are sedimented at 400 g for 10 min and resuspended in 40 ml of IMDM containing 0.2% BSA. The cells are loaded into the sedimentation chamber from the bottom followed by a steep gradient from 0.2 to 1% BSA and finally by the linear 1 to 2% BSA gradient. The linear BSA gradient which we use is a total of 1240 ml (620 ml of each BSA concentration). After 4.5 hr of sedimentation at 4°, the gradient is collected from the bottom. The first 200 ml is discarded, and the next 500 ml of the gradient is collected in 50-ml aliquots. Of these 10 50-ml aliquots, numbers 1 through 8 contain cells which sediment greater than 6 mm/hr and consist almost entirely of immature erythroblasts. Aliquot numbers 9 and 10 usually contain increasing numbers of erythrocytes and late erythroblasts and are discarded. Of the 10⁹ cells loaded, we routinely obtain a total of 1.1 to 1.5 × 10⁸ cells in the high purity fractions. Care must be taken not to overload the gradient and not to eliminate the steep buffer gradient between cells and the main gradient. Otherwise one may get a "cell streaming" effect which leads to poor separation results. One should consult the reference by Miller and Phillips[11] for more details of sedimentation method.

Koury and Krantz[12] have described a small-scale sedimentation procedure which uses a 100 ml, 35-mm-diameter syringe as the sedimentation chamber. We use these smaller gradients when fewer erythroblasts are desired. A continuous, linear gradient of 1 to 2% BSA (60 ml) is formed in the syringe using a small glass gradient maker (Southeastern Laboratory Apparatus, North Augusta, SC). Spleen cells from FVA-infected mice are loaded directly on the top of the gradient. Fewer than 10⁸ nucleated cells in IMDM free of BSA should be loaded to obtain an adequate separation of mature and immature erythroblasts. A typical gradient is loaded with 8 × 10⁷ nucleated cells and 1.5 × 10⁷ high purity immature erythroblasts are recovered after sedimentation. The duration of sedimentation at 4° can vary from 3 to 4.5 hr, and the gradient is fractionated in 5-ml aliquots from

[11] R. G. Miller and R. A. Phillips, *J. Cell. Physiol.* **73**, 202 (1969).
[12] M. J. Koury and S. B. Krantz, *Cell Tissue Kinet.* **15**, 59 (1982).

the bottom. After sedimentation, the pellets are judged for red color and only the white fractions are pooled to obtain the best preparation of immature erythroblasts. Routinely, after a 4 hr sedimentation, the erythroblasts in the first 30 to 35 ml collected (i.e., the first 6 to 7 aliquots) are used. These collections correspond to cells sedimenting greater than 7.2 mm/hr in the first 30 ml and greater than 6.2 mm/hr in the first 35 ml.

Erythroblast Culture. The culture medium which we use routinely for studying the differentiation of erythroid cells in response to EP is IMDM with 0.8% methylcellulose (Fisher, laboratory grade), 30% fetal bovine serum, 1% deionized BSA, 100 units/ml penicillin, 100 μg/ml streptomycin, and 10^{-4} M α-thioglycerol. The fetal bovine serum (Hy-Clone, Laboratories, Logan, Utah) is heated to 56° for 30 min before use, and the BSA (LEPTALB-7, Armour Pharmaceutical Co., Kankakee, IL) is deionized with a mixed bed resin, AG 501 × 8 (D) (Bio-Rad Lab., Richmond, CA). Cell cultures are initiated with 10^6 cells per ml or less and with varying amounts of EP, the standard dose for maximum effect being 0.2 units of EP per ml. Incubation is at 37° in humidified air plus 4% CO_2.

Liquid culture medium may be used for studying the differentiation effects of EP on the erythroid cells. A suitable liquid medium is IMDM supplemented with 20 to 30% fetal bovine serum. In addition, the medium should contain all of the ingredients of the semisold medium except methylcellulose.

Cells taken directly from the unit gravity sedimentation or cells taken from culture in the presence of erythropoietin can be held in serum-free culture for a short time (2 to 3 hr). Typically, these cells are washed and resuspended in 2% BSA in IMDM and can be held at 37° in a CO_2 incubator for up 3 hr or in a water bath for 30 min without appreciable cell death as judged by trypan blue exclusion.

Mouse Strain Variations. We have examined several mouse strains which are suitable for use in preparing EP-responsive erythroblasts. These include BALB/c, DBA/2, CBA, and (BALB/c × DBA/2)F₁ hybrid (CD₂F₁). The rate of development of erythroblastosis varies somewhat from strain to strain, a phenomenon which probably depends on the speed of replication and spread of the helper F-MuLV in the virus complex. Although the helper virus in our stock grows in mice which are *FV-1*ⁿⁿ or *FV-1*ᵇᵇ, it does appear to have some preference for *FV-1*ᵇᵇ mice and is thus somewhat "B-tropic." FVA stocks with different helper virus tropisms may differ with respect to the rate of the disease process in various mouse strains. Mouse strains which are *FV-2*ʳʳ, such as the C57BL strains, cannot be used as they are genetically resistant to development of Friend disease.

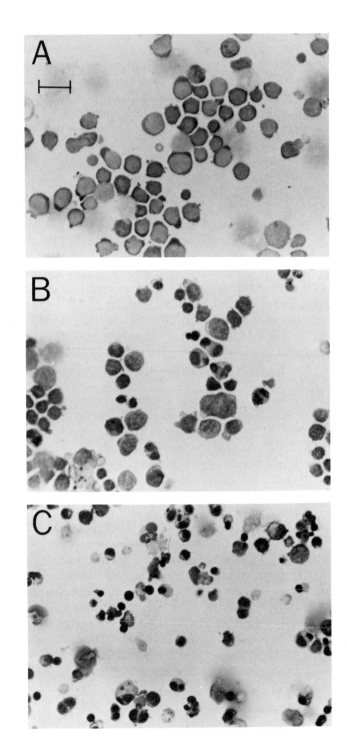

Characteristics of Cell Growth and Differentiation

Responsiveness to Erythropoietin. Figure 2 shows the morphological features of FVA-infected splenic erythroid cells immediately after isolation (Fig. 2A), after 24 hr of culture with EP (Fig. 2B), and after 48 hr of culture with EP (Fig. 2C). Initially, the cells are large with large nuclei containing diffuse chromatin. Fewer than 10% of the isolated erythroblasts typically stain with 3,3'-dimethoxybenzidine which detects the presence of heme. By 24 hr of culture, the cells become slightly smaller and the nuclei definitely stain more darkly with basic stains and are more compact. At 24 hr many cells exhibit faint benzidine staining revealing the beginning of heme synthesis. By 48 hr, the nuclei are very condensed, are often eccentrically located in the cell, and some enucleated reticulocytes are present. Greater than 85% of the erythroid cells exhibit strong benzidine staining at 48 hr. At 60 hr (not shown) reticulocytes and bare nuclei predominate. Upon culture with EP, the nucleated cell number typically more than triples by 48 hr and then gradually declines as the reticulocytes form. Studies of the proliferative potential of these erythroid cells showed that, although clusters of two to eight benzidine-staining cells were the most common progeny at 48 hr of culture with EP, many single cells and some colonies of 20 or more cells were present at the same time. Therefore, considering just proliferative capacity, these splenic erythroid cells include a range of erythroid precursors from the colony-forming units—erythroid (CFU-E)[13] to the late nonproliferating erythroblasts.

The incorporation of ^{59}Fe into heme is a very sensitive assay for the late erythroid maturation. EP-dependent stimulation of ^{59}Fe incorporation into heme peaks approximately 48 h after the cells are cultured, and the magnitude of this effect ranges from 50- to 100-fold over untreated cells. EP-dependent ^{59}Fe incorporation into heme is linear with cell number up to 10^6 cells plated per ml.[6] The ^{59}Fe incorporation increases with increasing EP up to 0.05 to 0.1 units of EP per ml and reaches a plateau at concentrations of EP higher than 0.1 to 0.2 units per ml (Fig. 3). We have used the ^{59}Fe incorporation into heme in response to EP to assay concentrations of EP in biological samples.

[13] A. A. Axelrad, D. L. McLeod, M. M. Shreeve, and D. S. Heath, *Proc. Int. Workshop Hemopoiesis Culture* **2nd**, 226 (1974).

FIG. 2. Photomicrographs of FVA-infected erythroblasts. (A) Spleen cells sedimenting at 7 mm/hr or greater at unit gravity. (B) Same cell population after 24 hr of culture with 0.2 units/ml EP, and (C) after 48 hr of culture with EP. Wright's stained, cytocentrifuged preparation. Bar equals 20 μm. From Koury *et al.*[6]

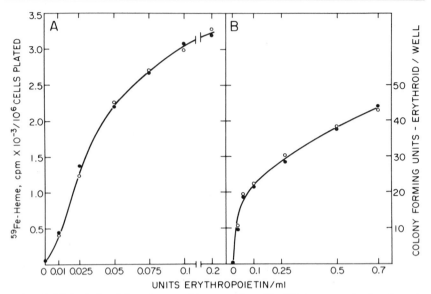

Fig. 3. Biological activity of mouse EP as quantitated by ^{59}Fe incorporation into heme in FVA-infected erythroid cells (A) and growth of CFU-E (B). Mouse EP (●) was compared to a standard preparation of human EP (○). EP-dependent ^{59}Fe incorporation into heme was determined in the interval from 24 to 48 hr after culture as described in the text. Normal marrow cells from an uninfected mouse were cultured in plasma clot by standard methods[13] with the indicated concentrations of EP. After 48 hr of culture, the clots were stained with benzidine and hematoxylin, and CFU-E were counted.

Biological Assay for Erythropoietin

The erythroid cell population we describe here is very well suited for the detection and quantitation of the biological activity of EP. These cells are very sensitive to EP in the range of 0.005 to 0.10 unit of EP per ml. Furthermore, the EP-dependent increase in ^{59}Fe incorporation into heme is up to 100-fold or greater over control levels. This low baseline response and a very large maximal response to EP are ideal factors for a highly accurate assay. In addition, to the present time we have not found any compound or a biological material other than erythropoietin which will trigger the maturation of these erythroblasts or increase the baseline level of ^{59}Fe incorporation into heme.

Method for the Assay of Erythropoietin. Splenic erythroid cells are prepared from FVA-infected mice and cultured in either liquid medium or semisolid, methylcellulose medium as described in the previous section. We routinely culture cells in plates containing 24, 16-mm-diameter wells. Each well is seeded with 5×10^5 cells in 0.5 ml of liquid medium. EP from a standardized preparation is added to some wells at a range of concentrations to generate a standard curve. Material from the source to be assayed for EP is added to other culture wells in a reasonable volume, usually 50 μl or less. However, if the material is a conditioned medium or present in a medium suitable for cell growth, volumes larger than 100 μl may be added as long as the cultures for the standard curve are diluted with a similar medium.

EP-dependent erythroid maturation is quantitated by the incorporation of ^{59}Fe into heme. The ^{59}Fe is first bound to transferrin. Human transferrin (Calbiochem-Behring, San Diego, CA) is dissolved in a buffer consisting of 250 mM Tris–HCl (pH 8.0) and 10 μM sodium bicarbonate (10 mg apotransferrin per ml). [59]FeCl$_3$ in 0.1 N HCl is added to this solution at 1 to 3 μCi of ^{59}Fe per 100 μg of transferrin. If the ^{59}Fe is added in a higher concentration of acid, the solution must be neutralized such that the pH does not drop below 5.0. The solution of ^{59}Fe and transferrin is incubated at 37° for 30 min, then IMDM is added such that the final concentration of transferrin in 2.5 mg/ml. At this stage the ^{59}Fe-transferrin can be added to the cultures, usually in a 50-μl aliquot (225 μg ^{59}Fe-transferrin per ml final concentration).

Two different time periods are commonly used in our laboratory to measure ^{59}Fe incorporation into heme: the first is to make the determination in the period from 40 to 48 hr after plating the cells, and the second is to determine ^{59}Fe incorporation into heme in the period from 24 to 48 hr after culturing the cells. However, one can obtain satisfactory results measuring the ^{59}Fe-heme synthesis during the entire 48-hr period or the first 24 hr of culture. The baseline level is slightly higher when ^{59}Fe-heme is determined during the first 12 hr of culture due to a small population of more mature erythroblasts which continue to synthesize heme independent of the action of EP.

[59]Fe-heme is extracted and counted by standard procedures. In our laboratory, the cells from culture wells are washed with 2 rinses of Hanks' balanced salt solution in 13 × 100-mm glass tubes. After sedimentation at 600 g for 10 min, the cell pellet is washed with 2 ml of Hanks' solution and sedimented again. The pellet is then resuspended in 2 ml of Drabkin's solution. Strong acid (0.2 ml of 1.0 N HCl) is added to release heme from hemoglobin. Following vortexing of the acidified cell extract, the heme is extracted into cyclohexanone (2 ml) with vigorous vortexing. Phases are separated by centrifugation of the mixture for 15 min at 1500 g. An aliquot of the upper cyclohexanone phase (usually 1.5 ml) is removed and counted in a gamma counter.

An example of an assay of the biological activity present in a preparation of mouse erythropoietin (Hy-Clone Laboratories, Logan, Utah) from a cell line which secretes EP is shown in Fig. 3. The concentration of the mouse EP had been previously estimated. An assay of the biological activity of EP based on the growth of erythroid colonies from normal mouse bone marrow (Fig. 3B) as well as our erythroblast assay (Fig. 3A) is compared. Both assays show identical responses to mouse EP compared to human EP (Cat-1, 1140 units EP/mg protein from the NIH, Bethesda, MD). We have also compared pure human EP obtained from recombinant methods (AMGen, Thousand Oaks, CA) and purified from human urine (from Dr. E. Goldwasser, University of Chicago) with preparations of EP from sheep urine (Hy-Clone Laboratories, Logan, Utah) and sheep serum (Connaught Laboratories, Willowdale, Ontario). All preparations of EP tested gave identical results in both our assay and the standard erythroid colony formation assay. We have also used this assay to verify that recombinant EP regained full biological activity after iodination.

We have found the EP-dependent incorporation of ^{59}Fe into heme in these erythroid cells is a highly reproducible assay for EP. In addition to being more sensitive than the colony forming assay, the assay is less time consuming since the ^{59}Fe-heme is quantitated by gamma counting rather than visual inspection of the cultures for erythroid colonies. Moreover, this assay method has advantages over other methods in which the level of EP is quantitated by measuring the incorporation of ^{59}Fe into heme. Fetal liver cells and marrow cells from erythropoietically stimulated animals have a large response in ^{59}Fe incorporation into heme to EP; however, these tissues have a heterogeneous mixture of developing erythroid cells, some of which have matured to a stage in which heme synthesis is no longer dependent on EP. Thus, the baseline level of ^{59}Fe incorporation into heme is very high compared to our assay method using a synchronous population of EP-sensitive cells. When EP-dependent ^{59}Fe incorporation into marrow cells explanted from erythropoietically suppressed animals is determined, the baseline response in the absence of EP is low, but the maximal response is also much less than our assay method since the number of responsive erythroid cells in a suppressed marrow is of the order of a few percent or less of the total cell population.

In view of the advantages of having a very large response to EP and a very low baseline response, the extra effort to procure purified EP-responsive erythroid cells for the assay of EP is justified in most cases. It is also possible that the purification of the cells from the enlarged spleens of the FVA-infected mice may be eliminated when assaying EP levels. We have shown that the baseline incorporation of ^{59}Fe into heme is greater in unseparated cells from the enlarged spleens compared to purified cells, yet this effect is minimal if heme synthesis is determined at a interval later than 36 hr after the cells are cultured (EP-dependent ^{59}Fe incorporation into heme is 20-fold or greater than control during the period 36 to 48 hr after culture of unseparated cells).[6] However, one should try to obtain cells only from intermediate sized spleens if the erythroid cells are to be used without further purification as the numbers of late erythroblasts are reduced in the smaller spleens earlier in the course of the FVA-induced erythroblastosis.

This method for the assay of biological activity of EP is affected by only a few interfering factors. High concentrations of transferrin in the material to be assayed can diminish the incorporation of ^{59}Fe into heme; however, the concentration of transferrin in the assay (30% fetal calf serum containing 2 to 3 mg transferrin per ml) minimizes this effect. Toxic heavy metals and other inhibitory substances in the material to be assayed may reduce the ^{59}Fe incorporation. In this event, the ^{59}Fe incorporation into heme will reach a lower plateau level. We have identified two compounds, dimethyl sulfoxide and calcium ionophore A23187, which enhance the effect of suboptimal levels of EP on ^{59}Fe incorporation into heme.[14] These agents have no effect in the absence of EP and with maximal concentrations of EP. We have tested sources of hemopoietic growth modulators such as erythroid burst promoting activity, interleukin-3, and conditioned medium from several sources for an enhancing effect on EP-dependent ^{59}Fe incorporation into heme in these erythroid cells. Thus far none of these factors has had an effect. Therefore, it seems unlikely that this assay for the biological activity of EP will give falsely elevated activity due to other biological modulators. While primitive erythroid precursor cells, such as the burst forming units–erythroid,[13] are probably greatly influenced by these regulatory factors, the more mature erythroid cells used in this assay are under the primary if not exclusive control of EP.

Concluding Comments

Our method for procuring a large population of synchronized, EP-responsive erythroid cells allows the investigation of the mechanism by which EP induces the maturation of erythroblasts. This had not been possible due to limitations of previously used cell systems. Continuous erythroid cell lines, such as the Friend erythroleukemia cells, have been used extensively in the study of erythroid maturation. The purity and developmental synchrony of these cells are well suited for the study of some events in erythroid maturation. However, these cells are induced by

[14] S. T. Sawyer and S. B. Krantz, unpublished data.

nonphysiological agents and not by EP. Many biochemical studies have employed primary explants of hematopoietic tissues such as bone marrow or fetal liver. The EP-responsive cells in these tissues are usually a minority of the cell population and are at many stages of development. This cellular impurity and lack of developmental synchrony seriously compromise the interpretation of data gathered using these tissues as a model for erythroid maturation.

We have established four criteria of a cell population which are necessary to study the mechanism by which EP acts on its target cell: responsiveness to the hormone, high purity, synchrony of differentiation, and adequate cell number for biochemical experiments. As far as we know, our erythroid cell population from spleens of mice infected with FVA is the only system which satisfies all of these criteria. Recently, a method to isolate a pure erythroid population of cells from mice bled while recovering from thiamphenicol treatment has been reported.[15] Although high purity and synchronized differentiation may exist in these cells, the bleeding of these animals leads to erythroid cells which have been exposed to high levels of EP prior to explantation. Moreover, to procure the same number of cells using this method, 10^7 cells/spleen,[15,16] which can be obtained by our method, 10 times the number of mice and much more time spent in cell purification are required.

With our method described here we have obtained data on the EP-dependent synthesis of heme, globins,[5] band 4.1, spectrin, and band 3.[17] In addition we have used these cells to study EP-dependent transcriptional regulation of globin message and the organization of chromatin around the globin gene during differentiation.[18] This cell system has also been used to identify binding sites for EP[19] as well as to study the regulation of transferrin receptors in these developing erythroid cells.[20] More recently, these cells were used to study the internalization of EP by receptor-mediated endocytosis[21] and the structure of the receptor for EP by cross-linking with ^{125}I-labeled EP.[22]

[15] W. Nijhof and P. K. Wierenga, *J. Cell Biol.* **96**, 386 (1983).
[16] W. Nijhof, P. K. Wierenga, G. E. J. Staal, and G. Jansen, *Blood* **64**, 607 (1984).
[17] M. J. Koury, M. C. Bondurant, and T. J. Mueller, *J. Cell. Physiol.* **126**, 259 (1986).
[18] M. C. Bondurant, R. N. Lind, M. J. Koury, and M. E. Ferguson, *Mol. Cell. Biol.* **5**, 675 (1985).
[19] S. B. Krantz and E. Goldwasser, *Proc. Natl. Acad. Sci. U.S.A.* **81**, 7574 (1984).
[20] S. T. Sawyer and S. B. Krantz, *J. Biol. Chem.* **261**, 9187 (1986).
[21] S. T. Sawyer, S. B. Krantz, and E. Goldwasser, *J. Biol. Chem.*, in press (1987).
[22] S. T. Sawyer, S. B. Krantz, and J. Luna, *Proc. Natl. Acad. Sci. U.S.A.*, in press (1987).

Section V

Techniques for the Study of Growth Factor Activity: Genetic Approaches and Biological Effects

[31] Isolation of Mitogen-Specific Nonproliferative Variant Cell Lines

By HARVEY R. HERSCHMAN

The mitogenic response of quiescent, nondividing cells following exposure to mitogens is accompanied by an extensive array of biochemical and physiological alterations. A large number of events—collectively termed the "pleiotypic response"[1]—occur prior to DNA synthesis in cells responding to mitogens, including (but not limited to) elevated transport of ions and small molecules, release of arachidonic acid, prostaglandin synthesis, induction of rate-limiting enzymes in anabolic pathways (e.g., ornithine decarboxylase, phosphofructokinase), a variety of phosphorylation reactions, receptor down-regulation, altered rates of protein degradation, elevated rates of RNA and protein synthesis, and induced expression of new gene products. Because tissues are a mixed population of different cell types, some of which are actively transversing the cell cycle, while others are in a quiescent, G_0 state, it is difficult to study mitogen induction of cellular proliferation in intact animals. In addition, in studies with whole animals it is difficult to distinguish primary effects of exogenously administered mitogenic agents from effects mediated via an intermediary tissue or cell. Finally, the amount of a biologically active agent necessary to obtain an effect *in vivo* is often prohibitive. For these reasons the majority of studies on the mechanism of action of mitogenic agents has been done in cell culture.

3T3 Cells as a Model System

Murine Swiss 3T3 cells exhibit "growth control" in culture; they reach a saturation density proportional to the amount of serum in the medium, then exit the cycle and enter a viable, nondividing state. If additional serum is added, after a 10–12 hr period the cells initiate DNA synthesis and subsequently divide.[2] Because Swiss 3T3 cells can be reversibly and synchronously shifted from noncycling to cycling, they have become an important research tool in the study of the regulation of cellular proliferation.

[1] A. Hershko, P. Mamont, R. Shields, and G. Tompkins, *Nature (London), New Biol.* **232**, 206 (1971).
[2] R. W. Holley, *Proc. Natl. Acad. Sci. U.S.A.* **60**, 300 (1968).

Quiescent 3T3 cells can be stimulated to reenter the cycle by a variety of agents, including (1) *polypeptide mitogens* such as epidermal growth factor (EGF), fibroblast growth factor (FGF), and platelet-derived growth factor (PDGF), (2) fatty acid derivatives such as *prostaglandin* $F_{2\alpha}$, and (3) the *tumor promoters* tetradecanoylphorbol acetate (TPA, a macrocyclic diterpene ester) and teleocidin (an indole alkaloid). In addition, the somatomedin-like molecules IGF-I and IGF-II, as well as insulin, play supportive roles by modulating the activity of a variety of mitogens.

Utility of Mitogen-Specific Nonproliferative Variants

Given the variety of agents able to stimulate the mitogenic response and the large number of phenotypic alterations that accompany mitogenesis, one is immediately confronted with two striking, fundamental questions: (1) Do mitogens of diverse structure stimulate the same set or different sets of causal events distal to receptor occupancy, i.e., are there common or distinct pathways to DNA synthesis and cell division for different mitogens? (2) What are the *causal*—as opposed to the *correlative*—events which occur in the transition from a noncycling to a cycling cell in response to a mitogen? To approach these questions we developed a procedure to select mitogen-specific, nonproliferative variants of Swiss 3T3 cells. We have used this procedure to select variant Swiss 3T3 cell lines unable to respond either to EGF,[3,4] or to TPA.[5]

An Overview of the Colchicine Selection Procedure for the Isolation of Mitogen-Specific Nonproliferative Variants

The procedure for isolating mitogen-specific nonproliferative variants is an adaptation of that used by Vogel *et al.*[6] for the selection of "flat" variants of transformed 3T3 cells. The selection takes advantage of the ability of colchicine to arrest cells in the mitotic phase of the cell cycle. The procedure is summarized in Fig. 1. Swiss 3T3 cells are grown to density arrest in medium containing 5% fetal calf serum. The 3T3 cells exit the proliferative phase of the cell cycle and enter a G_0, nondividing state. The mitogen for which nonproliferative variants are to be selected is added to the medium on the quiescent cells, along with a concentration of colchicine capable of arresting mitotic cells. Mitogen-responsive cells

---

[3] R. M. Pruss and H. R. Herschman, *Proc. Natl. Acad. Sci. U.S.A.* **74,** 3918 (1977).
[4] E. Terwilliger and H. R. Herschman, *Biochem. Biophys. Res. Commun.* **118,** 60 (1984).
[5] E. Butler-Gralla and H. R. Herschman, *J. Cell. Physiol.* **107,** 59 (1981).
[6] A. Vogel, R. Risser, and R. Pollack, *J. Cell. Physiol.* **82,** 181 (1973).

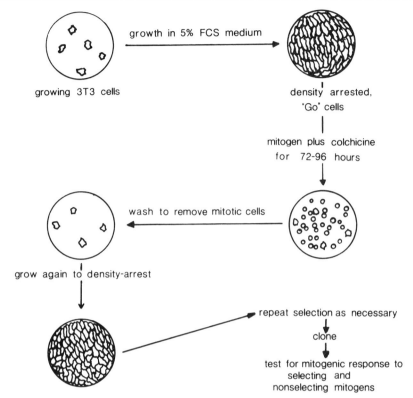

growing 3T3 cells

growth in 5% FCS medium

density arrested,
"Go" cells

mitogen plus colchicine
for 72-96 hours

wash to remove mitotic cells

grow again to density-arrest

repeat selection as necessary

clone

test for mitogenic response to
selecting and
nonselecting mitogens

FIG. 1. Protocol for isolation of mitogen-specific nonproliferative variants of murine 3T3 cells.

exit G_0, enter G_1, and proceed through the phases of the cycle until they enter mitosis. The colchicine in the medium prevents the mitogen-responsive cells from progressing through mitosis, and arrests their progression through the cell cycle at this point. The mitotic cells round-up, and are only loosely attached to the cell culture dish; in contrast, the interphase cells that have not responded to the mitogen treatment are tightly attached to the substratum. The mitotic cells can, therefore, be preferentially removed from the plastic surface by vigorous pipetting (or shaking, if the experiment is carried out in flasks). Three to four days are usually allowed between addition of mitogen/colchicine and the removal of arrested mitotic cells. After removal of mitotic, colchicine-arrested cells the remaining cells are detached with trypsin, replated, grown again to density arrest in medium containing 5% fetal calf serum, and subjected to the same selection procedure. Multiple rounds of enrichment are necessary to isolate mitogen-nonresponsive variants, since a significant portion of the initial population of quiescent cells does not, for unknown physiological reasons, respond to mitogen stimulation. The mitogen/colchicine procedure is continued until the response of the selected population is reduced to 50% of that observed for the parental population. At this point individual clones are isolated and tested for mitogenic responsiveness both to the selecting mitogen and to unselected mitogens.

In principle, this selection procedure should be applicable to any cell line that adheres well to the culture dish. It should also be possible to use any agent that kills dividing cells (e.g., tritiated thymidine suicide, bromodeoxyuridine plus UV light, etc.) to select against the mitogen-responsive cells (see Straus [32], this volume, for related methods). However, cells altered in a mitogen-driven, G_2-specific function would escape selection with an S phase selective agent, but would be unable to escape the colchicine selection.

The parental population of cells should exhibit the following characteristics: (1) the cells should demonstrate low saturation density, (2) they should exhibit a strong mitogenic response to the selecting mitogen and to a second mitogen (preferably elevated serum), and (3) they should be free from mycoplasma contamination.

If one wishes to develop multiple, independent, nonproliferative variants, it is necessary to first isolate independent, mitogen-responsive clones from the parental population and carry out independent selections for each variant, choosing only a single isolate from each selection. Since multiple rounds of selection are used, a variant could arise at any point in the selection, be enriched in the ensuing rounds, and be cloned a number of times in the final stages.

Characterization of the Mitogenic Response of the Parental Cell Population

The specific descriptions presented below will be for the selection of mitogen nonresponsive Swiss 3T3 cell lines. The parental cell population is first evaluated for mitogen response. Care should be taken to use cells that have not been allowed to grow to high density, since such conditions will select for transformed sublines. We routinely passage 3T3 cells at approximately three-quarters confluence. To characterize the mitogenic response, cells (5×10^4) are plated in 60-mm tissue culture plates in 5 ml of Dulbecco's modified Eagle's medium containing 5% fetal calf serum. Unless noted, all experiments are performed in this medium. Fifty plates are set up to demonstrate this initial mitogen response. Cell counts are

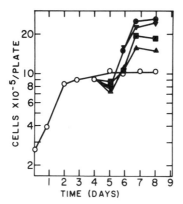

FIG. 2. Stimulation of cell division in quiescent, G_0 3T3 cells by various mitogens. Data are the averages of duplicate counts, 10 ng EGF/ml ($\bullet$), 100 ng TPA/ml ($\blacktriangledown$), 100 ng $PGF_{2\alpha}$/ml ($\blacktriangle$), 10 μg insulin/ml ($\blacksquare$), and control ($\bigcirc$). Data are from Herschman et al.[7]

done daily on two plates. For this purpose, cells are washed with phosphate-buffered saline and detached from the dish with a trypsin EDTA solution. Cell numbers can then be determined with a hemocytometer, or by using an electronic cell counter. On day 5 the medium on the remaining plates is replaced with fresh medium. Under these conditions Swiss 3T3 cells should reach saturation density on day 6 to 7. In order to test mitogen responsiveness the cells should be at saturation density for 2 to 3 days. Mitogen is then added to 10 of the remaining plates, using a small volume of appropriate vehicle. Additional fetal calf serum (0.27 ml) is added to a second set of 10 plates, to bring the final concentration to 10% fetal calf serum. A third set of plates receives only the control vehicle for the mitogen. On each successive day two plates for each group (mitogen-treated, serum-stimulated, control) are removed and use to determine cell number. Cells receiving vehicle only should remain viable and attached to the plate throughout this treatment; their cell number should not decline appreciably. Cells receiving additional serum should be stimulated to divide; their cell number 4 days after addition should be 1.7- to 2.2-fold greater than control cells. A vigorous response to a defined mitogen should be, in this context, 1.7- to 2.0-fold. If the optimal concentration of the mitogen is not known it should be determined at this point. A representative response curve for Swiss 3T3 cells to several mitogens is shown in Fig. 2.[7]

[7] H. R. Herschman, D. S. Passovoy, R. M. Pruss, and A. Aharonov, J. Supramol. Struct. **8**, 263 (1978).

Subcloning of Parental Cells Prior to Selection

Mitogens such as EGF, TPA, and FGF generally stimulate DNA synthesis in about 50–60% of a confluent, G_0 population of 3T3 cells. Since many cells in the population do not respond to the mitogen, multiple rounds of enrichment are necessary to isolate a mitogen nonresponsive variant. Once the response of the parental population has been established, independent clones should be isolated and used for the selection of only a single variant from each parental subclone. This will ensure that individual nonproliferative variants are the products of independent events. To clone Swiss 3T3 cells we simply dilute populations to one cell per milliliter in medium and plate 0.1 ml samples in 96-well plates. Wells are scored the following day for single cells, and individual clones are passaged, and frozen in multiple ampules. The subclones are then tested as described in the preceding section for saturation density and for mitogenic response, both to the defined mitogen to be used in the selection and to elevated serum. Once again, care should be taken to ensure that the subcloned populations are not permitted to grow to high densities.

Establishing a Concentration of Colchicine to Be Used for Selection

The ideal concentration of colchicine will be one that prevents mitosis, but has a minimal effect on the viability of noncycling cells. To establish an appropriate colchicine concentration for Swiss 3T3 cells 2×10^4 cells were plated in each of 20 60-mm dishes. Medium was changed every other day. Duplicate cell counts were performed on successive days. When the cells reached approximately 8×10^5 cells/dish, dilutions of colchicine bracketing a final concentration of 0.1–10 μg/ml (in 3-fold dilutions) were added to the remaining plates. The plates were observed after 24 and 48 hr, and the minimal concentration required to arrest the proliferating cells in mitosis was determined by inspection. For Swiss 3T3 cells this value is 1.5 μg/ml. Since the determination is simple and rapid, it should be done prior to beginning a new mitogen-selection series, in order to evaluate cells, drug, etc. If selections are to be done on other cell lines this preliminary evaluation is clearly required. Once an effective concentration of colchicine for inhibition of mitosis is established, its effect on density-arrested cells should be evaluated. Cells are grown to density arrest as described in the section on Characterization of the Mitogenic Response of the Parental Cell Population. Colchicine (at the concentration established above) is added 2–3 days after density arrest has been established. Three and four days later cells from control and colchicine-treated plates are removed with trypsin from their culture dishes, serially diluted in medium, and plated to evaluate cell viability by colony formation.

Enrichment of Mitogen Nonproliferative Variants

An appropriate mitogen concentration for stimulation of cell division has presumably been established by prior dose–response curves. For the initial round of selection 40 100-mm cell culture dishes containing 8×10^5 cells in 10 ml of medium are established for each parental cell clone. Duplicate cell counts are done daily as the cells approach confluence. One day prior to estimated density arrest, cells are fed with fresh medium. Two days after establishment of density arrest the following additions are made: (1) 8 plates of cells receive 0.55 ml of fetal calf serum (raising the serum concentration to 10%), (2) 8 plates of cells receive the mitogen for selection, (3) 2 plates receive colchicine, (4) eight plates receive control vehicle, and (5) 10 plates receive mitogen plus colchicine. Duplicate cell counts are done daily on the serum-stimulated, mitogen-stimulated, and control populations. Colchicine-treated plates (with and without mitogen) are observed with the microscope, to monitor the progress of the selection. Although the evaluation of the serum and mitogen responses in the absence of colchicine may seem tedious, it is comforting to know that an appropriate mitogenic response occurred in parallel cultures in the selection. The maximal mitogenic response for Swiss 3T3 cells is usually obtained by day 3 or 4 after mitogen addition. Although this may vary, comparison of the ongoing mitogen response data with prior responses allows one, in practice, to pick the appropriate day (e.g., day 3 or day 4) to harvest the cells. For cell lines other than Swiss 3T3, the prior characterization of the mitogen response will be of critical importance.

In the presence of colchicine plus mitogen the majority of the cells will be rounded, refractile, and only loosely attached to the plate. Cells receiving only colchicine, in contrast, will be well spread on the dish, and tightly attached to the plastic surface. Although some giant, multinucleated cells may be present and some bare areas may occur in the monolayer, the cells exposed to colchicine alone will resemble the control culture. Examples of colchicine and colchicine plus EGF treated Swiss 3T3 cells are shown in Fig. 3.

Mitogen-responsive cells are removed from the plate by repeated washing with a stream of medium from a pipet. If fresh medium is used after several minutes of washing, removal of mitotic cells can be monitored microscopically. When the experimenter is satisfied that all observable mitotic cells have been removed, the remaining cells from the 10

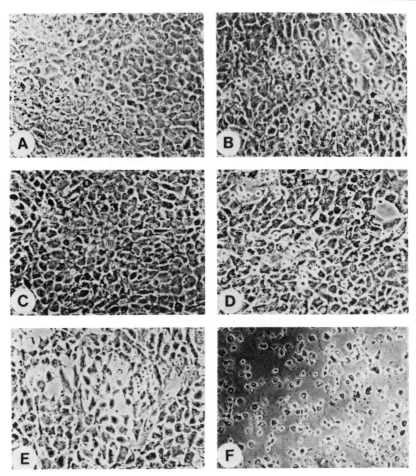

FIG. 3. 3T3 cells treated with colchicine (1.5 μg/ml) alone or colchicine and EGF (10 ng/ml). Cells treated with colchicine alone for (A) 1 day, (C) 2 days, and (E) 3 days. Cells treated with colchicine and EGF for (B) 1 day, (D) 2 days, and (F) 3 days. Data are reproduced from Pruss and Herschman.[3]

plates exposed to colchicine plus mitogen are fed and returned to the incubator. The next day the cells are removed with trypsin/EDTA, pooled, replated, and expanded. Once again, cells are subcultured at approximately three-quarters confluence, to prevent overgrowth of spontaneously transformed cells. The procedure described here thus selects for mitogen nonresponsive cells that have plating efficiencies and growth rates that permit them to compete with parental cells in mixed culture.

Repeated Rounds of Enrichment for Mitogen-Nonresponsive Variants

At this point the population will be enriched for mitogen nonrespond-ers, but will still have a substantial majority of parental, mitogen-respon-sive cells. Repeated enrichments, identical to the one described in the preceding section, are carried out. Analysis of the mitogenic responsive-ness of the enriched population to serum and to the selected mitogen is also carried out with each round, to monitor loss of specific vs general mitogenic response in the enriched population. Mitogenic responses to serum and the selected mitogen are also monitored on at least one paren-tal cell line. This control is valuable in determining whether reduced mitogen responsiveness during the rounds of enrichment is due to a change in the population of cells (i.e., increase in the number of variants) or to artifacts of the culture conditions or reagent preparation. Although the growth curves are tedious, they greatly facilitate evaluation of the progress of the selection. To prepare for the second round of enrichment the cells from the first selection are subcultured to a point where 40 plates of cells at 8×10^5 cells/100-mm dish can again be plated in medium containing 5% fetal calf serum. At this point cells should also be frozen, to provide a reference population and a source of cells for continued selec-tions, should a problem develop in the subsequent rounds of enrichment. The same additions are made at 2 days after density arrest: (1) 8 plates of cells surviving the first round of selection receive 0.55 ml of fetal calf serum, (2) 8 plates receive mitogen, (3) 8 plates receive vehicle, (4) 2 plates receive colchicine, and (5) 10 plates receive colchicine plus mito-gen. Parental cells are plated at the same density, and receive either elevated serum, mitogen, or vehicle at the appropriate time. Comparison of the serum response of the enriched versus the parental population is useful in evaluating the general mitogenic potential of the enriched popu-lation. Comparison of the mitogen response to the serum response in the selected population versus the parental population is useful in evaluating the change in mitogenic potential of the population with regard to the selecting mitogen. After the second selection cells surviving mitogen/colchicine exposure are once again pooled and expanded for additional selections.

Subsequent selections are carried out as described for the second round. Cells should be frozen after each round of selection, to provide a fresh starting source, should any problem arise in the subsequent selec-tions. Responses to elevated serum and to the selecting mitogen should be measured in each round of selection, both in the population under selec-tion and in a parental cell line, to evaluate the quality of the reagents, the general mitogenic responsiveness of the selected population, and the loss of specific mitogenic responsiveness.

Cloning and Initial Characterization of Potential Mitogen Nonresponsive Variants

As the selection progresses, a larger percentage of the mitogen/colchicine-treated cells will remain on the dish. Thus the growth and expansion of the resistant population become less demanding in reagents and time as the rounds of selection increase. When the mitogen response of the population to the selecting mitogen has been reduced 50% relative to the response of the parental population, cells are cloned by dilution in microtiter wells. The number of rounds of selection required will, of course, be a function of the mitogenic potency of the selecting mitogen. For EGF and TPA, both of which stimulate 40–60% of the quiescent Swiss 3T3 cells to enter DNA synthesis, three to five rounds of selection have been necessary.[3,5] Ten clones are usually isolated from each selected population. Cells are expanded and cryopreserved as soon as possible.

Most mitogens, in addition to stimulating cells to initiate a G_0–G_1 transition, also cause 3T3 cells to attain a higher saturation density if added to exponentially growing cells. If the selecting mitogen has this capability, it greatly facilitates evaluation of potential mitogen nonresponsive clones. To test the selecting mitogen for this property 30 35-mm plates of cells, each containing 10^5 cells in 2 ml of medium with 5% fetal calf serum, are prepared. Twenty-four hours after plating, cell counts are performed on the cells from two culture dishes. Ten of the remaining plates receive the selecting mitogen, 10 receive an additional 0.1 ml of serum, and 8 receive no addition. Cell counts are done at 2 day intervals on duplicate plates from each population. No medium changes are made throughout the course of the experiment. Mitogens are added 24 hr after plating in order to avoid any effect on plating efficiency. Swiss 3T3 cells grown in this fashion attain a saturation density in 10% serum, 5% serum plus 10 ng EGF/ml, or 5% serum plus 50 ng/ml TPA, about twice that of cells grown in medium supplemented with 5% serum. If the selecting mitogen demonstrates this characteristic, screening of clones is greatly facilitated, since entry into density arrest need not be monitored. Since clonal isolates often grow at different rates, it is difficult to coordinate mitogen response curves subsequent to density arrest for a variety of cell lines. Moreover, we have found that mitogen-nonresponsive lines may differ from their parent line in saturation density. Thus it is tactically much easier to carry out continuous growth curves of the type described here. In practice, three potentially mitogen-nonresponsive subclones from each parental line are tested at a time, since once a variant demonstrating a nonproliferative phenotype is observed we generally stop analysis of that population of subclones (at least temporarily), since our objective has been to isolate independent mitogen-nonresponsive variants.

Once clones have been provisionally identified as mitogen nonresponsive, their ability to escape mitogen responsiveness is tested in an assay that measures release from density arrest. Variant cell lines and parental lines are simultaneously tested as described above in the section on Characterization of the Mitogenic Response of the Parental Population. Responses both to the selecting mitogen and elevated serum are characterized, and appropriate variants are identified.

Additional Characterization of Mitogen-Specific Nonproliferative Variants

One potential explanation for the nonproliferative phenotype of a variant may be a reduced sensitivity to the selecting mitogen, although cells might still retain a maximal response potential at elevated mitogen concentrations. Cells producing an altered receptor of lowered affinity might have such a phenotype. This is easily established by carrying out a dose–response study of mitogenic response for the parental and variant cell lines. Mitogen response curves at differing concentrations of selecting mitogen are performed and the maximal cell number is plotted as a function of mitogen concentration. In practice, we have not found such variants; both EGF nonresponsive[3,4] and TPA nonresponsive[5] variants do not demonstrate an increase in cell number at mitogen concentrations two orders of magnitude greater than the parental cell line.

A second question of general importance is the ability of the variants to escape G_0 in response to the selecting mitogen. This may be approached in two ways. The ability of mitogen-nonresponsive variants to enter DNA synthesis may be monitored, relative to parental cells. To carry out such studies parental cells and mitogen-nonresponsive cells are grown to density arrest in 24-well culture plates in 1.0 ml of medium. Mitogens are added without a medium change to quiescent cells. At 0, 12, 18, 24, 30, and 36 hr after mitogen addition, triplicate sets of cultures are assayed for mitogen stimulated incorporation of tritiated thymidine. Fresh medium containing 0.2 μCi of tritiated thymidine per ml is added to each well at the appropriate time, and plates are incubated for 1 hr. Medium is removed and the cells are washed three times with phosphate-buffered saline solution. One milliliter of ice cold TCA (10%) is added, and the plates are incubated on ice for 15 min. The cells are again washed four times with PBS. Cells are solubilized with 1 ml of 0.5 N NaOH. After 2 hr 0.8 ml of the NaOH solution is neutralized with NaH_2PO_4 and the radioactivity is analyzed. Once again, the EGF and TPA nonproliferative variants isolated to date all are unable to initiate DNA synthesis in response to the selected mitogen.[3-5]

The ability of cells to exit the G_0 phase and enter the cell cycle can also be analyzed with a cytofluorograph. At appropriate times after treatment with mitogens cells can be removed with trypsin, collected by centrifugation, and resuspended in PBS containing 0.7 units/ml ribonuclease A (Sigma type IIIA) and 0.1% Nonidet P-40. After a 15 min incubation at 4°, the nuclei are pelleted and resuspended in an ice-cold solution containing (per liter) 1 g trisodium citrate, 0.58 g NaCl, 150 mg propidium iodide, and 1 ml of Nonidet P-40. DNA content per cell can, by this means, be measured in parental and variant cell lines after exposure to mitogen for various times.

The final general question to be addressed is the specificity of the nonproliferative phenotype in the variants. Is the nonproliferative phenotype unique to the selecting mitogen, or does the variant demonstrate a nonresponsive phenotype for a collection of mitogens? Since a variety of mitogens that are active in this paradigm are not commercially available (e.g., EGF, TPA, FGF, PDGF), the mitogenic specificity of the variant lines can be characterized by carrying out growth curve analyses of the type described above. Alternatively, entry into S can be monitored by tritiated thymidine synthesis after exposure of quiescent variant and parental cells to a phalanx of mitogens.

Once several independent variants are isolated, it is of interest to determine whether their phenotypes are dominant or recessive. In addition, complementation analysis is valuable in sorting the variants into functional classes. In the case of Swiss 3T3 mitogen-nonresponsive mutants the question of dominance may be able to be examined without inserting selectable markers in the variants. A Swiss 3T3 cell line exists that is both ouabain resistant (Ou^R) and thymidine kinase (TK^-) deficient.[8] Both normal Swiss 3T3 cells and the 3T3R5 Ou^RTK^- cell line will die in "ouabain-HAT" medium; ouabain will kill the former, "HAT" selective medium will kill the latter. Hybrids will survive; ouabain resistance will be conferred on the hybrid by the 3T3R5 parent while a functional thymidine kinase will be contributed by the other parent. If 3T3R5 has a mitogen response to the selecting mitogen (or if an alternative phe-

[8] E. Terwilliger and H. R. Herschman, *J. Cell. Physiol.* **123**, 321 (1985).

notype, such as growth factor binding, for the variant can be assayed) hybrids between 3T3R5 and the mitogen-nonresponsive variants can be analyzed for dominance. Single cell preparations of 3T3R5 and the variant cell line are prepared and 2×10^5 cells of each type are plated together in a 60-ml dish in medium containing 5% fetal calf serum. One day later the culture medium is removed and the cells are washed with serum-free medium. The mixed cultures are then treated with 2 ml of a 250 μg/ml solution of Bactohemagglutinin P (PHA, Difco) prepared in medium without serum. After removing the PHA solution the mixed cultures are treated with 1 ml of a solution of 5% dimethyl sulfoxide and 50% polyethylene glycol 1000 (BDH chemicals) prepared in serum-free medium. Plates are then washed three times with serum-free medium, fed with medium containing 5% serum, and incubated overnight. The following day cells are removed with trypsin and replated at 5×10^3 cells per 60-mm dish in selective medium (2×10^{-5} hypoxanthine, 10^{-5} M aminopterin, 3×10^{-5} M thymidine, 3×10^{-3} M ouabain). Plates are refed at intervals with selective medium, and resistant colonies are cloned and expanded. Mitogenic responses to serum and to the selecting mitogen are analyzed as described above in the parental cell lines and in several hybrids. To verify the hybrid nature of the fusion products DNA content should be measured by cytofluorography. Both dominant and recessive mitogen-nonproliferative variants of Swiss 3T3 cells have been described[8] (see Straus [32], this volume, for related methods).

To determine whether independent isolates of mitogen resistant variants fall into distinct functional classes, variants must be hybridized to one another in pair-wise fashion. This requires that each parent have a dominant, selectable trait. It is possible to do such experiments by first transfecting dominant selectable markers such as the *ecogpt*, *neo*, or amplified metallothionein genes into the variant cell lines, selecting resistant cells, verifying the mitogen-nonresponsive phenotypes, selecting hybrids, and testing the hybrid cell lines for mitogen responsiveness or for the presence or absence of an appropriate correlative phenotype, e.g., the presence of ligand binding sites.[8] The descriptions of transfection procedures to isolate cell lines carrying dominant selectable markers can be found in a variety of sources.[9] These procedures are not unique to mitogen-nonproliferative variants. Analysis of the hybrids between pairs of mitogen-nonresponsive variants for restoration of the parental phenotype (i.e., variant complementation) is carried out as described above, using an appropriate selective medium.

9 P. J. Southern and P. Berg, *J. Mol. Appl. Genet.* **1**, 327 (1982).

Additional Characterization of Mitogen-Nonproliferative Variants; Characteristics Unique to the Selecting Mitogen

The purpose of selecting mitogen-specific nonproliferative variants is to provide a tool for understanding pathways of mitogenesis. Once unique characteristics of the mitogens in question are known, a wide range of specific experiments are possible. Thus in our own work several EGF nonresponsive variants are unable to bind EGF[3,4] and have recently been shown to lack both a serologically recognizable antigen related to the EGF receptor and messenger RNA for the receptor.[10] These cells are, however, transformable by murine sarcoma viruses—suggesting that transformation by these viruses is not dependent on occupancy of the EGF receptor by transforming growth factors.[11] These experiments are cited to indicate the utility of such variants. Similarly, TPA nonproliferative variants have been demonstrated to retain c-kinase activity,[12] phorbol dibutyrate binding,[13] TPA-induced elevations in glucose transport,[14] and arachidonic acid release,[15] but be deficient in TPA-induced induction of ornithine decarboxylase activity.[14] Varshavsky proposed that TPA exerts its ability to enhance gene amplification via its activity as a mitogen.[16] TPA nonproliferative variants might well not demonstrate TPA-enhanced gene amplification if this hypothesis is correct; we have recently observed that the Swiss 3T3 TPA nonproliferative variant (3T3-TNR9) is, indeed, unable to demonstrate gene amplification in response to TPA, in contrast to the parental cell line.[17] Colburn et al.[18] have used this colchicine selection procedure to select TPA nonproliferative variants from the JB-6 murine epidermal cell line. This line normally cannot grow in soft agar, but will convert irreversibly to ability to grow in soft agar after exposure to TPA. It thus appears to be initiated, and subject to TPA-dependent tumor promotion. TPA is a potent mitogen for JB-b cells as well. One class of TPA nonproliferative variants is still sensitive to TPA-induced promo-

[10] C. A. Schneider, R. W. Lim, E. Terwilliger, and H. R. Herschman, Proc. Natl. Acad. Sci. U.S.A. 83, 333 (1986).
[11] R. M. Pruss, H. R. Herschman, and V. Klement, Nature (London) 274, 272 (1978).
[12] R. Bishop, R. Martinez, M. J. Weber, P. J. Blackshear, S. Beatty, R. Lim, and H. R. Herschman, Mol. Cell. Biol. 5, 2231 (1985).
[13] P. M. Blumberg, E. Butler-Gralla, and H. R. Herschman, Biochem. Biophys. Res. Commun. 102, 818 (1981).
[14] E. Butler-Gralla and H. R. Herschman, J. Cell. Physiol. 114, 317 (1983).
[15] E. Butler-Gralla, S. Taplitz, and H. R. Herschman, Biochem. Biophys. Res. Commun. 111, 194 (1983).
[16] A. Varshavsky, Cell 25, 561 (1981).
[17] H. R. Herschman, Mol. Cell. Biol. 5, 1130 (1985).
[18] N. H. Colburn, E. J. Wendel, and G. Abruzzo, Proc. Natl. Acad. Sci. U.S.A. 78, 6912 (1981).

tion. Thus selection of TPA nonproliferative variants has, in this case, dissociated TPA-induced mitogenesis from TPA-induced tumor promotion. Mitogen-nonproliferative variants provide research tools to investigate a wide (and ever increasing) range of questions concerned with receptor–ligand interactions, growth control, oncogenes, and gene expression.

Acknowledgments

Supported by Grant GM 24797 from the National Institutes of Health and Contract No. DE AC03 76 SF00012 from the Department of Energy. Drs. Rebecca Pruss, Edith Butler-Gralla, and Ernest Terwilliger contributed extensively to the development of the selection and characterization procedures described here.

[32] Use of Variant Cell Lines and Cell Hybrids for the Study of Hormone and Growth Factor Action

By DANIEL S. STRAUS

Introduction

Polypeptide hormones and growth factors regulate the metabolism, proliferation, and differentiation of mammalian cells via biochemical mechanisms that are at present poorly understood. Considerable progress has been made in the past few years in the biochemical characterization of receptors for polypeptide hormones and growth factors. A number of these receptors, including the receptors for insulin,[1-3] insulin-like growth factor I (IGF-I),[4,4a] epidermal growth factor (EGF),[5,5a] and platelet-derived growth factor (PDGF),[6,6a] have tyrosine-specific protein kinase activity that is activated by binding of the hormone or growth factor to the

[1] M. Kasuga, Y. Zick, D. L. Blithe, M. Crettaz, and C. R. Kahn, Nature (London) 298, 667 (1982).
[2] R. A. Roth and D. J. Cassell, Science 219, 299 (1983).
[3] M. A. Shia and P. F. Pilch, Biochemistry 22, 717 (1983).
[3a] A. Ullrich et al., Nature (London) 313, 756 (1985).
[4] S. Jacobs, F. C. Kull, Jr., H. S. Earp, M. E. Svoboda, and J. J. Van Wyk, J. Biol. Chem. 258, 9581 (1983).
[4a] A. Ullrich et al., EMBO J. 5, 2503 (1986).
[5] S. Cohen, H. Ushiro, C. Stoscheck, and M. Chinkers, J. Biol. Chem. 257, 1523 (1982).
[5a] A. Ullrich et al., Nature (London) 309, 418 (1984).
[6] C.-H. Heldin, B. Ek, and L. Ronnstrand, J. Biol. Chem. 258, 10054 (1983).
[6a] Y. Yarden et al., Nature (London) 323, 226 (1986).

receptor. Activation of the tyrosine-specific protein kinase activity appears to represent a key mechanism for transmembrane signaling by this class of hormones and growth factors. The major unsolved problem relating to the action of these growth factors remains the elucidation of the biochemical mechanisms that control their actions inside the cell. The application of genetics can play a key role in the elucidation of complex metabolic phenomena and pathways. In both prokaryotic and eukaryotic systems, the isolation, mapping, and characterization of mutants blocked in specific steps has in many cases provided crucial information regarding the number of genes involved in complex pathways, as well as the functional interaction of these genes. There is thus considerable interest at present in applying techniques of somatic cell genetics (isolation and characterization of variant cell lines, somatic cell hybridization) to a study of the mechanism of action of polypeptide hormones and growth factors. This chapter describes methods for isolating variant cell lines having altered responses to polypeptide hormones and growth factors, and methods for somatic cell hybridization that can be used for gene mapping and complementation analysis. Some recent applications of these techniques to the study of growth factors are also reviewed.

Isolation of Variant Cell Lines

Two different general strategies have been used to isolate variant cell lines with altered responses to hormones and growth factors. The first involves *positive selection* in favor of the variant, using culture conditions under which the variant can grow but the mass population of parental cells cannot. Variants isolated by positive selection frequently can be obtained in a single step. One important class of variants that can be isolated by positive selection are variants that can grow in the absence of a growth factor. The second type of strategy for isolation of variants is *negative selection,* or selection against cells that can grow under certain culture conditions. This approach is analogous to the penicillin selection used for isolation of auxotrophic mutants of bacteria.[7] Since the mass population of parental cells generally is not completely eliminated in a single round of negative selection, several rounds are usually required. A major class of variants that can be isolated by negative selection are variants that are unresponsive to a growth factor.

The frequency of variants arising from a mutational mechanism (i.e., a change in the primary nucleotide sequence of DNA) is greatly increased by treatment with chemical mutagens. Thus, it is generally desirable to

[7] W. Hayes, "The Genetics of Bacteria and Their Viruses," 2nd ed., pp. 211–213. Wiley, New York, 1968.

mutagenize cells before attempting to isolate variant cell lines (see also the chapter by Thompson, Vol. LVIII, this series).

Mutagenesis with Ethyl Methanesulfonate

Ethyl methanesulfonate (EMS) is a strong chemical mutagen that induces mutations primarily of the base pair substitution class.[8,9] A simple procedure for mutagenizing cells with EMS is described below. Since the optimal concentration of mutagen may vary with cell type and culture medium, it is desirable to perform controls to measure survival, and mutation frequency at a control locus (e.g., the Na^+,K^+-ATPase gene) to ensure that mutagenesis has in fact occurred and to optimize the concentration of mutagen.

Procedure

1. Best results are obtained with exponential phase, rapidly growing cultures. For monolayer cultures, good results are obtained with cultures that are approximately 25–50% confluent and observed to be growing rapidly by visual inspection for mitotic cells.

2. EMS (methanesulfonic acid, ethyl ester, Sigma) is added directly to the growth medium. In our experience, the optimal concentration of EMS for cells growing in minimal essential medium (MEM) with 10% fetal bovine serum is 200–400 μg/ml.[10] EMS is unstable in aqueous solution, thus any dilution into aqueous solution should be made immediately prior to use. The presence in culture media of compounds that react with EMS, for example, sulfhydryl compounds such as cysteine, may also affect its stability. It should also be noted that EMS is a carcinogen[11] and should be handled with appropriate care.

3. After 24 hr at 37°, the culture medium is discarded and cells are washed once with phosphate-buffered saline (PBS). At this point, survival can be determined for one set of control and treated cultures by trypsinizing cells and plating in 10-cm dishes at a clonal density (e.g., 100 cells per 10-cm dish for cells that clone with a high efficiency). For cultures to be used for variant isolation, fresh culture medium is added and incubation is continued until the cultures are confluent.

4. Cells are subcultured serially at a density of 1/4, one to three times to allow recovery from the mutagen and expression of mutations. The

[8] J. McCann, E. Choi, E. Yamasaki, and B. N. Ames, *Proc. Natl. Acad. Sci. U.S.A.* **72,** 5135 (1975).

[9] C. Coulondre and J. H. Miller, *J. Mol. Biol.* **117,** 577 (1977).

[10] D. S. Straus and D. L. Coppock, *J. Cell Biol.* **99,** 1838 (1984).

[11] J. V. Soderman (ed.), "CRC Handbook of Identified Carcinogens and Non-carcinogens: Carcinogenicity–Mutagenicity Data Base." CRC Press, Boca Raton, Florida, 1982.

longer expression period may be required for isolation of some variants with recessive mutations (see Ref. 12).

5. To verify that mutagenesis has occurred and to optimize the mutagen concentration, forward mutation frequency at the Na^+,K^+-ATPase locus is determined by measuring the frequency of ouabain-resistant (Oua^r) mutants[13-16] in control and mutagenized cultures. Sensitivity to ouabain varies considerably among different species. For example, Oua^r variants of human cells can be selected using ouabain concentrations of 10^{-7}–10^{-6} M.[13-15] In contrast, with rodent cells, a ouabain concentration of 1 mM is typically required.[13] The concentration of ouabain required also depends on cell density. With mouse cells growing in monolayer cultures, we have found that 1 mM ouabain can be used to isolate Oua^r variants if cells are plated at a density of 10^5 cells/10-cm dish.[10] In contrast, if cells are plated at 10^6/10-cm dish, a ouabain concentration of 10 mM is required to completely suppress growth of the background.[10] Operationally, it is advantageous to plate cells at 10^6/dish because this gives a larger sample size for estimation of mutation frequency, and fewer dishes need be used.

6. To count colonies for determination of survival and frequency of Oua^r mutants, culture dishes are rinsed twice with PBS and colonies are fixed with 0.25% glutaraldehyde in PBS for 15 min. The glutaraldehyde solution is discarded and colonies are fixed with 50% methanol. The methanol is discarded and the dishes are dried. Colonies are stained with crystal violet (1 mg/ml in PBS) for 15 min. Dishes are then rinsed thoroughly with water, and air dried.

7. Mutagenesis with EMS should yield a large increase in Oua^r mutants over the spontaneous mutation frequency.[10,13-16] Since ouabain resistance results from a missense mutation altering ouabain binding of the Na^+,K^+-ATPase, measurement of the forward mutation rate to ouabain resistance cannot be used for monitoring the induction of mutations by frameshift mutagens.[15,16]

Positive Selection Methods for the Isolation of Variant Cell Lines

One important class of variants that can be isolated by positive selection are hormone or growth factor-independent variants. The simplest way to isolate such variants is to plate cells in hormone-supplemented serum-

[12] L. A. Chasin, *J. Cell. Physiol.* **82,** 299 (1973).
[13] R. M. Baker, D. M. Brunette, R. Mankovitz, L. H. Thompson, G. F. Whitmore, L. Siminovitch, and J. E. Till, *Cell* **1,** 9 (1974).
[14] R. Mankovitz, M. Buchwald, and R. M. Baker, *Cell* **3,** 221 (1974).
[15] J. E. Lever and J. E. Seegmiller, *J. Cell. Physiol.* **88,** 343 (1976).
[16] U. Friedrich and P. Coffino, *Proc. Natl. Acad. Sci. U.S.A.* **74,** 679 (1977).

free medium lacking one growth factor or hormone that is required for growth. Cell density is of critical importance in the selection for growth factor-independent variants because metabolic cooperation or autostimulation frequently occurs in serum-free media, so that growth factor requirements tend to become more stringent at low cell densities. Therefore, a wide range of cell densities should be tried in any attempt to isolate growth factor-independent variants.

A good example of the use of this approach is the isolation of prostaglandin $E_1(PGE_1)$-independent variants of the Madin–Darby canine kidney (MDCK) cell line by Taub and co-workers.[17,18] These variants have provided evidence supporting the hypothesis that PGE_1 stimulates cell proliferation via a cyclic AMP-dependent mechanism and have also been useful in dissecting affects of PGE_1 on growth and the expression of differentiated functions.[17,18] Using a similar approach, Serrero[19] has isolated an insulin-independent variant of the teratoma-derived adipogenic cell line 1246. This variant has lost the ability to undergo adipose differentiation, is tumorigenic, and produces autostimulatory growth factors.[19]

Negative Selection Methods for the Isolation of Variants

Negative selection methods for the isolation of growth factor-insensitive variants using inhibitors of microtubule polymerization are described in detail by Herschman in this volume.[20] The use of inhibitors of microtubule polymerization has yielded density revertants of simian virus 40 (SV40)-transformed 3T3 cells,[21] variants of 3T3 that are unresponsive to EGF,[22] variants of 3T3 cells that have altered binding of insulin,[23] and variants of 3T3 cells that do not respond to the tumor promoter 12-O-tetradecanoylphorbol 13-acetate.[24] We have recently used this approach to isolate a variant cell line having a decreased growth response to platelet-derived growth factor.[10] If one uses colchicine for isolation of growth factor-unresponsive variant cell lines, it is important to bear in mind that one class of variants that might be isolated by this technique are variants that are resistant to colchicine rather than to the growth factor.[25,26] It is therefore important to test all variants for colchicine resistance.[10,22] In

[17] M. Taub, M. H. Saier, Jr., L. Chuman, and S. Hiller, *J. Cell. Physiol.* **114**, 153 (1983).
[18] M. Taub, P. E. Devis, and S. H. Grohol, *J. Cell. Physiol.* **120**, 19 (1984).
[19] G. Serrero, *In Vitro Cell Dev. Biol.* **21**, 537 (1985).
[20] H. R. Herschman, this volume [31].
[21] A. Vogel, R. Risser, and R. Pollack, *J. Cell. Physiol.* **82**, 181 (1973).
[22] R. M. Pruss and H. R. Herschman, *Proc. Natl. Acad. Sci. U.S.A.* **74**, 3918 (1977).
[23] Y. Shimizu and N. Shimizu, *Somatic Cell Genet.* **6**, 583 (1980).
[24] E. Butler-Gralla and H. R. Herschman, *J. Cell. Physiol.* **107**, 59 (1981).
[25] V. Ling and L. H. Thompson, *J. Cell. Physiol.* **83**, 103 (1974).
[26] V. Ling and R. M. Baker, *Somatic Cell Genet.* **4**, 193 (1978).

addition, treatment of cells with colchicine frequently causes mitotic failure and a consequent doubling of chromosome numbers.[27,28] Some variant cell lines isolated following selection with colchicine have twice the number of chromosomes as the parental cells; this change was presumably induced by the colchicine.[10] It can be determined whether a doubling of chromosome number has occurred by counting chromosomes (see below).

Another useful negative selection technique is [³H]thymidine suicide.[29] In this method, cells are plated in an appropriate culture medium and exposed to [³H]thymidine, which kills cells that go through a round of DNA synthesis during the interval of exposure. We have used this method to isolate variants of the NIL8 hamster cell line having an increased sensitivity to a polypeptide growth inhibitor produced by cultured liver cells.[29] To obtain good killing by this procedure, it is important to use [³H]thymidine having a high specific activity and to use a basal medium such as MEM that does not contain unlabeled thymidine. We have found that [³H]thymidine (40–60 Ci/mmol) gives good killing when added at a final concentration of 5 μCi/ml. After each round of selection, cells are washed thoroughly and allowed to recover in medium supplemented with unlabeled thymidine at a concentration of 10^{-6} M to chase the [³H]thymidine.[29]

Cloning of Variants

Before attempting to characterize putative variants, it is of great importance to establish that one is in fact studying a clonal cell population derived from a single variant cell. This is particularly important for variants isolated by negative selection techniques, which only enrich for, but do not absolutely select for variant cells. Typically, following several rounds of negative selection, cells are plated at clonal density, and several dozen clones are picked, expanded, and tested for the variant phenotype (e.g., resistance to a growth factor).[10,29] Techniques for single cell cloning are described in detail by Reid in an earlier volume of this series.[30] Variants obtained by positive selection also should be recloned unless there is an absolute certainty that a single, well isolated clone was picked in the initial selection. For variants isolated by positive selection, it is frequently desirable to propagate the variants for several passages in selective medium to be certain that all contaminating parental cells have been eliminated.

[27] M. Harris, *Exp. Cell Res.* **66**, 329 (1971).
[28] H. Salazar, E. Sidebottom, and H. Harris, *J. Cell Biol.* **67**, 378a (1975).
[29] D. W. Golub and D. S. Straus, *Exp. Cell Res.* **133**, 437 (1981).
[30] L. C. M. Reid, this series, Vol. 58, p. 152.

Independence of Variants

One possible problem that may be encountered in the isolation of variant cell lines is multiple isolation of the same variant. To increase the probability of isolating a number of different variants, typically several different mutagenized cultures are carried through selection. However, even if one follows this procedure, one cannot rigorously exclude the possibility that two variants are the same unless they were isolated from two different cultures that are independent in the sense that each has been expanded from a single clone.

Reversion Analysis

It is sometimes possible to gain information regarding the molecular nature of the lesion that produces a variant phenotype by studying the pattern of spontaneous and chemically induced reversion. For example, a very low spontaneous reversion frequency that is strongly enhanced by treatment of the variant with alkylating agents such as EMS suggests that the variant contains a base-pair substitution mutation. High revertability by 5-azacytidine is suggestive of gene inactivation and reactivation caused by changes in DNA methylation.[10,31–36] On the other hand, a very high spontaneous reversion frequency (e.g., $>5 \times 10^{-5}$) suggests that the variant can revert by a mechanism other than point mutation.[36–40] Even if one does not undertake a complete reversion analysis, it is important to periodically retest the phenotype of variants to be certain that the phenotype is stable.

Genetics Analysis Using Cell Hybridization

Somatic cell hybridization is a technique that is used for gene mapping and complementation analysis. The vast majority of human autosomal genes that have been mapped to date have been mapped by this method

[31] T. Mohandas, R. S. Sparkes, and L. S. Shapiro, Science 211, 393 (1981).
[32] S. J. Compere and R. D. Palmiter, Cell 25, 233 (1981).
[33] M. Groudine, R. Eisenman, and H. Weintraub, Nature (London) 292, 311 (1981).
[34] M. Harris, Cell 29, 483 (1982).
[35] R. D. Ivarie and J. A. Morris, Proc. Natl. Acad. Sci. U.S.A. 79, 2967 (1982).
[36] C. Steglich, A. Grens, and I. Scheffler, Somatic Cell Mol. Genet. 11, 11 (1985).
[37] D. S. Straus, Genetics 78, 823 (1974).
[38] D. S. Straus and L. D. Straus, J. Mol. Biol. 103, 143 (1976).
[39] R. G. Fenwick, Jr., J. C. Fuscoe, and C. T. Caskey, Somatic Cell Mol. Genet. 10, 71 (1984).
[40] T. D. Tlsty, P. C. Brown, and R. T. Schimke, Mol. Cell. Biol. 4, 1050 (1984).

(reviewed in Ref. 41). Methods for cell hybridization, and some application of these methods, are described below.

Selective Systems for the Isolation of Hybrids

HPRT$^-$ Mutants and the Use of HAT Medium. One of the most frequently used selective systems for the isolation of hybrid cell lines involves the utilization of a parental cell line containing a mutation in the hypoxanthine-guanine phosphoribosyltransferase (HPRT) gene, and selection of hybrid cells in HAT (hypoxanthine–aminopterin–thymidine) medium.[42] The HPRT$^-$ parental cells are unable to proliferate in HAT medium and are therefore selectively eliminated.

Isolation of HPRT$^-$ mutants. HPRT$^-$ mutants can be easily isolated by selecting for resistance to the base analog 6-thioguanine.[12] Stock solution of 6-thioguanine (Sigma) are prepared at a 100× concentration (1 mM) in dilute base, filter-sterilized, and diluted into culture medium to give a final concentration of 10^{-5} M. To select for HPRT$^-$ mutants, cells are mutagenized with EMS (see above) and plated into selective medium containing 10^{-5} M 6-thioguanine. Because metabolic cooperation occurs in this selective medium, it is important to keep the plating density at 3800 cells/cm^2 or lower.[12] Variant clones resistant to 6-thioguanine are picked and passaged several times in selective medium to ensure that any contaminating thioguanine-sensitive parental cells have been eliminated. To confirm that the thioguanine-resistant variants are actually HPRT$^-$ mutants, they are tested for sensitivity to killing by HAT medium.

Preparation of HAT medium. HAT medium[42] contains hypoxanthine (10^{-4} M), aminopterin (4×10^{-7} M), thymidine (1.6×10^{-5} M), and glycine (3×10^{-6} M). (Glycine is required in addition to the other components of HAT because aminopterin interferes with the biosynthesis of glycine.) HAT can be prepared as a 100× concentrate and is stored frozen prior to use.

Crosses between HPRT$^-$ mutants and nonproliferating cells. Hybrids between HPRT$^-$ mutants and nonproliferating cells or cells with a limited proliferation potential can be selected directly in HAT medium. Examples of nonproliferating cells that can be used as parents in such crosses are lymphocytes and macrophages. The proliferative life span of mouse embryo fibroblasts (MEFs) is very short,[43] thus MEFs can also be used as a parent in crosses performed in HAT medium.[44]

[41] Human Gene Mapping 7 (Seventh International Workshop on Gene Mapping), *Cytogenet. Cell Genet.* **37** (1984).
[42] J. W. Littlefield, *Science* **145,** 709 (1964).
[43] G. J. Todaro and H. Green, *J. Cell Biol.* **17,** 299 (1963).
[44] J. Jonasson, S. Povey, and H. Harris, *J. Cell Sci.* **24,** 217 (1977).

HPRT⁻ Ouaʳ Double Mutants and the Use of HAT Plus Ouabain Medium. Ouabain resistance (see above) is another selective marker that is useful for isolation of hybrid cells. Because ouabain resistance is dominant, hybrids between HPRT⁻ Ouaʳ double mutants and *any* ouabain-sensitive cell strain or line can be readily selected in medium containing HAT plus ouabain.[13,45,46] Techniques for isolation of HPRT⁻ and ouabain-resistant variants are outlined above. The ouabain resistance of hybrids formed by crossing Ouaʳ and Ouaˢ cells is generally somewhat lower than the Ouaʳ parent,[13,46] thus care must be taken not to use an excessively high ouabain concentration in the HAT plus ouabain medium.

Because of the higher sensitivity of human cells than rodent cells to killing by ouabain, it is possible to select for hybrids between HPRT⁻ ouabain-sensitive rodent cells and human cells in HAT plus ouabain medium by choosing a ouabain concentration that kills the human but not the rodent parent (see, for example, Refs. 47 and 48). This procedure has the advantage of avoiding the additional step of isolating a Ouaʳ variant of the parental HPRT⁻ cell line.

Isolation and Analysis of Hybrid Clones

The hybridization of somatic cells involves, as a first step, the fusion of cells and formation of a heterokaryon.[49] Mitotic division of the heterokaryon produces hybrid daughter cells containing chromosomes derived from both parental cells. The two agents used most commonly for promoting cell fusion are ultraviolet light-inactivated Sendai virus and polyethylene glycol (PEG). A detailed description of methods for cell fusion with Sendai virus and PEG has appeared in an earlier volume of this series.[50] Some of the hybrids that we have studied were isolated following Sendai virus-mediated cell fusion.[51] Following fusion, cells are plated into nonselective medium and allowed to recover for approximately 24 hr. The medium is then changed to HAT or HAT plus ouabain medium, and incubation is continued until hybrid clones appear. The hybrid clones are picked with cloning rings,[30] and passaged in selective medium for several passages to ensure that all contaminating parental cells have been eliminated.

[45] K. K. Jha and H. L. Ozer, *Somatic Cell Genet.* **2**, 215 (1976).
[46] C. M. Corsaro and B. R. Migeon, *Somatic Cell Genet.* **4**, 531 (1978).
[47] J. K. Rankin and G. J. Darlington, *Somatic Cell Genet.* **5**, 1 (1979).
[48] H. E. Schwartz, G. C. Moser, S. Holmes, and H. K. Meiss, *Somatic Cell Genet.* **5**, 217 (1979).
[49] H. Harris, "Cell Fusion." Harvard Univ. Press, Cambridge, Massachusetts, 1970.
[50] R. H. Kennett, this series, Vol. 58, p. 345.
[51] H. Harris and J. F. Watkins, *Nature (London)* **205**, 640 (1965).

Chromosomal Analysis

To ensure that clones isolated by the above methods are in fact hybrids (and not, for example, HPRT$^+$ revertants of one of the parental cell lines), it is necessary to perform a karyotypic analysis of the parental cell lines and hybrid clones. Chromosomal analysis also indicates the numbers of chromosomes derived from the two parental cell lines that are retained by each hybrid clone. All somatic cell hybrids are unstable to some degree, with the level of stability being determined largely by the species of the parental cells. For example, human × rodent hybrids are very unstable and preferentially lose human chromosomes.[52] On the other hand, intraspecies hybrids (e.g., mouse × mouse, human × human) can be relatively stable and lose chromosomes at a very slow rate. Methods for karyotypic analysis of cultured cells have been described in a paper in an earlier volume of this series.[53] Some simple methods that are useful for rapid analysis of hybrid cells are outlined below.

Chromosomal Preparation and Counting

An exponentially growing culture in a 75-cm^2 tissue culture flask is exposed to Colcemid (demecolcine, Sigma) at a concentration of 100 ng/ml for 30 min at 37°. Mitotic cells are dislodged by shaking the bottle, and collected by centrifugation at 800–1000 rpm for 7 min in a clinical centrifuge. Cells are treated for 10 min with a hypotonic solution (0.56% KCl in distilled water), collected again by centrifugation, and fixed three times with a 3 : 1 mixture of methanol/acetic acid. Cells are then applied to clean glass microscope slides and allowed to dry overnight. The slides may then be stained for 1 hr in freshly prepared Giemsa stain [Giemsa stock solution (Fisher), diluted 1 : 25 in 0.07 M phosphate buffer, pH 7.0]. The slides are thoroughly rinsed with the same phosphate buffer and then with water, and air dried. A coverslip is then applied to the slides, and metaphase cells are observed under 1250× magnification.

One can frequently derive a considerable amount of information about the origin of the hybrids using this conventional staining technique. For example, with this simple staining technique it is often possible to identify marker chromosomes derived from the two parental cell types that are present in the hybrid cells. This kind of analysis thus provides evidence that the cells are in fact hybrids. Chromosomal counts of the parental cells and hybrids can indicate whether the hybrids are 2 : 1 or 1 : 1 hybrids. (Hybrids constructed by crossing rapidly growing tumor cells with G$_1$-

[52] M. C. Weiss and H. Green, *Proc. Natl. Acad. Sci. U.S.A.* **58,** 1104 (1967).
[53] R. G. Worton and C. Duff, this series, Vol. 58, p. 322.

arrested or slowly growing normal cells frequently contain two chromo-some sets derived from the tumor cell parent and one from the normal cell parent[44].) Operationally, to do chromosome counts, one should count the chromosomes in approximately 50 well-spread metaphase cells. More detailed karyotypic analysis of the hybrid clones requires the use of chromosome banding techniques. Techniques for Giemsa banding[53] and fluorescence banding[44] of chromosomes are described elsewhere.

Applications of Cell Hybridization Techniques to the Study of Peptide Growth Factors and Their Mechanism of Action

One important application of cell hybridization techniques is gene mapping. For example, the structural genes for a number of polypeptide growth factors including insulin,[54] epidermal growth factor (EGF),[55] insu-lin-like growth factor I,[55,56] and insulin-like growth factor II[55,56] have been assigned to specific human chromosomes using cell hybrids. Human genes for a number of growth factor receptors, including the insulin,[58a,b] insulin-like growth factor I,[4a] platelet-derived growth factor,[6a] EGF[57,58] and transferrin receptors,[59] have also been mapped by this method.

Another application of cell hybridization techniques involves a com-plementation analysis of the response to growth factors or the require-ment for growth factors. For example, we have undertaken an analysis of insulin action using the mouse melanoma cell line PG19, and cell hybrids formed by crossing the melanoma cells with mouse embryo fibro-blasts.[60–63a] The PG19 melanoma cell line does not respond to the growth-stimulatory effects of insulin either under conditions of serum limitation

[54] D. Owerbach, G. I. Bell, W. J. Rutter, and T. B. Shows, *Nature (London)* **286**, 82 (1980).
[55] J. E. Brissenden, A. Ullrich, and U. Francke, *Nature (London)* **310**, 781 (1984).
[56] J. V. Tricoli, L. B. Rall, J. Scott, G. I. Bell, and T. B. Shows, *Nature (London)* **310**, 784 (1984).
[57] N. Shimizu, M. Ali Behzadian, and Y. Shimizu, *Proc. Natl. Acad. Sci. U.S.A.* **77**, 3600 (1980).
[58] R. L. Davies, V. A. Grosse, R. Kucherlapati, and M. Bothwell, *Proc. Natl. Acad. Sci. U.S.A.* **77**, 4188 (1980).
[58a] T. L. Yang-Feng, U. Francke, and A. Ullrich, *Science* **228**, 728 (1985).
[58b] D. S. Straus, K. J. Pang, F. C. Kull, Jr., S. Jacobs, and T. Mohandas, *Diabetes* **34**, 816 (1985).
[59] P. N. Goodfellow, G. Banting, R. Sutherland, M. Greaves, E. Solomon, and S. Povey, *Somatic Cell Genet.* **8**, 197 (1982).
[60] D. S. Straus and R. A. Williamson, *J. Cell. Physiol.* **97**, 189 (1978).
[61] D. L. Coppock, L. R. Covey, and D. S. Straus, *J. Cell. Physiol.* **105**, 81 (1980).
[62] D. L. Coppock and D. S. Straus, *J. Cell. Physiol.* **114**, 123 (1983).
[63] R. K. Kulkarni and D. S. Straus, *Biochim. Biophys. Acta* **762**, 542 (1983).
[63a] L. B. Hecht and D. S. Straus, *Endocrinology* **119**, 470 (1986).

or in hormone-supplemented serum-free medium, while melanoma × fibroblast hybrids exhibit a strong growth response to insulin under both conditions.[60,61] Insulin stimulates protein synthesis, inhibits protein degradation, and stimulates a rapid, large increase in phosphorylation of ribosomal protein S6 in the fibroblast × melanoma hybrid cells but not in the melanoma cells.[60–63a] A similar pattern of response to insulin-like growth factor II is observed with the melanoma and hybrid cells.[62] Both cell types have receptors for insulin and insulin-like growth factors.[61,62] Therefore, the melanoma cells are blocked or uncoupled in the "pathway" of insulin action at a step subsequent to binding of insulin to its receptor. Apparently, the fibroblast parent of the hybrids provides a gene or genes that enable the hybrid cells to respond to insulin and insulin-like growth factors. The identity of these gene(s) is the subject of current investigation.

Methodological Problems with Cell Hybridization

There are a number of potential methodological problems and pitfalls which should be considered before one attempts to use cell hybridization techniques. One such problem is the difficulty of isolating hybrids between certain combinations of cells. For example, we had great difficulty in isolating hybrids between an HPRT⁻ Ouaʳ variant of the AKR-SL2 mouse thymic lymphoma cell line[64] and IM-9 human lymphoblasts.[65] In eight different fusions between these two cell lines, only two stable HAT-resistant clones were obtained. One of these clones was studied in detail. It had the same number of chromosomes as the parental AKR-SL2 cells and contained no recognizable human chromosomes. However, analysis of the HPRT enzyme in these cells by isoelectric focusing revealed that it was human rather than mouse HPRT. Thus the clone was in fact a hybrid and seems to have arisen by a microtranslocation between a fragment of the human X chromosome carrying the HPRT gene and a mouse chromosome. The reason for the apparent incompatibility of the IM9 and AKR SL-2 cells in forming hybrids that retain human chromosomes is unclear. One possibility is that such hybrids are incapable of proliferating under the culture conditions employed.

Another methodological problem in hybridization experiments is the turn-off, or "extinction," of differentiated functions that occurs in hybrids between some types of cells. This phenomenon has been studied most extensively in hybrids formed by crossing hepatoma cells, which express liver-specific proteins such as serum albumin, and fibroblasts. Expression of the liver-specific proteins is turned off in such hybrids but

[64] D. S. Straus and K. J. Pang, *Mol. Cell. Biochem.* **47**, 161 (1982).
[65] P. De Meyts, A. R. Bianco, and J. Roth, *J. Biol. Chem.* **251**, 1877 (1976).

may be reexpressed in subclones of the hybrids (reviewed in Ref. 66). The "extinction" of expression of the liver-specific proteins occurs at the level of mRNA synthesis.[67,68] The molecular basis of the extinction phenomenon is not known but appears to involve trans-acting regulatory factors.[69] The extinction phenomenon can create a problem if one wishes to obtain hybrid cells expressing a differentiated function.

Conclusions and Perspective

It has become evident in the past few years that the receptors for a number of polypeptide growth factors, including insulin,[1-3] insulin-like growth factor I,[4] epidermal growth factor,[5] and platelet-derived growth factor,[6] each possess a tyrosine-specific protein kinase activity. However, the biological effects of these growth factors vary considerably. In particular, the synergistic effects of insulin and insulin-like growth factors in combination with PDGF, as well as the possible differences in their temporal sequence of action, suggest that there must be some differences in the pathway of action of the different growth factors[70] (see also Harrington and Pledger [36], this volume). It is possible that this difference may be explained ultimately by a difference in substrate specificity of the different receptor kinases. Alternatively, it is possible that the different hormones act via branched pathways, with some branch or branches being shared by all of the different growth factors. The isolation and genetic characterization of variants that are blocked or uncoupled in the postreceptor pathways of action of these growth factors may help elucidate their mechanism(s) of action.

Acknowledgment

This work was supported by NIH Grant AM21993.

[66] M. C. Weiss, in "Somatic Cell Genetics" (T. C. Caskey and D. C. Robbins, eds.), pp. 169–179. Plenum Press, New York, 1982.
[67] D. Cassio, M. C. Weiss, M. O. Ott, J. M. Sala-Trepat, J. Fries, and T. Erdos, Cell 27, 351 (1981).
[68] W. K. Church, J. Papaconstantinou, S.-W. Kwan, A. Poliard, C. Szpirer, and J. Szpirer, Somatic Cell. Mol. Genet. 10, 541 (1984).
[69] A. M. Killary and R. E. K. Fournier, Cell 38, 523 (1984).
[70] C. D. Stiles, G. T. Capone, C. D. Scher, H. N. Antoniades, J. J. Van Wyk, and W. J. Pledger, Proc. Natl. Acad. Sci. U.S.A. 76, 1279 (1979).

[33] Conjugation of Peptide Growth Factors with Toxin to Isolate Cell Variants Involved in Endocytosis and Mitogenic Response

By NOBUYOSHI SHIMIZU

Peptide growth factors can be made cytotoxic without loss of receptor-recognizing specificity by cross-linking them to the A fragment of diphtheria toxin (Dta)[1] or the A subunit of toxic ricin (Rica).[2] The resulting hybrid proteins (conjugates) can be used as an agent to select resistant cell variants that are unable to bind, take up, or process the growth factor. The general strategy of this conjugate technique has been described.[2–4] Two different cross-linking techniques, both through disulfide linkages, are described: one[5] that modifies carboxyl groups and one[6] that modifies amino groups of growth factors such as epidermal growth factor (EGF) and insulin. Procedures for isolating conjugate-resistant cell variants are also described.

Purification of Toxic Fragments or Subunits

Dta. Purified diphtheria toxin is purchased as a solution from Connaugh Laboratories, Canada. Toxin protein is precipitated by adding ammonium sulfate (ultrapure grade, Schwartzman) to 70% saturation. The precipitate is collected by centrifugation at 10,000 g for 15 min, dissolved in a small volume of 10 mM sodium phosphate (pH 7.0), and dialyzed against the same buffer. The dialyzed toxin (35 mg/ml) is mixed with two volumes of 150 mM Tris–HCl (pH 8.0) containing 3 mM EDTA. It is then treated with 1 μg/ml trypsin (Sigma) at 25° for 30 min. The reaction is terminated by adding soybean trypsin inhibitor (Sigma) at 1.5 μg/ml. The nicked toxin is reduced by incubation with 100 mM dithiothreitol (DTT)

[1] A. M. Pappenheimer, Jr., *Annu. Rev. Biochem.* **46**, 69 (1977).
[2] S. Olsnes and A. Pihl, *Phamacol. Ther.* **15**, 355 (1982).
[3] N. Shimizu, in "Receptors and Recognition" (P. N. Goodfellow, ed.), Series B, Vol. 16, p. 109. Chapman and Hall, London, 1984.
[4] N. Shimizu, in "Cell Culture Methods for Molecular and Cell Biology" (G. Sato, D. Barnes, and D. Sirbasku, eds.), Vol. 3, p. 233. Liss, New York, 1984.
[5] D. Gilliland, R. J. Collier, J. M. Moehring, and J. T. Moehring, *Proc. Natl. Acad. Sci. U.S.A.* **75**, 5319 (1978).
[6] J. Carlsson, H. Drevin, and R. Axen, *Biochem. J.* **173**, 723 (1978).

and 4 M urea at 25° for 60 min, and applied onto a Sephadex G-100 column (1.5 × 110 cm) equilibrated with 50 mM Tris–HCl (pH 8.0) containing 4 M urea, 1 mM EDTA, and 1% 2-mercaptoethanol.[7] Every other fraction is analyzed by polyacrylamide (10%) gel electrophoresis (PAGE) in the presence of sodium dodecyl sulfate (SDS).[8] The fractions containing DTa (M_r = 21,000) are pooled, concentrated by ammonium sulfate precipitation, and dissolved in a minimum volume of 20 mM N-tris(hydroxymethyl)methyl-2-aminoethane sulfonate (TES, Sigma). The solution is dialyzed against the same buffer containing 50% glycerol and kept at −20° until use. The absorbance of 1 mg/ml DTa at 280 nm is 1.52. Activity of DTa for ADP-ribosylation is measured with partially purified mouse liver elongation factor 2 (EF-2).[9]

RICa. Decorticated castor beans (*Ricinus communis,* 200 g) are stirred overnight in 500 ml of 5% acetic acid and homogenized in a Waring blender. This is centrifuged at 10,000 g for 20 min. The oil is scraped off and ammonium sulfate is added to the supernatant to 80% saturation. The precipitate is collected by centrifugation, dissolved in distilled water, and dialyzed extensively against distilled water. The resulting yellow solution is passed through a DE-52 (Whatman) column (3.5 × 25 cm) equilibrated with 5 mM sodium acetate (pH 5.8). The flow-through fractions are collected and treated with ammonium sulfate at 80% saturation. The precipitate is collected by centrifugation and dissolved in a minimum volume of 10 mM sodium phosphate (pH 7.4). An aliquot of this solution is applied to a Sephadex G-100 column (3.5 × 60 cm) equilibrated with 10 mM sodium phosphate (pH 7.4). The second of the two sharp major peaks is collected as purified toxic ricin. This ricin solution is then applied to a Sepharose 4B column (2.0 × 60 cm) equilibrated with 0.1 M Tris–HCl (pH 7.6) containing 0.1 M NaCl. After thorough washing with the column buffer, the RICa is eluted with 5% 2-mercaptoethanol in the same buffer.[10] The eluate is dialyzed against 5 mM sodium phosphate (pH 6.5) and applied to a CM52 column (4.0 × 40 cm) equilibrated with the same buffer. RICa (M_r = 30,000) is eluted with a 0–0.1 M NaCl gradient in the same buffer. Activity of RICa to modify ribosomes is assayed by an *in vitro* protein synthesis system.[11] The RICa preparation is kept at −20° in 5 mM sodium phosphate (pH 6.5) containing 50% glycerol until use. The absorbance of 1 mg/ml RICa at 280 nm is 0.77.

---

[7] J. Kandel, R. J. Collier, and D. W. Chung, *J. Biol. Chem.* **249**, 2088 (1974).
[8] U. K. Laemmli, *Nature (London)* **227**, 680 (1970).
[9] T. Honjo, Y. Nishizuka, O. Hayashi, and I. Kato, *J. Biol. Chem.* **243**, 3533 (1980).
[10] T. T.-S. Lin and S. S.-L. Li, *Eur. J. Biochem.* **105**, 453 (1980).
[11] K. Sandvig and S. Olsnes, *Exp. Cell Res.* **121**, 15 (1979).

Conjugation of Growth Factors with DTa or RICa

Insulin–DTa Conjugate. Cystamine is used to activate insulin's carboxyl groups.[5] To a 2 mg/ml porcine insulin solution (Eli Lilly, in 0.01 N HCl) is added cystamine–diHCl (Sigma) at a final concentration of 1 M, and the pH is adjusted to 4.7. Then, 1-ethyl-3-(dimethylaminopropyl)-carbodiimide–HCl (EDAC, Sigma) is added at a concentration of 2 mM and the pH is adjusted at 4.7 for 10 min. The solution is gently stirred during these periods of time. The reaction is stopped by adjusting the pH to 8.8 by carefully adding 10 N NaOH. The reaction mixture is applied onto a BioGel P-4 column (1.6 × 18 cm) equilibrated with 10 mM TES (pH 8.8) to remove unreacted cystamine–diHCl and EDAC. Excess reagents can also be removed by dialysis. The resulting cystaminyl insulin is charged onto a DE-52 column (2.0 × 30 cm) equilibrated with 20 mM TES (pH 8.8). When a 0–0.3 M NaCl gradient in the same buffer is applied, multiple peaks of cystaminylinsulin are eluted from the column. Unmodified insulin is eluted last. The extent of cystaminylation is determined with S,S'-dithiobis(2-nitrobenzoic acid) (Sigma).[12,13] The fractions corresponding to a 1 : 1 molar ratio of cystamine to insulin are concentrated, if necessary, by Amicon membrane (YM5) filtration and dialyzed against 20 mM TES (pH 8.0).

Prior to conjugation, the DTa is completely reduced by incubation with 0.1 M DTT for 4 hr and then dialyzed against 20 mM TES (pH 8.0). The cystaminylinsulin is then mixed with the reduced DTa at a molar ratio of 8 to 1. Cross-linking of the two protein species is spontaneously achieved via thiol–disulfide exchange. The resulting conjugate is purified by a Sephadex G-75 column (2.0 × 60 cm) equilibrated with 20 mM TES (pH 8.0) containing 0.1 M NaCl. The insulin–DTa conjugate is eluted first, separating from DTa and cystaminylinsulin. The purified conjugate is dialyzed against 20 mM TES (pH 8.0) containing 50% glycerol and stored as aliquots at −20°. The absorbance of 1 mg/ml insulin–DTa at 280 nm is 1.42.

Insulin–RICa Conjugate. N-Succinimidyl-3-(2-pyridyldithio)propionate (SPDP)[6] is used to modify B29 lysine of insulin. It is necessary to block the two amino-termini of insulin because modification of these groups decreases affinity for the insulin receptor. For this, citraconic anhydride is used, since it can be removed after conjugation.[14]

Porcine insulin is dissolved in 10 mM sodium carbonate at a final

[12] W. K. Miskimins and N. Shimizu, *Biochem. Biophys. Res. Commun.* **91,** 143 (1979).
[13] A. F. S. A. Habeeb, this series, Vol. 25B, p. 457, 1972.
[14] Y. Shechter, J. Schlessinger, S. Jacobs, K. J. Chang, and P. Cuatrecasas, *Proc. Natl. Acad. Sci. U.S.A.* **75,** 2135 (1978).

concentration of 2 mg/ml, and the pH is adjusted to 6.9 with NaOH. Two aliquots (5 μl each) of citraconic anhydride (Sigma) are added to the insulin solution (20 ml) with gentle stirring over a period of 1 hr, and the pH is maintained at 6.9 for 15 min. The mixture is then applied to a Sephadex G-25 column (1.0 × 15 cm) equilibrated with 10 mM sodium phosphate (pH 7.8) containing 0.1 M NaCl. The blocked insulin is eluted in the void volume. This is then reacted with a 5-fold molar excess of SPDP for 2 hr at room temperature. Excess reagent is removed by passing through a BioGel P-6 column (1.6 × 18 cm) equilibrated with 20 mM TES (pH 7.8).

The derivatized insulin is mixed with freshly reduced RICa (1 mg in 20 mM TES, pH 7.8) at a molar ratio of 20 to 1 for cross-linking. The mixture is dialyzed overnight against 10 mM sodium phosphate (pH 7.8) containing 0.1 M NaCl. Then the pH of the solution is lowered for deblocking the amino-termini by dialyzing against distilled water adjusted to pH 2.0 with HCl. The insulin–RICa conjugate is separated free of insulin and RICa by a Sephadex G-75 column (2.0 × 60 cm) equilibrated with 10 mM sodium phosphate (pH 7.8) containing 0.1 M NaCl. The peak corresponding to the insulin–RICa conjugate is collected, dialyzed against 50% glycerol in 20 mM TES (pH 8.0), and stored as aliquots at −20°. The absorbance of 1 mg/ml insulin–RICa at 280 nm is 0.81.

EGF–DTa Conjugate. This is similar to the insulin–DTa conjugate formation. Briefly, mouse EGF (0.8 mg) is dissolved in 1 ml 50 mM acetic acid and dialyzed against 2 changes of distilled water containing 1 μg phenylmethylsulfonyl fluoride (PMSF)/ml, 500 ml each. Cystamine–diHCl (225 mg) is added, dissolved, and the pH is adjusted to 4.7 with 1 N HCl. EDAC is then added and the pH maintained at 4.7 for 10 min with gentle stirring. The reaction mixture is then dialyzed against 2 changes of 20 mM TES (pH 6.5), 500 ml each, to remove unreacted reagents. Completely reduced DTa (1 mg/ml) is added to 1 ml derivatized EGF and the pH is adjusted to 7.4. The mixture is dialyzed against 20 mM TES (pH 7.4) containing PMSF overnight at 4° to allow formation of spontaneous disulfide bonds. The resulting conjugate is applied to a BioGel P-10 column equilibrated with 0.15 M NaCl in 50 mM HCl. The EGF–DTa conjugate is eluted with 20 mM TES (pH 7.4). The absorbance of 1 mg/ml EGF–DTa at 280 nm is 1.16.[15]

EGF–RICa Conjugate. This is similar to the insulin–RICa conjugate formation. Briefly, mouse EGF (2.8 mg) in 2 ml of 0.1 M sodium phosphate (pH 7.8) containing 0.1 M NaCl is mixed with 2.88 mg of SPDP and incubated for 2 hr at room temperature. Excess reagent is removed by

[15] N. Shimizu, W. K. Miskimins, and Y. Shimizu, *FEBS Lett.* **118,** 274 (1980).

chromatography on a Sephadex G-25 column equilibrated with the same buffer. The derivatized EGF is mixed with completely reduced RICa at a 7 : 1 molar ratio. The mixture is concentrated by Amicon YM5 membrane filtration and dialyzed against 20 mM TES (pH 8.0). The resulting conjugate is purified on a Sephadex G-75 column equilibrated with the same buffer.[16]

Activities of the Conjugates

The conjugates prepared by these methods proved to be toxic to a variety of cultured cells, such as human A431 epidermoid cells, Swiss/ 3T3 mouse fibroblasts, 3T3-L1 preadipocytes, Chinese hamster ovary (CHO) cells, and rat hepatoma H35 cells with an ED$_{50}$ of 7.0–7.7 $\times$ 10^{-9} M.[3,12,15–18]

Isolation of Cell Variants Resistant to Growth Factor–Toxin Conjugates

Variants are selected from cells with or without mutagenesis. The general procedures for mutagenesis have been described.[4,19,20] Cells are grown to about 70% confluency in 75-cm^2 flasks. Conjugates are mixed with Dulbecco's modified Eagle's medium (DME medium) containing 0.1% bovine serum albumin (BSA), and sterilized through a Millipore filter (0.22 μm). The sterilized conjugate is added directly to the 1 to 2 day conditioned media at a final concentration of 2–3 times the ED$_{50}$. The medium is changed after 3 days to remove dead cells and debris. To ensure the specific binding of conjugate to receptors, cells are incubated with conjugate in serum-free DME medium at 4° for 60 min and washed free of unbound conjugate, and the medium is replaced in serum-containing DME medium. Incubation is continued at 37° for 60 min in a CO$_2$ incubator to allow internalization of the cell-bound conjugates. Then, cultures are maintained in the serum-containing DME medium. Treatment with conjugate is repeated 2 days after the medium change. The cells are usually treated with the same concentration of conjugate two more times and with 2–3 times higher concentrations one time. Two weeks after the last treatment, surviving colonies are isolated with cloning rings

[16] N. Shimizu, Y. Shimizu, and W. K. Miskimins, *Cell Struct. Funct.* **9,** 302 (1984).
[17] N. Shimizu, W. K. Miskimins, S. Gamou, and Y. Shimizu, *Cold Spring Harbor Conf. Cell Proliferation* **9,** 397 (1982).
[18] W. K. Miskimins and N. Shimizu, *Proc. Natl. Acad. Sci. U.S.A.* **78,** 445 (1981).
[19] Y. Shimizu and N. Shimizu, *J. Biol. Chem.* **261,** 7342 (1986).
[20] Y. Shimizu and N. Shimizu, *Somatic Cell Genet.* **6,** 583 (1980).

and allowed to expand in DME medium containing 10% fetal calf serum (FCS) for further analysis.

The conjugates of peptide growth factors and toxins not only allow selection of cell variants specifically related to the biological activity of each growth factor but also provide variants with a wide array of properties. The reasons for this variety may be 2-fold: the properties of the conjugates and the nature of the target cell's endocytotic mechanisms. Possible modification of the method and various applications have been discussed in detail.[3,4]

[34] Transmembrane Delivery of Polypeptide Growth Factors Using Their Conjugates with Cell Surface Binding Proteins

By Fumiaki Ito and Nobuyoshi Shimizu

Polypeptide growth factors are delivered into cells through their binding to specific membrane receptors. In the course of this internalization pathway, growth factors induce a broad range of early and delayed biological responses leading to cell growth.[1-3] However, the signals that initiate and maintain these events have not been identified. The functional significance of the internalized growth factors and/or the receptors has not been completely understood. In this chapter, we describe the methods to deliver polypeptide growth factors across the plasma membrane bypassing their own receptors. For this, EGF or insulin is cross-linked to a human serum protein α_2-macroglobulin (α_2-M) because α_2-M binds to specific cell surface receptors and undergoes intracellular processing in a manner similar to those growth factors.[4] Methods to construct radioactive conjugates and techniques to analyze their intracellular processing using a Percoll density gradient are also described.

[1] A. C. King, L. Hernaez-Davis, and P. Cuatrecasas, *Proc. Natl. Acad. Sci. U.S.A.* **78**, 717 (1981).

[2] G. Carpenter and S. Cohen, *Annu. Rev. Biochem.* **48**, 193 (1979).

[3] A. B. Schreiber, T. A. Libermann, I. Lax, Y. Yarden, and J. Schlessinger, *J. Biol. Chem.* **258**, 846 (1983).

[4] F. R. Maxfield, J. Schlessinger, Y. Schechter, I. Pastan, and M. C. Willingham, *Cell* **14**, 805 (1978).

Purification of α_2-M and EGF

α_2-M was prepared from human plasma as described by Swenson and Howard[5] with some modifications. Briefly, α_2-M was isolated using polyethylene glycol (4–12%), ammonium sulfate precipitation (40–55%), ion-exchange chromatography using DE-52 cellulose with a 0 to 0.15 M NaCl gradient, concentration by an Amicon PM10 membrane, and Sephadex G-200 gel filtration. EGF was isolated from the submaxillary glands of male mice as previously described[6] with the exception that fractions obtained from the acid BioGel P-10 column were dialyzed against 5 mM ammonium acetate (pH 5.6), and then purified on a column of DEAE-Sephacel with a 5 to 200 mM ammonium acetate gradient.

Cross-Linking of Insulin or EGF with α_2-M. Two different cross-linking techniques, both forming intermolecular disulfide bonds, are utilized. The first technique involves modification of carboxyl groups and is used for the preparation of insulin–α_2-M conjugates. The second technique involves modification of amino groups, which is used for the preparation of EGF–α_2-M conjugates.

Insulin–α_2-M Conjugate. Porcine insulin (Eli Lilly, 10 mg) is dissolved in 5 ml of 0.01 N HCl and cystamine dihydrochloride (Sigma, 1.13 g) is added. 1-Ethyl-3-(3-dimethylaminopropyl)carbodiimide–HCl (EDAC, 3.83 mg) is added and the pH is maintained at 4.7 for 10 min with gentle stirring. The reaction is stopped by adjusting the pH to 8.8 with 10 N NaOH. The reaction mixture is applied onto a column of BioGel P-4 equilibrated with 10 mM N-tris(hydroxymethyl)methyl-2-aminoethanesulfonic acid (TES) (pH 8.8) to remove unreacted cystamine dihydrochloride and EDAC. The cystaminylinsulin (320 μg) is mixed with α_2-M (2 mg) at 23° for 2 hr in a solution of 10 mM TES (pH 8.0). Since insulin contains six free carboxyl groups, multiple species of cystaminylinsulin are produced. These multiple species can be separated by applying the mixture onto a column of DE52 equilibrated with 20 mM TES (pH 8.8), and eluting with a 0 to 0.3 M NaCl gradient in the same buffer.[7] The mixture of derivatized insulin is used directly for the conjugation. The reaction mixture is then applied onto a column of Sephadex G-100 equilibrated with the same buffer. Fractions eluted at the void volume are collected as the insulin–α_2-M conjugates, dialyzed against phosphate-buffered saline and stored at $-20°$.

[5] R. P. Swenson and J. B. Howard, *J. Biol. Chem.* **254**, 4452 (1979).
[6] C. R. Savage, Jr., and S. Cohen, *J. Biol. Chem.* **247**, 7609 (1972).
[7] N. Shimizu, *in* "Cell Culture Methods for Molecular and Cell Biology" (G. Sato, D. Barnes, and D. Sirbasku, eds.), Vol. 3, p. 233. Liss, New York, 1984.

EGF–α_2-M Conjugate. A mixture of unlabeled mouse EGF (1.56 mg) and a trace amount of ^{125}I-labeled EGF is reacted with 1.5 mg of *N*-succinimidyl-3-(2-pyridyldithio)propionate (SPDP, Pierce) in 2 ml of 0.1 *M* sodium phosphate buffer (pH 7.4) containing 0.1 *M* NaCl for 2 hr at room temperature. SPDP is dissolved in ethanol at a concentration of 30 m*M* just prior to use and the reaction with SPDP is carried out with occasional stirring. Excess reagent is removed by chromatography on a Sephadex G-25 column equilibrated with the same buffer. The derivatized EGF is concentrated by Amicon YM5 membrane filtration and then dialyzed against 20 m*M* TES (pH 8.0). The degree of substitution with 2-pyridyl disulfide is determined by measuring the absorbance at 343 nm which quantifies pyridine 2-thione released upon reduction with dithiothreitol.[8] The modified EGF is mixed with α_2-M at a 5 : 1 molar ratio. The mixture is reacted for 1 hr at room temperature and then for 18 hr at 4°. The resulting conjugate is purified on a column of Sephadex G-75 equilibrated with 20 m*M* TES (pH 8.0) containing 0.1 *M* NaCl and 0.1% bovine serum albumin (Sigma, BSA). Based on the radioactivity recovered in the conjugates, we calculate that each α_2-M molecule carries about 0.5 EGF molecule.

Iodination of Conjugates. For studying the binding of conjugates to cells and the intracellular fate of conjugates, the insulin portion of insulin–α_2-M conjugates and EGF portion of the EGF–α_2-M conjugates are iodinated. The procedures for preparing the labeled conjugates are basically the same as those described under cross-linking of insulin or EGF with α_2-M.

The cystaminylinsulin (10 μg) is reacted with 1 mCi of ^{125}I using 0.16 μg of chloramine-T. After 2 min at room temperature, tyrosine (8 μg) is added to stop further iodination and the reaction mixture is kept overnight at 4°. The mixture is then incubated with α_2-M (60 μg) in 10 m*M* TES (pH 8.0) and fractionated on a column of Sephadex G-100 equilibrated with 10 m*M* TES (pH 8.0) containing 0.1% BSA. The specific activity of the resulting ^{125}I-labeled insulin–α_2-M conjugates was 1.47 μCi/μg protein. The conjugates are analyzed on a column of Sephadex G-50 equilibrated with 4 *M* urea, 1 *M* acetic acid, and 0.1% Triton X-100. Most of the radioactivity appears at the position of void volume at which α_2-M is eluted. When the conjugates are treated with 50 m*M* dithiothreitol (DTT), the high-molecular-weight peak disappears and another peak eluting with insulin and its chains appears. These data show that ^{125}I-labeled insulin is linked to α_2-M via a disulfide bond and the conjugate preparation contains

---

[8] T. Stuchburg, M. Shipton, R. Norris, J. Paul, G. Malthouse, K. Brocklehurst, J. A. L. Herbert, and H. Suschitzky, *Biochem. J.* **151**, 417 (1975).

no free insulin. From the specific radioactivity of the labeled insulin preparation used, we calculate that each α_2-M molecule carries about 2 insulin molecules.

The 2-pyridyl disulfide-containing EGF (5 μg) is reacted with 1 mCi [125]I (15.7 mCi/μg) using Iodo-Beads (Pierce). One Iodo-Beads, which had been washed with 50 mM sodium phosphate buffer (pH 7.4), is then added to a ^{125}I solution diluted with 0.4 M sodium phosphate buffer (pH 7.4) and left at room temperature for 5 min. The modified EGF is then incubated with the preloaded Beads for 1 min at room temperature. The Iodo-Beads is removed and the mixture is incubated with 10 μg of tyrosine for 16 hr at 4° to stop further iodination. The labeled and activated EGF is then cross-linked to α_2-M at a 5 : 1 molar ratio as described above. The purity of the conjugates is estimated to be >80%, which is based on the densitometric analysis of autoradiographs of SDS–PAGE.[9] The specific activity of the conjugates is ~200 μCi/mg.

The following method may be used to avoid iodination of the α_2-M portion of the conjugates. After iodination of the modified EGF with ^{125}I, tyrosine (10 μg) is added and the mixture kept for 2 hr at 4°. The mixture is then applied on a column of BioGel P-4 to remove free [125]I. The column is eluted with 30 mM sodium phosphate buffer (pH 7.5) containing 0.1 M NaCl and 0.1% myoglobin (Sigma). The fraction eluted in the void volume is used for the conjugation to α_2-M at a 5 : 1 molar ratio. Myoglobin is used because it contains no cystein residues.

Binding of Conjugates. To successfully apply conjugates to transmembrane delivery of many substances, the structure of their constituents should be retained in terms of the abilities to recognize their own specific receptors and to exert the appropriate biological activities. For this, binding properties of the conjugates are tested using mouse Swiss/3T3 cells, CI-3 cells,[10] and NR-6 cells.[11] The CI-3 and NR-6 cells are deficient in insulin receptor and EGF receptor, respectively (see also chapter by Herschman [31], this volume). The cell variants, however, retain normal binding ability for α_2-M. The time course of binding of the conjugates at 23° is usually tested. We also examine whether the binding of the conjugates is inhibited by adding cold insulin (or cold EGF) and/or α_2-M in excess. These results usually show that the insulin–α_2-M conjugates have binding capacity for both insulin receptors and α_2-M receptors and that insulin is delivered to the insulin receptor-deficient cells through α_2-M

[9] U. K. Laemmli, *Nature (London)* **227**, 680 (1970).
[10] W. K. Miskimins and N. Shimizu, *Proc. Natl. Acad. Sci. U.S.A.* **78**, 445 (1981).
[11] R. M. Pruss and H. R. Herschman, *Proc. Natl. Acad. Sci. U.S.A.* **74**, 3918 (1977).

receptors.[12] Under the same experimental conditions, about 2 to 3% of input insulin bind to 3T3 cells. Similarly, the EGF–α_2-M conjugates are capable of binding to both EGF receptors and α_2-M receptors in Swiss/3T3 cells and they can deliver EGF into EGF receptor-deficient NR-6 cells through α_2-M receptors.[13]

Internalization of Conjugates. Percoll density gradient centrifugation is used for studying the subcellular localization of both insulin–α_2-M and EGF–α_2-M conjugates.[12–14] Cells grown to confluence in 100-mm dishes are incubated with labeled conjugates in the appropriate binding buffer at desired temperature and time. The dishes are then placed on ice and the cells are rinsed and scraped from the dishes in 10 ml of 0.25 M sucrose. After centrifugation at 800 rpm for 10 min, the pellet is then resuspended in 1 ml of 0.25 M sucrose containing 1 mM EDTA, 10 mM acetic acid, and 10 mM triethanolamine, pH 7.4 (SEAT buffer). The cells are lysed by pipetting 40 times with a P1000 Gilson pipetman and the lysates centrifuged for 10 min at 1000 rpm. The resulting postnuclear supernatant is layered on a 9 ml 20% Percoll in SEAT buffer and centrifuged at 30,000 g for 90 min. The gradients are fractionated from the bottom. The density of each fraction is determined using an Abbe refractometer or density marker beads (Pharmacia Fine Chemicals). The positions of subcellular particles fractionated on a Percoll gradient are determined by assaying UDP-galactose-glycoprotein galactosyltransferase (EC 2.4.1.38) for Golgi, acid phosphatase (EC 3.1.3.2), and β-galactosidase (EC 3.2.1.23) for lysosomes and NADPH–cytochrome-c reductase (EC 1.6.2.4) for endoplasmic reticulum.[15] The position of plasma membrane in a gradient is determined by binding of ^{125}I-labeled α_2-M at 4°. These results show that the conjugates are able to deliver insulin or EGF into cells through the internalization pathway of α_2-M receptors.[12–14]

We cross-linked EGF to α_2-M using SPDP. This cross-linking technique has also been used for the conjugation of insulin to α_2-M with some modifications.[7] Since modification of the two amino-termini of insulin causes a decrease in affinity for the insulin receptor, the amino-termini are blocked with citraconic anhydride prior to the conjugation procedures described above. Two different cross-linking techniques, which we describe here, are generally useful to conjugate polypeptide growth factors to other cell surface binding ligands.

---

[12] F. Ito, S. Ito, and N. Shimizu, *Mol. Cell. Endocrinol.* **36,** 165 (1984).
[13] F. Ito, S. Ito, and N. Shimizu, *Cell Struct. Funct.* **9,** 105 (1984).
[14] F. Ito and N. Shimizu, *FEBS Lett.* **152,** 131 (1983).
[15] W. K. Miskimins and N. Shimizu, *J. Cell. Physiol.* **118,** 305 (1984).

α_2-M is the tetrameric plasma glycoprotein, consisting of four identical subunits of $M_r = 180,000$.[16] The subunits contain 24 cysteinyl residues, in which 23 residues form intramolecular disulfide bridges and one SH group of cystein is thiol esterified.[17] We have not identified which cysteinyl residues are used for preparing these conjugates. α_2-M–proteinase complex formation has been found to cleave the thiol ester.[18] It is most likely that free thiol groups utilized for cross-linking are formed upon thiol ester cleavage.

Dickson *et al.* have introduced sulfhydryl groups into α_2-M molecules by modifying lysyl residues on α_2-M with methyl 4-mercaptobutyrimidate.[19] Since this derivatization of α_2-M does not affect its binding to cell surface receptors, it may be applied to introduce many sulfhydryl groups into α_2-M and to cross-link more growth factors to α_2-M.

[16] J. M. Jones, J. M. Creeth, and R. A. Kekwick, *Biochem. J.* **127**, 187 (1984).
[17] L. Sottrup-Jensen, T. M. Stepanik, T. Kristensen, D. M. Wierzbicki, C. M. Jones, P. B. Løblad, S. Magnusson, and T. E. Petersen, *J. Biol. Chem.* **259**, 8318 (1984).
[18] G. S. Salvesen, C. A. Sayers, and A. J. Barrett, *Biochem. J.* **195**, 453 (1981).
[19] R. B. Dickson, J.-C. Nicolas, M. C. Willingham, and I. Pastan, *Exp. Cell Res.* **132**, 488 (1981).

[35] Kinetic Analysis of Cell Growth Factor–Nutrient Interactions

By W. L. McKeehan and K. A. McKeehan

Principle

The external nutrient supply directly controls the proliferation of prokaryotic cells. However, proliferation of different mammalian cells in a relatively constant and limiting nutritive environment for cell proliferation is under control of hormone-like growth factors.[1-3] Growth factors promote mammalian cell proliferation by modifying the cellular requirement and metabolism of nutrients such that the constant nutritive environment is temporarily permissive. Similar to enzymes, mammalian cells can be isolated and their interaction with substrates and effectors can be studied

[1] W. L. McKeehan, in "Molecular Interrelations of Nutrition and Cancer" (M. S. Arnott, J. van Eys, and Y. M. Wang, eds.), p. 249. Raven, New York, 1982.
[2] W. L. McKeehan, *Cold Spring Harbor Conf. Cell Proliferation* **9**, 65 (1982).
[3] W. L. McKeehan, *Fed. Proc., Fed. Am. Soc. Exp. Biol.* **43**, 113 (1984).

quantitatively under defined and controlled conditions *in vitro*.[4] Similar to the three-way interaction between allosteric effectors of enzymes, enzyme–substrate requirements, and enzyme reaction rate, the three-way interaction of cellular effectors, cell nutrient requirements, and cell proliferation rate can be studied. Employing the principles used for kinetic analysis of enzymes, we assume the cell is a catalyst that converts substrates to new cells. The number of cell catalysts doubles during each cycle of reaction. We assume that nutrient substrates interact with cellular sites to form cell–nutrient complexes equivalent to the enzyme–substrate complex of first-order enzyme reactions. Cells have many nutritive substrates; however, we assume conditions can be established where all nutrients are constant and in excess such that the formation of a single cell–nutrient complex can be considered a first-order function of a single nutrient concentration. The rate of cell multiplication can then be considered a first-order function of the concentration of a single cell–nutrient complex and, in turn, a single nutrient concentration. We assume the equilibration among cells, a single nutrient, and cell–nutrient complex is rapid relative to rate of cell division (often greater than 24 hr) or other processes that deplete cell–nutrient complexes. Lastly, experimental conditions must be designed where cell multiplication rates can be determined and all effector and nutrient concentrations remain in excess of cell–substrate complexes. This can be achieved by two means: (1) at a cell density sufficiently low where a multiplication rate of the cell population can be measured, but cells have no significant impact on nutrient or effector concentration during the process; or (2) perfusion of higher cell density cultures at a rate where the external environment remains effectively constant. For large numbers of replicate assays, the latter is technically cumbersome. Similar to most investigators who study enzyme kinetics, we chose the former approach to study cell growth kinetics under near-steady-state conditions.

Reagents

Purified cell type[5–7]
Optimal nutrient media[8]

[4] W. L. McKeehan and K. A. McKeehan, *J. Supramol. Struct. Cell. Biochem.* **15**, 83 (1981).
[5] Methods for isolation, purification, and characterization of specific cell types is beyond the scope of this chapter. See Refs. 6 and 7 for recent representative approaches.
[6] W. B. Jakoby and I. H. Pastan (eds.), this series, Vol. 59.
[7] D. W. Barnes, D. A. Sirbasku, and G. H. Sato (eds.), "Methods for Preparation of Media, Supplements, and Substrata for Serum-Free Animal Cell Culture," Vols. 2–4. Liss, New York, 1984.
[8] R. G. Ham and W. L. McKeehan, this series, Vol. 59, p. 44.

Standardized set of growth factors[9,10]
Optimal culture substrate or assay vessel[11,12]
5% glutaldehyde or 1% formalin or 3:1 methanol/acetic acid for fixation
0.1% crystal violet or Harris hematoxylin and 1 mM lithium carbonate for cell staining

Procedure

The following is applicable to anchorage-dependent cells attached to a clear substratum as a monolayer. Nutrient media minus the nutrient under analysis is prepared from high quality stock solutions.[13] Media in 1-ml aliquots are distributed among the 24 flat bottom wells of a 10 × 15 cm Linbro tissue culture plate (Flow Laboratories, McLean, VA). Well bottoms are coated with applicable reagent as needed for the individual cell type before addition of assay media.[11,12] Usually triplicate wells are used per condition. The nutrient under test is added as a separate sterile solution to reach the desired final concentration. At least 5 concentrations of nutrient are used in each analysis. At least three concentrations of growth factor are used at each nutrient concentration. If crude serum protein is used as a mixture of growth factors, then standardized serum protein is added at empirically determined levels.[14] If the effect of a single defined growth factor is studied, then a mix of all required defined growth factors minus the one under consideration is added to the assay media. At least three concentrations of single growth factor at five concentrations of nutrient are evaluated. Cells are harvested from a stock culture, serially diluted with assay medium, and then added to each assay to yield a minimum cell concentration still adequate to result in measurable increases in cell number within a 7- to 14-day incubation period. If necessary, the status of nutrients and growth factors during incubation is measured to ensure maintenance of steady-state conditions.[4,15] The multiwell

[9] D. Barnes and G. Sato, *Anal. Biochem.* **102**, 255 (1980).

[10] W. L. McKeehan, *In Vitro* **5**, 48 (1984).

[11] D. W. Barnes, D. A. Sirbasku, and G. H. Sato (eds.), "Methods for Preparation of Media, Supplements, and Substrata for Serum-Free Animal Cell Culture," Vol. 1, Chaps. 11–19. Liss, New York, 1984.

[12] M. Rojkind, Z. Gatmaitan, S. Mackensen, M. A. Giambrone, P. Ponce, and L. Reid, *J. Cell Biol.* **87**, 255 (1980).

[13] W. L. McKeehan, K. A. McKeehan, S. L. Hammond, and R. G. Ham, *In Vitro* **13**, 399 (1977).

[14] W. L. McKeehan, D. P. Genereaux, and R. G. Ham, *Biochem. Biophys. Res. Commun.* **80**, 1013 (1978).

[15] W. L. McKeehan and K. A. McKeehan, *In Vitro* **16**, 475 (1980).

plates are then incubated in a humidified atmosphere at 37° and 5% carbon dioxide. Oxygen-sensitive cell types are incubated in an atmosphere of 7% oxygen and 5% carbon dioxide. After 7 to 14 days, media is removed, the cells are fixed with 5% glutaraldehyde, 1% formalin or a solution of methanol : acetic acid (3 : 1). Fixed cells are then stained with 0.1% crystal violet or Harris hematoxylin and 1 mM lithium carbonate. The fixation and staining procedure depends on cell type.

Videometric Measurement of Cell Number

In our laboratory, the number of fixed and stained cells is estimated *in situ* by videometry.[16] We employ an Artek Model 980 cell counter (Artek Systems Corp., Farmingdale, NY) equipped with an Artek Model 890 light box. The cell counter is interfaced with a Franklin Ace 1000 microcomputer with the Artek Omni-Interface. A data analysis program in Applesoft Basic was developed by Dr. Richard Jones (Director, Institute for Computer Studies, University of New Haven, West Haven, CT) to process and plot data from the cell counter. The program accepts groups of assay data, averages the data, and calculates mean and standard error among replicates. Videometric measurement of stained cells allows a nondestructive measurement of cell number and has the following advantages[16]: (1) it is practical for large numbers of replicate assays, (2) it allows cell counts at cell densities too low for accurate counts by Coulter counter or hemocytometer, and (3) it is independent of harvesting procedure for cell counting which is often variable and may not yield single cells. The laboratory set-up is shown in Figs. 1 and 2.

Analysis of Data

Figure 3 shows the effect of different Ca^{2+} concentrations on multiplication of human diploid fibroblasts in a model system[17] designed to fit the assumptions described earlier. Cell multiplication rates determined from the slope of the log–linear curves appear uniform over a period of 8 to 16 days and regress to the same point on the ordinate. Cell population multiplication rate is determined by the following formula:

$$r = 3.32(\log C_t - \log C_0)/t$$

where r is cell generations per day, 3.32 is the reciprocal log 2, C_t is the stained cells in area units determined by videometry after t days of incu-

[16] W. L. McKeehan, P. S. Adams, and M. P. Rosser, *Cancer Res.* **44**, 1998 (1984).
[17] W. L. McKeehan and K. A. McKeehan, *Proc. Natl. Acad. Sci. U.S.A.* **77**, 3417 (1980).

FIG. 1. Instrumentation for videometric measurement of cell number. From left to right, the unit consists of the light box with a 25-well plate (A) under the camera (B). The counter shows an area count of 255. On top of the counter is the interface and a monitor that displays and flags the images to be analyzed. At right is the computer. Optimum settings for each application should be determined experimentally to achieve best sensitivity and a linear relationship between area measurements and actual cell number.[16]

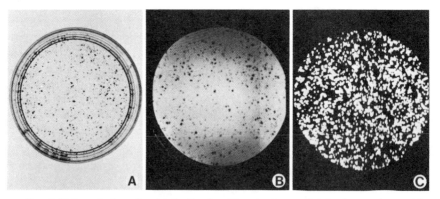

FIG. 2. Enlarged view of stained cells, the video monitor, and stained areas detected by the counter. A 60-mm Petri dish containing rat ventral prostate epithelial cells was fixed with methanol : acetic acid (3 : 1) and then stained with a solution of Harris hematoxylin followed by 1 mM lithium carbonate (A). (B) is the positive image seen on the monitor in Fig. 1 and (C) shows the image on the monitor in area measurement mode. The counter flags stained areas on the monitor that it measures which appear as enhanced white images.

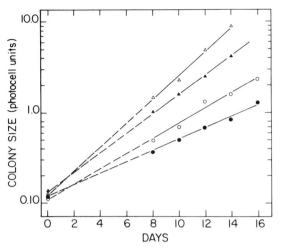

FIG. 3. Time course of increase in number of human fibroblasts at different concentrations of $CaCl_2$. The multiplication of 100 cells was measured in a $CaCl_2$-deficient nutrient medium containing 500 $\mu g/ml$ fetal bovine serum protein (FBSP) as a source of growth factors.[14] In this example, cell number was estimated densitometrically by a manual colony sizing method before the automated system in Figs. 1 and 2 was designed.[13,15] $CaCl_2$ was added at the following concentrations: (●) 30 μM, (○) 100 μM, (▲) 300 μM, (△) 1 mM. (Data reproduced from McKeehan and McKeehan.[17])

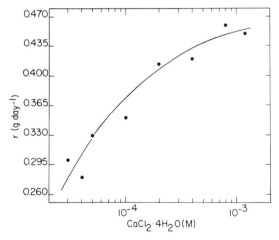

FIG. 4. Multiplication rate as a function of increasing $CaCl_2$ concentration. Multiplication of 100 human fibroblasts was analyzed after 14 days of incubation. Multiplication rate (r) in cell generations per day was calculated as described in the text. (Data reproduced from McKeehan and McKeehan.[17])

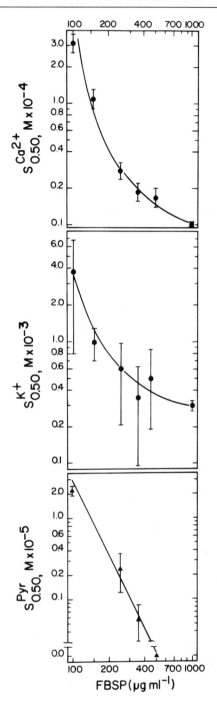

bation, C_0 is the stained area when $t = 0$, and t is the time of incubation in days.

Multiplication rate, r, at various concentrations of nutrients was substituted for V (velocity) and S (substrate) in the Henri–Michaelis–Menten formula used in enzyme kinetic analysis.[18] Data were fitted to a right rectangular hyperbola (Fig. 4). Kinetic parameters, $S_{0.50}$ and R_{max}, for each nutrient and level of growth factor are determined by nonlinear least-squares analysis using the Marquadt strategy.[19] The Basic language nonlinear least-squares program (NLLSQ) and plotting program for Apple II and Franklin computers was employed (CET Research Group, LTD, Norman, OK).[19] The $S_{0.50}$ value is the nutrient concentration required to promote a half-maximal cell proliferation rate. R_{max} is the maximal proliferation rate.

Effect of Growth Factor Effectors on Cell Nutrient Requirements

Using the kinetic analysis, we showed that serum-derived growth factors as a group specifically reduce the requirement of human fibroblasts for Ca^{2+}, K^+, Mg^{2+}, phosphate ions (P_i), and 2-oxocarboxylic acids out of over 30 nutrients analyzed.[1–4,15,17,20] Figure 5 shows the effect of serum proteins on the $S_{0.50}$ value for Ca^{2+}, K^+, and pyruvate. Further analysis revealed that transformation of the normal fibroblasts with oncogenic simian virus made the $S_{0.50}$ value for Ca^{2+}, K^+, and Mg^{2+} independent of growth factor effectors.[20] A similar kinetic analysis revealed that the single defined growth factor, epidermal growth factor (EGF), modifies the $S_{0.50}$ value for Ca^{2+}, Mg^{2+}, and 2-oxocarboxylic acids, but not K^+ and phosphate ions for normal human fibroblasts.[21,22] Using a similar kinetic analysis, Lechner and Kaighn demonstrated a relationship similar to fibroblasts among EGF, Ca^{2+}, and the multiplication rate of normal and tumorigenic human prostate cells.[23] The kinetic analysis described here has also been extended to analysis of cell attachment/survival rate in

[18] I. H. Segel, "Enzyme Kinetics." Wiley, New York, 1975.
[19] S. D. Christian and E. E. Tucker, *Am. Lab.* **14**, 36 (1982).
[20] W. L. McKeehan, K. A. McKeehan, and D. Calkins, *J. Biol. Chem.* **256**, 2973 (1981).
[21] W. L. McKeehan and K. A. McKeehan, *Exp. Cell Res.* **124**, 397 (1979).
[22] W. L. McKeehan, K. A. McKeehan, and D. Calkins, *Exp. Cell Res.* **140**, 25 (1982).
[23] J. F. Lechner and M. E. Kaighn, *Exp. Cell Res.* **121**, 432 (1979).

FIG. 5. Effect of fetal bovine serum protein concentration on requirement for Ca^{2+}, K^+, and pyruvic acid. The $S_{0.50}$ value, the concentration of each nutrient required to promote a half-maximal multiplication rate, was determined at the indicated concentrations of serum protein (FBSP). (Reproduced from McKeehan and McKeehan.[17])

addition to multiplication rate of the cell population.[4] The robust quantitative analysis described here is applicable to many isolated cell types and can be easily extended to analyze interactions among environmental effectors of cells other than nutrients and hormone-like growth factors (e.g., adhesion factors, inhibitors, drugs, and xenobiotics).

[36] Characterization of Growth Factor-Modulated Events Regulating Cellular Proliferation

By Maureen A. Harrington and W. Jackson Pledger

Introduction

Specific cellular events in the G_1 phase of the cell cycle, modulated by serum-derived growth factors, regulate the rate of cellular proliferation. Delineation of these events at the biochemical and molecular level may be achieved or aided by an experimental system in which cells are synchronized at various growth factor-modulated regulatory points within G_0/G_1. While several pharmacologic agents are available that can be used to synchronize populations of cells, these agents are limited by a lack of G_0/G_1 specificity and/or toxic side effects. An alternative approach to chemical synchronization of cells is to synchronize cells by limiting the supply of required growth factors and/or other required physiological substances.

Traverse of density-arrested fibroblastic cells through G_0/G_1 is mediated by multiple growth factors with sequential or interdependent actions. Platelet-derived growth factor (PDGF) and other known competence factors [fibroblast growth factor, $Ca_3(PO_4)_2$] initiate the proliferative response of BALB/c 3T3 cells and program cells to respond to the progression factors required for G_0/G_1 traverse and entry into S phase. Progression activity can be supplied by platelet-poor plasma (PPP) or by a combination of epidermal growth factor (EGF) and somatomedin C (SmC, insulin-like growth factor I). While competence formation is nutrient independent and requires only a transient exposure to PDGF, G_0/G_1 traverse is contingent upon the continued availability of nutrients, such as amino acids, and progression activity. A continued and sufficient supply of progression activity is necessary for cells to modulate the two G_1 control points: the V point, which bisects the 12 hr G_0/G_1 phase and the W point, which approximates the G_1/S border.

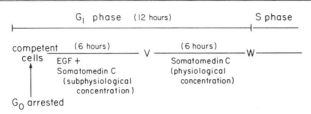

FIG. 1. Sequence of growth factors required for density-arrested BALB/c 3T3 cells to traverse G_0/G_1.

The sequence of growth factors required for density-arrested BALB/c 3T3 cells to traverse G_0/G_1 can be summarized as follows: PDGF is required to initiate a proliferative response, EGF and SmC are required for the first 6 hr of G_0/G_1 (from G_0 to the V point), whereas SmC alone is required for the last 6 hr of G_1 and entry into S phase (from the V point to S phase) (Fig. 1). In this chapter various methods employing requirements for a specific growth factor(s) and essential nutrients will be described that can be used to synchronize BALB/c 3T3 cells at the various regulatory points within G_0/G_1. Cells synchronized in this fashion can be used to (1) study the mechanism of action of certain growth factors at the specific cell cycle time when they are required and (2) delineate and order those cellular events required for G_0/G_1 traverse.

Cell Culture and Growth Factors

The competence–progression cell cycle model is based on the serum-derived growth factor requirements of density-arrested BALB/c 3T3 cells, an embryonic mouse fibroblastic cell line exhibiting density-dependent inhibition of growth. The reader may refer to Aaronson and Todaro[1] for a description of the isolation and characterization of the cell line. The competence–progression model can also be applied to a variety of mesenchyme-derived cell lines. However, consideration should be given to the potential for (1) constitutive production of growth factors which would mask an apparent requirement for a competence or progression factor, and (2) less stringent requirements for exogenously supplied growth factors due to reasons other than endogenously produced growth factors. For example, human embryonic fibroblasts produce both a competence-like factor[2] and somatomedin C.[3] Alternatively, the 10T1/2 cell line differs significantly from BALB/c 3T3 cells in that EGF-supplemented PPP-con-

[1] S. A. Aaronson and G. J. Todaro, *J. Cell. Physiol.* **72**, 141 (1972).
[2] D. R. Clemmons, *J. Cell. Physiol.* **114**, 61 (1983).
[3] D. R. Celmmons, L. E. Underwood, and J. J. VanWyk, *J. Clin. Invest.* **67**, 10 (1981).

taining medium will elicit a mitogenic response in a significant portion of a 10T1/2 cell population. However, parallel to the BALB/c 3T3 cell line, prior treatment of 10T1/2 cells with PDGF reduces the concentration of EGF required and maximizes the population of responding 10T1/2 cells.[4]

Stock cultures of BALB/c 3T3 cells are maintained at subconfluent densities in the high glucose formulation of Dulbecco's modified Eagle's medium (GIBCO) containing 4.0 mM glutamine, 50 units/ml penicillin, and 50 μg/ml streptomycin (DMEM) supplemented with 10% calf serum (CS) (Colorado Serum Co.) in a humidified 5% CO_2/95% air atmosphere. For experimental purposes, stock cultures of cells allowed to grow to confluent densities are removed by trypsinization (0.1% trypsin in phosphate-buffered saline) and are reseeded at subconfluent densities (1–3 $\times$ 10^4 cells/cm^2) in serum-supplemented DMEM. Depending on the size of culture dish employed, plated cells may require refeeding 3 days after plating with fresh serum-supplemented medium. After an additional 3 days, confluent monolayers of cells will be present. These cells are referred to as quiescent and reside in the G_0 portion of the cell cycle. Physical and biochemical parameters that distinguish G_0 from G_1 cells have been described by Pardee et al.[5] and Baserga.[6]

Procedures describing the isolation of platelet-derived growth factor can be found in Antoniades et al.,[7] Heldin et al.,[8] Deuel et al.,[9] and Raines and Ross,[10] epidermal growth factor in Savage and Cohen,[11] and somatomedin C in Svoboda et al.[12] Bovine insulin (Sigma Chemical Company) used at pharmacological concentrations can substitute for physiological concentrations of SmC, since at μM concentrations insulin binds to SmC receptors and functions as a somatomedin-like peptide.[13] The amount of

---

[4] W. Wharton, E. Leof, N. Olashaw, E. J. O'Keefe, and W. J. Pledger, *Exp. Cell Res.* **147,** 443 (1983).

[5] A. B. Pardee, R. Dubrow, J. L. Hamlin, and R. F. Kletzien, *Annu. Rev. Biochem.* **47,** 715 (1978).

[6] R. Baserga, "Multiplication and Division in Mammalian Cells," pp. 175–188. Dekker, New York, 1976.

[7] H. N. Antoniades, C. D. Scher, and C. D. Stiles, *Proc. Natl. Acad. Sci. U.S.A.* **76,** 1809 (1979).

[8] C.-H. Heldin, B. Westermark, and A. Wasteson, *Proc. Natl. Acad. Sci. U.S.A.* **76,** 3722 (1979).

[9] T. F. Deuel, J. S. Huang, R. T. Proffitt, J. U. Baenziger, D. Chang, and B. B. Kennedy, *J. Biol. Chem.* **256,** 8896 (1981).

[10] E. W. Raines and R. Ross, *J. Biol. Chem.* **257,** 5104 (1982).

[11] C. R. Savage and S. Cohen, *J. Biol. Chem.* **247,** 7609 (1972).

[12] M. E. Svoboda, J. J. VanWyk, D. G. Klapper, R. E. Fellows, and R. J. Schlueter, *Biochemistry* **19,** 790 (1980).

[13] J. J. VanWyk, L. E. Underwood, J. B. Baseman, R. L. Hintz, D. R. Clemmons, and R. W. Marshall, *Adv. Metab. Dis.* **8,** 127 (1975).

PDGF necessary for conducting experiments is related to the number of cells in the confluent monolayer. Consequently, the concentration of PDGF will increase disproportionately as the size of the cell monolayer increases. Therefore, the appropriate PDGF concentration should be determined for the cells and experimental conditions used. When PDGF is added to DMEM, a small amount of protein (0.1–0.2% PPP or other nonmitogenic carrier protein) should also be present as PDGF binds tenaciously to plastic in protein-free medium.

Platelet-Poor Plasma

Platelet-poor plasma (PPP) is prepared as follows: drawn venous blood is quickly placed into prechilled 50.0-ml conical tubes embedded in ice. Chilled blood is centrifuged (2500 g, 4°, 30 min) to remove formed elements. The plasma, down to an area immediately above the buffy coat (platelet layer), is carefully removed and recentrifuged (2500 g, 4°, 30 min) to ensure full removal of formed elements. Plasma is then transferred to a sterile glass bottle, and incubated at 37° for 3–5 hr after which the fibrin clot is removed. Plasma is then heat inactivated at 56° for 30 min, aliquoted, and stored at −20°.

The key to identifying regulatory points within the cell cycle and establishing temporal relationships between cellular events is the ability to quantitate movement of cells through the cell cycle. Several techniques are available that can be used to monitor cell cycle traverse, and in general they are based upon establishing a temporal coordinate between a given event and a measurable part of the cell cycle. For G_0/G_1 specific events, the time lag between a given event and (1) S phase can be measured by quantitating [³H]thymidine incorporation (cpms or labeled nuclei), (2) mitosis can be measured by adding colchicine and quantitating the rate of appearance of mitotic figures, and (3) subsequent cell division, by quantitating cell number.

The primary interests of this laboratory are focused on identifying unique G_0/G_1 events which are required for the initiation of proliferation. S phase represents the next measurable part of the cell cycle; consequently the temporal location of cells within G_0/G_1 in this system is based upon the time lag prior to entry into S phase. To accurately gauge the response of the cell population a method to determine entry into S phase is required. Methods such as autoradiography or microflow cytometry are advantageous because they can be used to quantitate the response. Autoradiography has been the method used in this laboratory and a description of the physical principles and procedures can be found in Vol. LVIII of this series.

Competence Formation

Density-arrested BALB/c 3T3 cells apparently reside in G_0, which is temporally 12 hr from the G_1/S border or the initiation of DNA synthesis, and are synchronized by a requirement for PDGF. This type of arrest can be accomplished by (1) growing cells in DMEM supplemented within 10% CS until growth-arrested confluent cultures are evident, and (2) changing proliferating cultures into DMEM supplemented with 5% PPP or suboptimal CS (0.1–0.5%) and allowing cultures to become quiescent. Release of cells from a G_0 block can be accomplished by the addition of DMEM supplemented with PDGF, FGF, or $Ca_3(PO_4)_2$. For competent cells to traverse G_0/G_1 and enter S phase, G_0 cells may be treated transiently (3–4 hr) with an appropriate quantity of competence factor, followed by replacement of the medium with DMEM supplemented with 5% PPP or EGF (10 ng/ml) plus SmC (20 ng/ml).

The addition of PDGF-supplemented medium to G_0 arrested cells can be used to investigate cellular reactions that may be required for the initiation of cellular proliferation. Since plasma is biologically active, but does not stimulate the proliferation of density-arrested cells, the addition of PPP to quiescent cells serves as a useful negative control for monitoring basal fluctuations in cellular responsiveness.

Early G_0/G_1

Competence and progression activities interact synergistically to promote G_0/G_1 traverse; increasing the amount of either activity will reduce the requirement for the corresponding activity. Consequently, pharmacologic doses of either type of activity alone could stimulate a portion of the cell population to traverse G_0/G_1 and enter S phase. However, at physiological concentrations, both competence and progression activities are necessary for G_0/G_1 traverse. In the absence of progression activity (5% PPP, EGF and SmC, EGF and insulin), PDGF-treated cells initiate the proliferative cycle but do not traverse G_1. Competence is a stable cellular state, that decays with a $t_{1/2}$ of 15–18 hr.[14] Therefore, PDGF-treated cells can be transferred into medium supplemented with a suboptimal concentration of progression activity (0.2% PPP) which does not support traverse of the cell cycle. If at any time before the total loss of competence, medium is resupplemented with an optimal concentration of PPP (5%) or progression factors, the remaining competent cells will traverse G_1 and

---

[14] J. P. Singh, M. A. Chaiken, W. J. Pledger, C. D. Scher, and C. D. Stiles, *J. Cell Biol.* **96,** 1497 (1982).

entry into S phase will begin 12 hr later. As the lag time between removal of PDGF and addition of PPP increases, the number of cells entering S phase will decrease.[15] This protocol can be used to correlate the stability of cellular components or reactions with the initiated competent state.

G_1 Arrest Points

Traverse of competent cells through the 12-hr G_1 lag requires the continued presence of both nutrients and progression activity. Movement of cells through G_1 can be arrested at 2 key points: (1) the V point, which bisects G_1 and is temporally located 6 hr before the G_1/S border and (2) the W point which lies immediately prior to the G_1/S border. Arrest of cells progressing through G_1 at the V point can be accomplished in several ways.

V Point Arrest

Timed PPP Removal.[15] Density-arrested cells can be transiently exposed to PDGF (3–5 hr) and then placed into medium supplemented with 5% PPP. After 8–10 hr this medium is removed and medium containing 0.2% PPP is added. Cells that have not traversed beyond the V point (50–60% of the population) will arrest at the V point. After appropriate incubation cells can be restimulated with optimal plasma to study late G_1 and entry into S phase.

Limited Somatomedin C.[16] Density-arrested cultures can be transiently exposed to PDGF and then changed into a medium supplemented with EGF (10 ng/ml) and subnanogram amounts of SmC (0.25–0.75 ng/ml) or EGF plus a low concentration of insulin (10^{-8} M). After 12 hr, approximately 50% of the population will be arrested at the V point. Alternatively, cultures may be transiently exposed to PDGF, switched into medium supplemented with 5% PPP-derived from hypopituitary animals, and cells will arrest at the V point. The addition of an optimal concentration of SmC allows progression to continue through G_1 and into S phase.

Elevated cAMP.[17] Density-arrested cultures of BALB/c 3T3 cells can be transiently exposed to PDGF and then switched into medium supple-

[15] W. J. Pledger, C. D. Stiles, H. N. Antoniades, and C. D. Scher, *Proc. Natl. Acad. Sci. U.S.A.* **74**, 4481 (1977).
[16] E. B. Leof, J. J. VanWyk, E. J. O'Keefe, and W. J. Pledger, *Exp. Cell Res.* **147**, 202 (1983).
[17] E. B. Leof, W. Wharton, E. O'Keefe, and W. J. Pledger, *J. Cell. Biochem.* **19**, 93 (1983).

mented with 5% PPP, cholera toxin (1 μg/ml), and isobutylmethylxanthine (IBMX; 0.1 mM) for 11 hr. Greater than 80% of the population will be arrested at the V point. The removal of IBMX and the presence of progression activity (PPP or SmC) allows continuation of cell cycle traverse from the V point arrest.

Amino Acid Arrest.[18] Traverse through G_1 is both growth factor and nutrient dependent. Consequently under conditions where the supply of amino acids is limited, competent cells undergoing progression through G_1 will arrest at the V point. For this procedure, DMEM deficient in either Group A or Group B amino acids is required.[18] Group A contains arginine, glycine, histidine, methionine, serine, threonine, tryptophan, valine, and glutamine, whereas Group B contains cystine, isoleucine, leucine, lysine, phenylalanine, and tyrosine. Additionally, dialyzed PPP must be used and is prepared by extensive dialysis of plasma against phosphate-buffered saline.

To achieve an amino acid arrest, density-arrested cells are prestarved for at least 24 hr (longer prestarvation times may be required) in DMEM minus either Group A or B amino acids supplemented with 5% dialyzed PPP. Cells are then exposed to DMEM deficient in Group A (or B) amino acids supplemented with PDGF and 5% dialyzed PPP to begin G_1 traverse. Approximately 12–18 hr later, cells will be uniformly arrested at the V point. The addition of required amino acids allows progression to continue. Progression activity must also be present and can be supplied by plasma, SmC, or insulin.

Cells arrested at the V point are synchronized by a requirement for somatomedin C. Release of V point-arrested cells can be achieved by placing cells in DMEM (with complete amino acid formulation) supplemented with somatomedin C (20 ng/ml) or pharmacological concentrations of insulin (10^{-5}–10^{-6} M) or 5% PPP. In comparing the kinetics of S phase entry of the cells arrested by these four procedures, the condition where amino acids are limited has provided the most uniform, complete, and stable arrest.

W Point Arrest[16]

Traverse between the V point and the G_1/S border requires physiological concentrations of SmC. Thus far we have defined only one method that is capable of arresting cells at the W point. Density-arrested BALB/c 3T3 cells are transiently exposed to PDGF, and then transferred into 5%

[18] C. D. Stiles, R. R. Isberg, W. J. Pledger, H. N. Antoniades, and C. D. Scher, *J. Cell. Physiol.* **99**, 395 (1979).

PPP containing medium for 15 hr. Replacement of the 5% PPP containing medium with medium containing 0.2% PPP will result in approximately 50% of the cells arrested at the W point. After incubation in medium with limiting amounts of PPP, the addition of high concentration of PPP (5%) or SmC allows W point arrested cells to enter the S phase.

A key feature of the synchronization methods described here is the potential for studying the mechanism of action of specific growth factors at discrete intervals during the cell cycle when they are known to be required. Under these conditions, the mechanisms by which growth factors modulate gene expression, cellular biochemical events, and alteration of cellular responsiveness to other growth factors can be analyzed.

[37] Assays of Growth Factor Stimulation of Extracellular Matrix Biosynthesis

By WILLIAM R. KIDWELL

The epithelium of organs such as the mammary gland is separated from the surrounding stromal tissue by an extracellular matrix scaffolding called the basement membrane. This structure, which is synthesized by the epithelium,[1] appears to be essential for the growth and survival of these glandular cells[2] and it probably modulates their differentiation as well. In primary culture a strong, positive correlation between the growth stimulating potencies of growth factors and their abilities to differentially amplify basement membrane protein (laminin, collagen IV, and proteoglycan) synthesis has been established.[3] Additionally, when mammary cells are cultured on different types of extracellular matrix proteins, they display differing requirements for growth factors.[3] These observations suggest that at least part of the mechanism by which growth factors act is by enhancing the biosynthesis of extracellular matrix proteins. This is especially true in the case of primary cell cultures in which cell selection for culture conditions has not taken place. In the present report we outline

[1] L. A. Liotta, M. S. Wicha, S. I. Rennard, S. Garbisa, and W. R. Kidwell, Lab. Invest. 41, 511 (1980).
[2] M. S. Wicha, L. A. Liotta, B. K. Vonderhaar, and W. R. Kidwell, Dev. Biol. 80, 253 (1980).
[3] W. R. Kidwell, D. Salomon, and M. Bano, in "Culture Methods for Molecular and Cell Biology" (D. H. Barnes, D. A. Sirbasku, and G. H. Sato, eds.), p. 7. Liss, New York, 1984.

briefly some of the more useful methods for the quantitative and qualitative measurements of extracellular matrix components affected by growth factors.

General Considerations

In culture extracellular matrix components are not only deposited at the basal cell surface but are also secreted into the growth medium. Thus quantitation of growth factor requires analysis of cells and growth medium. Growth factors have been shown to alter both the biosynthetic and degradative rates of matrix proteins, especially of collagen, the component that has been studied in greatest detail.

Collagen is composed of a triple helix that is stabilized by posttranslational hydroxylations of proline (see Vol. 80, this series). Intracellular degradation of collagen is very rapid in the prehelix assembly phase.[4] Whether this rate of degradation is affected by growth factors has not yet been determined, but it is not unlikely. The collagen biosynthetic rate appears to be constant throughout the growth cycle in some cells[5] in contrast to total cell protein synthesis which is usually reduced in confluency. For this reason growth factor effects on collagen and possibly other matrix proteins should be assessed in log-phase cells. Collagen degradation also occurs via cell-produced procollagenases, which are converted to active enzymes by proteases.[6] Growth factors, hormones,[6] and agents that affect cell shape and cell–cell interaction can alter the production of collagenases, the enzymes likely to be responsible for most of the extracellular degradation of collagen.

Intra- and extracellular degradation processes are best assessed by pulsing cells for 15 min with a precursor amino acid followed by a chase without labeled precursor. Quantitation of labeled collagen at the end of the pulse and at various times during the chase period will reveal whether growth factors modulate turnover. Alternatively, one can measure the ratio of collagen degradation products to intact collagen in the presence and absence of growth factors. In the case of collagen, this can be done by measuring the amount of acid-soluble and acid-insoluble hydroxyproline present in the medium and cell layer.[7] If the ratio is constant in the presence and absence of growth factor, then turnover is unaffected. If

[4] P. Bruckner, E. F. Eikenberry, and D. J. Prockop, Eur. J. Biochem. 118, 607 (1980).

[5] R. A. Berg, M. L. Schwartz, and R. G. Crystal, Proc. Natl. Acad. Sci. U.S.A. 77, 4746 (1980).

[6] D. Salomon, L. A. Liotta, and W. R. Kidwell, Proc. Natl. Acad. Sci. U.S.A. 78, 382 (1981).

[7] M. Bano, D. Salomon, and W. R. Kidwell, J. Biol. Chem. 258, 2729 (1983).

different ratios are seen, the growth medium should be examined for collagenase activity using labeled, native collagen of the type the cells make.[6]

Dual-isotope labeling has also provided valuable information on the net biosynthesis, the extent of proline hydroxylation, and the degradation rates for collagen.[8] Cells are labeled with [^{14}C]proline and trans-4-[^{3}H]proline. The amount of each label in collagen and the amount of ^{3}H$_2$O are determined. If the ratio of the two isotopes in collagen is constant, then growth factors have no effect on hydroxylation of proline in collagen. That being the case, alterations in the ratio of labeled collagen to tritiated water would indicate a change in the collagen degradation rate, either intra- or extracellularly.

Collagen Assays Using Protease-Free Collagenase

Cells are plated at a density such that log-phase growth is maintained throughout the labeling period. For long-term labeling (2–3 days), 2–5 μCi [U-^{14}C]proline (200 μCi/mmol) or [^{3}H]lysine (50 Ci/mmol) is added. For pulse-chase studies, cells are labeled for 15 min with 100 times as much [^{3}H]lysine. Media and cells are separately analyzed for the presence of acid-insoluble, labeled protein that is converted to an acid-soluble form on digestion with protease-free collagenase.[9] Cells are scraped from the dishes into 5 ml of 0.5 M Tris–HCl, pH 7.4, containing 0.11 M NaCl and 1 mM unlabeled lysine or proline. Cells are centrifuged down for 10 min at 1500 g and resuspended in 1 ml 50 mM Tris–HCl, pH 7.4, containing proline or lysine and then sonicated briefly. Proteins are precipitated for 30 min with cold 20% TCA and the precipitates collected by centrifugation for 15 min at 3000 g. Precipitates are resuspended and washed 3 times in cold 5% TCA. The resultant precipitate is dissolved in 0.5 ml of 0.2 M NaOH and 0.1 ml removed for counting. This represents the amount of label in total cell protein. Then 0.2 ml of the remaining sample is placed into two tubes containing 0.16 ml of 0.15 M HCl and 0.1 ml of 1 M HEPES, pH 7.3. Ten microliters of 25 mM CaCl$_2$ and 20 μl of collagenase (Form 111, Advanced Biofactures, 3200 U/ml 50 mM Tris–HCl, pH 7.6, 5 mM CaCl$_2$) are added and the samples incubated for 90 min at 37°. After cooling on ice, 0.5 ml of 10% TCA and 5% tannic acid are added and samples stored for 30 min in the cold. Acid-insoluble material (noncollagen protein) is pelleted by centrifugation for 15 min at 4000 g. The supernatant is removed and saved. The pellet is wahed once with TCA/tannic

[8] M. Chojker, B. Peterkofsky, and J. Bateman, Anal. Biochem. **108**, 385 (1980).
[9] B. Peterkofsky, Arch. Biochem. Biophys. **152**, 318 (1972).

acid and recentrifuged. Combined supernatents are analyzed for radioactivity. The second tube is processed in an identical way except that collagenase is replaced with buffer. Differences in the counts with and without collagenase digestion represent the labeled collagen in the sample. The procedure is based on methods developed by Peterkofsky.[9]

Quantitation of Hydroxyprolines

The abundance of 3- and 4-hydroxyproline in cultures may be used as a rough estimate of collagen[10] and as an index of collagen type synthesized. Since these imino acids are nearly uniquely present in collagens, their quantitation in acid-soluble form (peptides and free amino acids) in the culture medium is a useful measure of collagen degradation that is sometimes altered by growth factors.[7] Labeled 4-hydroxyproline can be quantitated readily because of its unique chemistry. Acid-hydrolyzed samples are oxidized with chloramine-T and extracted with toluene then heated. The 4-hydroxyproline which is ultimately converted to pyrrole is recovered in a second toluene extraction.[11]

Radioactive 3-hydroxyproline can be resolved from other amino acids in a protein hydrolyzate on an automated amino acid analyzer and counted.[12] Quantitation of the mass of this amino acid is most readily accomplished with the small amounts likely to be encountered in cultured cells as follows. Culture medium and cells, separately or combined, are hydrolyzed to free amino acids with constant boiling HCl and concentrated on a rotoevaporator. Following desalting, the amino acids are concentrated by lyophylization and then treated with nitrous acid.[13] This procedure converts all the amino acids except proline, and 3- and 4-hydroxyproline into 2-hydroxy acids which can be separated from the imino acids by partitioning between aqueous 1 N HCl and diethyl ether.[13] This results in as much as a 3000-fold enrichment in 3-hydroxyproline when using cultures of rat mammary cells. The imino acids are then derivitized by reacting with 7-chloro-4-nitrobenz-2-oxa-1,3-diazole (NBZ). Fifty microliters of imino acid mix in water is mixed with 50 μl of 10 mM sodium borate, pH 9, and 50 μl of 30 mM NBZ in methanol. After standing for 30 min at 25° 200 liters of 0.1 N HCl and 1.3 ml ethyl acetate are added. The sample is mixed vigorously and centrifuged to separate the two layers. The derivatized imino acids recovered in the top layer are

[10] H. Stegemann, Hoppe Seyler's Z. Physiol. Chem. 311, 41 (1958).
[11] B. Peterkofsky and D. J. Prockop, Anal. Biochem. 4, 400 (1962).
[12] M. Bano and W. R. Kidwell, Cancer Res. 44, 3055 (1984).
[13] D. D. Dziewiatkowski, V. C. Haskall, and R. L. Riolo, Anal. Biochem. 49, 550 (1972).

concentrated to dryness with a stream of nitrogen and dissolved in a small volume of methanol and separated by thin-layer chromatography on silica gel 60 plates (EM Labs) using methanol : toluene : acetone : trimethylamine (20 : 40 : 35 : 5). The fluorescent derivatives, localized with a UV lamp, are scraped into tubes and recovered from the silica gel by adding 0.2 ml 0.1 N HCl and 1.3 ml ethyl acetate. Following vortexing, the sample is centrifuged and the ethyl acetate recovered. Fluorescence is determined with excitation at 460 nm and emission at 525 nm. Standards prepared with 4-hydroxyproline can be used since the quantum yields of 3- and 4-hydroxyproline are equivalent. Fluorescence is linear from 0 to 300 pM. The assay is suitable for as little as 10^6 mammary cells. The R_f values are 5.9, 4.2, and 3.2 for proline, 3- and 4-hydroxyproline, respectively. A 3-hydroxyproline : 4-hydroxyproline ratio of about 1 : 10 is seen for type IV collagen whereas ratios of 1 : 100 to 1 : 1000 are indicative of other collagens.[14]

Quantitation of Glycosaminoglycans

Glycosaminoglycans (GAGs) are labeled for 1–3 days with 50–100 μCi [^{3}H]glucosamine. Cell layer and medium, separately or combined, are digested for 48 hr with Pronase (15 mg/ml) that is previously predigested for 1 hr at 37° to inactivate any contaminating GAG-degrading activity. Residual acid-insoluble material is removed by precipitating with 10% cold TCA. After centrifugation for 15 min at 4000 g, the supernatant is collected and the TCA removed with three extractions with water-saturated diethyl ether. About 20 μg carrier heparin, hyaluronic acid, chondroitin sulfates A and C, and dermatin sulfate is added followed by 3 vol 1.5% potassium acetate in ethanol. GAGs are precipitated overnight at $-20°$ and pelleted at 100,000 g for 1 hr. The pellets are resuspended in 50 mM unlabeled glucosamine and reprecipitated 3 times with potassium acetate/ethanol. The various GAG classes are then resolved by ion-exchange chromatography on a DEAE-cellulose column[15] or divided into aliquots which are digested with specific hydrolases[15] or with nitrous acid to cleave heparan sulfate.[16] Nondigested GAGs are then precipitated with alcohol and the radioactivities in the pellets determined. The type of culture substratum can dramatically affect the relative amounts of the GAG classes,[17] the extent to which they are distributed into the cell layer

[14] N. A. Kefalides, R. Alper, and C. C. Clark, *Int. Rev. Cytol.* **61**, 167 (1979).
[15] S. R. Baker, D. L. Blithe, C. Buck, and L. Warren, *J. Biol. Chem.* **255**, 8719 (1980).
[16] H. Saito, T. Yamagata, and S. Suzuki, *J. Biol. Chem.* **243**, 1536 (1968).
[17] G. Parry, E. Y.-H. Lee, D. Farson, M. Koval, and M. Bissell, *Exp. Cell Res.* **156**, 487 (1985).

and medium,[17] and the rate of degradation of the GAGs.[18] It is clear, however, that growth factors, such as epidermal growth factor and mammary tumor factor, act by enhancing GAG precursor biosynthesis.[19]

Immunological Methods

Antibodies against most of the common matrix components have been developed and offer a specific means for quantitative and qualitative analysis. Mass determination is rapidly accomplished by means of enzyme-linked immunoassay (ELISA) whereas growth factor effects on labeled precursor incorporation can be assessed by immunoprecipitation and gel electrophoresis.

ELISA Assays. Detailed procedures for this assay are outlined for matrix components by Rennard *et al.*[20] In the assay purified antigen is coated onto microtiter wells. Solubilized cell layer or medium is mixed with suitable dilutions of specific antiserum plus protease inhibitors and incubated overnight in the cold then placed into the microtiter wells. Any unbound antibody present binds to the antigen on the well. Following removal of the solution and washing the well, the bound antibody is quantitated with peroxidase coupled to a second antibody with specificity for the first antibody. Methods for the suitable purification of collagens, laminin, and fibronectin needed for the assays have been described.[3,20] Using this procedure we have demonstrated that most growth factors that enhance the rate of mammary cell growth also differentially enhance the amount of laminin and type IV collagen associated with the cells.[3,21]

Immunoprecipitation and Gel Electrophoresis. A suitable procedure for quantitating laminin, fibronectin, and collagen from the same sample is as follows. Cells are labeled for 5–10 hr with [35S]methionine (5 μCi/ml) in methionine-free medium containing 50 μg/ml β-aminopropionitrile (to prevent collagen cross-linking). Matrix proteins in the medium are precipitated for 24 hr with ammonium sulfate (283 mg/ml) for 24 hr. Phenylmethylsulfonyl fluoride (PMSF, 0.2 mM) and N-ethylmaleimide (NEM, 1 mM) are present during this step to limit proteolysis. Precipitates are collected by centrifugation for 30 min at 20,000 g and dissolved in a small volume of 50 mM Tris–HCl, pH 7.4, containing PMSF and NEM. The

[18] G. David and M. Bernfield, *Proc. Natl. Acad. Sci. U.S.A.* **76,** 876 (1979).
[19] Y. Kumegawa, M. Hiramatsu, T. Yajima, M. Hatakeyama, S. Hosoda, and M. Namba, *Endocrinology* **110,** 607 (1982).
[20] S. I. Rennard, R. A. Berg, G. R. Martin, P. G. Robey, and J.-M. Foidart, *Anal. Biochem.* **104,** 205 (1980).
[21] M. Bano, D. Salomon, and W. R. Kidwell, *J. Biol. Chem.* **260,** 5745 (1985).

samples are dialyzed against 1000 volumes of the same solution. Cell layers are scraped into a minimum volume of 6 M urea–4% sodium dodecyl sulfate–20 mM dithiothreitol and extensively dialyzed against Tris–HCl buffer–PMSF–NEM. Two hundred microliters of cell lysate or medium protein preparation is mixed with 100 μl of bovine serum albumin (10 mg/ml in 0.15 M NaCl, 5 mM EDTA, 50 mM Tris–HCl, pH 7.4, 0.05% Nonidet P-40). Antiserum against the particular matrix component is added at 1 : 50 to 1 : 100 dilution and immune complex allowed to form for 16 hr at 4°. Immunobeads with the appropriate coupled anti-γ-globulin is added and the samples incubated for 2 hr at 37°. The complexes formed are then centrifuged down in a microfuge and washed 3–5 times with the albumin–Tris–EDTA–Nonidet P-40 buffer. The final pellet is boiled for 3 min in 4% sodium dodecylsulfate–6 M urea–50 mM Tris–HCl, pH 8.8, and electrophoresed on acrylamide gels according to Davidson and Bornstein.[22] Nonimmune serum controls and samples predigested with protease-free collagenase are used to confirm the identity of radioactive bands localized on gels by autoradiography.[23]

Indirect Immunofluorescence. If possible, cells should be grown on glass coverslips to reduce background fluorescence. Cells are fixed for 5 min in absolute methanol at -20° followed by 5 min in acetone at -20°. After air drying, 250 μl antibody suitably diluted with phosphate-buffered saline containing 1% bovine serum albumin (PBS–BSA) is added. Samples are incubated overnight at 4° and washed 5 times with PBS–BSA. The second antibody, with conjugated fluorescein or rhodamine (Cappel labs), diluted about 1/100 in PBS–BSA, is placed on the coverslips for 1 to 24 hr in the cold. Coverslips are then extensively washed with PBS–BSA and examined for fluorescence. When photographing, the exposure time should be kept short (15–30 sec) and equivalent for antiserum and prebleed controls. The film ASA number should be set at 1/2 when photographing rhodamine fluorescence. This procedure will detect matrix proteins localized both intra- and extracellularly.

Messenger RNA Quantitation

Increased collagen synthesis is sometimes,[21] but not always,[24] associated with increased collagen mRNA levels. We have recently found that

[22] J. M. Davidson and P. Bornstein, *Biochemistry* **14**, 5188 (1975).
[23] W. M. Lewko, L. A. Liotta, M. S. Wicha, B. K. Vonderhaar, and W. R. Kidwell, *Cancer Res.* **41**, 2855 (1981).
[24] P. Tolstoshev, R. Haber, B. C. Trapnell, and R. G. Crystal, *J. Biol. Chem.* **256**, 9672 (1981).

one factor, MDGF1, enhances collagen synthesis 4- to 10-fold in a rat kidney cell line and gives a comparable increase in collagen mRNA levels.[21] Procedures for the isolation of cellular RNA and quantitation of collagen mRNA using nick-translated collagen cDNAs and slot–blot hybridization have been detailed in Bano *et al.*[21] The procedure is very useful for confirming conclusions of growth factor effects on biosynthesis vs degradation drawn from pulse-chase studies.

Culture Conditions Affecting Matrix Protein Production

While it is widely known that a number of growth factors can modulate the biosynthesis of matrix proteins, it is less well appreciated that extracellular matrix components can modulate growth factor responses.[3,6,21] Preliminary studies suggest that the differences in cellular responses to growth factors are due to the effects of the substratum on growth factor receptor expression.[25] It has also been suggested that matrix components can feedback regulate their own biosynthesis[26] as well as regulate certain nutritional requirements. In the latter case a proline auxotrophy has been observed in primary cultures of mouse mammary and rat liver cells grown in primary culture on various types of collagen substrata.[26,27] These results suggest that much effort is needed to unravel the complex relationships between growth regulation, differentiation, and the extracellular matrix. It is probable that studies with primary cultures will reveal regulatory processes that, because of cell selection, have been lost in established cell lines. It is also probable that synergistic effects of growth factors on matrix production will be detected, especially with increasing use of serum-free, defined media for cell culture.

---

[25] S. Monaham and W. R. Kidwell, unpublished observations, 1984.
[26] W. R. Kidwell, G. Smith, and B. K. Vonderhaar, *in* "Growth and Differentiation of Cells in Defined Environments" (H. Murakami, ed.), pp. 103–108. Springer-Verlag, Hamburg, 1985.
[27] T. Nakamura, H. Teramoto, Y. Tomita, and A. Ichihara, *Biochem. Biophys. Res. Commun.* **122**, 884 (1984).

[38] Measurement of Growth Factor-Induced Changes in Intracellular pH

By DAN CASSEL and PAUL ROTHENBERG

Introduction

Addition of growth factors to quiescent cells leads to a rapid activation of Na^+/H^+ exchange and consequently to an elevation of the intracellular pH (pH_i). The activation of Na^+/H^+ exchange can be an important component of the mitogenic response of cells to growth factors. Thus, in cultured Chinese hamster fibroblasts, it has been demonstrated that a variant lacking Na^+/H^+ exchange activity is unable to emerge from quiescence upon addition of growth factors in a medium at neutral pH which does not contain bicarbonate buffer.[1,2] However, this variant shows a mitogenic response to growth factors similar to the wild-type cells when pH_i is elevated by other means such as relatively alkaline extracellular pH or the presence of a bicarbonate buffer, suggesting that pH_i elevation by growth factors may serve a permissive (albeit essential) role rather than a causal role in the stimulation of cellular proliferation in hamster fibroblasts. Yet, the precise role of Na^+/H^+ exchange in the general mechanism of action of growth factors remains to be fully elucidated (see also Mendoza and Rozengurt [37] and Lever [36], Vol. 146 of this series).

Methods for pH_i Determination in Cultured Animal Cells

Determination of growth factor-induced elevation of pH_i depends on the availability of methods to reliably determine the cytoplasmic pH in cells which grow in monolayer culture. Two major techniques are currently available to measure pH_i of cells grown in monolayers, the weak acid distribution methods[2,3] and the intracellularly trapped fluorescent pH indicator method[4–6] (for general review, see Ref. 7). The weak-acid distri-

[1] J. Pouyssegur, C. Sardet, A. Franchi, G. L'Allemain, and S. Paris, *Proc. Natl. Acad. Sci. U.S.A.* **81**, 4833 (1984).

[2] G. L'Allemain, S. Paris, and J. Pouyssegur, *J. Biol. Chem.* **259**, 5809 (1984).

[3] S. Schuldiner and E. Rozengurt, *Proc. Natl. Acad. Sci. U.S.A.* **79**, 7778 (1982).

[4] J. A. Thomas, R. N. Bucksbaum, A. Zimniak, and E. Racker, *Biochemistry* **18**, 2210 (1979).

[5] W. H. Moolenar, R. Y. Tsien, P. T. Van der Saag, and S. W. de Laat, *Nature (London)* **304**, 645 (1983).

[6] P. Rothenberg, L. Glaser, P. Schlesinger, and D. Cassel, *J. Biol. Chem.* **258**, 4883 (1983).

bution method has some disadvantages due to the time required to reach a steady-state distribution of the weak acid and the high ratio of extracellular to intracellular volume. Inherently slow, this method is also subject to errors resulting from subcellular compartmentalization, as well as from incomplete separation of extracellular medium from intracellular volumes.

Two methods for intracellular trapping of fluorescent pH indicators have been described. In one method[4,5] an esterified fluorescein derivative is taken up by cells and is hydrolyzed in the cytoplasm to release additional negative charges which render the dye less permeable. Although this is a readily applied method, dye leakage from cells is still sufficiently rapid (depending in part on cell type) to limit pH_i measurements to short times (minutes)—a problem exacerbated at physiological (37°) temperatures. In the second method, to be described in this chapter,[6,7] the fluorescent pH indicator is coupled to an inert high-molecular-weight carrier, dextran, and this dye conjugate is then introduced into the cell cytoplasm by osmotic lysis of the pinocytic vesicles. Although technically more elaborate, this method stably traps the fluorophore within the cytoplasm, permitting continuous nondestructive pH_i measurements over long periods of time. Small adjustments in the experimental conditions are necessary with each cell type. This method is easier to apply to some cell types than to others but yields cells with a stable intracellular fluorescent signal, since the coupling to dextran prevents any leakage of the dye from the cell.

DMFD Loading Method for pH_i Measurement

The fluorescent pH indicator 4′,5′-dimethylfluorescein coupled to dextran (DMFD) (pK_a 6.75)[8] is introduced into the cell cytoplasm by the method of Okada and Rechsteiner,[9] for the introduction of macromolecules into cells, as outlined in Fig. 1 (see Appendix for synthesis of DMFD). Cell monolayers grown on glass slides are first exposed to a *hypertonic* solution containing the dye, resulting in the uptake of the dye in a hypertonic solution into pinocytic vesicles within the cytoplasm. Subsequently, the cell monolayer is washed and exposed to a *hypotonic* solution resulting in the preferential osmotic lysis of the pinocytic vesicles and the release of DMFD into the cytoplasm. After this procedure, the cells are incubated overnight to allow recovery from any deleterious effects of the osmotic shock before they are used for measurements of pH_i.

---

[7] L. Reuss, D. Cassel, P. Rothenberg, B. Whiteley, D. Mancuso, and L. Glaser, *Curr. Top. Membr. Transp.*, in press (1985).

[8] P. Rothenberg, L. Glaser, P. Schlesinger, and D. Cassel, *J. Biol. Chem.* **258**, 12644 (1983).

[9] C. Y. Okada and M. Rechsteiner, *Cell* **29**, 33 (1982).

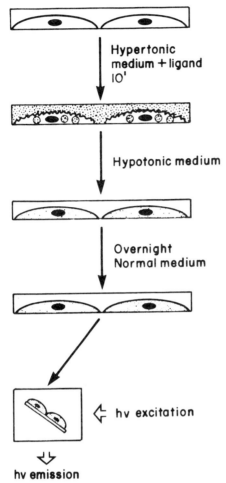

Hypertonic
medium + ligand
10'

Hypotonic medium

Overnight
Normal medium

⟵ hv excitation

hv emission

FIG. 1. Schematic model of the introduction of dimethylfluorescein-dextran (DMFD) into cytoplasm. Cells are incubated in hypertonic medium containing high concentrations of DMFD for a short period of time to allow uptake of medium components into primary endocytic vesicles, but not enough time to reach lysozomes. The cells are then treated in hypotonic medium to lyse the primary endocytic vesicles releasing DMFD into the cytoplasm. After an overnight recovery, the fluorescence of the cells can be used to measure pH_i.

Methods

Salt Solution

A physiological salt solution for incubating cells during pH_i measurements (referred to as solution A) contains 120 mM NaCl, 5.4 mM KCl, 1 mM MgSO$_4$, 1.8 mM CaCl$_2$, 1 mM NaP$_i$, 10 mM glucose, and 25 mM HEPES or MOPS buffer. The solution is brought to the desired pH with NaOH. CaCl$_2$ is added last following pH adjustment to prevent the formation of calcium phosphate precipitate. The solution is filter sterilized.

Preparation and Clean-Up of Glass Slides

High-quality microscope slides (0.1 × 2.5 × 5 cm) are cut into 0.9 × 2.5 cm rectangular pieces. Oversized or scratched slides are discarded. The cut slides are cleaned by manually scrubbing each slide in 7× detergent solution, followed by extensive washing with distilled water. The slides are wiped with tissue paper to remove excess water and are individually placed in a beaker containing fuming nitric acid and equipped with a glass cover. (Caution: All procedures involving fuming nitric acid should be carried out in a fume hood and the slides should be handled with a long pair of forceps.) After the slides have been immersed in the nitric acid for at least 24 hr, the nitric acid is diluted with at least one volume of distilled water. The slides in the beaker are washed by several changes of distilled water, and then washed by consecutive transfer through three 500-ml beakers containing glass distilled water, using a clean pair of fine forceps. Slides should be separated during the washing procedure to prevent trapping of nitric acid between the planar surfaces. Finally, the slides are placed in 100% A.R. ethanol and are stored covered until used.

Slides which have been used for experiments are collected and stored in a 7× detergent solution. They can be regenerated by the clean-up procedure described above.

Growing Cells on Glass Slides

The slides are individually removed from the beaker with ethanol in which they are stored and are dried by flaming. They are placed in 35-mm culture dishes (two slides per dish) and are covered with 2.5 ml of trypsinized cell suspension in growth medium. Tissue culture as well as bacteriological culture dishes can be used. The latter dishes appear to be advantageous for growing certain cell types which adhere much more strongly to tissue culture plastics than to glass, such as NR6 cells.[10] After plating, the cells are propagated by standard tissue culture techniques.

Preparation of DMFD Loading Solution

The fluorescent indicator dye, dimethylfluorescein, is synthesized and coupled to dextran as described by Rothenberg et al.[8] The 5000 MW

[10] D. Cassel, P. Rothenberg, Y. X. Zhuang, T. F. Deuel, and L. Glaser, *Proc. Natl. Acad. Sci. U.S.A.* **80**, 6224 (1983).

dextran used in the original procedure may no longer be commercially available. However, the readily available 10,000 MW dextran can be substituted without any change in the procedures. We have successfully used an indicator dye coupled to a 10,000 MW dextran for pH_i measurement. While both the 5000 MW and the 10,000 MW dextrans give a similar degree of substitution with dimethylfluorescein, the fluorescence intensity of the 10,000 MW DMFD solution is approximately 60% of that of 5000 MW DMFD solution (D. Cassel, unpublished observations) (see Appendix). DMFD prepared from a 9000 MW dextran has recently become commercially available from Sigma. While the chemical characteristics of the commercial DMFD appear similar to those reported by Rothenberg *et al.*, we have not tested its compatibility for pH_i measurement by the method described herein.

A solution that is suitable for loading A431 cells with DMFD contains 1 M sucrose, 10% polyethylene glycol (PEG), 2% fetal calf serum, and 15% (w/v) DMFD. To prepare 10 ml of this solution, 1 g PEG [Carbowax PEG 1500 (MW 400–600), Fisher Scientific] is spread with a spatula on the bottom part of a 50-ml conical centrifuge tube (commercially available PEG is often cytotoxic; several lots may need to be tested to obtain satisfactory results). A.R. sucrose (3.42 g) and 8 ml warm solution A are added and the tube vortexed until all material is dissolved. The pH of the solution is brought to 7–7.2 with 2 N NaOH. DMFD (1.5 g) is added and the tube is incubated at 37° with occasional tilting to wet the DMFD. When most of the DMFD has dissolved (10–15 min), the tube is vortexed for approximately 15 sec, 0.2 ml fetal calf serum is added, and the tube is centrifuged at 1000 g for 10 min to remove residual insoluble DMFD. The pH of the supernatant is checked and is readjusted to 7–7.2 if necessary. Finally the solution is filter sterilized, using a 12-ml syringe equipped with a 0.22-μm Millipore filter, and the filtrate is divided into sterile freezing tubes and stored in liquid nitrogen. The solution can be used for loading cells for at least 6 weeks. Storage at more elevated temperatures is not recommended since it results in a gradual decrease in the pH of the solution which is somehow generated by the PEG.

Loading Protocol for A431 Cells

All procedures are carried out under sterile conditions in a laminar flow hood. Cells are grown on glass slides to near confluency.[6,8] The slides are taken from the dish by holding the sides of the slide with forceps. Excess medium is removed by first holding the slide vertically and absorbing the drop of medium at the bottom with sterile tissue paper, and then laying the slide on the paper with the cells facing upward. This procedure is carried out quickly to avoid overdraining. The slides are placed into the middle of an empty 35-mm dish (one per dish) and are immediately covered with 50 μl of sterile DMFD loading solution added dropwise with an automatic pipettor. The dish is tilted several times to facilitate complete coverage of the slide with the solution. The cells are incubated for 10 min at 37° in a humidified air incubator. Subsequently, the slides are washed twice with 3–4 ml of hypotonic solution prewarmed to 37°, and are covered with 3 ml of this solution. Hypotonic solution is prepared as a dilution of serum-free tissue culture medium [Dulbecco's modified Eagle's (DME)-HEPES buffered, pH 7.2]. A 6 parts H_2O : 4 parts DME ratio has been used with A431 carcinoma cells; a 1 : 1 dilution has been employed for NR-6 fibroblasts. Care should be taken to achieve mixing of the viscous DMFD with hypotonic solution during washing. The dishes are left in the incubator or laminar flow hood for 2 min. The slides are then lifted with forceps and gently swirled in the solution several times to remove residual DMFD which may be adhered to the bottom of the slide. The slides are transferred into new 35-mm dishes containing 2 ml of HEPES-buffered DME supplemented with 1.5% FCS (two slides per dish). The dishes are incubated overnight in a CO_2-free humidified incubator. The cells can also be used 2 days after loading.

Loading of Other Cell Types

Loading with DMFD was successfully employed with the mouse 3T3 cell variant NR6.[10] Loading solution for NR6 cells contains 0.5 M rather than 1 M sucrose and, following incubation with DMFD, the cells were washed very gently to avoid detachment of the cells from the glass surface. Swiss mouse 3T3 cells come off the glass slides even with gentle washing.[10] When attempting to load other cell types with DMFD it may be necessary to retest the loading conditions such as the sucrose and DMFD concentrations in the loading solution and the washing method. Loading of cells that have been grown on different surfaces and to different final densities can be attempted. Another alternative is to load the cells in suspension by the method described by Okada and Rechsteiner[9] followed by plating on glass slides. This procedure has been used successfully for PC12 cells.[11]

Fluorometer Setup

The slides bearing loaded cells are inserted into square 1 cm glass or quartz cuvettes using a specially designed slide holder.[12] Both the cuvette

---

[11] C. Chandler, E. Cragoe, Jr., and L. Glaser, *J. Cell. Physiol.* **125**, 367 (1985).
[12] S. Ohkuma and B. Poole, *Proc. Natl. Acad. Sci. U.S.A.* **75**, 3327 (1978).

and the slide holder should be fixed in place in the fluorometer to prevent any change in the position of the slide during solution changes. Ordinary rubber bands looped around the top of the cuvette stem of the slide holder can be used for this purpose. The side of the slides on which the cells are growing is facing the excitation beam at a 30–35° angle (Fig. 2). For the Aminco SPF500 fluorometer, fluorescence is measured using excitation wavelength and bandwidth of 510 and 0.5–1 nm, respectively, and emission wavelength and bandwidth of 545 and 20 nm, respectively. These parameters are set to maximize the fluorescent signal and minimize light scattering. The excitation bandwidth should be selected to avoid rapid photobleaching, and thus should be adjusted for individual fluorometers depending on the light intensity of the excitation beam. If excitation bandwidths larger than 2 nm are required to reduce the noise, excitation and emission wavelengths may have to be moved apart to reduce light scattering.

Solutions in the cuvette are changed by injecting 10–20 ml of the new solution through a small hole in the slide holder using overflow suction

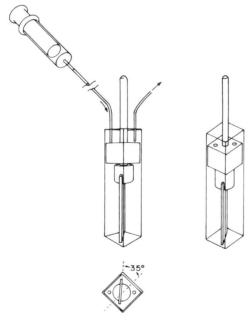

FIG. 2. Arrangement of cells in fluorometer cuvette. The diagram shows the arrangement of the glass slide in a cuvette and the procedure to hold it firmly in place. The block that holds the glass slide is machined from Teflon, and the glass slide is held in place in a slot with a rubber gasket. Note: arrangement for changing solutions should be made so that liquid is not drained from cells and at least 10 ml of new solution warmed to 37° is injected into the cuvette.

through another hole. Injections are done with a syringe fitted with short plastic tubing, with a diameter just slightly smaller than that of the hole in the slide holder. The tubing is just inserted into the hole and should not be inserted deeply into the cuvette (Fig. 2).

Preliminary Steps in Fluorescence Measurements

At the beginning of each set of experiments, an empty slide and a slide with loaded cells are inserted into two cuvettes. First, cellular fluorescence is measured and instrument gear is adjusted to register the signal at about the middle of the recorder scale. Light scattering from the empty slide is then measured and the reading is offset to zero. Subsequently, cellular fluorescence is measured and adjusted once more to about the initial value. More accurate light scattering values from each slide are obtained after quenching the DMFD fluorescence by injecting 0.1 M HCl[8,10]; these values should be used as the blank in cellular fluorescence measurements.

Following the insertion of cells into the cuvettes, there may be a transient period (10–30 min) of fluorescence instability. To avoid repeated delays it is advisable to have, during the experiments, at least one extra cuvette into which loaded cells have been introduced and maintained at 37°.

Determination of pH$_i$ Values

Determination of the initial pH$_i$ value relies on methods for the equilibration of pH$_i$ with solutions at various extracellular pH values. When equilibration is achieved, a calibration curve of fluorescence versus pH is constructed by exposing the cells to solutions at different pH values, and pH$_i$ is determined from the intersection of the initial cellular fluorescence with the calibration curve.[8,10] Two equilibration methods have been used: (1) Na$^+$ loading. In this method, cells are incubated in solution A containing ouabain (0.5 mM for A431 cells) to inhibit the Na$^+$/K$^+$ pump for at least 2 hr. This incubation results in the equilibration of Na$^+$ across the cytoplasmic membrane as well as in activation of Na$^+$/H$^+$ exchange by unknown mechanisms.[6,8] In the absence of a transmembrane Na$^+$ gradient, the Na$^+$/H$^+$ exchange activity rapidly equilibrates the intracellular pH with the external pH. A certain disadvantage of this method is that some drift in baseline fluorescence may occur during the prolonged incubation required to equilibrate Na$^+$ across the cell membrane. Photobleaching during this period can be eliminated either by closing the excitation shutter or by using a very narrow excitation bandwidth. The equilibration time can also be decreased by using Li^+ instead of Na^+ in the medium.[1,13] (2) Equilibration by the nigericin-high K^+ method.[4,10] In this method, the K^+/H^+ exchange ionophore, nigericin, rapidly equilibrates the pH across the cell membrane in the presence of external $[K^+]$ which approximates the intracellular $[K^+]$ (about 120 mM). A shortcoming of this method is that nigericin also affects the internal pH of cellular organelles. In some cells, such as A431 cells,[8] a fraction of the loaded DMFD is present in acidic organelles which become more basic in the presence of nigericin. When this fraction is high, the nigericin method should not be relied on for standardization of cytoplasmic pH. The nigericin technique has been successfully employed to determine pH_i in mouse NR6 cells, in which DMFD does not appear to be trapped in acidic organelles following loading with DMFD at a reduced temperature.[10]

It is quite impractical to measure pH_i for each set of experiments, particularly so with the Na^+ (or Li) equilibration method. Rather, it can usually be assumed that pH_i is always the same for a certain cell type under certain initial experimental conditions. Due to the limitations of the above-mentioned pH_i determination methods, absolute pH_i values should be treated with some caution. It may also be useful to measure pH_i in cells that have not been loaded with DMFD using a pH_i indicator dye which does not appear to be trapped in intracellular organelles.[4,5]

Measurement of Growth Factor-Induced pH_i Elevation

Once the conditions for pH_i measurement for a particular cell type have been set, as described in the previous paragraphs, measurement of growth factor-induced changes in pH_i is quite simple. DMFD-loaded cells are preincubated at 37° in solution A at the desired pH for at least 1 hr, and are then inserted into the cuvettes and the solution is changed once. After fluorescence has stabilized, the growth factor or serum is injected into the cuvette in 10 ml of the same solution and the changes in cellular fluorescence are recorded. Fluorescence is translated into intracellular pH units using the pH_i vs fluorescence calibration curve.

The exact conditions of the experiment should be pretested as follows. Before the addition of the growth factor, the cells must be depleted from serum growth factors by preincubation in a serum-free solution. The length of the preincubation period is determined as that which gives maximal pH_i elevation upon subsequent addition of the growth factor. However, prolonged preincubation in nutrient-depleted physiological saline,

[13] B. Whiteley, D. Cassel, Y. X. Zhuang, and L. Glaser, *J. Cell Biol.* **99**, 1162 (1984).

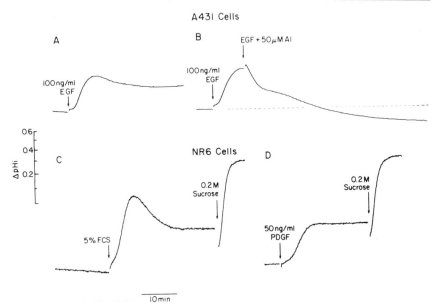

FIG. 3. Mitogen-induced changes in intracellular pH. The figure shows the time course of the pH_i elevation by EGF in A431 (A and B) cells and by PDGF in NR6 (C and D) cells. As shown in A, the addition of the potent Na^+/H^+ exchange inhibitor dimethylamiloride (A1) to stimulated A431 cells results in a reversal of the pH_i elevation. After mitogen addition to NR6 cells, further alkalization due to activation of the Na^+/H^+ exchange can be observed by the addition of hypertonic medium.[16] The data shown are from unpublished experiments by B. Whitely and L. Glaser.

such as solution A, may reduce cellular viability. The magnitude of pH_i elevation by growth factors may also depend on the extracellular pH. In both A431 cells[8] and NR6 cells,[10] pH_i elevation by growth factors and serum is maximal at pH_o 6.8, and is greatly reduced at pH_o 7.4. However, the optimal pH_o may differ in different cell types.

A representative curve of pH_i elevation by growth factors is shown in Fig. 3. In all cases examined to date[8,10,13–15] pH_i elevation by growth factors, as well as by other activators of Na^+/H^+ exchange, is preceded by a lag period lasting between 1 and 3 min. In some cases (NR6 cells) a decrease in pH_i is observed during this period.[10] Whether this initial decrease in pH_i is transient or whether it also affects the final pH_i elevation is unclear. pH_i elevation is complete within 5–10 min and subsequently pH_i may either stay constant or partially decline.[8,10]

[14] D. Cassel, Y. X. Zhuang, and L. Glaser, *Biochem. Biophys. Res. Commun.* **118,** 675 (1984).

[15] D. Cassel, B. Whiteley, Y. X. Zhuang, and L. Glaser, *J. Cell. Physiol.* **122,** 178 (1985).

The role of Na^+/H^+ exchange in growth factor-induced pH_i elevation may be verified by two types of experiments. (1) The cells are exposed to Na^+-free solution (with isotonic KCl, choline chloride, or tetramethylammonium chloride replacements), in the presence or absence of the growth factor. The growth factor is not expected to induce pH_i elevation in the absence of Na^+. (2) Inhibition of Na^+/H^+ exchange by amiloride and its analogs is expected to inhibit growth factor-induced pH_i elevation. Although amiloride itself can be used, it is advantageous to employ some of the more potent 5-N substituted amiloride analogs.[16,17] The greater potency of these analogs at low concentrations eliminates possible complications due to quenching of DMFD fluorescence which may occur at the relatively high concentrations of amiloride that are required to inhibit Na^+/H^+ exchange.[10] The efficacy of inhibition of Na^+/H^+ exchange by these analogs parallels their ability to inhibit growth factor stimulated pH_i elevation.

Due to the relative stability of the fluorescent signal in DMFD-loaded cells, growth factor-induced changes in pH_i can be monitored for an extended period. However, as already mentioned, prolonged incubation (several hours) in solution A results in cellular damage, as evident by morphological alterations and by a decrease in intracellular DMFD fluorescence. Thus, prolonged measurements require incubation in a medium containing additional components as required for the maintenance of cellular viability. A comprehensive review of the effects of mitogens on Na^+/H^+ exchange is presented by Reuss et al.[7]

Appendix: Synthesis of Dimethylfluorescein-Dextran[8]

Synthesis of 4',5'-Dimethylfluorescein Isothiocyanate. The preparation of 4',5'-dimethylfluorescein as first described by H. Burton and F. Kurzer [*J. Soc. Chem. Ind., Lond.* **67**, 345 (1948)] was modified to yield 4',5'-dimethylnitrofluorescein. Two-tenths mole of 4-nitrophthalic acid (Sigma), 0.4 mol of 2-methylresorcinol (Aldrich), and 0.2 mol of anhydrous, granular $ZnCl_2$ were melted together at 185° for 1 hr followed by 20 min at 200°. The cooled, glass-like product was powdered, boiled in 800 ml of 3% HCl for 15 min, filtered, washed with 500 ml of hot 3% HCl, washed with water, and dried to give crude 4',5'-dimethylnitrofluorescein.

[16] G. L'Allemain, A. Franchi, E. J. Cragoe, Jr., and J. Pouyssegur, *J. Biol. Chem.* **259**, 4313 (1984).
[17] Y. X. Zhuang, E. J. Cragoe, Jr., T. Shaikewitz, L. Glaser, and D. Cassel, *Biochemistry* **23**, 4481 (1984).

Reduction of the nitro group to an amine and recrystallization from HCl was carried out using the method of McKinney *et al.* [R. M. McKinney, J. T. Spillane, and G. W. Pearce, *J. Org. Chem.* **27**, 3986 (1962)]. The partially purified 4′,5′-dimethylfluoresceinamine (2.5 g) was dissolved in 30 ml of dry acetone and added with stirring during 30 min to 20 ml of dry acetone also containing 2.5 ml of thiophosgene (Baker). The suspension was stirred for 1 hr at 25° and the product was filtered, washed with acetone, and then with H_2O to remove HCl. The product was dried *in vacuo* for 18 hr at 25°. Thin-layer chromatography of this material on silica gel (Kieselguhr F254, Merck, Darmstadt) using chloroform: methanol: acetic acid (74:25:1, v/v/v) revealed the presence of the two expected isomers together comprising about 85–90% of the spotted material.

Preparation of Hydrazine-Dextran. Four grams of dextran (M_r 5000) was dissolved in 8 ml of water. Two millimoles of sodium metaperiodate was added with continuous stirring, the pH maintained at 4.5 by adding 5 N NaOH. The solution of oxidized dextran was applied to a column containing 30 ml of Dowex anion-exchange resin (AG2-X8) in the chloride form and the dextran eluted with distilled water. The eluate was added dropwise into 20 ml of 0.1 mol of hydrazine hydrochloride, pH 5.0. One hour after the addition of dextran was complete, the pH of the solution was adjusted to 6.0 with NaOH. Four millimoles of sodium cyanoborohydride in 4 ml of 1 M Mes buffer, pH 6.0, was then added and the solution was incubated for 16–20 hr at room temperature under nitrogen. Subsequently, the pH was brought to 9.0 with NaOH, 2 mmol of sodium borohydride in 2 ml of 1 M potassium phosphate buffer, pH 9.0, was added, and the solution was incubated for 5 hr at room temperature.

The solution was added slowly into 3 volumes of 95% ethanol with shaking to precipitate the dextran. After settling, the supernatant was decanted and the sticky precipitate was dissolved in 20 ml of water and dialyzed (in Spectrapor No. 3 tubing, M_r = 3500 cutoff) at 4° against 4 liters of water for 36 hr with two changes, and then lyophilized. The yield was 2.5–2.8 g. The colored reaction product of free amino groups with trinitrobenzene sulfonate was used to estimate the presence of about 200–240 μmol of free NH_2 per g of dextran product. The hydrazine dextran was stored in a desiccator.

Coupling of Dimethylfluorescein Isothiocyanate to Hydrazine-Dextran. Hydrazine-dextran (2.5 g) was dissolved in 50 ml of dry dimethyl sulfoxide with gentle heating, 400 mg of dimethylfluorescein isothiocyanate was dissolved in 6 ml of dimethyl sulfoxide, and the two dimethylsulfoxide solutions were combined and incubated for 15 min at room temperature. The dextran was precipitated by adding the dimethyl sulfoxide solution to 200 ml of ethanol with rapid stirring. The precipitate was filtered with suction and washed with ethanol until the filtrate was clear. The product was dissolved in 20 ml of 1 M ammonium acetate, pH 8.0, and dialyzed against 2 liters of 0.1 M ammonium acetate, pH 8.0, at 4° overnight. Dialysis was continued against excess volumes of double-distilled water for 24 hr with one change and the product then was lyophilized. Substitution of the free amino groups on the dextran with dimethylfluorescein isothiocyanate was nearly quantitative as assessed by fluorimetry. About 1 mol of dimethylfluorescein isothiocyanate was coupled per mol of dextran (average M_r = 5000).

Acknowledgments

The research on which this chapter is based was initiated while the authors were at the Department of Biological Chemistry, Washington University School of Medicine, and was supported by a Grant from the Monsanto Company to L. Glaser. Paul Rothenberg was supported by Medical Scientist Training Grant GM 02016. The research of Dan Cassel is supported by Grant GM 34659 from the National Institutes of Health, by grant from the United States–Israel Binational Science Foundation, and by a grant from the Bat-Sheva de Rothschild Fund for the Advancement of Science and Technology. We thank B. Whiteley for help in preparation of the manuscript, and especially P. Schlesinger for advice and assistance in the development of these procedures.

[39] Detection of Proteins Induced by Growth Regulators

By Marit Nilsen-Hamilton and Richard T. Hamilton

Introduction

Based on our studies of the induction of proteins and glycoproteins by peptide growth factors and a peptide growth inhibitor we have established a method for detecting growth regulator-induced proteins synthesized by cells in culture. With this method we have studied the induction of proteins in five different cell lines and various oncogenic transformants of these cell lines; and we have also examined at least 14 other types of cells and their transformed counterparts. The advantage of the method is that both intracellular and extracellular proteins can be studied in the same population of cells. This is important because many secreted proteins are induced by growth regulators.

Examples of the results from the use of this method to detect the effects of growth regulators on the proteins secreted by some cells are shown in Fig. 1. The cells were labeled with a radiolabeled amino acid, then the cellular fractions separated, and the proteins in each fraction resolved by polyacrylamide gel electrophoresis. Each type of cell responds to growth regulators by producing a unique set of one to six secreted proteins. The pattern of secreted proteins and the effect of growth regulators on that pattern is so distinctive that it can be used to identify a particular cell type. Growth factors can induce different proteins than do growth inhibitors. In many cases the proteins produced in response to growth regulators appear transiently. Also, the response to growth factors is a function of the growth state of the cells. While the

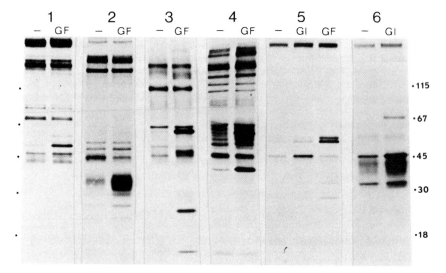

Fig. 1. The effect of growth regulators on the pattern of proteins secreted by several cell types. The proteins secreted by several cell types and the effects of either a growth factor (GF) or a growth inhibitor (GI) on the pattern of secreted, [^{35}S]methionine-labeled proteins are shown. In each case the fluorogram is of proteins resolved by SDS–polyacrylamide gel electrophoresis through 7.5 to 15% gradient gels. The types of cells, treatments, and the molecular weights of the major induced secreted proteins are (1) quiescent human fibroblasts with and without FGF (52,000 M_r), (2) growing Swiss mouse 3T3 cells with and without FGF (a heterogeneously glycosylated protein with a median M_r of 34,000), (3) quiescent BALB/c 3T3 cells with and without EGF with 1 μg/ml cycloheximide present during the preincubation period only (12,000, 24,000, 48,000, and 62,000 M_r), (4) growing BALB/c 3T3 cells with and without FGF (37,000 M_r), (5) growing BSC-1 cells with and without EGF (59,000 and 62,000 M_r) or the BSC-1 growth inhibitor (48,000 M_r), (6) growing CCL64 mink lung epithelial cells with and without BSC-1 growth inhibitor (48,000 and 67,000 M_r). To the right of the figure the approximate positions of molecular weight standards are marked (115,000, β-galactosidase; 67,000, bovine serum albumin; 45,000, ovalbumin; 30,000, carbonate dehydratase; 18,000, myoglobin).

most pronounced increases in DNA synthesis are found in quiescent cells that have been stimulated with growth factors, the most pronounced selective increases in protein synthesis can occur in growing cells[1] or in quiescent cells.[2]

Several different methods can be used to resolve a protein from a mixture of radiolabeled proteins. One of the best techniques is sodium dodecyl sulfate (SDS)–polyacrylamide gel electrophoresis either in one dimension or combined with isoelectric focusing to give a two-dimensional pattern of labeled proteins. Hundreds of proteins in a cell extract can be clearly resolved by two-dimensional gel electrophoresis.[3] It is however difficult to unambiguously identify specific changes in the patterns of proteins distributed by two-dimensional polyacrylamide gel electrophoresis, and it is even more difficult to quantitate those differences. Nevertheless, procedures for quantitating the amount of radiolabel incorporated into proteins resolved by two-dimensional gel electrophoresis are available and constantly being improved. When a much smaller number of proteins in a mixture needs to be resolved (for example, a mixture of proteins that has been secreted into the extracellular medium) the simpler one-dimensional SDS–polyacrylamide gel electrophoresis is the method of choice: many samples can be run on one gel and the amount of protein in each band can be quantitated more easily. Because of the advantages of one-dimensional over two-dimensional polyacrylamide gel electrophoresis, it is desirable to reduce the number of proteins in a sample so that they can be run on the one-dimensional gel. This can be done, for example, by fractionating the cell lysate before resolving the proteins on one-dimensional gels.

Although we have identified many secreted proteins that are selectively induced by growth regulators, we have rarely been able to detect selective changes in the synthesis of intracellular proteins using one-dimensional SDS–polyacrylamide gels. Only a few intracellular proteins reported to be induced by growth factors can be readily detected by one-dimensional gel electrophoresis—they are major intracellular proteins that can be detected in a protein band even though there may be several other proteins in the same position on the gel. Whereas intracellular mitogen-induced proteins have been difficult to identify, we have found that for almost every cell line that we have studied, we have been able to identify one or more secreted proteins induced by growth regulators.

[1] M. Nilsen-Hamilton, J. M. Shapiro, S. L. Massoglia, and R. T. Hamilton, *Cell* **20,** 19 (1980).
[2] R. T. Hamilton, M. Nilsen-Hamilton, and G. Adams, *J. Cell. Physiol.* **123,** 201 (1985).
[3] P. H. O'Farrell, *J. Biol. Chem.* **250,** 4007 (1975).

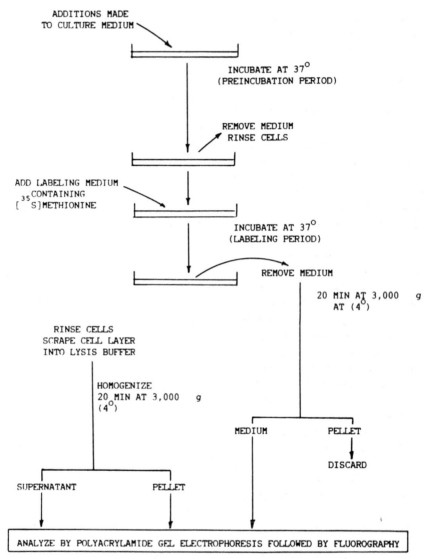

FIG. 2. Procedure for analyzing the synthesis of proteins by cells in culture in response to growth regulators. Additions of growth regulators are made to the medium in which the cells are growing. The cells are then incubated for a period of time (preincubation period) that can be varied according to the cells and the experiment and is determined by the time course of induction of the particular protein(s) to be studied. After the preincubation period the medium is removed and the cells rinsed with TBS. For cells in suspension the cells are rinsed by resuspending and centrifuging them. The centrifugal force should be kept to a minimum during the rinses to avoid damage to the cells. The cells are then incubated with

Methods

Solutions

All pH values given below have been measured at room temperature.
Electrode buffer: 0.384 M glycine, 0.1% SDS, and 49.5 mM Tris base (pH 8.35)

Gel electrophoresis sample buffer: this buffer is made up as a twice-concentrated solution containing 4% (w/v) SDS, 4% (v/v) 2-mercaptoethanol, 20% (v/v) glycerol, 0.008% bromphenol blue, and 120 mM Tris-PO$_4$ (pH 6.9). The samples are diluted 1 : 1 with this buffer for SDS–polyacrylamide gel electrophoresis. To inhibit proteases 200 kallikrein inactivator units/ml aprotinin and 2 mM phenylmethylsulfonyl fluoride (PMSF) are also added to the sample buffer directly before mixing with the sample.

Lysis buffer: 0.5% Nonidet P-40, 10 mM Tris, pH 7, 100 kallikrein inactivator units/ml aprotinin, and 1 mM PMSF. The buffer is usually made up and frozen in aliquots with all ingredients except the PMSF. Because PMSF has a half-life in water at pH 7 of only 110 min, the PMSF is stored as a stock solution in ethanol (0.1 M) and is added to the lysis buffer just before the lysis buffer is to be used.

TBS: 0.137 M NaCl, 6.7 mM KCl, 0.68 mM CaCl$_2$, 0.5 mM MgCl$_2$, 0.7 mM Na$_2$HPO$_4$, and 25 mM Tris–HCl (pH 7.4).

Radiolabeling Cells in Culture

Although the assay was originally designed for cells attached to the substratum, it can also be used, with minor modifications, for cells grown in suspension. A general outline of the method is shown in Fig. 2.

---

labeling medium containing a radiolabeled amino acid for a labeling period with a length that varies depending on the cells and the type of experiment. A convenient length for the labeling period is 4 hr. The medium used to label the cells is the medium in which the cells have been grown except that the methionine concentration is reduced to 20 μM or less (see the text for precautions regarding the reduced concentration of methionine in the labeling medium). Serum which is present at a concentration of 1% or less during labeling is the other component of the medium that is often reduced in the labeling medium compared to its concentration in the medium in which the cells were cultured. After the debris in the medium is removed by centrifugation and the Nonidet P-40 lysate is separated into two fractions (pellet and supernatant) the three fractions are resolved by SDS–polyacrylamide gel electrophoresis or two-dimensional gel electrophoresis. The gels are then impregnated with an enhancing agent and exposed to preflashed film at $-70°$. Fluorograms can be scanned using a densitometer to determine the amount of [^{35}S]methionine incorporated into each band.

Cells are plated on 24-well tissue culture dishes (diameter of each well = 1.6 cm) in their growth medium containing the appropriate growth requirements such as serum (1 ml medium/well). The minimum number of cells per well necessary for detecting an induced protein can vary considerably. Many fewer cells are needed to radiolabel the major intracellular proteins than are needed to radiolabel secreted proteins or minor intracellular proteins. For example, to detect the secreted proteins and glycoproteins that are shown in Fig. 1, a minimum of 10^4 cells per well is sufficient, whereas about one-tenth that number of cells would be needed to incorporate enough radiolabel into the major intracellular proteins to see them easily by fluorography. The number of cells is only one determinant of the amount of radiolabel incorporated. Other factors are the concentration of radioisotope and specific activity of the [^{35}S]methionine in the labeling medium and the length of the labeling period. These factors can be balanced against one another to achieve a desirable level of incorporation of radiolabel into cellular and secreted proteins.

Once the cells have grown to the required density and have achieved the desired growth state, they can be treated with growth regulators and other agents. In general, it is not desirable to change the medium over the cells before adding the growth regulator. The fresh serum usually added with the medium contains growth factors and growth inhibitors that can interfere with the effect of the purified growth regulator being tested. In most cases the effect of growth regulators on protein synthesis is transient, completed within about 48 hr after addition of the growth regulator (Fig. 3). Therefore, it is usually better to change the medium over the cells several days before starting the assay. After the growth regulators have been added the cells are incubated under normal growth conditions for the chosen length of time (preincubation period in Fig. 2). If several preincubation times are being tested, it is better to add the growth regulators at staggered times so that the period of radiolabeling starts at the same time for all samples.

For the labeling period, the medium is replaced with medium containing a radiolabeled amino acid. Before adding the labeling medium, the culture medium is sucked off with a pasteur pipet and the cells are rinsed once with TBS, an isoosmotic, buffered salt solution. The culture dish is then tilted to allow the remaining TBS to drain to one edge; the TBS is then carefully sucked off. It is important in this step to remove all liquid above the cell layer because only a small amount of labeling medium will be added to each well. If this labeling medium were diluted randomly by residual TBS then the concentration of radioisotope would vary from one well to the next. It is also important to avoid disturbing the cell layer while sucking off the fluids. The pasteur pipet tip is therefore not touched to the

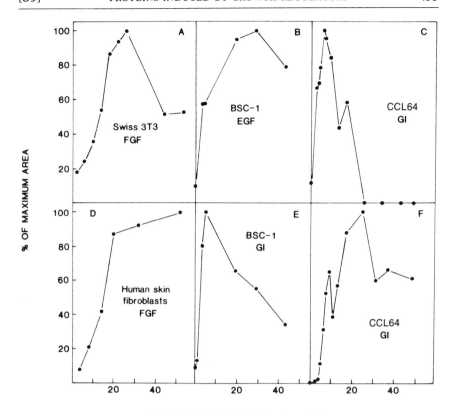

TIME, h after addition of growth regulator

FIG. 3. Time courses for the induction of secreted proteins after the addition of growth regulators. The amount of a particular growth regulator-induced protein secreted into the medium was determined from the area under the peak of densitometric scans of fluorograms of gels as a function of time after adding the growth regulator. After adding the growth regulator, the cells were labeled at the times indicated on the figure. The labeling period was either 2 hr (BSC-1 and CCL64 cells) or 4 hr (Swiss mouse embryo 3T3 cells, human skin fibroblasts). The names and molecular weights of the proteins whose time courses of induction are shown are (A) MRP (median M_r of 34,000) secreted by mouse embryo 3T3 cells in response to FGF; (B) MIP59 and MIP62 (59,000 and 62,000) secreted by BSC-1 African green monkey kidney epithelial cells in response to EGF; (c) IIP48^{mink} (48,000) secreted by CCL64 mink lung epithelial cells in response to the BSC-1 GI; (D) MIP52 (52,000) secreted by human skin fibroblasts responding to FGF; (E) IIP48 (48,000) secreted by BSC-1 cells in response to the BSC-1 growth inhibitor (GI); (F) IIP70 (70,000) secreted by CCL64 cells in response to the BSC-1 GI.

cell layer but placed a short distance above the bottom of the well against the side of the well and as far away from vertical with respect to the bottom of the well as possible. Touching the cell layer could damage the cells and so release cellular contents into the medium.

The cells are labeled with 150 μl of medium per 1.6 cm diameter well. The labeling medium consists of the medium in which the cells were grown but with 20 μM [^{35}S]methionine (100–200 μCi/ml). For those cells that are grown in serum, 0.2 to 1% serum is also usually included in the labeling medium. The growth regulators and other agents added to the medium for the preincubation period are usually also added to the labeling medium. Substances that may be toxic to the cells but that act rapidly (for example, actinomycin D or tunicamycin) can be present during the labeling period only. The cells are then incubated in the labeling medium with gentle rocking at the desired temperature in a humidified atmosphere containing the concentration of CO_2 in air necessary for maintaining the appropriate pH of the labeling medium. After 2–4 hr, the medium is removed and spun for 20 min at 3000 g to remove cellular debris.

For some cells a small amount of serum is needed in the labeling medium to ensure that protein synthesis continues at a constant rate throughout the period of labeling. Serum contains growth factors and growth inhibitors. Therefore if a protein is induced rapidly (within a time period similar to the length of the labeling period) the presence of serum may be a complication and an attempt should be made to remove it from the labeling medium. When serum is included in the labeling medium, its concentration is kept at 1% or below if the secreted proteins are to be analyzed by gel electrophoresis. At serum concentrations above about 1%, the serum albumin band on SDS–polyacrylamide gels is so large that it distorts the pattern of labeled proteins in its lane and in neighboring lanes. For many cells, the concentration of serum in the growth medium is greater than 1%. Therefore, it is necessary to show that (1) lowering the serum concentration for the short period of radiolabeling does not alter the pattern of proteins made by the cells and (2) the cells maintain a constant rate of incorporation of radiolabel into proteins over the labeling period. It is also important to determine the concentration of serum below which the rate of incorporation of radiolabel decreases. This lower limit will vary from one type of cell to another. Even though the cells may require serum concentrations of 5–10% or more for continued growth, they do not require as high a serum concentration during the short labeling period. For the types of cells described in Fig. 1, 0.2% serum is sufficient to maintain a constant rate of incorporation of radiolabel into protein for 8 hr. Many cells can be labeled in the absence of serum (for example, BSC-1 cells, CCL64 cells, primary hepatocytes, and some fibroblasts); but some cells are sensitive to the removal of serum during radiolabeling (for example, Swiss 3T3 cells).

In assays where several different variables are being tested, it is convenient to add the labeling medium in two different portions. Portion 1: 135 μl containing the [^{35}S]methionine and any other components that are to be added to most of the wells. Portion 2: 15 μl containing those additions that are to be made to only a few wells. With this arrangement the radioisotope is contained in fewer tubes resulting in less waste. For portion 1, a solution is first prepared that contains the medium with low methionine, [^{35}S]methionine, serum, and any other component that is to be added to all wells. Solutions for portion 2 are made up separately and contain the agent to be added at a concentration 10-fold above the final concentration.

The specific activity of [^{35}S]methionine necessary for sufficient incorporation into protein depends on the length of the labeling period and the type of cells that are being labeled. "Sufficient incorporation of label" refers to the ability to obtain a readily visible image on film after the enhancer-impregnated gel has been exposed to X-ray film for a period of 2 to 10 days at $-70°$. To increase the incorporation of radiolabel into proteins we have raised the specific activity of [^{35}S]methionine by reducing the concentration of methionine to 20 μM. This is 10% of the concentration normally found in Dulbecco and Vogt's modified Eagle's medium. Because the methionine concentration varies greatly between different growth media this is one aspect of the method that can be varied. For many cells the methionine concentration can be lower than 20 μM without detrimental effects. But for each cell line it is important to test that methionine is not depleted during labeling. This is done by showing that the [^{35}S]methionine is incorporated linearly for a period longer than the labeling period. We have found that for the cells described in Fig. 1 and also for primary hepatocytes, the incorporation of [^{35}S]methionine is linear for at least 8 hr. It is also necessary to check that the pattern of labeled proteins is not altered by lowering the concentration of the amino acid. For some cells it is not even necessary to add unlabeled amino acid with the radiolabeled amino acid to achieve linear incorporation and an unaltered labeling pattern, although it is probably prudent to include some methionine apart from that provided with the radiolabel. For 3T3 cells, for example, we have found that removing all the methionine in the medium did not alter the secretion pattern, although it did alter the secretion pattern of esophageal epithelial cells. Other amino acids or sugars labeled with ^{3}H or ^{14}C can also be used. In each case there are the same considerations as we have described for an assay where the radiolabeled amino acid is methionine. If the proteins have been labeled with ^{3}H-labeled precursors the X-ray films need to be exposed for longer times because of the weaker emissions from this radioisotope. The specific activity of the [^{35}S]methionine is determined by the amount of radiolabeled methionine added to the mixture since the specific activity of commercially available [^{35}S]methionine (>500 Ci/mmol) is high enough that its addition does not appreciably affect the concentration of methionine in the labeling medium. For the results shown in Figs. 1 and 3, with labeling periods of 2 or 4 hr, the concentration of ^{35}S was 50–150 μC/ml. For intracellular proteins, the amount of radiolabel incorporated is directly proportional to the length of the labeling period. For secreted proteins the amount of radiolabel found in proteins in the medium is proportional to the length of the labeling period minus the time lag between synthesis and secretion of the protein. For most cells this time lag varies between 30 and 60 min.

A convenient labeling period is 4–5 hr. But it is sometimes necessary to shorten the labeling period; for example, to determine the time course of induction of a protein. The minimum period of labeling for secreted proteins is determined by the time between synthesis and secretion. The time between synthesis and secretion of a protein seems to vary between different types of cells. The lag time for 3T3 cells is about 35 min, whereas it is about 60 min for BSC-1 cells. Because of the small volume of labeling medium over the cells, the incubator must contain an atmosphere that is saturated with water and the dishes must be rocked gently back and forth during labeling. A lightweight rocker of plexiglass can be made by connecting a 2 rpm electric clock motor (Synchron brand; supplied by Empire Clock Co., St. Paul, MN) through an arm to a platform that rocks about its central axis making a maximum angle of about 10°. The advantages of the small volume of labeling medium are that the least amount of [^{35}S]methionine is used per well and the secreted proteins are at the highest possible concentration when the medium is harvested. With volumes of labeling medium of 300 μl or larger in each well rocking can be omitted.

For resolution by one-dimensional polyacrylamide gel electrophoresis, samples of the medium and supernatant are diluted (1 : 1) with twice-concentrated electrophoresis sample buffer and pellets from the Nonidet P-40 cell lysate are resuspended in 100 μl of electrophoresis sample buffer at its final sample concentration. The samples are heated for 2 min at 98° then resolved by SDS–polyacrylamide gel electrophoresis. If more than about 10^6 cells are present in each well the pellets might be viscous because the concentration of DNA is high. In this case the DNA can be sheared by passing the sample rapidly through a syringe needle of high gauge (27 gauge). If sheared or sufficiently dilute the DNA will not decrease the resolution of the proteins on the gel. After electrophoresis the gels are impregnated with an enhancing agent, dried, and the [35S]methionine-labeled bands are visualized by fluorography.[4] We have compared two different methods of enhancing the signal from the radioisotope incorporated into protein. We found that gels impregnated with diphenyloxazole (PPO)[5] gave slightly sharper images upon fluorography than those impregnated with salicylic acid.[6] However the salicylic acid method takes much less time and does not involve the use of toxic organic solvents such as dimethyl sulfoxide, the solvent for PPO. Commercial preparations of enhancers for fluorography are also available.

Collecting the Labeled Cell Fractions

After the cells have been labeled, the medium is transferred to glass test tubes (10 × 75 mm) and centrifuged to remove cellular debris [5000 rpm (3000 g) for 20 min in a Sorvall SM24 rotor or an SS34 rotor with adaptors for the small tubes]. Alternatively, the medium can be transferred to microcentrifuge tubes (1.5 ml) and spun at 12,000 g for 5 min in a microcentrifuge. Usually, no pellet can be detected after centrifugation. However, a detectable pellet signifies that an appreciable number of cells were dislodged from the dish during labeling and that one should be wary of the medium being contaminated with the products of cell lysis. The supernatant is removed and may be stored frozen before resolving the radiolabeled proteins by gel electrophoresis. Because the specific activity of the [35S]methionine is high, the medium need not be concentrated before the proteins are resolved.

After the medium has been removed, the cell layer is rinsed gently with TBS and then 100 µl of lysis buffer at 4° is added to each well. From here on the samples are kept at 4° unless it is specified otherwise. The cells are scraped into the lysis buffer using a rubber policeman with an end cut at right angles to the handle. The scraped cells are transferred to test tubes, mixed on a vortex mixer for 5 sec, then centrifuged as for the medium. The pellet after lysis with Nonidet P-40 contains nuclei, cytoskeletal, and extracellular matrix material. The supernatant contains the cytoplasmic, organellar, and membrane proteins. All of the samples can be stored at −20° until resolved by SDS–polyacrylamide gel electrophoresis.

---

[4] R. A. Laskey and A. D. Mills, *Eur. J. Biochem.* **56,** 335 (1975).

[5] W. M. Bonner and R. A. Laskey, *Eur. J. Biochem.* **46,** 83 (1974).

[6] J. P. Chamberlain, *Anal. Chem.* **98,** 132 (1979).

Gel Electrophoresis to Resolve Labeled Proteins

The [^{35}S]methionine-labeled proteins in the medium and the cellular fractions can be resolved by SDS–polyacrylamide slab gel electrophoresis. We have found that we get the best resolution with a procedure based on the discontinuous gel electrophoresis system described by Ornstein[7] and Davis.[8] The changes that we have made to the system described by Ornstein and Davis are to add 0.1% SDS to all buffers and to increase the concentration of Tris buffer in the running gel. The running gels contain a linear gradient, from top to bottom, of 7.5 to 15% acrylamide. The ratio of acrylamide to bisacrylamide (N,N'-methylenebisacrylamide) is 30:0.8 (w/w) in both the running and stacking gels. The purity of the acrylamide is important. It should be electrophoresis grade or, if a lesser grade, the acrylamide should be recrystallized from a solvent such as acetone. A gradient of acrylamide in the running gel gives sharper bands and resolves over a larger molecular weight range than a gel of a single acrylamide concentration. The running gel buffer is 600 mM Tris–HCl, pH 8.9, with 0.1% SDS. The stacking gel contains 5% acrylamide, 0.1% SDS, and 59 mM Tris–PO$_4$, pH 6.9. Both stacking and running gels are polymerized using 0.05% TEMED (N,N,N',N'-tetramethylethylenediamine) and ammonium persulfate (0.008% for the running gels and 0.1% for the stacking gel). The same electrode buffer is used for both anode and cathode. The pH values of all solutions are critical to the resolution. The gels are run at a constant amperage of 10–12 mA/gel. It takes about 5 hr for the bromphenol blue to travel 10 cm. The gels are then removed, stained with 0.1% Coomassie blue in 25% (v/v) isopropanol and 10% (v/v) acetic acid for a minimum of 2 hr, and then destained with 10% (v/v) isopropanol and 10% (v/v) acetic acid.

Quantitating the Amount of Radiolabel Incorporated into Protein

The amount of [^{35}S]methionine incorporated into a protein can be determined from the gels in two ways. To identify the radiolabeled protein bands, the dried gel, impregnated with an enhancer, is exposed to preflashed film at −70° according to the method of Laskey and Mills.[4] If sufficient [^{35}S]methionine has been incorporated into a protein band then the band can be cut out of the gel and the amount of [^{35}S]methionine in the protein can be determined using a liquid scintillation spectrometer. We have found that to have sufficient radioactivity (100 cpm or more) in a

---

[7] L. Ornstein, *Ann. N.Y. Acad. Sci.* **121**, 321 (1964).
[8] B. J. Davis, *Ann. N.Y. Acad. Sci.* **121**, 404 (1964).

band to be able to measure it by scintillation spectrometry, the band must be clearly visible on a film that has been exposed to the dried gel for 24 hr or less. Although it is relatively easy to label major intracellular proteins such as actin to this specific activity, it is often difficult to achieve a high enough specific activity of secreted proteins and minor intracellular proteins to be able to directly determine the radioisotope content of the band. If there is not sufficient radiolabel incorporated into the protein for directly measuring the cpm, the fluorogram can be scanned with a densitometer and the area under the protein peak determined. If film exposures are maintained within the linear range of a standard curve, then accurate values that are proportional to the amount of [^{35}S]methionine incorporated into the protein can be obtained by scanning film exposures. A standard curve can be created to estimate the amount of radioactivity in each protein band by exposing a series of strips of dried acrylamide gel containing known amounts of radioisotope along with the gel to the same X-ray film. These standard strips are prepared by allowing strips of dried enhancer-impregnated gel to absorb a known amount of radioisotope. The swollen gel strip is again dried, this time between dialysis tubing along with other strips containing calibrated amounts of radioisotope. The density of the bands produced by these calibrated strips can then be plotted against the amount of radioactivity in each strip, and the amount of radioactivity incorporated into each protein band on the gel read from the standard curve.

Normalizing the Data to the Rate of Protein Synthesis or to the Number of Cells

As well as selectively inducing certain proteins, growth factors sometimes also nonselectively increase the rate of protein synthesis (reviewed in Ref. 9) and decrease the rate of protein degradation.[10] Because of this, the amount of radiolabel incorporated into every protein will often increase in response to a growth factor. Therefore, to determine the extent to which a protein is specifically induced it is necessary to normalize a value for the amount of [^{35}S]methionine incorporated into a protein to the general effect of a growth regulator on the radiolabeling of all proteins synthesized by the cells. One procedure that is often used is to load an equal amount of acid-insoluble radioactivity on each gel channel for electrophoresis. We have found that this is not the best approach, particularly for secreted proteins where the amount of radioactivity incorporated into

---

[9] G. Thomas and J. Gordon, *Cell Biol. Int. Rep.* **3**, 307 (1979).
[10] F. J. Ballard, S. E. Knowles, S. S. C. Wong, J. B. Bodner, C. M. Wood, and J. M. Gunn, *FEBS Lett.* **114**, 209 (1980).

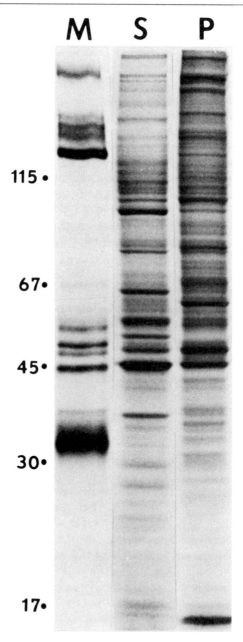

FIG. 4. Comparison of the patterns of proteins distributed on SDS–polyacrylamide gel electrophoresis from different cellular fractions. The pattern of [^{35}S]methionine-labeled proteins found in the medium (M) is compared with the patterns of [^{35}S]methionine-labeled

single proteins that are major components of the mixture can vary greatly. A simpler approach, and one less prone to error, is to load equal volumes of medium on the gels, then later normalize the results from densitometer scans of films to either number of cells, acid-insoluble cpm, or to the density of a protein band that is not altered by the growth regulator. The supernatant samples from the Nonidet P-40 lysate can be used to obtain values for the acid-insoluble cpm incorporated by the cells. We have sometimes found the last procedure of normalizing to the levels of another protein very useful, particularly with secreted proteins, because normalized in this way the data are corrected for errors in sample preparation and differences in the overall rate of secretion as well as in the rate of protein synthesis. However, because one can never be sure that the growth regulator does not influence the levels of a particular reference protein under all conditions, this procedure should be used with caution and in conjunction with one of the other two procedures.

Cell Lysis and Proteolytic Activity

Most cells secrete only a few percent of all the protein that they synthesize. Thus the possibility of contamination of the secreted protein fraction with cellular protein is a major concern. We have found that lysis of only about 2% of the cells provides enough protein in the medium to be detected as contaminants of the secreted protein fraction. In most cases, substantial cell lysis can be easily avoided but it needs to be recognized when it occurs. Changing the medium over the cells frequently results in cell lysis, releasing radiolabeled proteins into the medium that obscure the secreted protein pattern. Thus we found it more difficult to obtain preparations of secreted proteins contaminated with few or no intracellular proteins when we tried to use methods in which the cells are first labeled with radioisotope followed by a change of medium to an unlabeled collection medium into which the secreted proteins are released by the prelabeled cells. Because of the potential of cell lysis, it is important to establish that the radiolabeled proteins in the medium have actually been secreted by the cells rather than being released by lysed cells. This can be done in two ways. First, the pattern of secreted proteins can be examined as it is quite different from the pattern of intracellular proteins (Fig. 4).

proteins found in the pellet (P) and the supernatant (S) of an Nonidet P-40 lysate of 3T3 cells that have been treated for 16 hr with FGF and then labeled for 4 hr with [^{35}S]methionine in the presence of FGF. It can be seen that the pattern of secreted proteins is simpler than that of intracellular proteins and that the bands formed by the secreted proteins tend to be broader and more diffuse than those formed by the intracellular proteins. This is probably because many of the secreted proteins are glycosylated.

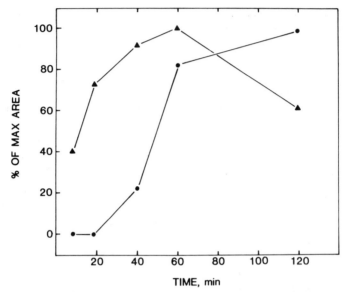

FIG. 5. Typical time courses of the appearance in the medium of a radiolabeled secreted and cellular protein. The levels of two proteins were measured in the medium as a function of time after the cells had been labeled with an 8 min pulse of [^{35}S]methionine. After the 8 min labeling period, the labeling medium was removed and replaced with medium containing 200 μM methionine and no radioisotope. Samples of medium were removed at various times after this change of medium, the proteins in the medium were resolved by SDS–polyacrylamide gel electrophoresis and the amount of radioisotope incorporated into each protein determined from a fluorogram by scanning with a densitometer. The proteins whose time courses are depicted are fibronectin (●) and a 45,000 M_r protein (▲) that is probably actin.

FIG. 6. The effect of proteolysis on the pattern of proteins on a one-dimensional gel. Proteases can sometimes be introduced with preparations of growth regulators. The effect of proteolysis is to reduce the average size of the protein bands distributed by SDS–polyacrylamide gel electrophoresis such as shown. A secretion pattern from human fibroblasts (channel 1) is compared with another sample from the same experiment which shows evidence of proteolysis (channel 2). Similar results can also be observed in the cell lysate if the particular protease is not inhibited by the protease inhibitors added to the lysis buffer, and if sufficient protease remains with the cell layer after removing the medium and rinsing the cells. Every cell type that we have examined secretes some proteins with molecular weights of 100,000 and above. For the secretion pattern of human fibroblasts shown, the high-molecular-weight proteins, fibronectin (broad arrow) and collagen (arrowheads) are indicated.

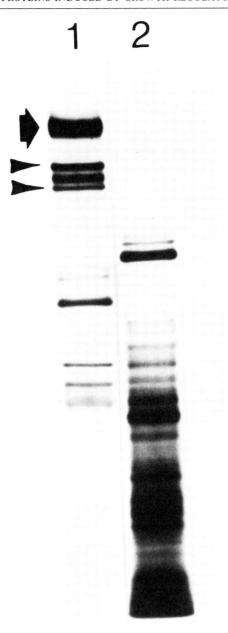

FIG. 6.

Also, whereas secreted proteins are often glycosylated and therefore appear as relatively broad bands after electrophoresis through SDS–polyacrylamide gels, intracellular proteins appear as much sharper bands. There are also many fewer secreted proteins than there are intracellular proteins. Even though the pattern of secreted proteins is different from and simpler than the pattern of intracellular proteins, these characteristics do not allow one to unambiguously distinguish between cellular and secreted proteins; some of the minor proteins in the medium often prove to be intracellular proteins. One way to determine whether a particular protein is an intracellular protein derived from lysed cells or a bona fide secreted protein is to determine whether, after labeling the cells for a short time (5–10 min), there is a lag before the protein appears in the medium. Proteins that are secreted will not appear in the medium until 35 to 60 min after the pulse label and their levels will increase steadily until they reach a plateau after about 60 to 90 min (fibronectin in Fig. 5). The time course of appearance of secreted proteins depends on the type of cell. Proteins that appear because of cell lysis will appear immediately (45K protein in Fig. 5).

Although infrequently encountered, another potential problem in identifying proteins induced by growth regulators is proteolysis. This often results in the loss of higher molecular weight protein bands on the gels and the concomitant gain of lower molecular weight protein bands (Fig. 6). This problem can usually be avoided by adding the appropriate protease inhibitor. We include two protease inhibitors (PMSF and aprotinin) in the lysis buffer and in the electrophoresis sample buffer. These inhibitors are generally effective although they do not inhibit all proteases. Proteases can also be introduced from the plastic combs that are used to prepare the wells in the stacking gel if the combs have been stored wet and uncleaned for an extended period.

We have found that by taking the precautions discussed above we have been able to apply this method to every cultured or primary cell type that we wished to study. The method is reproducible, inexpensive, and easy to use. It is an excellent method for identifying proteins, particularly secreted proteins, whose levels are induced or reduced by growth regulators. The assay has also proved extremely useful for studying regulation of the synthesis and secretion of proteins by growth factors and growth inhibitors in cultured cells.

Acknowledgments

We thank Fred Thalacker for providing the data showing the time course of appearance of IIP48 and IIP70 from CCL64 cells. This work was supported by grants from the NIH (GM33528) and the American Cancer Society (CD-242).

Author Index

Numbers in parentheses are footnote reference numbers and indicate that an author's work is refered to although the name is not cited in the text.

Subject Index